Unraveling the Exposome

Sonia Dagnino • Anthony Macherone
Editors

Unraveling the Exposome

A Practical View

Springer

Editors
Sonia Dagnino
Department of Epidemiology
and Biostatistics, MRC-PHE Centre
for Environment and Health, School
of Public Health
Imperial College London
London, UK

Anthony Macherone
The Johns Hopkins School of Medicine
Agilent Technologies
Baltimore, Maryland, USA

ISBN 978-3-030-07739-6 ISBN 978-3-319-89321-1 (eBook)
https://doi.org/10.1007/978-3-319-89321-1

This Springer imprint is published by the registered company Springer Nature Switzerland AG.
The registered company address is: Gewerbestrasse 11, 6330 Cham, Switzerland

Foreword

With the completion of the human genome project in 2003, it was anticipated that new genomic methods would find causes as well as treatments for the chronic diseases that account for most human mortality (Guttmacher and Collins 2003). Certainly, new omic technologies fostered genome-wide association studies (GWAS) that can glean causal factors from essentially all genetic loci. But results from roughly 2000 GWAS have found only modest effects of common genetic variants (Welter et al. 2014). Indeed, GWAS findings are consistent with heritable cancer risks from entire genotypes that are typically about eight percent in families and twins (Hemminki and Czene 2002; Rappaport 2016). Thus, it is reasonable to infer that exposures and gene–exposure interactions are the major causes of cancer and other chronic diseases. But in contrast to high-tech GWAS, searches for causal exposures have been limited to only a few hundred chemicals or mixtures and have relied upon low-tech methods, primarily questionnaires.

This contrast of technologies used to characterize genetic and "environmental" factors led Christopher Wild to promote an "exposome"—representing all exposures experienced during life—as a complement to GWAS in etiological investigations of cancer (Wild 2005). In this context, Wild considered "environmental exposures" with the implication that the exposome was comprised of substances derived from external sources, such as pollutants, food contaminants, and lifestyle factors. Wild also recognized that, whereas the genome is essentially static, the exposome varies throughout life and, therefore, presents challenges for epidemiologic studies. Yet, he was optimistic that nascent omic technologies (in 2005), i.e., transcriptomics, proteomics, and metabolomics, could someday identify "signatures or fingerprints" of causal exposures (Wild 2005).

I was intrigued with Wild's speculations and, with colleague Martyn Smith, functionalized the exposome as the totality of chemicals that can be measured in blood (Rappaport and Smith 2010). In doing so, we expanded Wild's view of exposure to include all chemicals that reach the "internal chemical environment," regardless of their origins from external or internal sources. This extended the exposome umbrella to cover endogenous molecules like hormones as well as substances generated by human and microbial metabolism, lipid peroxidation, inflammation, etc.

A compilation of about 1600 circulating molecules and metals, derived from studies of normal humans, led to further insights into the nature of the "blood exposome" (Rappaport et al. 2014). On the one hand, this dataset revealed that chemicals derived from foods, drugs, and endogenous sources covered the same dynamic range of blood concentrations (pM–mM), while those from pollution sources were present at 1000-fold lower levels (fM–μM). But on the other hand, chemical-similarity maps, representing the chemical space and chronic disease space of these circulating chemicals, were surprisingly similar. This indicates that all sources of exposure (foods, drugs, pollution, and endogenous processes) generate chemicals that contribute to chronic diseases, and thus are worthy of investigation.

The vast chemical space and dynamic range of the blood exposome emphasize the importance of untargeted methods for discovering causes of disease. In fact, state-of-the-art liquid chromatography–mass spectrometry can now detect thousands of small molecules and proteins in a few microliters of archived blood from prospective cohorts. Likewise, preprocessing methods for metabolomics and proteomics are rapidly evolving to filter and normalize untargeted proteomics and metabolomics data that are appropriate for statistical analyses. Thus we can now conduct exposome-wide association studies (EWAS) as complements to GWAS in epidemiologic studies (Rappaport 2012) and thereby identify causal "signatures and fingerprints" of exposure that fueled Wild's speculation in 2005.

Although the exposome paradigm was originally proposed and developed for etiologic research, it has grown to encompass other avenues of inquiry. That is, our conceptualization of "exposure" has evolved from single substances, derived primarily from air and water pollution and the diet, to the myriad chemicals emanating from arrays of external and internal sources (Rappaport and Smith 2010; Wild 2012). This has led, in turn, to the realization that the exposome is perhaps the biggest of the big data entities, whose characterization will require advanced computation, bioinformatics, and statistics. And finally, as combined GWAS/EWAS lead to discoveries of causal exposures and gene–exposure interactions, public health professionals will be able to develop interventions for reducing disease risks.

In this book, *Unravelling the Exposome: A Practical View*, the editors Anthony Macherone and Sonia Dagnino have assembled a diverse team of authors who delve into the many channels of research and practice emerging from the exposome paradigm. By blending the multidisciplinary endeavors of epidemiologists, exposure scientists, toxicologists, statisticians, physicians, chemists, and omics scientists, this book offers an essential overview of "exposomics" for health professionals.

Environmental Health
University of California
Berkeley, CA, USA

Stephen M. Rappaport

References

Guttmacher AE, Collins FS (2003) Welcome to the genomic era. N Engl J Med 349 (10):996–998

Hemminki K, Czene K (2002) Attributable risks of familial cancer from the family-cancer database. Cancer Epidemiol Biomark Prev 11(12):1638–1644

Rappaport SM (2012) Biomarkers intersect with the exposome. Biomarkers 17 (6):483–489

Rappaport SM (2016) Genetic factors are not the major causes of chronic diseases. PLoS One 11(4):e0154387

Rappaport SM, Smith MT (2010) Epidemiology. Environment and disease risks. Science 330(6003):460–461

Rappaport SM, Barupal DK, Wishart D, Vineis P, Scalbert A (2014) The blood exposome and its role in discovering causes of disease. Environ Health Perspect 122(8):769–774

Welter D, MacArthur J, Morales J, Burdett T, Hall P, Junkins H, Klemm A, Flicek P, Manolio T, Hindorff L, Parkinson H (2014) The NHGRI GWAS catalog, a curated resource of SNP-trait associations. Nucleic Acids Res 42(Database issue):D1001–D1006

Wild CP (2005) Complementing the genome with an "exposome": the outstanding challenge of environmental exposure measurement in molecular epidemiology. Cancer Epidemiol Biomark Prev 14(8):1847–1850

Wild CP (2012) The exposome: from concept to utility. Int J Epidemiol 41(1):24–32

Preface

In 2013, I was asked to write a chapter for Imma Ferrer and E. Michael Thurman in the next volume of Comprehensive Analytical Chemistry, Vol. 61 (Elsevier B.V. Amsterdam: The Netherlands. ISBN: 978-0-444-62623-3), entitled The Future of GC/Q-TOF in Environmental Analysis. Lacking possession of a crystal ball, I began to research exactly what I could espouse about the future of time-of-flight mass spectrometry in the environmental sciences. In doing so, I came across a topic I had heard of only once before from Frank David of the Research Institute of Chromatography in Belgium—Exposomics. Through the efforts of writing the chapter, it became very clear that the concept of the exposome filled an immense gap for the understanding of how nongenetic chemical exposures from both within and outside of the body contributed to the onset and progression of chronic disease.

Several years later, I presented "Exploring the exposome: a critical quantity for determining the causes of chronic human disease" as the Plenary Speaker at a meeting of the London Biological Mass Spectrometry Discussion Group at University College, London. While there, I ran into Sonia Dagnino whom I had met previously when she was a post-doctoral fellow in Shane Snyder's lab at the University of Arizona. Sonia spoke to me about an opportunity to edit a book for Springer about the exposome and we agreed to partner as co-editors on the project—thus the journey began.

Through both Sonia's and my exposome networks, we secured contributions from the top experts in the field of exposomics. We thank the authors for their contributions, and truly believe this volume will be a principal reference for the field of exposomics.

Baltimore, MD, USA — Anthony Macherone

Contents

Part I
The Utility of the Exposome Paradigm

Chapter 1
Using Exposomics to Assess Cumulative Risks from Multiple Environmental Stressors

Martyn T. Smith, Cliona M. McHale, and Rosemarie de la Rosa

Abstract Humans are exposed to a milieu of environmental stressors of a chemical, physical, and social nature that may change over time. Interaction of these stressors with various intrinsic factors such as genetics, sex, life stage, and health status determines susceptibility to related diseases. Cumulative risk assessment seeks to determine the combined risks to health from exposures to multiple agents or stressors. This can be achieved by expanding beyond a G × E approach—where "G" represents genetic susceptibility and "E" (environment) represents a limited range of exposures—to an I × E approach—where "I" (intrinsic) represents the many inter-related biological factors that contribute to disease susceptibility and "E" (extrinsic) represents all nongenetic factors including the exposome. Exposomics is poised to advance this concept and make significant advances in environmental health science and our understanding of the causes of chronic diseases. The internal exposome can be assessed using targeted and untargeted exposomics tools to measure individual chemicals, groups of chemicals, or the totality of chemicals acting on a particular receptor or biological pathway in a functional assay. Comprehensive data on the internal, external, and public health components of the exposome together could inform risk assessment and ultimately guide risk management. These approaches could be applied in vulnerable populations such as migrants or those burdened with multiple types of stressor simultaneously as identified through map- or indicator-based approaches. Development and refinement of additional exposomics tools that can be applied in prospective human epidemiology studies should be a focus of future studies.

Keywords Cumulative risk · Environmental stressors · Internal exposome · Stressogens

---

M. T. Smith (✉) · C. M. McHale · R. de la Rosa
Superfund Research Program, Division of Environmental Health Sciences, School of Public Health, University of California, Berkeley, CA, USA
e-mail: martynts@berkeley.edu; cmchale@berkeley.edu; rmd1025@berkeley.edu

S. Dagnino, A. Macherone (eds.), *Unraveling the Exposome*,
https://doi.org/10.1007/978-3-319-89321-1_1

Cumulative Risk and the Exposome

The US EPA defines cumulative risk assessment as "Combined risks from aggregate exposures to multiple agents or stressors, where agents or stressors may include chemical and nonchemical stressors" (US EPA 2003). This definition fits well with the exposome paradigm where all nongenetic factors contributing to disease are considered to be "environmental" including industrial chemicals, drugs, infectious agents, and psychosocial stress (Rappaport and Smith 2010). All of these nongenetic factors may be considered cumulatively as environmental stressors. Exposomics can be conceptualized through the comprehensive analysis of exposure to all environmental stressors and should yield a more thorough understanding of chronic disease development. It can also be used in the context of cumulative risk assessment. Since the goal of cumulative risk assessment is to analyze, characterize, and quantify the combined risks to health or the environment from exposures to multiple agents or stressors, it seems that exposomics is perfectly poised to advance this important area of environmental health science.

Chronic disease is rarely caused by a single factor, e.g., an individual gene or a single chemical exposure. Chemical exposures, be they occupational or environmental, do not occur in isolation and can be modified by other lifestyle and biological factors, such as stress, obesity, concurrent tobacco smoking, and chronic infections (e.g., hepatitis B virus). Collectively, these factors play a role in determining a person's susceptibility to chemical exposures. As such, their cumulative risk can vary from very high to very low depending on how many susceptibility factors they possess (Fig. 1.1). This is illustrated by our recent findings on the health

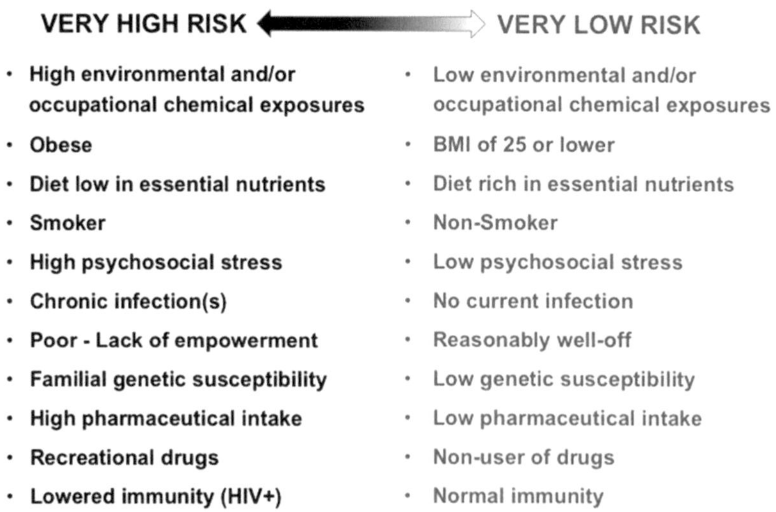

Fig. 1.1 High vs. low cumulative risk factors. Combinations of factors underlie the degree of risk of disease that an individual has at a given time

effects of arsenic that show that arsenic's carcinogenic effects are synergistically higher in obese individuals, smokers, and in those with concurrent occupational exposures (Steinmaus et al. 2015). Thus, to understand human susceptibility to disease we must look beyond genetics and account for all forms of exposure as well as intrinsic factors that confer additional vulnerability, including sex, life stage, and health status.

Cumulative risk is particularly high among ethnic minorities and low-income communities, as these groups are disproportionately exposed to environmental pollutants and psychosocial stressors that may interact to increase disease susceptibility (Morello-Frosch and Shenassa 2006; Solomon et al. 2016). For example, asthma prevalence is higher in Puerto Rican children than in any other ethnic group in the United States and this population also experiences the largest proportion of families below the poverty level, high exposure to allergens and community violence, poor dietary habits, and limited access to health care (Szentpetery et al. 2016). A study conducted in a Mexican American population from the Salinas Valley found that the association between prenatal organophosphate pesticide exposure and childhood IQ was stronger in children that experienced greater adversity (Stein et al. 2016). The adverse effects of air pollution, such as asthma exacerbations in children (Lee et al. 2006; Lin et al. 2004; Neidell 2004) and respiratory deaths in elderly adults (Martins et al. 2004), are worse for individuals of poor socioeconomic status (SES). Similar associations were observed at the community level. Low neighborhood-level SES is associated with elevated risk of air pollution-related asthma (Shmool et al. 2014), preterm births (Ponce et al. 2005), and adult mortality (Finkelstein et al. 2003). The social environment may promote vulnerability to physical exposures through the induction of chronic psychological stress. Allostatic load is a composite measure of the "wear and tear" on physiological systems caused by chronic stress. High allostatic load was reported to amplify the impact of lead on diastolic blood pressure in adults (McEwen 1998; Zota et al. 2013). Life stress and exposure to violence also increase asthma risk and related symptoms in children exposed to traffic-related air pollution (Chen et al. 2008; Clougherty et al. 2007). Prenatal exposure to psychosocial stress may further enhance fetal susceptibility to environmental pollutants. Maternal stress during pregnancy was found to interact with PAH exposure to impact childhood IQ (Vishnevetsky et al. 2015) and neurobehavioral development (Perera et al. 2013). In addition to chronic stress, the lack of adequate coping mechanisms and limited health promoting resources (e.g., green-space and recreational programs) in communities without social and economic resources may further increase susceptibility to chemical exposures (Morello-Frosch et al. 2011). Exposomics may help to inform researchers and policy makers of the cumulative risk in these communities and could be used to monitor the impact of interventions aimed at mitigating inequities in exposure to environmental stressors. More importantly perhaps, exposomics could provide clues as to the mechanisms behind these adverse interactions allowing for effective interventions.

I $\times$ E Is a Recent Concept Incorporating Multiple Environmental Stressors

As researchers have historically viewed disease susceptibility as being mainly genetic, genomic-era research has largely focused on genome-wide association studies and to a lesser degree on gene–environment (G $\times$ E) interaction studies. In these studies, "environment" (E) has typically comprised a limited range of factors such as occupational chemical exposures, infections, and lifestyle factors and reported mainly modest associations with adverse outcomes (Simonds et al. 2016). However, as mentioned earlier, environmental stressors such as chemical exposures, diet, and infection interact with each other against a background of multiple intrinsic factors that modulate risk, such as preexisting health conditions. Consequentially, the causes of disease are complex and people exhibit varying degrees of vulnerability to the adverse health effects of chemical exposures and other stressors. In the context of environmental health and policy, Morello-Frosch et al. (2011) described how intrinsic susceptibility, related to biological traits such as age, genetics, and preexisting health conditions, and extrinsic social vulnerability, related to social constructs such as race and class, interact with environmental hazard inequalities to exacerbate health disparities. Recently, we expanded this concept to encompass a broader range of intrinsic factors (I) that respond to and modulate susceptibility to multiple extrinsic (E) factors including traditional environmental stressors such as occupational and nonoccupational chemical exposures, as well as nutritional, lifestyle, and socioeconomic factors (McHale et al. 2018). We described a scenario whereby the interaction of combinations of I and E factors (I $\times$ E) determines where an individual falls on the spectrum of health and disease throughout their lifetime. We proposed expansion of a G $\times$ E-based approach to an I $\times$ E-based approach that would require the consideration of all extrinsic factors, including the exposome, and the many interrelated intrinsic biological factors that contribute to disease susceptibility (Gonzalez-Bulnes et al. 2016; Simonds et al. 2016; Lill 2016), e.g., allostatic load, nutritional status, immune status, and even age. Exposomics tools may enable the measurement of many I and E parameters and allow improved quantification of individual and population risk (McHale et al. 2018).

Using Exposomics to Measure the Environmental Factors Underlying Cumulative Risk

Classical risk assessment largely focuses on measuring singular extrinsic environmental exposure factors via traditional exposure assessment, and estimates interindividual variability from intrinsic factors by imprecise variability factors for toxicokinetics and toxicodynamics (Hattis et al. 1999; Renwick and Lazarus 1998). Various approaches have been taken to perform cumulative risk assessments such as using tiered or phased approaches to identify relevant factors (Menzie et al. 2007;

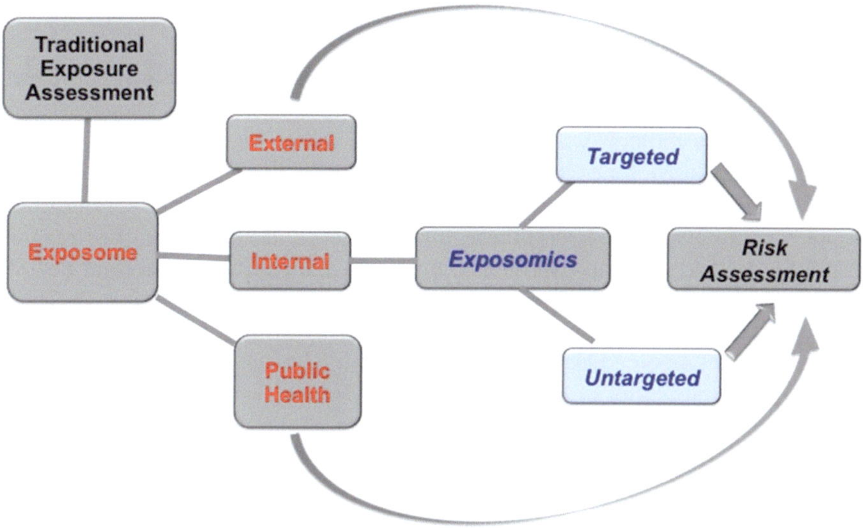

Fig. 1.2 The exposome in practice. Analysis of the components of the exposome by targeted and untargeted exposomics methods can complement traditional exposure assessment and improve risk assessment

Moretto et al. 2017; National Research Council 2009), considering interaction among factors that affect common adverse outcomes such as lead/methylmercury/polychlorinated biphenyls, nutritional factors, social deprivation and IQ (National Research Council 2008) or multiple endocrine disruptors and reproductive tract development (Rider et al. 2010). However, these methods do not comprehensively account for the complex exposures that an individual faces over a lifetime. New approaches are needed to adequately measure the combined effects of multiple environmental stressors on human health.

All chemical and nonchemical stressors mediate their effects on the body via signaling of small molecules. For example, acute psychological stress triggers the secretion of adrenaline and other hormones into the bloodstream that have physiological effects on breathing, heart rate, and blood pressure. Thus, the amount of biologically active chemicals (small molecules) present in the body's internal environment stem from both extrinsic and intrinsic sources (Rappaport and Smith 2010). The new field of exposomics should attempt to measure as many small molecules as possible in human bodily fluids. This would be a measure of the internal exposome (Fig. 1.2). Further, it should attempt to link the presence of these small molecules with functional changes in biology leading to chronic illnesses. The internal measurements made in exposomics could be of individual chemicals, groups of chemicals, or the totality of chemicals acting on a particular receptor or biological pathway in a functional assay. Examples of current targeted and untargeted exposomic methods and promising future strategies are described in the following sections. Hence, exposomics can be operationalized by studying all the small molecules in the body and their influence on biological pathways that lead to impaired health.

Another, completely different approach to examine the relationships between environmental exposures and disease is to measure exposures to various environmental stressors through wearable and regional sensors and survey instruments. These are being used, for example, to measure exposure to air pollution and drinking water contaminants; to better assess the diet through smartphone capture of dietary habits; and, to evaluate exercise through pedometers and other sensor devices that could be used to measure exposure to air pollutants over extended time periods (Turner et al. 2017). These are measures of the external exposome (Fig. 1.2). These tools are, of course, useful for improving exposure assessment in targeted epidemiology studies of specific risk factors and for avoiding known risks. They are limited, however, in their ability to identify novel environmental causes of disease, but in combination with internal exposomics tools they could be a useful approach to assessing risk (Fig. 1.2). The Public Health Exposome framework proposed by Juarez et al. (2014) recognizes and assesses both Individual and Community level vulnerability and espouses a population-based, exposure science approach to health disparities research. The multi-disciplinary methods, tools, and terms used by this approach are discussed in further detail elsewhere in this book.

Cumulative risk assessment goals overlap with those of exposomics. Both fields pursue common goals: (1) to redefine "exposures" as "stressors," encompassing not just chemical but biological, physical, and psychosocial elements; (2) to assess multiple stressor sources, exposure routes, and endpoints; (3) to assess mixture or synergistic effects; and (4) to prioritize vulnerable populations. The exposome encompasses the totality of internal and external human exposures that occur at both the individual and population level. Therefore, a comprehensive approach to risk assessment could come from the analysis of data on the internal exposome generated using exposomics tools, together with information obtained from the external and Public Health exposomes (Fig. 1.2). This integrative approach would advance the end goal of cumulative risk assessment, which is to inform risk management decisions that aim to protect public health.

Targeted and Untargeted Methods to Measure the Internal Exposome

Modern mass spectrometry and nuclear magnetic resonance (NMR) now allow us to measure pharmaceuticals, lipids, and other dietary components with relative ease and are being expanded to untargeted methods which can measure thousands of molecular ions (Rappaport et al. 2014) (Table 1.1). Psychosocial stress could be measured by various markers including telomere length, cortisol and amylase levels, and activity through stress response pathways, such as the glucocorticoid receptor pathway (Table 1.1). It is also important to measure current and prior exposures to

Table 1.1 Targeted and untargeted methods to measure the internal exposome

	Untargeted		Targeted	
Measurement	Method	No. Analytes	Method	No. Analytes
Metabolomics	MS & NMR	300,000	MS & NMR	100s
Adductomics	LC-MS/MS	100–1000	LC-MS/MS	20–30
Infectome	Subtractive sequencing	100–1000s	Targeted arrays	100s
Transcriptome/methylome	Sequencing	100,000s	Arrays, qPCR	10s-1000s
Inflammation			Cytokine panels	20–30
Stress measures	CALUX bioassay	Total glucocorticogenic activity	Individual assays	Cortisol, amylase, telomere length
Metals			ICP-MS	~20
Known pollutants			MS	100s
Recepter-ome	CALUX bioassays	Total ER, AR, AhR	MS	Individual hormones

MS mass spectrometry, *NMR* nuclear magnetic resonance, *LC-MS/MS* liquid chromatography-MS/MS, *CALUX* Chemical-Activated LUciferase gene eXpression, *ICP-MS* inductively coupled plasma mass spectrometry, *ER* estrogen receptor, *AR* androgen receptor, *AhR* Aryl hydrocarbon receptor

infectious agents, as they can play an important role in chronic disease development. Various methods for the targeted and untargeted analysis of these and other endpoints are shown in Table 1.1.

Metabolomics

The concentrations of circulating small molecules in human blood span 11 orders of magnitude (from fM to nM) and include over 100 chemical classes (Rappaport et al. 2014). Pollutant-derived molecules and metals are present at concentrations three orders of magnitude lower than molecules derived from foods, drugs, and endogenous processes. Though untargeted high-resolution metabolomics (HRM) is becoming more affordable and routine, chemical identification and the detection of molecules that are at low abundance in a population remain challenging (Jones 2016). Agilent Technologies has developed a sensitive, targeted method using a quadrupole GC-MS/MS system that can measure 60 persistent organic pollutants (POPs) in a small amount of human plasma or serum, including polychlorinated biphenyls (PCBs), polybrominated diphenyl ethers (PBDEs), organochlorine pesticides, polycyclic aromatic hydrocarbons (PAHs), furans, and dioxins (Macherone et al. 2015). The human salivary metabolome was recently been shown to be a useful source of relevant molecular information by mass spectrometry (Bessonneau et al. 2017). As saliva is easier and more practical to collect than blood, it could inform intraindividual variation in biomarker levels in longitudinal studies.

Adductomics

Reactive electrophiles are a toxic subset of the blood exposome that damages DNA and functional proteins (Liebler 2008). Various electrophilic molecules, including reactive oxygen species (ROS), are generated through the metabolism of exogenous chemicals, such as benzene, polycyclic aromatic hydrocarbons and omega-6 polyunsaturated fatty acids (PUFAs), and endogenous processes such as lipid peroxidation (Rubino et al. 2009; Liebler 2008). Small reactive electrophiles cannot be readily measured directly in vivo as they are short-lived. Adducts of reactive electrophile adducts can, however, be detected in abundant relatively long-lived blood proteins, particularly hemoglobin (Hb) and human serum albumin (HSA) (Rubino et al. 2009; Ehrenberg et al. 1996). Measurement of these adducts can be used to identify and infer the systemic concentrations of electrophilic precursor molecules (Rubino et al. 2009; Granath et al. 1992). For example, the Rappaport Research Group developed a "Cys34 blood adductomics" method (Rappaport et al. 2012), targeting the highly reactive Cys34 residue in HSA (Aldini et al. 2008; Carballal et al. 2003). An untargeted analytical scheme and bioinformatics pipeline can be used to detect, quantitate, and annotate Cys34 adducts in tryptic digests of human serum or plasma (Grigoryan et al. 2016). This untargeted method was used to identify adducts in plasma that could distinguish smokers from nonsmokers (Grigoryan et al. 2016) and users of smoky coal from users of smokeless coal and wood (Lu et al. 2017). Some of the products identified included Cys34 oxidation products, methylation and addition products of Cys34 representing cyanide, benzaldehyde, benzoquinone, methoxybenzoquinone, acetonitrile, crotonaldehyde, and acrylonitrile arising from diverse endogenous sources, including production of ROS, lipid peroxidation, and products of microbial metabolism. As these and other products are validated, subsets of them can be analyzed in targeted adductomic approaches. Recently, the same group identified clusters of adducts associated with occupational exposure to benzene in non-smoking Chinese factory workers (Grigoryan et al. 2018).

Infectome

As blood-based RNA sequencing (RNA-seq) captures data on both the human and nonhuman transcriptomes present, subtractive computational methods have been developed to characterize the nonhuman transcripts (Kostic et al. 2011; Feng et al.

2007; Moore et al. 2011; Naeem et al. 2013; Weber et al. 2002; Xu et al. 2003). For example, RNA CoMPASS simultaneously analyzes transcriptomes and metatranscriptomes from diverse biological specimens (Xu et al. 2014). It enabled the detection of Epstein-Barr Virus (EBV) in Burkitt's lymphoma associated with EBV infection. VirusSeq is an algorithmic method that accurately detects known viruses and their integration sites with high sensitivity and specificity in a variety of cancer types (Chen et al. 2013; Khoury et al. 2013). VirScan detects antiviral antibodies in just 1 μL of serum with high sensitivity and specificity (Xu et al. 2015). The method uses immunoprecipitation and massively parallel DNA sequencing of a bacteriophage library that displays peptides from all human viruses.

Transcriptome/Methylome

Transcriptomic signatures represent both biomarkers of exposure and effect, reflecting the unique interaction between genetic background and exposure levels in an individual. Transcriptomes can be analyzed in an untargeted manner by RNA-Seq or in a targeted manner by microarray, L1000/S1500, NanoString platforms, and quantitative or digital PCR (McHale et al. 2016). Epigenetic mechanisms such as DNA methylation modulate the gene expression response to environmental exposures (Feil and Fraga 2012). An altered DNA methylome may represent a stable signature of past environmental exposure (Feil and Fraga 2012; Herceg et al. 2013). DNA methylomes can be analyzed in an untargeted manner using whole genome bisulfite sequencing (majority of CpG sites) or in a targeted manner using the more affordable reduced representation bisulfite sequencing (RRBS, one to two million sites) or array technology (Yong et al. 2016). Illumina's Infinium MethylationEPIC BeadChip Array interrogates DNA methylation at 850,000 CpG sites. Array-based methods include a hybridization step and are more susceptible to batch effects than the sequencing-based methods. Integrative or cross-omics approaches leverage the power of analyzing multiple omic datasets to inform mechanisms (Rotroff and Motsinger-Reif 2016; Yan et al. 2017). For example, a transcriptome-metabolome wide association study design was used to explore complex cellular responses to a combined exposure to a fungicide, maneb (MB), and an herbicide, paraquat (PQ) (Roede et al. 2013). Data showed that key mechanistic hubs could be identified based on the strength of correlations of genes and metabolites.

Inflammation

Pro-inflammatory cytokine/chemokine profiling in serum or plasma can be performed by comprehensive, validated bead-based immunoassays. For example, xMAP assays generally demonstrate sensitivity at physiologically relevant levels <10 pg/mL (based on detectable signal >2 standard deviations above background);

have inter- and intra-assay CVs of <10% (Dossus et al. 2009; Dupont et al. 2005); >90% correlation with data obtained by more conventional enzyme-linked immunosorbent assay (ELISA) assays (Dossus et al. 2009; Dupont et al. 2005; Elshal and McCoy 2006); and high equivalency between recombinant standards and natural samples ($R^2 > 0.99$).

Stress

Cortisol, the endogenous glucocorticoid present in humans, is released in response to stress. Total glucocorticogenic activity in plasma or serum can be measured using a Chemical-Activated LUciferase gene eXpression (CALUX) glucocorticoid receptor (GR)-based reporter bioassay, which measures the overall net effect of both endogenous and exogenous molecules acting on the glucocorticoid receptor simultaneously (Fejerman et al. 2016). This assay links a luciferase reporter gene to one or more glucocorticoid response elements, thus providing a light-based readout that is proportional to the degree of receptor activity. Targeted stress-response assays include the measurement of salivary alpha amylase (sAA) and salivary cortisol, as noninvasive markers of the sympathetic nervous system (SNS) and hypothalamic–pituitary–adrenal system (HPA) axis stress responses, respectively. sAA activity can be measured using an enzymatic assay and salivary cortisol can be measured by immunoassay. Telomere shortening is an established marker of chronic psychological stress and cellular aging (Blackburn and Epel 2012). Relative mean telomere length (TL) can be measured by a monochrome multiplex quantitative real-time PCR assay (Zalli et al. 2014; Steptoe et al. 2011).

Metals

Metals are a component of the blood exposome with a broad dynamic range (Rappaport et al. 2014). They can be measured in plasma by sensitive inductively coupled plasma mass spectrometry (ICP-MS) (Guan et al. 2017; Yu et al. 2013).

Recepter-Ome

Exposure to endocrine disrupting chemicals (EDCs) interferes with a variety of endocrine pathways, including estrogen, androgen, thyroid, retinol, aryl hydrocarbon, and peroxisome proliferator-activated receptor pathways and likely contributes substantially to multiple diseases and dysfunction (Trasande et al. 2015). We have developed sensitive and reproducible CALUX assays to measure the activity of aryl hydrocarbon receptor (AhR), estrogen receptor α and β (ERα and ERβ), androgen receptor (AR), and glucocorticoid receptor (GR) in human serum and plasma (Fejerman et al. 2016). In the future, CALUX assays can be developed for additional endocrine receptors such as PPAR and RAR to expand the receptor-ome.

The Utility of Exposome Studies in Vulnerable Populations

Smith et al. (2015) have proposed using exposomics tools to quantify cumulative risk and suggested engaging impacted communities in participatory exposome research. Potential populations of interest include migrant populations, newborn infants, and communities located near multiple sources of pollution who are exposed to multiple stressors.

Studies in Migrant Populations

The world is a constantly evolving place and people migrate to other countries for many different reasons. Migrant populations are ideal for exposomics research. These populations were exposed to different environmental stressors in early-life and may have made changes in behavior due to acculturation as compared to the native populations in their new and former residences. This leads to profound differences in disease incidence rates that could be driven by both environmental and genetic factors. For instance, Hispanic immigrants have a longer life expectancy and lower infant mortality rates than their US-born counterparts, a phenomenon often referred to as the "Hispanic Paradox (Markides and Coreil 1986)". Compared to Caucasian populations, South Asian Indians are at fourfold higher risk of type 2 diabetes (T2D) and begin to develop insulin resistance at a relatively lower BMI and younger age of onset (reviewed in Bakker et al. 2013). While various socio-demographic and lifestyle factors, such as social support, access to healthcare, the "healthy immigrant effect," and the acculturation process are believed to contribute to differences in immigrant health status, the biological processes underlying these disparities in disease vulnerability remain unknown (Singh and Miller 2004). Furthermore, these societal factors may enhance susceptibility to the adverse affects of chemical exposures that are more prevalent among these ethnic groups (e.g., pesticides) (Cox et al. 2007; Jaacks and Staimez 2015). Using an exposomics approach to conduct cumulative risk assessments allows us to examine differences in disease onset in immigrant populations. For example, this approach may help to resolve enigmas such as the "Hispanic Paradox," and the "South Asian Phenotype" of diabetes. We are currently taking an exposomics approach in an attempt to explain these two unique phenotypes in immigrant Latino and Asian Indian populations.

Prenatal Period as a Window of Susceptibility

Another population that may be well suited to exposomic analysis is pregnant women and their newborn infants. The prenatal period is a critical window of susceptibility since most organ systems are formed in utero. Individuals born during the Dutch "hunger winter" of the Second World War report increased risk of cardiovascular and metabolic disease in adulthood (Ravelli et al. 1976). From these observations emerged the Developmental Origins of Health and Disease (DOHaD) theory, which hypothesizes that prenatal exposures impact health later in life (Barker 2004). Chemical exposures that occur during pregnancy can also have persistent effects on the offspring. For example, women exposed to diethylstilbestrol (DES) in utero are at increased risk for rare cancers and adverse reproductive effects (Newbold 1995). Furthermore, there is evidence demonstrating that maternal stress may amplify the negative associations between prenatal chemical exposures and fetal growth (Vesterinen et al. 2017). In the past decade, there has been increasing interest in understanding the mechanism linking prenatal exposures to adverse health effects in adulthood (Haugen et al. 2015). Exposomics tools can be applied to biosamples from existing prospective cohort studies (e.g., mid-pregnancy maternal blood, cord blood, Guthrie card blood spots) and analyzed in relation to disease incidence later in life. This approach would be able to assess the simultaneous effects of multiple stressors during this critical period of development and elucidate biological pathways associated with disease etiology.

Identification of Vulnerable Communities Using Indicator- and Map-Based Approaches

Communities located near multiple sources of pollution are often disproportionately exposed to social stressors. Indicator and map-based approaches can identify cumulative impacts to such communities. These methods rely on indicators of extrinsic environmental exposures and population vulnerability and use models based upon basic risk assessment concepts. For example, the California Communities Environmental Health Screening Tool, CalEnviroscreen was developed by the California Environmental Protection agency (CalEPA) as part of the agency's Environmental Justice Action Plan (Meehan August et al. 2012). The tool, which was developed in conjunction with extensive community input, includes indicators that assess pollutants (exposures), environmental threats and adverse conditions (environmental effects), populations with biological traits that may exacerbate the effects of pollutants (sensitive populations), and community characteristics that may increase vulnerability to pollutants (socioeconomic factors). Updated in 2017 (OEHHA 2017), the CalEnviroScreen 3.0 tool calculates a community disadvantage score (CES Score) based on the assessment of the weighted contribution of 19 individual indicators related to pollution burden and population characteristics at the census tract level. Studies using CalEnviroscreen have shown that environmental health hazards such as pesticide use and toxic chemical releases disproportionately burden communities of color in California (Cushing et al. 2015) and that air pollution (ozone levels) and socioeconomic disadvantage contributed to geographic disparities in ovarian cancer survival in California (Vieira et al. 2017).

Semi-quantitative indicator-based approaches can incorporate multiple factors that are precluded from traditional risk assessments and the resulting tools paint a broad picture of the cumulative burdens and vulnerabilities faced by different communities. Indicator-based approaches can incorporate an almost unlimited number of stressors for which data are available, and can combine very different types of stressors (chemical, psychosocial, health) into one analysis. Rather than producing estimates of health risk for any given population, such approaches are intended to facilitate comparisons of communities differentially impacted by intrinsic and environmental factors and generate hypotheses for further evaluation. Exposomic approaches could be then applied.

Strategies for Future Exposomics Research

There is some debate over the best strategies to use for exposomics research, given the limitations of both targeted and untargeted methods (Rappaport et al. 2014). While untargeted methods provide promise in examining 1000s of molecules simultaneously, some sensitivity is sacrificed and this indicates the additional need for targeted methods with improved sensitivity. Therefore, it is important to incorporate both approaches in exposomics research.

The advantage of untargeted methods is the potential for discovery of novel analytes while measuring hundreds to thousands of compounds simultaneously. To improve upon characterization of "active" molecules in the blood by metabolomics, the method could be paired with other assays that quantify the net effect of endogenous and exogenous compounds in human serum on specific biological pathways. These preliminary screening methods may allow discrimination between analytes of interest and background noise that are measured using untargeted approaches (i.e., metabolomics). An example of such methods is use of receptor-binding reporter assays in response to chemicals in human blood samples (Brouwers et al. 2011). Our group uses sensitive CALUX receptor-based reporter bioassays to measure the overall net effect of both endogenous and exogenous molecules present in human plasma on the estrogen, androgen, and glucocorticoid receptors. Recently, we measured total estrogenic (E) activity in plasma from 90 Mexican American women using a CALUX assay and found associations with Indigenous American ancestry (19% decrease in E activity per 10% increase in ancestry, $P = 0.014$) and with length of US residency (28% increase in E activity for every 10 years, $P = 0.035$) (Fejerman et al. 2016). These results suggest that differences in E activity may contribute to higher breast cancer incidence among women with low Indigenous American ancestry compared to those with high

Indigenous American ancestry and in US-born compared to foreign-born women. Moving forward, we plan to profile plasma from subjects with extremes of estrogenic activity by untargeted high-resolution mass spectrometry (HRMS) to determine which compounds may be responsible for the differing levels of receptor activity. This coupled approach will provide a list of candidate of novel estrogenic compounds that can be further studied in relation to breast cancer risk and other endocrine-related health outcomes.

Conclusions

Cumulative risk assessment can be improved in two ways: 1) by moving from a G × E paradigm to a more comprehensive I × E paradigm and (2) by applying exposomics to characterize and quantify the internal exposome. Targeted and untargeted exposomics tools can be used to measure individual chemicals, groups of chemicals, or the totality of chemicals acting on a particular receptor or biological pathway in a functional assay. To complement the existing suite of exposomics tools, the development of additional targeted, semi-targeted, and untargeted methods is needed. These methods could be initially applied in high priority vulnerable populations such as migrant populations or communities burdened by socioeconomic and environmental stressors. Data on the internal exposome along with data on the external and public health components of the exposome have the potential to improve cumulative risk assessment, and ultimately inform risk management.

Acknowledgments We thank our lab colleagues Sarah Daniels, Sylvia Sanchez, Fenna Sille, Phum Tachachartvanich, Luoping Zhang, and Felicia Castriota and collaborators Laura Fejerman, Stephen Rappaport, Esther John, Anthony Macherone, Paul Elliott, Jaspal Kooner, John Chambers, Michele La Merrill, Craig Steinmaus, Allan Smith, Daniel Nomura, Jen-Chywan Wang, Kurt Pennell, Michael Denison, and Catherine Thomsen for their collaboration in pursuing exposomics. This work was supported by NIH grant P42 ES004705 from the National Institute of Environmental Health Sciences and award 21UB-8009 from the California Breast Cancer Research Program.

References

Aldini G, Vistoli G, Regazzoni L, Gamberoni L, Facino RM, Yamaguchi S, Uchida K, Carini M (2008) Albumin is the main nucleophilic target of human plasma: a protective role against pro-atherogenic electrophilic reactive carbonyl species? Chem Res Toxicol 21(4):824–835

Bakker LE, Sleddering MA, Schoones JW, Meinders AE, Jazet IM (2013) Pathogenesis of type 2 diabetes in South Asians. Eur J Endocrinol 169(5):R99–R114. https://doi.org/10.1530/EJE-13-0307

Barker DJ (2004) The developmental origins of adult disease. J Am Coll Nutr 23(6 Suppl):588S–595S

Bessonneau V, Pawliszyn J, Rappaport SM (2017) The saliva exposome for monitoring of individuals' health trajectories. Environ Health Perspect 125(7):077014. https://doi.org/10.1289/EHP1011

Blackburn EH, Epel ES (2012) Telomeres and adversity: too toxic to ignore. Nature 490 (7419):169–171. https://doi.org/10.1038/490169a

Brouwers MM, Besselink H, Bretveld RW, Anzion R, Scheepers PT, Brouwer A, Roeleveld N (2011) Estrogenic and androgenic activities in total plasma measured with reporter-gene bioassays: relevant exposure measures for endocrine disruptors in epidemiologic studies? Environ Int 37(3):557–564. https://doi.org/10.1016/j.envint.2010.11.001

Carballal S, Radi R, Kirk MC, Barnes S, Freeman BA, Alvarez B (2003) Sulfenic acid formation in human serum albumin by hydrogen peroxide and peroxynitrite. Biochemistry 42 (33):9906–9914. https://doi.org/10.1021/bi027434m

Chen E, Schreier HM, Strunk RC, Brauer M (2008) Chronic traffic-related air pollution and stress interact to predict biologic and clinical outcomes in asthma. Environ Health Perspect 116 (7):970–975. https://doi.org/10.1289/ehp.11076

Chen Y, Yao H, Thompson EJ, Tannir NM, Weinstein JN, Su X (2013) VirusSeq: software to identify viruses and their integration sites using next-generation sequencing of human cancer tissue. Bioinformatics 29(2):266–267. https://doi.org/10.1093/bioinformatics/bts665

Clougherty JE, Levy JI, Kubzansky LD, Ryan PB, Suglia SF, Canner MJ, Wright RJ (2007) Synergistic effects of traffic-related air pollution and exposure to violence on urban asthma etiology. Environ Health Perspect 115(8):1140–1146. https://doi.org/10.1289/ehp.9863

Cox S, Niskar AS, Narayan KM, Marcus M (2007) Prevalence of self-reported diabetes and exposure to organochlorine pesticides among Mexican Americans: Hispanic health and nutrition examination survey, 1982–1984. Environ Health Perspect 115(12):1747–1752. https://doi.org/10.1289/ehp.10258

Cushing L, Faust J, August LM, Cendak R, Wieland W, Alexeeff G (2015) Racial/ethnic disparities in cumulative environmental health impacts in California: evidence from a statewide environmental justice screening tool (CalEnviroScreen 1.1). Am J Public Health 105(11):2341–2348. https://doi.org/10.2105/AJPH.2015.302643

Dossus L, Becker S, Achaintre D, Kaaks R, Rinaldi S (2009) Validity of multiplex-based assays for cytokine measurements in serum and plasma from "non-diseased" subjects: comparison with ELISA. J Immunol Methods 350(1–2):125–132. https://doi.org/10.1016/j.jim.2009.09.001

Dupont NC, Wang K, Wadhwa PD, Culhane JF, Nelson EL (2005) Validation and comparison of luminex multiplex cytokine analysis kits with ELISA: determinations of a panel of nine cytokines in clinical sample culture supernatants. J Reprod Immunol 66(2):175–191 https://doi.org/10.1016/j.jri.2005.03.005

Ehrenberg L, Granath F, Tornqvist M (1996) Macromolecule adducts as biomarkers of exposure to environmental mutagens in human populations. [Review] [38 refs]. Environ Health Perspect 104(Suppl 3):423–428

Elshal MF, McCoy JP (2006) Multiplex bead array assays: performance evaluation and comparison of sensitivity to ELISA. Methods 38(4):317–323. https://doi.org/10.1016/j.ymeth.2005.11.010

Feil R, Fraga MF (2012) Epigenetics and the environment: emerging patterns and implications. Nat Rev Genet 13(2):97–109. https://doi.org/10.1038/nrg3142

Fejerman L, Sanchez SS, Thomas R, Tachachartvanich P, Riby J, Gomez SL, John EM, Smith MT (2016) Association of lifestyle and demographic factors with estrogenic and glucocorticogenic activity in Mexican American women. Carcinogenesis 37(9):904–911. https://doi.org/10.1093/carcin/bgw074

Feng H, Taylor JL, Benos PV, Newton R, Waddell K, Lucas SB, Chang Y, Moore PS (2007) Human transcriptome subtraction by using short sequence tags to search for tumor viruses in conjunctival carcinoma. J Virol 81(20):11332–11340. https://doi.org/10.1128/JVI.00875-07

Finkelstein MM, Jerrett M, DeLuca P, Finkelstein N, Verma DK, Chapman K, Sears MR (2003) Relation between income, air pollution and mortality: a cohort study. CMAJ 169(5):397–402

Gonzalez-Bulnes A, Astiz S, Ovilo C, Garcia-Contreras C, Vazquez-Gomez M (2016) Nature and nurture in the early-life origins of metabolic syndrome. Curr Pharm Biotechnol 17(7):573–586
Granath F, Ehrenberg L, Tornqvist M (1992) Degree of alkylation of macromolecules in vivo from variable exposure. Mutat Res 284:297–306
Grigoryan H, Edmands WMB, Lan Q, Carlsson H, Vermeulen R, Zhang L, Yin S-N, Li G-L, Smith MT, Rothman N, Rappaport SM (2018) Adductomic signatures of benzene exposure provide insights into cancer induction. Carcinogenesis 39(5):661–668
Hasmik Grigoryan, William M B Edmands, Qing Lan, Henrik Carlsson, Roel Vermeulen, Luoping Zhang, Song-Nian Yin, Gui-Lan Li, Martyn T Smith, Nathaniel Rothman, Stephen M Rappaport, (2018) Adductomic signatures of benzene exposure provide insights into cancer induction. Carcinogenesis 39 (5):661-668
Guan C, Dang R, Cui Y, Liu L, Chen X, Wang X, Zhu J, Li D, Li J, Wang D (2017) Characterization of plasma metal profiles in Alzheimer's disease using multivariate statistical analysis. PLoS One 12(7):e0178271. https://doi.org/10.1371/journal.pone.0178271
Hattis D, Banati P, Goble R (1999) Distributions of individual susceptibility among humans for toxic effects. How much protection does the traditional tenfold factor provide for what fraction of which kinds of chemicals and effects? Ann N Y Acad Sci 895:286–316
Haugen AC, Schug TT, Collman G, Heindel JJ (2015) Evolution of DOHaD: the impact of environmental health sciences. J Dev Orig Health Dis 6(2):55–64. https://doi.org/10.1017/S2040174414000580
Herceg Z, Lambert MP, van Veldhoven K, Demetriou C, Vineis P, Smith MT, Straif K, Wild CP (2013) Towards incorporating epigenetic mechanisms into carcinogen identification and evaluation. Carcinogenesis 34(9):1955–1967. https://doi.org/10.1093/carcin/bgt212
Jaacks LM, Staimez LR (2015) Association of persistent organic pollutants and non-persistent pesticides with diabetes and diabetes-related health outcomes in Asia: a systematic review. Environ Int 76:57–70. https://doi.org/10.1016/j.envint.2014.12.001
Jones DP (2016) Sequencing the exposome: a call to action. Toxicol Rep 3:29–45. https://doi.org/10.1016/j.toxrep.2015.11.009
Juarez PD, Matthews-Juarez P, Hood DB, Im W, Levine RS, Kilbourne BJ, Langston MA, Al-Hamdan MZ, Crosson WL, Estes MG, Estes SM, Agboto VK, Robinson P, Wilson S, Lichtveld MY (2014) The public health exposome: a population-based, exposure science approach to health disparities research. Int J Environ Res Public Health 11(12):12866–12895. https://doi.org/10.3390/ijerph111212866
Khoury JD, Tannir NM, Williams MD, Chen Y, Yao H, Zhang J, Thompson EJ, Network T, Meric-Bernstam F, Medeiros LJ, Weinstein JN, Su X (2013) Landscape of DNA virus associations across human malignant cancers: analysis of 3,775 cases using RNA-Seq. J Virol 87 (16):8916–8926. https://doi.org/10.1128/JVI.00340-13
Kostic AD, Ojesina AI, Pedamallu CS, Jung J, Verhaak RG, Getz G, Meyerson M (2011) PathSeq: software to identify or discover microbes by deep sequencing of human tissue. Nat Biotechnol 29(5):393–396. https://doi.org/10.1038/nbt.1868
Lee JT, Son JY, Kim H, Kim SY (2006) Effect of air pollution on asthma-related hospital admissions for children by socioeconomic status associated with area of residence. Arch Environ Occup Health 61(3):123–130. https://doi.org/10.3200/AEOH.61.3.123-130
Liebler DC (2008) Protein damage by reactive electrophiles: targets and consequences. Chem Res Toxicol 21(1):117–128. https://doi.org/10.1021/tx700235t
Lill CM (2016) Genetics of Parkinson's disease. Mol Cell Probes 30(6):386–396. https://doi.org/10.1016/j.mcp.2016.11.001
Lin M, Chen Y, Villeneuve PJ, Burnett RT, Lemyre L, Hertzman C, McGrail KM, Krewski D (2004) Gaseous air pollutants and asthma hospitalization of children with low household income in Vancouver, British Columbia, Canada. Am J Epidemiol 159(3):294–303
Lu SS, Grigoryan H, Edmands WM, Hu W, Iavarone AT, Hubbard A, Rothman N, Vermeulen R, Lan Q, Rappaport SM (2017) Profiling the serum albumin Cys34 adductome of solid fuel users

in Xuanwei and Fuyuan, China. Environ Sci Technol 51(1):46–57. https://doi.org/10.1021/acs.est.6b03955
Macherone A, Daniels S, Maggitti A, Churley M, McMullin M, Smith MT (2015) Measuring a slice of the exposome: targeted GC-MS/MS analysis of persistent organic pollutants POPs) in small violumes of human plasma. In: 63rd AMS conference on mass spectrometry and allied topics, St. Louis, MO, p Abstract TP 309
Markides KS, Coreil J (1986) The health of Hispanics in the southwestern United States: an epidemiologic paradox. Public Health Rep 101(3):253–265
Martins MC, Fatigati FL, Vespoli TC, Martins LC, Pereira LA, Martins MA, Saldiva PH, Braga AL (2004) Influence of socioeconomic conditions on air pollution adverse health effects in elderly people: an analysis of six regions in Sao Paulo, Brazil. J Epidemiol Community Health 58 (1):41–46
McEwen BS (1998) Stress, adaptation, and disease. Allostasis and allostatic load. Ann N Y Acad Sci 840:33–44
McHale CM, Smith MT, Zhang L (2016) Application of toxicogenomics in exposed human populations: benzene as an example. Toxicogenomics in Predictive Carcinogenicity, vol 28, The Royal Society of Chemistry, Cambridge, UK
McHale CM, Osborne G, Morello-Frosch R, Salmon AG, Sandy M, Solomon G, Zhang L, Smith MT, Zeise L (2018) Assessing health risks from multiple environmental stressors: a paradigm shift from GxE to IxE. Mutat Res Rev 775:11–20. https://doi.org/10.1016/j.mrrev.2017.11.003
Meehan August L, Faust JB, Cushing L, Zeise L, Alexeeff GV (2012) Methodological considerations in screening for cumulative environmental health impacts: lessons learned from a pilot study in California. Int J Environ Res Public Health 9(9):3069–3084. https://doi.org/10.3390/ijerph9093069
Menzie CA, MacDonell MM, Mumtaz M (2007) A phased approach for assessing combined effects from multiple stressors. Environ Health Perspect 115(5):807–816. https://doi.org/10.1289/ehp.9331
Moore RA, Warren RL, Freeman JD, Gustavsen JA, Chenard C, Friedman JM, Suttle CA, Zhao Y, Holt RA (2011) The sensitivity of massively parallel sequencing for detecting candidate infectious agents associated with human tissue. PLoS One 6(5):e19838. https://doi.org/10.1371/journal.pone.0019838
Morello-Frosch R, Shenassa ED (2006) The environmental "riskscape" and social inequality: implications for explaining maternal and child health disparities. Environ Health Perspect 114 (8):1150–1153
Morello-Frosch R, Zuk M, Jerrett M, Shamasunder B, Kyle AD (2011) Understanding the cumulative impacts of inequalities in environmental health: implications for policy. Health Aff 30 (5):879–887. https://doi.org/10.1377/hlthaff.2011.0153
Moretto A, Bachman A, Boobis A, Solomon KR, Pastoor TP, Wilks MF, Embry MR (2017) A framework for cumulative risk assessment in the 21st century. Crit Rev Toxicol 47(2):85–97. https://doi.org/10.1080/10408444.2016.1211618
Naeem R, Rashid M, Pain A (2013) READSCAN: a fast and scalable pathogen discovery program with accurate genome relative abundance estimation. Bioinformatics 29(3):391–392. https://doi.org/10.1093/bioinformatics/bts684
National Research Council (2008) Phthalates and cumulative risk assessment. National Academy Press, Washington, DC
National Research Council (2009) Science and decisions: advancing risk assessment. committee on improving risk analysis approaches used by the U.S. EPA, Board on Environmental Studies and Toxicology, Division on Earth and Life Studies, Washington DC
Neidell MJ (2004) Air pollution, health, and socio-economic status: the effect of outdoor air quality on childhood asthma. J Health Econ 23(6):1209–1236. https://doi.org/10.1016/j.jhealeco.2004.05.002
Newbold R (1995) Cellular and molecular effects of developmental exposure to diethylstilbestrol: implications for other environmental estrogens. Environ Health Perspect 103(Suppl 7):83–87

OEHHA (2017) California Communities Environmental Health, Version 3.0 (CalEnviroscreen 3.0) screening tool. OEHHA, California Environmental Protection agency. https://oehha.ca.gov/calenviroscreen/report/calenviroscreen-30

Perera FP, Wang S, Rauh V, Zhou H, Stigter L, Camann D, Jedrychowski W, Mroz E, Majewska R (2013) Prenatal exposure to air pollution, maternal psychological distress, and child behavior. Pediatrics 132(5):e1284–e1294. https://doi.org/10.1542/peds.2012-3844

Ponce NA, Hoggatt KJ, Wilhelm M, Ritz B (2005) Preterm birth: the interaction of traffic-related air pollution with economic hardship in Los Angeles neighborhoods. Am J Epidemiol 162 (2):140–148. https://doi.org/10.1093/aje/kwi173

Rappaport SM, Smith MT (2010) Epidemiology. Environment and disease risks. Science (New York, NY) 330(6003):460–461. https://doi.org/10.1126/science.1192603

Rappaport SM, Li H, Grigoryan H, Funk WE, Williams ER (2012) Adductomics: characterizing exposures to reactive electrophiles. Toxicol Lett 213(1):83–90. https://doi.org/10.1016/j.toxlet.2011.04.002

Rappaport SM, Barupal DK, Wishart D, Vineis P, Scalbert A (2014) The blood exposome and its role in discovering causes of disease. Environ Health Perspect 122(8):769–774. https://doi.org/10.1289/ehp.1308015

Ravelli GP, Stein ZA, Susser MW (1976) Obesity in young men after famine exposure in utero and early infancy. N Engl J Med 295(7):349–353. https://doi.org/10.1056/NEJM197608122950701

Renwick AG, Lazarus NR (1998) Human variability and noncancer risk assessment—an analysis of the default uncertainty factor. Regul Toxicol Pharmacol 27(1 Pt 1):3–20

Rider CV, Furr JR, Wilson VS, Gray LE Jr (2010) Cumulative effects of in utero administration of mixtures of reproductive toxicants that disrupt common target tissues via diverse mechanisms of toxicity. Int J Androl 33(2):443–462. https://doi.org/10.1111/j.1365-2605.2009.01049.x

Roede JR, Uppal K, Park Y, Lee K, Tran V, Walker D, Strobel FH, Rhodes SL, Ritz B, Jones DP (2013) Serum metabolomics of slow vs. rapid motor progression Parkinson's disease: a pilot study. PLoS One 8(10):e77629. https://doi.org/10.1371/journal.pone.0077629

Rotroff DM, Motsinger-Reif AA (2016) Embracing integrative multiomics approaches. Int J Genomics 2016:1715985. https://doi.org/10.1155/2016/1715985

Rubino FM, Pitton M, Di Fabio D, Colombi A (2009) Toward an "omic" physiopathology of reactive chemicals: thirty years of mass spectrometric study of the protein adducts with endogenous and xenobiotic compounds. Mass Spectrom Rev 28(5):725–784. https://doi.org/10.1002/mas.20207

Shmool JL, Kubzansky LD, Newman OD, Spengler J, Shepard P, Clougherty JE (2014) Social stressors and air pollution across New York City communities: a spatial approach for assessing correlations among multiple exposures. Environ Health 13:91. https://doi.org/10.1186/1476-069X-13-91

Simonds NI, Ghazarian AA, Pimentel CB, Schully SD, Ellison GL, Gillanders EM, Mechanic LE (2016) Review of the gene-environment interaction literature in cancer: what do we know? Genet Epidemiol 40(5):356–365. https://doi.org/10.1002/gepi.21967

Singh GK, Miller BA (2004) Health, life expectancy, and mortality patterns among immigrant populations in the United States. Can J Public Health 95(3):I14–I21

Smith MT, de la Rosa R, Daniels SI (2015) Using exposomics to assess cumulative risks and promote health. Environ Mol Mutagen 56(9):715–723. https://doi.org/10.1002/em.21985

Solomon GM, Morello-Frosch R, Zeise L, Faust JB (2016) Cumulative environmental impacts: science and policy to protect communities. Annu Rev Public Health 37:83–96. https://doi.org/10.1146/annurev-publhealth-032315-021807

Stein LJ, Gunier RB, Harley K, Kogut K, Bradman A, Eskenazi B (2016) Early childhood adversity potentiates the adverse association between prenatal organophosphate pesticide exposure and child IQ: the CHAMACOS cohort. Neurotoxicology 56:180–187. https://doi.org/10.1016/j.neuro.2016.07.010

Steinmaus C, Castriota F, Ferreccio C, Smith AH, Yuan Y, Liaw J, Acevedo J, Perez L, Meza R, Calcagno S, Uauy R, Smith MT (2015) Obesity and excess weight in early adulthood and high risks of arsenic-related cancer in later life. Environ Res 142:594–601. https://doi.org/10.1016/j.envres.2015.07.021

Steptoe A, Hamer M, Butcher L, Lin J, Brydon L, Kivimaki M, Marmot M, Blackburn E, Erusalimsky JD (2011) Educational attainment but not measures of current socioeconomic circumstances are associated with leukocyte telomere length in healthy older men and women. Brain Behav Immun 25(7):1292–1298. https://doi.org/10.1016/j.bbi.2011.04.010

Szentpetery SE, Forno E, Canino G, Celedon JC (2016) Asthma in Puerto Ricans: lessons from a high-risk population. J Allergy Clin Immunol 138(6):1556–1558. https://doi.org/10.1016/j.jaci.2016.08.047

Trasande L, Zoeller RT, Hass U, Kortenkamp A, Grandjean P, Myers JP, DiGangi J, Bellanger M, Hauser R, Legler J, Skakkebaek NE, Heindel JJ (2015) Estimating burden and disease costs of exposure to endocrine-disrupting chemicals in the European union. J Clin Endocrinol Metab 100(4):1245–1255. https://doi.org/10.1210/jc.2014-4324

Turner MC, Nieuwenhuijsen M, Anderson K, Balshaw D, Cui Y, Dunton G, Hoppin JA, Koutrakis P, Jerrett M (2017) Assessing the exposome with external measures: commentary on the state of the science and research recommendations. Annu Rev Public Health 38:215–239. https://doi.org/10.1146/annurev-publhealth-082516-012802

US EPA (2003) Framework for cumulative risk assessment. vol EPA/600/P-02/001F. U.S. Environmental Protection Agency, Office of Research and Development, National Center for Environmental Assessment, Washington, DC

Vesterinen HM, Morello-Frosch R, Sen S, Zeise L, Woodruff TJ (2017) Cumulative effects of prenatal-exposure to exogenous chemicals and psychosocial stress on fetal growth: systematic-review of the human and animal evidence. PLoS One 12(7):e0176331. https://doi.org/10.1371/journal.pone.0176331

Vieira VM, Villanueva C, Chang J, Ziogas A, Bristow RE (2017) Impact of community disadvantage and air pollution burden on geographic disparities of ovarian cancer survival in California. Environ Res 156:388–393. https://doi.org/10.1016/j.envres.2017.03.057

Vishnevetsky J, Tang D, Chang HW, Roen EL, Wang Y, Rauh V, Wang S, Miller RL, Herbstman J, Perera FP (2015) Combined effects of prenatal polycyclic aromatic hydrocarbons and material hardship on child IQ. Neurotoxicol Teratol 49:74–80. https://doi.org/10.1016/j.ntt.2015.04.002

Weber G, Shendure J, Tanenbaum DM, Church GM, Meyerson M (2002) Identification of foreign gene sequences by transcript filtering against the human genome. Nat Genet 30(2):141–142. https://doi.org/10.1038/ng818

Xu G, Strong MJ, Lacey MR, Baribault C, Flemington EK, Taylor CM (2014) RNA CoMPASS: a dual approach for pathogen and host transcriptome analysis of RNA-seq datasets. PLoS One 9(2):e89445. https://doi.org/10.1371/journal.pone.0089445

Xu GJ, Kula T, Xu Q, Li MZ, Vernon SD, Ndung'u T, Ruxrungtham K, Sanchez J, Brander C, Chung RT, O'Connor KC, Walker B, Larman HB, Elledge SJ (2015) Viral immunology. Comprehensive serological profiling of human populations using a synthetic human virome. Science 348(6239):aaa0698. https://doi.org/10.1126/science.aaa0698

Xu Y, Stange-Thomann N, Weber G, Bo R, Dodge S, David RG, Foley K, Beheshti J, Harris NL, Birren B, Lander ES, Meyerson M (2003) Pathogen discovery from human tissue by sequence-based computational subtraction. Genomics 81(3):329–335

Yan J, Risacher SL, Shen L, Saykin AJ (2017) Network approaches to systems biology analysis of complex disease: integrative methods for multi-omics data. Brief Bioinform. https://doi.org/10.1093/bib/bbx066

Yong WS, Hsu FM, Chen PY (2016) Profiling genome-wide DNA methylation. Epigenetics Chromatin 9:26. https://doi.org/10.1186/s13072-016-0075-3

Yu LL, Davis WC, Nuevo Ordonez Y, Long SE (2013) Fast and accurate determination of K, Ca, and Mg in human serum by sector field ICP-MS. Anal Bioanal Chem 405(27):8761–8768. https://doi.org/10.1007/s00216-013-7320-4

Zalli A, Carvalho LA, Lin J, Hamer M, Erusalimsky JD, Blackburn EH, Steptoe A (2014) Shorter telomeres with high telomerase activity are associated with raised allostatic load and

impoverished psychosocial resources. Proc Natl Acad Sci U S A 111(12):4519–4524. https://doi.org/10.1073/pnas.1322145111
Zota AR, Shenassa ED, Morello-Frosch R (2013) Allostatic load amplifies the effect of blood lead levels on elevated blood pressure among middle-aged U.S. adults: a cross-sectional study. Environ Health 12(1):64. https://doi.org/10.1186/1476-069X-12-64

Chapter 2
The Public Health Exposome

Paul D. Juarez

Abstract The exposome has been described by Wild as the measure of all of the exposures an individual has in a lifetime and how those exposures relate to health. The exposome represents the totality of exogenous (external) and endogenous (internal) exposures from conception onwards, simultaneously distinguishing, characterizing, and quantifying etiologic, mediating, moderating, and co-occurring risk and protective factors and their relationship to disease.

The exposome provides a systems science approach to bringing together and organizing data needed to model the relationships, mechanisms, and pathways among and between external exposures, endogenous exposures, health outcomes, and population-level health disparities. It holds promise for identifying completed exposure pathways from source of exposure in the natural, built, social, and policy environments to route of entry into the body, biomarkers of exposure, biomarkers of disease, disease phenotype, clinical outcomes, and population level disparities, across the lifespan, and between generations. This chapter proposes a new taxonomy for sequencing the public health exposome first described by Wild as the eco-exposome. The aim of this chapter is to identify a common taxonomy for conceptualizing and operationalizing environmental exposures as an important step towards articulating a science of health disparities.

Keywords Public health exposome · Universal exposure-tracking framework

P. D. Juarez (✉)
Department of Family and Community Medicine, Health Disparities Research Center of Excellence, Meharry Medical College, Nashville, TN, USA
e-mail: pjuarez@mmc.edu

S. Dagnino, A. Macherone (eds.), *Unraveling the Exposome*,
https://doi.org/10.1007/978-3-319-89321-1_2

Introduction

The causes of health disparities are complex and have been identified as including genetics, epigenetics, "omics," environmental exposures, and gene × omics × environment interactions, at key life stages, over the life course, and across generations. Despite the increased attention and prioritization that have been given to health disparities over the past 30 years, many measures of population health outcomes for racial/ethnic minority groups and other subpopulations remain stubbornly intractable, while some have actually increased (Noonan et al. 2016). Much remains to be learned about how exposures in the natural, built, social, and policy environments get under the skin to affect personal health outcomes and contribute to population-level health disparities.

Modeling the contributions of multiple environmental exposures to poor health and health disparities, as experienced in the real world, is in its formative stage. Its amorphous nature has precluded prior efforts to create a common nomenclature and taxonomy to operationalize it. This is an initial effort to create a standard taxonomy of external environmental exposures.

History of Health Disparities Research

After 2003, when the identification and final sequencing of the human genome were completed, there was a strong belief that identification of the causes of disease would be shortly forthcoming followed by personalized medical interventions that could be tailored to address a patient's individual genetic sequence (International Human Genome Sequencing Consortium 2004). There was great enthusiasm and support for discovering the single-nucleotide polymorphisms (SNPs) underlying differences in a person's susceptibility to disease and that this would contribute to our knowledge of the underlying causes of racial health disparities. Despite some successes, it became readily evident that the causes of most diseases and of disparities were more complex than genetics alone could explain.

Since that time, it has become widely recognized that the regulation of genetic expression is influenced by both different intrinsic signals and stimuli in response to complex exogenous and endogenous conditions (Handy et al. 2011). In recent years, increased attention has been given towards understanding the role of epigenetics and other cellular and biochemical mechanisms on health, such as the proteome, exome, transcriptome, epigenome, metabolome, microbiome, connectome, and exposome (Pećina-Šlaus and Pećina 2015; Sun and Shi 2015). While promising to increase our understanding of the pathways and mechanisms through which chemical and nonchemical stressors affect health, this approach too has offered little to understanding the complex causes of population-level disparities.

People have many different contaminants in their body at any one time arising from the multitude of chemical and nonchemical insults people are exposed to in their daily lives, all of which can affect a person's health and cause disease, over time (Institute of Medicine (US) Committee on Assessing Interactions Among Social B, and Genetic Factors in Health 2006; National Research Council 2012; Goodson et al. 2015). There is a growing recognition, however, of the importance for understanding the effects of cumulative and multiple, chemical and nonchemical, at critical life stages, over the life course and across generations (Goodson et al. 2015; USEPA 2003). To date, however, much of the research emanating from environmental health sciences has been limited to identifying cause and effect relationships associated with single chemical exposures and specific health outcomes (National Research Council 1983; National Research Council (US) Committee on Applications of Toxicogenomic Technologies to Predictive Toxicology 2007). The complex biological mechanisms and pathways underlying diseases and health conditions suggest the need for applying a systems approach to increase our understanding of the complex relationships between multifactorial, environmentally induced, health-related symptoms and disorders (Wimalawansa and Wimalawansa 2016; GBD 2013 Risk Factors Collaborators et al. 2015).

Measuring the totality of exposures that a person experiences from conception to death, along with the associated biological response for how exposures get under the skin, has emerged as a recent field of study within exposure science, now referred to as exposomics (Buck Louis and Sundaram 2012; Brunekreef 2013; Juarez 2013). An exposomics approach portends the development of complex models of cumulative risk which can account for and measure the effects and interactions between multiple environmental exposures, biological perturbations, and epigenetic variations, both simultaneously, and across space and time (Lentz et al. 2015; Williams et al. 2012; Sexton and Linder 2011). Completing exposure pathways will require a comprehensive understanding of both external and internal environmental mechanisms and pathways and their relationship to personal and population measures of adverse health outcomes. New life course methods and cumulative risk models are needed that are capable of integrating exposure mechanisms and pathways across these different domains. Novel mathematical and computational models and analytics also are needed to study the complex associations between etiological, moderating, and co-occurring exposures and disease phenotypes, health outcomes, and population-level disparities (Langston et al. 2014b).

Environmental Exposures

Wild (2012) identified two categories of external environmental exposures (general and unique). He identified unique external exposures that an individual experiences to include radiation, infectious agents, chemical contaminants, and exposures arising from individual behaviors such as smoking, diet, and lifestyle factors. He distinguished these from more general external exposures found in the world in which we live, such as air pollution and community stressors, to which everyone in a geographic area is potentially exposed.

No standardized taxonomy for classifying the full range of external chemical and nonchemical, environmental exposures has yet been developed. The purpose of this chapter is to propose a taxonomy for classifying general external exposures. We propose that the external environment can be captured in four broad domains: the natural, built, social, and policy environments (Ruiz et al. 2016; Juarez et al. 2014; Langston et al. 2014a; Coughlin and Smith 2015).

Natural Environment

The natural environment consists of climate, weather, and natural resources that affect human survival and economic activity. It includes exposures from the air, water, and land that humans interact with on a daily basis. Studies have shown that exposure to different components of the natural environment can have an independent, positive or negative effect on health and health-related behaviors (Lobdell et al. 2011). While much is known about the adverse effects of chemical emissions that are released into the air, water, and land, there is growing evidence that exposure to positive attributes of the natural environment such as green space and state and national parks and forests can have a positive impact on health and health-related behavior.

Air

Air toxics are chemicals which have been released into the air that have the potential to cause adverse health effects, such as respiratory disease, cardiovascular disease, cerebrovascular disease, cancer, and birth defects (Atkinson et al. 2015; Beelen et al. 2015; Wang et al. 2014; Samet et al. 2000). Sources of air pollutants include those that are emitted from point (Kibble and Harrison 2005), nonpoint (United States Comptroller General 1977), mobile (National Research Council 2004), and non-road mobile sources (Zhang et al. 2010). A point source is a stationary facility that emits a significant amount of air pollution during manufacturing, power generation, heating, incineration, or other such activity. Point sources include power plants, refineries, municipal waste incinerators, and other industrial sources. Area or nonpoint sources are defined as small sources of air pollution which by themselves may not emit very much pollution, but when added together, account for a significant portion of emissions. These include small industrial sources that often are too numerous to be inventoried individually, such as auto body shops, gas stations, and dry cleaners. This category also includes smaller amounts of emissions from consumer product usage (USEPA) and residential heating. Mobile sources can be divided into two categories, on-road and non-road. On-road mobile sources include cars, trucks, buses, and motorcycle while non-road mobile sources include aircraft, trains, lawnmowers, boats, and construction and farm vehicles and equipment.

The Clean Air Act of 1970 (42 U.S.C. §7401 et seq.) identified six common air pollutants, also known as criteria air pollutants. They include ground-level ozone, carbon monoxide, sulfur dioxide, particulate matter, lead, and nitrogen dioxide. In order to determine the potential health effects of monitored or modeled air levels, "health benchmarks" based on chemical-specific toxicity values have been developed by the US Environmental Protection Agency (USEPA) and other government agencies. For non-cancer health effects, the benchmark is called a reference concentration, and it indicates the level to which someone can be exposed without harm. For chemicals suspected of causing cancer (carcinogens), the health benchmark is derived from the chemical's potency and the estimated likelihood of contracting cancer over a lifetime of exposure. A carcinogenic benchmark concentration is set at a risk level of one-in-one-million. Cumulative cancer risk and total hazard indices are calculated for hazardous air pollutants exceeding health benchmark levels.

Ground level ozone is created by chemical reactions between oxides of nitrogen (NOx) and volatile organic compounds (VOC) in the presence of sunlight (USEPA). It is not emitted directly into the air. Major sources of NOx and VOC include emissions from industrial facilities and electric utilities, motor vehicle exhaust, gasoline vapors, and chemical solvents. Breathing ozone can trigger a variety of health problems, particularly for children, the elderly, and people of all ages who have lung diseases such as asthma.

Carbon monoxide (CO) is a colorless and odorless gas that is released when something is burned and can be harmful when inhaled in large amounts (USEPA 2017a). The greatest sources of CO are released in outdoor air, and include cars, trucks, and other vehicles that burn fossil fuels. A variety of common sources that can affect air quality, however, also are found indoors, such as unvented kerosene and gas space heaters, leaking chimneys and furnaces, and gas stoves. Breathing air with a high concentration of CO can be dangerous because it reduces the amount of oxygen that can be transported in the blood stream to critical organs like the heart and brain. At very high levels, in indoor environments, CO can cause dizziness, confusion, unconsciousness, and even death.

Sulfur dioxide (SO_2) is the most unhealthful of an entire group of sulfur oxides (SO_x) that are found in the atmosphere (USEPA 2017e). The largest sources of SO_2 emissions are from fossil fuel combustion at power plants and other industrial facilities. Smaller sources of SO_2 emissions include: natural sources such as volcanoes; vehicles that burn fuel with a high sulfur content such as locomotives and ships; and industrial processes such as extracting metal from ore. Short-term exposures to SO_2 can be harmful to the human respiratory system and make breathing difficult. Children, the elderly, and those who suffer from asthma and other respiratory illnesses are particularly sensitive to effects of SO_2.

Particulate matter (PM) is the sum of all solid and liquid particles suspended in air, many of which are hazardous (USEPA 2017d). This complex mixture includes both organic and inorganic particles such as dust, pollen, soot, smoke, and liquid droplets. Some particles, such as dust, dirt, soot, or smoke are large or dark enough to be seen with the naked eye. Others are so small they are only detectable by using an electron microscope. Particulate matter includes three types: PM10: inhalable particles, with diameters that are generally 10 μm and smaller; PM2.5: fine inhalable particles, with diameters that are generally 2.5 μm and smaller; and ultrafine particles that are less than 100 nm in diameter.

PM2.5 has been linked to a host of health problems including respiratory disease (Dong et al. 2012; Chen et al. 2007; Miller et al. 2007), cardiovascular disease (Breitner et al. 2011; Brook et al. 2010; Crouse et al. 2012; Dehbi et al. 2017), and cancer. Particles less than 10 μm in diameter pose the greatest problems, because they can get deep into the lungs, and may even get into the bloodstream (Lepeule et al. 2012; Atkinson et al. 2015; Levy and Hanna 2011). Once inhaled, these particles can affect the heart and lungs and cause serious health effects. Regulations currently do not exist for ultrafine PM which is believed to have more aggressive health implications than larger particulates.

Lead exposures occur from different sources (such as paint, gasoline, solder, and consumer products) and through different pathways (such as air, food, water, dust, and soil) (Occupational Safety and Health Administration 2017). Lead-based paint, which was widely used in homes up until the 1950s and banned for residential use in 1978, remains the most widespread and dangerous high-dose source of lead exposure and is still found in many older homes today. Sources of lead in the air include ore and metals processing and piston-engine aircraft that operate on leaded aviation fuel. Other sources of lead include waste incinerators, utilities, and lead-acid battery manufacturers. The highest air concentrations of lead usually are found near lead smelters.

Once taken into the body, lead is distributed throughout the body via the blood and accumulates in the bones. Depending on the level of exposure, lead can adversely affect the nervous system, immune system, and reproductive, developmental, and cardiovascular systems and cause kidney and brain damage. The most common effects of lead encountered in current populations are neurological effects in children and cardiovascular effects (e.g., high blood pressure and heart disease) in adults. Infants and young children are especially sensitive to even low levels of lead, which may affect neuro-cognitive development and contribute to behavioral problems, learning deficits, and lowered IQ.

Nitrogen dioxide (NO_2) is one of a group of highly reactive gases known as nitrogen oxides (NO_x), which also includes nitrous acid and nitric acid. NO_2 (along with other NOx) reacts with other chemicals in the air to form both particulate matter and ozone (Kornartit et al. 2010). NO_2 primarily gets into the air from the burning of fossil fuels: coal, oil, and gas. Leading sources of emissions are from cars, trucks and buses, power plants, and off-road equipment.

Nitrogen oxides can significantly impact human health when inhaled due to its effects on the respiratory system. Exposures over short periods of time can aggravate respiratory diseases, particularly asthma, leading to respiratory symptoms (such as coughing, wheezing, or difficulty breathing), hospital admissions, and frequent visits to emergency rooms. Exposures to elevated concentrations of NO_2 over longer periods of time can contribute to the development of asthma and potentially increase susceptibility for respiratory infections. People with asthma, as well as children and the elderly are generally at greater risk for the health effects of NO_2.

Water

Surface waters are the natural water resources of the earth and include oceans, streams, lakes, and rivers (USEPA 2017b). The main sources of freshwater pollution include storm water waste, discharge of untreated waste, dumping of industrial effluent, mining operations, and run-off from agricultural fields. All wastewater can cause pollution of lakes and rivers. Domestic waste water, agricultural run-off, and industrial effluents contain phosphorus and nitrogen, fertilizer run-off, manure from livestock operations, which increase the level of nutrients in water bodies and can cause eutrophication in the lakes, rivers, and coastal areas. Because storm water runoff carries nutrient pollution directly into rivers, lakes, and reservoirs, which serve as sources of drinking water for many people, it can result in water-caused diseases and health problems.

Persistent organic pollutants (POPs) found in industrial chemicals and agricultural pesticides represent the most harmful effects of wastewater for the ecosystem and human health. These chemicals can get into the food chain and cause serious damage to human health over time. Disinfectants used to treat drinking water have been found to react with toxic algae creating harmful chemicals called dioxins, which have been linked to reproductive and developmental health risks and even cancer. Infants who drink water too high in nitrates also can become seriously ill and even die.

Ground water is the water found beneath the earth's surface and provides drinking water for more than one-half of the nation's population. It is the sole source of drinking water for up to 90% of persons residing in rural communities. Groundwater pollution occurs when pollutants are released into the ground and make their way down into underground aquifers. Threats to groundwater have been increasing due to population growth, agricultural practices, and widespread use of chemicals and road salts. Sources of groundwater pollution include sewage and wastewater, septic systems, underground storage tanks, uncontrolled hazardous waste, landfills, fracking, chemicals and road salts, and atmospheric contaminants (Barret 1997).

Chemical pollutants in groundwater may result from point sources such as sewage disposal systems and livestock facilities or nonpoint sources such as fertilized cropland, parks, golf courses, lawns, and gardens, or naturally occurring sources of nitrogen. When it rains, these chemical pollutants seep into the ground and eventually into the aquifers. Arsenic, fluoride, total inorganic nitrogen, volatile organic compounds, phosphorous, iron, manganese, boron, most heavy metals, and pathogens are among the contaminants that are of greatest concern.

Drinking contaminated water can have serious effects on a person's health. Acute side effects of groundwater contamination may include nausea and vomiting, diarrhea, headaches, Legionnaire's disease, and eye and nose irritation. Contaminated water can also cause various waterborne diseases, such as typhoid, dysentery, hepatitis, and cholera. Exposure to chemicals in untreated groundwater can cause kidney damage and damage to the brain and nervous system.

Land

Land Cover

Land cover refers to the biophysical coverage of land. The USGS identifies 92 land cover classifications in 9 broad categories: *Water* (open water, perennial ice/snow); *Barren* (bare rock/sand/clay, quarries/strip mines/gravel pits, transitional); *Shrub land, Herbaceous upland natural/semi-natural vegetation* (grasslands/herbaceous); *Wetlands* (woody wetlands, emergent herbaceous wetlands); *Developed* (low intensity residential, high intensity residential, commercial/industrial/transportation); *Forested upland* (deciduous forest, evergreen forest, mixed forest); nonnatural Woody (Orchards/Vineyards/others); and *Herbaceous Planted/cultivated* (pasture/hay, row crops, small grains, fallow, urban/recreational, grasses) (USGS 2017).

Decisions about how and where we build our communities have significant impacts on the natural environment and on human health. There is increasing evidence that green spaces have a range of health benefits and are important to physical activity (Lee and Shaddick 2010; Hunter et al. 2015), mental health (Bodin and Hartig 2003; Bowler et al. 2010; Cohen-Cline et al. 2015), birth outcomes (Hystad et al. 2014), and cognitive development (Dadvand et al. 2015).

Land Use

Land use is a description of how people utilize the land and of socioeconomic activity. Land pollution and soil contamination deteriorate the earth's land surfaces, often directly or indirectly, as a result of man's activities and use of natural resources (Atlanta Regional Health Forum and Atlanta Regional Commission 2006). The major sources of land pollution are the result of an increase in urbanization, agriculture, commerce, and industry which have resulted in increases in domestic waste, solid waste, pesticides, fertilizers, and chemicals. Domestic waste includes household products like paper, plastic containers, bottles, cans, food, used cars, broken furniture, broken electronic goods, and hospital waste. Some of these are biodegradable (meaning they easily rot or decay into organic matter) while others are not. Solid waste includes solid, liquid, semi-solid, or contained gaseous material, resulting from industrial, commercial, pharmaceutical, and mining operations. Pesticides, insecticides, and fertilizers are used in many large-scale farming practices. Chemicals and nuclear power plants produce waste materials, byproducts, and residue that need to be stored in secure environments.

Land use is generally classified into seven categories for zoning purpose: rural and agricultural, recreation, forestry, residential, commercial/industrial, public/semi-public, and undeveloped. Rural and agriculture land is used for the production of crops and rearing of livestock. The preservation of open space and natural features for recreational purposes is seen as a way to promote public health and safety and provide visual enjoyment. Forestry regulations are used not only as a way to promote conservation of a natural resources but increasingly as an important and relatively cost-effective way of offsetting air pollution by increasing the removal of greenhouse gases from the atmosphere. Residential zoning typically includes single family residences (SFR) and any number of other designation which cover homes, apartments, duplexes, trailer parks, co-ops, and condominiums. Residential zoning can cover a broad range of issues such as whether mobile homes can be placed on property, and the number of structures allowed on certain property, the types of animals that are allowed at a residence, and the nature of home-based businesses.

Land use, community design, and transportation systems substantially impact local air quality, water quality and supply, traffic safety, physical activity, and exposure to contaminated, industrial "brownfields" (Atlanta Regional Health Forum and Atlanta Regional Commission 2006). Local government designates various zones for different uses of land, such as industrial, agricultural, commercial, and residential. Land zoned for public/semi-public use include institutional, park and recreation, open space, roadway rights of way, utility, railway, and airports. Commercial zoning has several categories which are dependent upon the business use of the property and the number of business patrons. Office buildings, shopping centers, nightclubs, hotels, certain warehouses, some apartment complexes can all be zoned as commercial. Industrial zoning includes manufacturing sites and storage facilities but is generally determined by the type of business, the total land area covered by all buildings on a lot, and building height. Environmental factors, including noise, usually are issues in determining into which industrial level a business falls. Lands zoned as undeveloped areas include rivers, lakes, forests, streams, wetlands, and undeveloped open spaces. Preservation of undeveloped areas is often seen as providing multiple social benefits, including ensuring habitat preservation, protecting water quality, and preserving open space.

Contaminated lands and environments can cause problems in the human respiratory system, problems of the skin, and various kinds of cancers. Land development patterns have disproportionately exposed children, the elderly, racial and ethnic minorities and people of lower socioeconomic status, and other vulnerable populations to the health effects of pollution. In addition, these groups often have fewer options for where to live and work. Land use is now addressed as an environmental justice concern.

Built Environment

The built environment includes all of the physical parts of where we live, work and play (e.g., homes, buildings, streets, open spaces, and infrastructure). The EPA estimates that people spend up to 90% of their time in buildings (USEPA 2017c). Buildings are built almost entirely with materials that are synthetic, chemically processed, or treated. The materials we build with can affect our well-being as much as the food we eat, the water we drink, and the air we breathe. For the most part, we can't see the toxic chemicals that leak into our indoor air.

Patterns and types of development, building location and design, and transportation infrastructure are different features of the built environment that are likely to have direct and indirect impacts on health. The degree to which neighborhood is important to health may differ, depending on the broader social context. For instance, planning decisions that influence the location of supermarkets, fast-food eateries, farmers markets, and convenience stores can have profound effects on people's diets and their health (Wells 2017). Inaccessible or nonexistent sidewalks and bicycle or walking paths also contribute to sedentary habits and a person's level of physical activity and can lead to poor health outcomes such as obesity, cardiovascular disease, diabetes, and some types of cancer.

Places You Live

Residence

There is a growing body of research that has found that characteristics of a home can have a major impact on the health of persons who live there over their life span. For instance, a home that is cold and damp or has allergens may cause respiratory illnesses and asthma in the residents (Mendell et al. 2011). The height and size of housing also have been found to affect health of residents. High-rise housing has been found to be associated with psychological stress, particularly among low-income mothers of young children (Evans 2003). There also is a great deal of evidence about the toxic effects of building materials that are used in residential construction. Among them are polyvinyl chloride (PVC), lead, mercury, asbestos, formaldehyde, chromated copper arsenate, perfluorinated compounds (PFCs), phthalates, polybrominated diphenyl ethers (PBDEs), short-chain chlorinated paraffins (SCCPs), and halogenated flame retardants (Song et al. 2015; Korpi et al. 1998; Volchek et al. 2014).

Polyvinyl Chloride (PVC)

PVC is used to make pipes, pipe fittings, pipe conduits, vinyl flooring, and vinyl siding. It is used to make wire and cable coatings, packaging materials, wrapping film, gutters, downspouts, door and window frames, gaskets, electrical insulation, hoses, sealant liners, paper and textile finishes, thin sheeting, roof membranes, swimming pool liners, weather-stripping, flashing, molding, irrigation systems, containers, and automotive parts, tops, and floor mats. PVC releases phthalates during its use phase of life and releases dioxin when burned. Dioxin is one of the most toxic substances known to exist. PVC and PVC-byproducts contain known carcinogens (Yang et al. 2014). PVC is made from vinyl chloride, which is listed as a human carcinogen. Exposure to PVC dust may cause asthma and affect the lungs.

Lead

Lead was previously described under criterion air pollutants. The main toxic effects of lead are on the brain and nervous system. In adults, high levels of lead can cause headaches and problems with mood, thinking, and memory. Lead exposure also has been found to increase risk for kidney tumors and cancer of the brain, lungs, and other organs. In children, the results can be even more profound because of the developmental growth of the brain and central nervous system. Lead poisoning in children has been found to lead to lower intelligence, stunted growth, impaired hearing, and behavioral and learning problems.

Mercury

While mercury is still found in some electronics and thermostats, its most significant source in the residential sector is in lighting. Mercury is a known developmental toxicant, and it's suspected of many other health effects.

Asbestos

Asbestos fibers are strong, heat resistant, chemical resistant, and useful in providing heat insulation. Their most common uses include floor and ceiling tiles, plasters, insulations, adhesives, wallboard, roofing materials, fireproofing materials, and cement products. Asbestos is a known carcinogen, and inhalation of asbestos fibers is known to cause respiratory problems and lung diseases such as asbestosis, mesothelioma, or lung cancer (Goswami et al. 2013; Case et al. 2011). All three of these diseases experience long lag times and the diseases may not manifest for 10–40 years after the initial asbestos exposure.

Formaldehyde

Formaldehyde is present in substantial concentrations both indoors and outdoors. It is widely used in the manufacturing of building materials and numerous household products. It is used to add permanent-press qualities to clothing and draperies, as a component of glues and adhesives, and as a preservative in some paints and coating products. In homes, the most significant sources of formaldehyde are pressed wood products made using adhesives that contain urea-formaldehyde (UF) resins. Pressed wood products made for indoor use include particleboard, hardwood plywood paneling, and medium density fiberboard, which is generally recognized as being the highest formaldehyde-emitting pressed wood product. Formaldehyde is a known respiratory irritant and carcinogen. It can cause watery eyes, burning sensations in the eyes and throat, nausea, and difficulty in breathing in when persons are exposed at elevated levels (Dannemiller et al. 2013; Golden 2011).

Chromated Copper Arsenate (CCA)

CCA is a chemical wood preservative containing chromium, copper, and arsenic. It is used to prevent rotting in lumber designed for outdoor use such as play sets, decks, and picnic tables. CCA was widely used for residential purposes in the United States from the 1970s until EPA phased it out in 2003. Itching, burning rashes, neurological symptoms, and breathing problems have been associated with handling chromated arsenical wood preservatives (Hamula et al. 2006; Barraj et al. 2007).

Perfluorinated Compounds (PFCs)

PFCs include a family of fluorine-containing chemicals that are used to make materials stain and stick resistant. PFCs are used in wide array of consumer products and food packaging, such as microwave popcorn bags, pizza boxes, and cleaning and personal-care products like shampoo, dental floss, and denture cleaners. There are many forms of PFCs, but the two most notorious are: PFOA or perfluorooctanoic acid, used to make Teflon products. PFOA is broadly toxic. PFOS or perfluorooctane sulfonate was used on carpet, furniture, and clothing until 2002 in the manufacture of 3M's Scotchgard treatment. It does not break down in the environment and has a half-life in the body of more than 4 years. PFOA is a likely human carcinogen; it causes liver, pancreatic, testicular, and mammary gland tumors in laboratory animals.

Phthalates

Phthalates are a group of industrial chemicals used to make plastics like polyvinyl chloride (PVC) more flexible or resilient. Building materials are the largest end use for PVC. Major uses of flexible PVC in buildings include carpet backing, resilient flooring, wall coverings, acoustical ceiling surfaces, upholstery textiles, roof membranes, waterproofing membranes, and electrical cord insulation. Phthalates are nearly ubiquitous in modern society, found in a broad array of consumer products such as toys, food packaging, hoses, raincoats, shower curtains, vinyl flooring, adhesives, detergents, hair spray, and shampoo. Certain phthalates are known or suspected endocrine disruptors, meaning they impact and alter the human hormone system. Phthalates are also suspected to be potent reproductive toxins, especially in boys (Zimmer et al. 2012; Lovekamp-Swan and Davis 2003).

Polybrominated Diphenyl Ethers (PBDEs)

PBDEs are used as flame retardants in plastic building materials and are particularly widespread in polyurethane foam products (insulation and cushions). Key routes of human exposure are thought to be from their use in household consumer products, and their presence in house dust, and not from dietary routes. PBDEs have been associated in animal studies with liver toxicity, thyroid toxicity, developmental and reproductive toxicity, and developmental neurotoxicity (Zhao et al. 2015; Dishaw et al. 2014).

Halogenated Flame Retardants (HFRs)

HFRs are added to many types of building materials. During a fire, HFRs release significant amounts of smoke and toxic gases that can harm/kill occupants and firefighters. The European Union has already banned some HFRs, but the USA lags behind. There is currently a concerted effort in the green building movement to remove HFRs from materials when there is no added fire safety benefit (Stapleton et al. 2011).

Neighborhoods

Characteristics about the neighborhood one lives in, apart from the dwelling, have been found to independently affect people's physical and mental health. Neighborhood design factors such as whether streets are laid out in intersecting grids, have sidewalks, and shared recreation spaces have been shown to promote walking and have other health and mental health benefits. Research by Wells and her colleagues, for instance, revealed that nearby green space buffers the impact of stress on children and promotes their resilience (Wells 2017). Research further suggests that higher levels of access to nature have the greatest protective effect on children dealing with stressful life events. The buffering effect of green space has been found to be greatest for the most vulnerable children. Similarly, there is an increasing body of research that has shown that the availability of healthy foods in a neighborhood may influence the dietary behaviors of individuals, and individual behaviors may in turn affect food availability.

Places You Work (Occupational)

The work environment exposes workers to a wide variety of risk factors. Occupational hazards that have been found to affect health include biophysical, chemical, biological, ergonomic, nanoparticles, and psychosocial exposures. Respiratory disease (asthma, COPD, silicosis and coal workers pneumoconiosis, farmer's lung, allergic rhinitis, and byssinosis), cardiovascular diseases (arrhythmia, ischemic heart disease), neurodegenerative diseases (Alzheimer's disease (AD) and other dementias, Parkinson's disease (PD), and multiple sclerosis), cancer (bladder, bone, brain/CNS, breast, colon and rectal, kidney, laryngeal, leukemia, liver and biliary, lung), psychosocial stress (substance abuse, suicide, depression, anxiety, burnout) all have been linked to occupational exposures (Allen et al. 2015; Boers et al. 2005; GBD 2013 Risk Factors Collaborators et al. 2015).

Injury

The leading causes of fatal occupational injuries (2015) are transportation incidents (2054), falls, slips, and trips (800), contact with objects and equipment (722), violence and other injuries by persons or animals (703), and exposure to harmful substances or environments (424). The highest rates per 100 workers of total nonfatal occupational injury and cases are aquaculture (13.6), nursing and residential care facilities (12.0), police protection (local government) (11.3), household furniture manufacturing (except wood and metal) (10.8), manufactured (mobile) home (10.2), fire protection (10.2), and veterinary services (10.0) (Smith and DeJoy 2012).

Respiratory Disease

The leading causes of death from work-related respiratory condition are pneumoconiosis or occupations associated with inhaling mineral dust during mining, processing, or manufacturing. Occupational lung diseases include asbestosis, coal workers' pneumoconiosis, malignant mesothelioma, and respiratory tuberculosis. Deaths from asbestosis and mesothelioma have been linked to construction occupations: plumbers, pipefitters, and steamfitters, insulation workers, carpenters, and electricians (Chuang et al. 2016; Fay and Ashford 1961).

Cancer

Estimates of the percent of cancer deaths that are attributable to occupational exposure are: mesothelioma (85–90% in men; 23–90% in women), sino-nasal and nasopharyngeal (31–43% in men), lung (6.3–13%), bladder (3–19%), laryngeal, 1–20% (men), skin cancer (non-melanoma, 1.5–6% in men), and leukemia 0.08–2.8%). Cancers which previously have been associated with occupational exposure include:

- Bladder cancer (exposures include: arsenic, aromatic amines, coal tars and pitched, diesel engine exhaust; metalworking fluids and mineral oils; occupations include work as a hairdresser or barber, painter; in the rubber industry) (Latifovic et al. 2015).
- Bone cancer, brain and other central nervous system cancers (exposures include: ionizing radiation; occupations are: health care facilities, research institutions, nuclear reactors, nuclear weapon production facilities, and other various manufacturing settings) (Brown et al. 2012).
- Breast cancer (exposures are nonionizing radiation exposure, pesticides, polycyclic aromatic hydrocarbons and metals; occupations are night shift workers, seamstress, dressmakers, and tailor) (Sobel et al. 1995).
- Colon and rectal cancer (exposures are: asbestos, dioxin, wood dust, organic solvents and metal-working fluids; occupations include textile industry, automobile industry, and beverage industry) (Oddone et al. 2014).
- Kidney cancer (exposures are asbestos and crystalline silica exposures, trichloroethylene glass fibers, mineral wood fibers, and brick dust; occupations include: concrete/terrazzo finishers; pottery, manufacturers of motor vehicles, trailers and semitrailers, manufacturers of fabricated metal products, except machinery and furniture, manufacturers of electric components, computer, radio, television, and communication equipment and apparatuses, and manufacturers of machinery and equipment) (Kim et al. 2014).
- Laryngeal cancer (exposures include asbestos; strong inorganic acid mists, including sulfuric acids; occupations include semiskilled and unskilled workers, persons who work in the rubber industry, workers exposed to dust, out of doors workers, drivers, and people working in the cement industries and port services (Olsen and Sabroe 1984).
- Leukemia (exposures include benzene, ethylene oxide, formaldehyde, ionizing radiation, non-arsenical insecticides; occupations include plumbing, heating and air conditioning industries, and sales of nondurable goods (such as paints and varnishes) (Blair et al. 2001).
- Liver and biliary cancer (exposures include ionizing radiation; trichloroethylene, vinyl chloride methylene chloride; occupations include health care workers; smelting of ores containing arsenic; vinyl chloride production; and wood preservation) (Cancer Sites Associated with occupational exposures) (Canadian Centre for Occupational Health and Safety 2017).

- Lung cancer (exposures include arsenic; asbestos; beryllium; cadmium; chromium; coal gasification, coal tar and pitches, cobalt; coke production, diesel engine exhaust, dioxins; inorganic lead; iron and steel foundry work, mineral oils; nickel; natural radon in workplaces, ionizing radiation, rubber production; silica; strong inorganic acid mists; occupations include work as a welder, aluminum production; asphalt workers; coal gasification; copper smelting; hematite mining (underground) with radon exposure; iron and steel founding; isopropanol manufacture (strong acid process); occupations include painters; printing processes; roofers; rubber production; uranium mining; and vineyard workers) (Hancock et al. 2015; Lacourt et al. 2015; Malhotra et al. 2015).
- Mesothelioma (exposures include asbestos, talc containing asbestiform fibers; occupations include blasters; boilermakers; bricklayers; construction workers; drillers; electricians; machinists; mechanics; miners; pipefitters; plumbers; sheet metal workers; shipbuilding workers; welders) (Canadian Centre for Occupational Health and Safety 2017; Nielsen et al. 2014).
- Nasal and sinus cancer (exposures include: chromium, formaldehyde, leather dust, nickel, textile; occupations include boot and shoe manufacturing and repair; carpenters; furniture and cabinet making; isopropanol manufacture (strong acid process); occupations include: miners; plumbers; pulp and paper mill workers; textile workers; and welders) (Greiser et al. 2012; Pesch et al. 2008).
- Non-Hodgkin's disease (exposures include: tetrachloroethylene, trichloroethylene; occupations include farmers, pesticide applicators, drivers, hairdresser or barber, non-arsenical insecticides, work as a painter) (Alicandro et al. 2016).
- Non-melanoma skin cancer (exposures include: coal tars and pitches, mineral oils, solar radiation; occupations include agriculture, military service, construction, transport, and forestry) (Surdu et al. 2013).
- Esophageal cancer (exposures: silica dust, metals, endotoxin, a contaminant of cotton dust, sulfuric acid and carbon black tetrachloroethylene; occupations include cotton textile industry) (Parent et al. 2000; Astrakianakis et al. 2007).
- Pharyngeal cancer (exposures include: asbestos, aliphatic and alicyclic hydrocarbons, pesticides and alcohol; occupations include: lawyers, authors, journalists, performing artists, musicians, electronics and telefitters, painters, dockers, unskilled laborers and hotel porters in males; and private secretaries, dressmakers, shoemakers and cobblers, waiters, pursers and stewardesses in females) (Tarvainen et al. 2008).
- Stomach cancer (exposures include: asbestos; lead compounds, inorganic; ionizing radiation; occupations include: asbestos mining; insulation material production (pipes, sheeting, textiles, clothes, masks, asbestos cement products); insulators and pipe coverers; rubber production industry; shipyard and dockyard workers) (Canadian Centre for Occupational Health and Safety 2017).

Pesticides can be classified according to their target within three main categories and their numerous chemical groups. (1) Insecticides: organochlorines, organophosphates, carbamates, pyrethroids, and newer chemical groups such as neonicotinoids and phenylpyrazoles. Occupation deaths associated with pesticides include:

(2) Herbicides: amides, chlorophenoxy, bipyridyls, dinitroanilines, triazines, urea herbicides, and aminophosphonates. (3) Fungicides: inorganic, dithiocarbamates, anilides, dicarboximides, strobilurin, aromatic, (benz)imidazoles, and conazoles. There also are other categories based on target organisms, i.e., nematicides, acaricides, rodenticides, and fumigants (USEPA 2017f).

Workers exposed to organophosphate pesticides and certain organochlorines (lindane products, DDT) exhibit an increased risk of lymphatic and bone marrow cancers. Agricultural workers, workers in pesticide production plants, and rural populations all exhibit an increased risk of prostate cancer. An increased risk of Parkinson's disease has been observed among individuals occupationally exposed to pesticides, and some studies also point to an increased risk of Alzheimer's disease. The literature suggests that pregnant women occupationally exposed to pesticides face an increased risk of miscarriage and that their children are at greater risk of birth defects and impairment of fine motor skills, visual acuity, or short-term memory. It has recently been shown that children exposed to pesticides are also at significantly higher risk of developing leukemia and brain tumors. The children of women living near agricultural zones or exposed to pesticides in household settings are likewise at increased risk of birth defects (GreenFacts Scientific Board 2017).

Places You Play

The degree to which the built environment supports or constrains the adoption of a physically active lifestyle has emerged as an important public health consideration in recent years reflecting the critical role that the design and configuration of transport and land use systems may play, alongside personal and social factors, in influencing levels of physical activity.

Recreational Facilities

Access to recreational facilities has been demonstrated to have a significant impact on health and individual human development. Recreational facilities that are easily accessible for families with children greatly increase the likelihood of regular physical activity being undertaken which can promote stronger muscles and bones and have a positive effect on their physical development. Undertaking physical exercises also reduces the risk of obesity and cardiovascular disease and on their mental health (Malambo et al. 2016; Elwell Bostrom et al. 2017).

Green Space

There has been considerable work done in recent years exploring the value of urban green space for health and well-being. Urban green spaces provide environmental benefits through their effects on negating urban heat, offsetting greenhouse gas emissions, and attenuating storm water. They also have direct health benefits by providing urban residents spaces for physical activity and social interaction, and allowing psychological restoration to take place (Liu et al. 2017; Lee et al. 2015).

Transportation

The connections between public health and transportation are varied and dependent on mode, access, and proximity. Where transportation infrastructure is designed to accommodate or even encourage nonmotorized transportation, such as through complete streets policies, it can have a positive effect on public health.

Mode of Transportation

Mode of transportation (e.g., automobiles, motorcycles, bicycles, and walking) all are leading causes of injury morbidity and mortality in subpopulation groups. African-American, Native American, and Latino drivers all face higher traffic-related risks related to seatbelt use, impaired driving, and pedestrian safety (UC Berkeley Safe Transportation Research and Education Center 2017).

Access

There is growing body of evidence that supports the importance of access to transportation on health, particularly for those with lower incomes or who are under/uninsured (Syed et al. 2013). Research has shown that limited access to transportation creates health inequities, as well as decreased access to education, employment, and opportunities for recreational activities for older adults and people with disabilities. Access to health care services, fresh fruits and vegetables, and recreational facilities all have been associated with health.

Proximity

Residential proximity to major roads and highways also has been found to be associated with poor health outcome. Transportation is a source of pollution, generating air, soil, water, and noise pollutants, including particulate matter, carbon monoxide, nitrogen oxide, and carcinogens. Many studies show that living in proximity to freeways and other high traffic roadways leads to adverse health effects beyond those associated with regional air pollution.

Reports by the American Public Health Association and others have linked air pollution to negative health outcomes, including asthma, respiratory illness, heart disease, poor birth outcomes, cancer, and premature death. A number of studies that focused on children have found slower lung development and significant increases in the incidence of lung disease, such as asthma, bronchitis, and decreased lung function, in children who live or attend school near heavily travelled roadways. In addition to children, seniors, and people with heart and lung conditions are considered particularly sensitive to effects of air pollution (Atkinson et al. 2015; Balluz et al. 2007; Beelen et al. 2015; Cao et al. 2011).

Social

The social determinants of health (SDH) constitute the conditions in which people are born, grow, work, live, and age. They are shaped by the distribution of money, power, and resources throughout local communities, nations, and the world and determine the conditions of daily life. These forces and systems include economic policies and systems, development agendas, social norms, social policies, and political systems (World Health Organization 2017). According to the CDC, the complex, integrated, and overlapping social structures and economic systems are responsible for most health inequities (Centers for Disease Control and Prevention 2017).

The US Census and in recent years, the American Community Survey (ACS) provide the majority of the data used to estimate the demographic, social, education, housing, and occupational characteristics of the United States population. The ACS is a series of surveys designed to produce critical information on small areas and small population groups previously collected on the decennial long form. One year estimates now are available for populations of geographic areas of 65,000 or more, and 5-year estimates are available for geographic areas of 20,000 or less at a census block and block group level. This ACS 5-year period estimates are now available as an alternative to the decennial census long-form sample for nearly all geographic areas recognized by the Census Bureau, including census tracts and block groups (U.S. Census Bureau 2014).

The ACS addresses 35 topics: acreage and agricultural sales; age; ancestry; citizenship; place of birth; year of entry; class of worker; commute/journey to work; commuter and internet use; cost of utilities, condominium fee; disability; educational attainment, bachelor's field of degree; family, relationships; fertility; food stamps benefit; grandparents as caregivers; health insurance coverage; Hispanic origin; home heating fuel; income; industry, occupation, labor force status; language spoken at home; marital history, marital status; mortgage, taxes, insurance; name; home ownership, property value, rent; plumbing facilities, kitchen facilities, telephone service; race; residence 1 year ago; school enrollment; sex; units in structure, rooms, bedrooms; vehicles available; veteran status; work status last year; year unit built, year moved into unit.

Demographic

Population demographics are derived from the ACS. Demographic data include the number and characteristics of people who live in a particular area or form a particular group. It includes a count of the age, sex, education level, income level, marital status, occupation, religion, birth rate, death rate, average size of a family, and average age at marriage of people in a particular area. Demographic data are now available down to the block group level for 5-year periods from the ACS.

Social/Cultural

Sociocultural factors are customs, lifestyles, and values that characterize a society or group and are perpetuated through its institutions. Social factors include reference groups, family, role, and status in society, time, and available resources. Cultural factors are the established beliefs, values, traditions, laws and languages of the various racial, ethnic, religious, and social groups. Social/cultural institutions include education, language, law, politics, religion, health, social organizations, values, and attitudes.

Data on sociocultural factors are largely available through the US Census. However, more detailed data on a broad range of sociocultural topics also are available from other federal agencies, including the FBI (Uniform Crime Reports, crime by county, arrests by county).

Political

The majority of data on political conditions and climate are provided by national polling organizations such as Gallup, Harris, and the Pew Research Center and media groups, such as the New York Times, CNN, Fox News, and MSNBC and are available only at a national or state level. Geo-political district boundaries of elected officials also are used to assess political factors. "My Congressional District" (U.S. Census Bureau 2017) provides access to detailed demographic, social, economic, and housing statistics every year by congressional district. In addition, ACS also provides data on voting and voter registration.

Economic

Socioeconomic differences in health outcomes have been widely documented to be associated with most health conditions. Many of the effects of socioeconomic status (SES) on health outcomes are indirect, affected by a variety of life experiences, opportunities, or choices related to SES, beginning in early life and either cumulating or being tempered by later life situations (Crimmins et al. 2004). People who are poorer and who have less education are more likely to suffer from diseases, to experience loss of functioning, to be cognitively and physically impaired, and to experience higher mortality rates (Adler et al. 1993, 1994; Marmot 2005; Marmot et al. 1997; Williams 1990).

Area-level measures of SES have different patterns of association with health outcomes than individual-level SES measures (Pardo-Crespo et al. 2013). Neighborhood or area-level measures of SES can be an independent contextual factor influencing health outcomes and should be considered in health studies independent of individual socioeconomic measures (Pardo-Crespo et al. 2013). While SES can be applied to an individual, we use it to describe characteristics of a community or of a geographic area.

Economic data comes from a variety of sources, including: Department of Housing and Urban Development; Federal Reserve Home Mortgage Disclosure Act Data on high cost loans; Office of Federal Housing Enterprise Oversight Data on falling home prices; and Bureau of Labor Statistics data on place and county unemployment rates; Low-Income Housing Tax Credit (LIHTC) Qualified Census Tract (QCT); 50th Percentile Rent Estimates; Fair Market Rents For The Section 8 Housing Assistance Payments Program; US Postal Service (business and residential vacancy data); Department of Labor (labor force, unemployment).

Policy

Federal, state, and local laws and regulations can have a direct or indirect impact on health by affecting the environments where one lives, works, and plays. While federal laws and regulations have a national scope, their impact may be attenuated by how states implement them and by how rigorously they are enforced. State, and local laws and regulations, meanwhile vary by jurisdiction, both in content and enforcement. It is possible to identify when a law took effect and assess its impact by comparing changes on a health indicator after the law was enacted.

Federal Laws

At the federal level, some laws impact health and health equity by providing health services directly to eligible persons. There are also many laws which contribute to health and health equity by addressing it indirectly, including those which seek to remediate toxic environments, including low levels of education, homelessness, unsafe working conditions, crime, and unemployment.

There are a number of federal programs that provide direct health care services including the Veterans Administration which provides hospital care and outpatient services and the Indian Health Service which runs federal direct care service facilities, contract health service operations, tribally administered programs, and urban Indian health programs. Other federal programs promote increased access to care through a health insurance mechanism including: Medicare, Medicaid, children's health insurance program (CHIP), CHAMPVA, and the Affordable Care Act.

Additional federal laws have been passed to promote health and health equity through laws intended to promote access to care, food, and nutrition, improve workplace conditions, reduce exposure to toxic chemicals, improve quality of care, develop, and train culturally competent providers, increase the diversity of the health care workforce, collect better race/ethnicity health data, and make a greater investment in public health. These programs are administered through various departments and agencies including the US Department of Health and Human Services, Department of Agriculture, Department of Energy, Environmental Protection Agency, Department of Justice, Department of Labor, Department of Transportation, etc.

Federal agencies also promote health and health equity through collection and dissemination of data on health care, health access, and health finance. Health-related data include measure of health behaviors, morbidity and mortality, vital statistics, utilization, access, cost, and facilities. Sources and types of health-related data include: Centers for Disease Control & Prevention (Wonder: natality, detailed mortality, compressed mortality, multiple cause of death, infant deaths, BRFSS, Community Health Status Indicators, National Program of Cancer Registries Cancer Surveillance System (NPCR-CSS)); Vital Statistics: online (birth, linked birth-infant deaths, birth cohort linked birth-infant death, mortality multiple cause, and fetal death); Center for Medicare and Medicaid (all beneficiaries, beneficiaries over 65, beneficiaries under 65; chronic conditions, Alzheimer's Disease and Related Dementia, Heart Failure, Arthritis (Osteoarthritis and Rheumatoid), Hepatitis (Chronic Viral B & C), Asthma, HIV/AIDS, Atrial Fibrillation, Hyperlipidemia (High cholesterol), Autism Spectrum Disorders, Hypertension (High blood pressure), Cancer (Breast, Colorectal, Lung, and Prostate), Ischemic Heart Disease, Chronic Kidney Disease, Osteoporosis, Chronic Obstructive Pulmonary Disease, Schizophrenia and Other Psychotic Disorders, Depression, Stroke); utilization and spending; Quality Improvement Evaluation System (hospitals, skilled nursing facilities, home health agencies, and other types of facilities); and Health Resources and Services Administration (Area Health Resource Files, health center service delivery and look-alike sites, organ donation and transportation, primary care service areas, health professions shortage areas, dental health shortage areas, mental health shortage areas, medical underserved areas/populations). In addition, many surveys exist on a wide array of sociocultural factors that are published annually but their results generally are only available at a national or state level, primarily due to the prohibitive sampling costs.

Due to the HIPAA Privacy Rule [Pub. L. 104–191], individually identifiable health information, including demographic data, is considered "protected" and requires individual consent to access, share, or use. Only de-identified protected information can be used by researchers without consent. This requires the removal of "All geographic subdivisions smaller than a State, including street address, city, county, precinct, zip code, and their equivalent geocodes, except for the initial three digits of a zip code if, according to the current publicly available data from the Bureau of Census (1) the geographic units formed by combining all zip codes with the same three initial digits contains more than 20,000 people; and (2) the initial three digits of a zip code for all such geographic units containing 20,000 or fewer people is changed to 000" (USDHHS 2017).

State Laws

State laws that have a direct or indirect impact on health and health equity cut across the scope of work of multiple states agencies. They include agencies that address agriculture and rural development, civil and criminal justice, education, energy, environment and natural resources, health and human services, immigration, labor and employment, and financial services and commerce.

Health-related activities of state government traditionally have included public health functions and the financing and/or delivery of personal health services. Duties of a state health department typically include (1) disease surveillance, epidemiology, and data collection; (2) state laboratory services; (3) preparedness and response to public health emergencies; (4) population-based primary prevention; (5) health care services; (6) regulation of health care providers and other licensed professions; (7) environmental health; and (8) technical assistance and training. The most common prevention services of state health departments include tobacco (87%), HIV/AIDS (85%), sexually transmitted disease counseling and partner notification (85%), nutrition (79%), and physical activity (77%) (National Health Policy Forum 2010).

Health care services are provided either directly through delivery of categorical services for high risk groups populations such as persons with developmental disabilities or mental health issues through public hospitals or indirectly through insurance programs such as Medicaid, and CHIP. Despite a wide range of governance structures, each state's health department is generally the primary public health authority within the state and plays a key role in supporting the delivery of public health services.

Differences in what health issues are addressed and how they are addressed are constrained by state revenue generating mechanisms. Level of state revenues vary dramatically across states and provide the parameters of what state government can and cannot do. The most common revenue generating mechanisms used by states include: business property taxes; sales and excise taxes paid by businesses on their input purchases and capital expenditures; gross receipts taxes; corporate income and franchise taxes; business and corporate license taxes; unemployment insurance taxes; individual income taxes paid by owners of noncorporate (pass-through) businesses; and other state and local taxes that are the statutory liability of business taxpayers (Cost Council on State Taxation 2015). In addition, states generate a significant amount of their resources through intergovernmental transfers with the federal government.

Local laws

Local governments include county governments, municipalities, townships, special districts (such as water and sewage authorities), and school districts. Overall, revenue from property tax makes up the largest percent of local government general revenue (30%) and is the largest source of local government expenditures. Revenue from sales taxes (7%), individual income taxes (2%), and other taxes (such as stadium taxes and business license taxes) is just over 2%. Charges and fees, on water, sewerage, and parking meter fees provided local governments with about 23% of general revenue. Since 1977, the portion from intergovernmental transfers declined from 43% of general revenue in 1977 to 36% in 2013, while revenue from charges and fees increased from 15% to 23%. Aid to school districts account for more than half of all state government transfers to localities. Housing programs make up 40% of federal transfers to local governments (Urban Institute and Brookings Institution 2017).

The responsibility for the health of its citizens ultimately is the responsibility of local government. Whether policies and standards are set by the federal and/or state governments, local governments deliver many health services, especially those directed at vulnerable populations. The largest direct expenditures on health care by local governments include locally owned and operated hospitals, employee health care, retiree health care, and public health services. Like state government, local governments provide a broad array of services that have an indirect impact on health. This has led many local governments to adapt a broad approach to address "health in all policies."

Local health departments typically are responsible for the recording and analysis of reports of births, deaths, marriages, divorces, and notifiable diseases, maintenance of disease registries, conduct of special surveys to determine the prevalence and outcomes resulting from various diseases, collection and interpretation of morbidity data, health planning activities, and periodic evaluation of community health needs and services (American Public Health Association 1950).

Regulations

When Congress passes laws, they rarely contain enough specific language to guide their implementation completely. Regulations/rules are used to clarify definitions, authority, eligibility, benefits, and standards. It is the responsibility of the federal administrative agencies to fill in the details of new or amended laws with rules and regulations. The Code of Federal Regulations (CFR) is the codification of the regulations (sometimes called administrative law) published in the Federal Register by the executive departments and agencies of the federal government of the United States (USGPO 2017).

Sequencing the Public Health Exposome

While a great deal is known about the underlying mechanisms and pathways through which individual chemical exposures "get under the skin," much less is known about the cumulative and interactive effects of multiple chemical exposures affect health at key developmental periods or across the life course. Even less is known about how the mechanisms and pathways associated with nonchemical exposures found in the built, social, and policy environments cause poor health outcomes or mediate and/or moderate the effects of chemical exposures.

The public health exposome can be considered analogous to the human genome, not just an extension of the interface between genes and environment (Juarez 2013). The Human Genome Project (HGP) identified and mapped approximately 20,000–25,000 genes of the human genome from both a physical and functional standpoint. In contrast, the building blocks of the public health exposome are those factors which comprise the natural, built, social, and policy environments. While the number of data elements we have curated currently is about 30,000, this number is likely to increase dramatically as more persons from different backgrounds become involved and contribute to the refinement of the taxonomy.

A taxonomy of the Public Health Exposome is presented in Figs. 2.1, 2.2, 2.3, and 2.4. The four domains include: natural (see Fig. 2.1), built (see Fig. 2.2), social (see Fig. 2.3), and policy environments (see Fig. 2.4). Within each domain, sources of exposures are identified. Subdomains within each of the four domains are identified. For the natural environment, the subdomains are air, water, and land; for the built environment subdomains are places you live, work, and play; for the social environment, subdomains are demographic, social, political, and economic factors; and for the policy domain, subdomains are federal, state, and local laws and regulations. For each subdomain, category, exposures, and health outcomes are described.

DOMAIN	Subdomains	Source	Category	Exposure	Health Outcomes
Natural					
	Air	Motor vehicle	Fossil Fuels (coal, oil, gas)	Ground level Ozone	Cardiovascular disease
		Power plants	Emissions	Carbon Monoxice	Unconsciousness/death
		Refineries		Sulfur Dioxides	Respiratory disease
		Industrial		Particulate Matter	Asthma
		Utilities		Lead	
		Unvented space		Volatile organic	
		heaters		compounds	
		Leaking		Nitrogen Dioxide	
		Furnaces		Min/max temperature	
		Volcanos		Heat Index	
		Locomotives		Precipitation	
		Ships		Solar radiation	
		Off-road		Persistent Organic	
				Dioxins	
				Nitrates	
	Water	Surface Water	Storm water waste	Persistent organic	Reproductive health
			Discharge of untreated waste	Dissinfectants	Developmentl health
			Dumping of industrial effluent	Dioxins	Cancer
			Mining		
			Agriculture runoff		
		Ground Water	Agriculatural practice	Arsenic	Nausea
			Population growth	Fluoride	Vomiting
			Sweage disposal systems	Total Inorganic nitrogen	Diarrhea
				Volatile organic	
			Livestock facilities	compounds	Headaches
			Croplands	Phosphorous	Legionnaire's disease
			Golf courses	Iron	eye and nose irritation
			lawns and gardens	Manganese	Typhoi
				Boron	Hepatitis
				Heavy metals	Cholera
					Kidney damage
					Damage to brain and CNS
	Land	Land Cover	Barren	Exercise	Birth outcomes
			Water	Diet	Mental health
			Shrub land	Recreational use	Cognitive development
			Grasslands/Herbaceoius		Obesity
			Wetlands		
			Developed		
			Forested upland		
			Herbaceious		
		Land Use	Rural/agriculture	Pesticides	Respiratory illness
			recreation	Insecticides	Skin problems
			Forestry	fertlizers	Cancers
			Residential	Solid wastes	
			Commercial/industrial	Chemicals	
			Public/semi-public	Household products	
			Undeveloped	nuclear byproducts	
			Buildings	Protecting water quality	
			Streets		
			Open Spaces		
			Infrastructure		
			Transportation		
			Community design		
			Urbanization		

Fig. 2.1 Natural Environment

DOMAIN					
Built	**Subdomains**	**Source**	**Category**	**Exposure**	**Health Outcomes**
	Live	Neighborhood	Roads	Stress	Mental health
			Sidewalks	Exercise	Stress
			Green space	Diet	Obesity
			Food availability	Access to health care	
				Access to education	
				Access to employment	
				Access to resources	
				Social networks	
		Residence	Building materials	Polyvinyl Chloride (PVC)	Respiratory illness
				Lead	Asthma
				Mercury	Psycho-social stress
				Asbestos	Cancers
				Formaldehyde	Cognitive development
				Chromate copper	Asbestosis
				PFCs	Mesothelioma
				Pthalates	Eye nose irritation
				PBDEs	Neurological symptoms
				Short-chain chlorinated	Rashes/itching
				Halogenated flame	Endocrine disruptors
					Reproductive problems in
				Toxic gases	boys
					Liver/thyroid toxicity
					Developmental
					neurotoxicity
	Work	Physical	Injury	Transportation	Falls, slips, trips
		Ergonomic		Contact with objects	Homicide/assault
		Nanoparticles		Violence	Psycho-social stress
		Psychosocial		Harmful substances	
		Biological	Respiratory		Asthma
					COPD
					Pneuomoconiosis
					Respiratory tuberculosis
					Coal workers'
					pneumoconiosis
		Chemical	Cancer	Lung	Bladder cancer
					Bone cancer, brain cancer
				Esophogeal	and other CNS cancers
				bladder	Breast Cancer
				Larynx/Pharynx	Colon and rectal cancer
				Skin	Kidney cancer
				Nasal/sinus	Laryngeal cancer
				Breast	Lung cancer
				Colon/rectal	Nasal and sinus cancer
				Bone	Non-Hodgkin's disease
				Kidney	Non-melanoma skin cancer
				Laryngeal	Esophogeal cancer
				Lung	Pharyngeal cancer
				Stomach	Stomach cancer
					Leukemia
		Pesticides	Insecticides	Organochlorines	Parkinson's disease
				Organophosphates	Miscarraige
				carbamates	Birth defects
					Impairment of fine motor
				Pyrethroids	skills
				Neonicotinoids	Visual accuity
				Phenylpyrazoles	Alzheimers disease
					Short term memory

Fig. 2.2 Built environment

		Herbicides	Chlorphenoxy Bphyridyls Dinitroanilines Triazines Urea herbicides Aminophosphonates	Leukemia Brain tumors
		Fungicides	Inorganic Dithiocarbamates Anilides, Dicarboximides Strobilurin Aromatic (benz)imidazoles Conazoles	
Play	Recreational Facilities	Parks Sidewalks Bikelanes Facilities	Exercise Social interaction	Obesity Mental health Muscles/bones Stress
	Green Spaces		Urban heat Greenhouse gas Storm water attenuation	
Transportation	Access	Automobiles Walking/cycling Public transportation	Recreational facilities Health care Education Employment	Injury Cardio-vascular health Mental health
	Proximity	Air Water Land Noise	Particulate matter Carbon monoxide Nitrogen oxide Carcinogens	Asthma, bronchitis, decreased lung function Respiratory disease Heart disease Poor birth outcomes Cancer Premature death

Fig. 2.2 (continued)

Discussion

The PHE represents the general external environment and is not separate or apart from Wild's broader description of the exposome. As part of the exposome, the PHE is one part of a new risk assessment methodology which can be used to better understand complete exposure pathways. The PHE provides public health researchers with an opportunity to harness publically available secondary data and to combine the use of both data-driven and hypothesis-driven approaches to consider the complex role of environmental exposures on personal health and population-level health disparities. Adapting an exposome approach, however, will challenge a number of traditional public health practice, policy, research methods, training, and funding mechanisms and will require a significant retooling effort if it is to be adopted.

DOMAIN

Social	**Subdomains**	**Source**	**Category**	**Exposure**	**Health Outcomes**
	Demographic	Age	Children	Marital history	Health Disparities
		Race	Teens & young adults	Immigration status	Health status
		Gender	Older adults	Educational attainment	Access to health care
		SES	African American	School enrollment	Costs of health care
		Residence	Latino	Labor force status	
			American Indian	Commute time	
			API	Work status last year	
			LGBT	Home ownership	
			Gender	# Units in structure	
			Disability	# rooms	
			Low income	# of bedrooms	
			Rural	vehicles available	
			Urban	Year unit built	
				Year moved into unit	
				Home heating fuel	
				Mortgage	
				Taxes	
				Insurance	
				Rent	
	Socio-cultural	Personal	Ancestry	Social supports	Premature death
		Family	Role & status in society	Support networks	Health status
		Reference			
		groups	Time and available resources	Primary language at home	Mental health status
		Social groups	Education	veteran status	Health disparities
				Grandparents as	
			Language	caregivers	
			Laws	Religiosity	
			Religion	Criminal history	
			Social organizations	Community involvement	
			Social Norms	Values & attitudes	
	Political		Politics	Geo-political boundaries	Acess to health care
				Voting history	Health care financing
	Economic		Employment opportunities	Food stamp recipient	Increased morbidity
			Health	Cost of food	Loss of functioning
			Income	Cost of housing	cognitively impaired
				Cost of utilities	Physicially impaired
				Cost of health care	
				Health Insurance	

Fig. 2.3 Social environment

First, the exposome challenges the categorical, discipline-specific manner in which investigators are trained and supports the need for transdisciplinary training. The exposome spans a wide array of traditional disciplines from basic to social and behavioral science, computer science and engineering, public health and environmental health, urban planning to geography and health economics, and biostatistics, epidemiology and computational methods to mixed effects, Bayesian, spatial-temporal, and predictive models. No one person can have the breadth of knowledge and skills needed to apply an exposome approach. This will only be achieved by embracing transformative, transdisciplinary team science.

Second, the exposome likewise challenges the disease/organ focus currently used to fund research. Research funding that uses a disease or organ-specific approach is less suitable for transdisciplinary team science. Instead, research that focuses on

DOMAIN Policy	Subdomains	Source	Category	Exposure	Health Outcomes
	Federal	USDHHS	Direct Care	Indian health services	Health care financing
			Insurance	Medicare	Health disparities
				Medicaid	
				Affordable Care Act	Access to care
				CHIP	Children's health
				FQHCs	Health professions shortage areas
		CDC, NCHS, HRSA	Data collection and dissemination	Medicare, Medicaid, Wonder, BRFSS, HPSAs/MUAs, etc.	Health status and equity
		Department of Agriculture		Supplemental Nutrition Assistance Program	Food stamps
				Women's Infants and Children Program	Food and nutrition for pregnant women & children
				National School Lunch Program	Food and nutrition for children
		Department of Labor	Other Services	Disclocated workers training	
				Occupational Safety and Health	Worker health and safety
				Mine Safety Health Admnistration	Mine Safety
				Office of Disability Employment	Worker disability
		Department of Education		Discretionary grants	Student education
				Pell grants/Loan repayment	Student education
				Agricultural programs	Chemical emissions control
		USEPA		Air quality programs	Clean air
				Energy efficiency	Pollution prevention
				Pollution prevention	Toxic substances
				Product labeling	Consumer safety
				Superfund site clean up	Environmental remediation
				Waste management	Conservation
				Water quality programs	Clean water
		Department of Housing & Urban Devpt		Community Development	
				Homeless assistance	
				HOPWA	
				Rural housing	
				Multifamily Housing	
				Healthcare programs	
				Public & Indian housing	
		Veterans Administration	Champus/VA	Health insurance	
			VA Hospital & outpatient care	Access to care/direct	

Fig. 2.4 Policy domain

	Department of Interior	Bureau of Indian Affairs	Land Management	Conservation
			Environmental enforcement	Conservation
			Surface mining	Conservation
			Fish & wildlife	Conservation
	Department of Justice	Federal Bureau of Investigation	Criminal justice	Safety
State	Health Policy	varies	Health and Human services	Food
	Other policies	varies	Agriculture & rural development	Health status
		varies	Criminal justice	Medicaid
		varies	Education	Affordable Care Act
		varies	Energy	Mental health status
		varies	Labor & Employment development	Disability
		varies	Financial Services & Commerce	Prevention
Local	Health policy	Varies	Sanitation	Morbidity & mortality
	Other policies	Varies	Public Health	Clean environment
			Environmental health	Safe work place
			Occupational health	Infectious disease
				Health promotion/disease prevention
				STIs/HIV

Fig. 2.4 (continued)

cross-cutting pathways and mechanisms, whether molecular or environmental or both, will require funders to rethink how research goals are to be achieved.

Third, the exposome promotes the use of new predictive models, methods, and analytics that can be used to handle "big data." Traditional biostatics and epidemiologic methods will need to be coupled with data-driven, scalable combinatorial analytics which can be used to analyze the complex relationships between multiple environmental exposures, pathways, and health outcomes.

Fourth, collecting large amounts of heterogeneous data on environmental exposures across space and time dramatically increases the amount of data that investigators will be using, requiring investigators to be adept in bioinformatics. The recent advent of data that from personal monitors, social media, and electronic health records are now available in streaming or near real time, increases the demands on researchers to be knowledgeable about data capture, curation, and storage technologies.

Little is currently known about how or why some people who are exposed to similar exposures experience different personal health outcomes or why health disparities appear so intractable over time among certain vulnerable populations. An exposome approach presents a new paradigm for conceptualizing the causal pathways through which environmental exposures affect both personal health trajectories and population level health disparities and in so doing, provides a clear pathway for establishing a science of health disparities.

References

Adler NE, Boyce W, Chesney MA, Folkman S, Syme S (1993) Socioeconomic inequalities in health: no easy solution. JAMA 269(24):3140–3145. https://doi.org/10.1001/jama.1993.03500240084031

Adler NE, Boyce T, Chesney MA, Cohen S, Folkman S, Kahn RL, Syme SL (1994) Socioeconomic status and health: the challenge of the gradient. Am Psychol 49(1):15–24

Alicandro G, Rota M, Boffetta P, La Vecchia C (2016) Occupational exposure to polycyclic aromatic hydrocarbons and lymphatic and hematopoietic neoplasms: a systematic review and meta-analysis of cohort studies. Arch Toxicol 90(11):2643–2656. https://doi.org/10.1007/s00204-016-1822-8

Allen EM, Alexander BH, MacLehose RF, Nelson HH, Ramachandran G, Mandel JH (2015) Cancer incidence among Minnesota taconite mining industry workers. Ann Epidemiol 25 (11):811–815.e811. https://doi.org/10.1016/j.annepidem.2015.08.003

American Public Health Association (1950) The local health department: services and responsibilities. Am J Public Health Nations Health 40:67–72

Astrakianakis G, Seixas NS, Ray R, Camp JE, Gao DL, Feng Z, Li W, Wernli KJ, Fitzgibbons ED, Thomas DB, Checkoway H (2007) Lung cancer risk among female textile workers exposed to endotoxin. J Natl Cancer Inst 99(5):357–364. https://doi.org/10.1093/jnci/djk063

Atkinson RW, Mills IC, Walton HA, Anderson HR (2015) Fine particle components and health—a systematic review and meta-analysis of epidemiological time series studies of daily mortality and hospital admissions. J Expo Sci Environ Epidemiol 25(2):208–214. https://doi.org/10.1038/jes.2014.63

Atlanta Regional Health Forum, Atlanta Regional Commission (2006) Land use planning for public health: the role of local boards of health in community design and development. National Association of Local Boards of Health, Bowling Green, OH

Balluz L, Wen XJ, Town M, Shire JD, Qualter J, Mokdad A (2007) Ischemic heart disease and ambient air pollution of particulate matter 2.5 in 51 counties in the U.S. Public Health Rep 122 (5):626–633

Barraj LM, Tsuji JS, Scrafford CG (2007) The SHEDS-wood model: incorporation of observational data to estimate exposure to arsenic for children playing on CCA-treated wood structures. Environ Health Perspect 115(5):781–786. https://doi.org/10.1289/ehp.9741

Barret M (1997) Initial tier screening of pesticides for groundwater concentration using the SCI-GROW model. USEPA, Washington, DC

Beelen R, Hoek G, Raaschou-Nielsen O, Stafoggia M, Andersen ZJ, Weinmayr G, Hoffmann B, Wolf K, Samoli E, Fischer PH, Nieuwenhuijsen MJ, Xun WW, Katsouyanni K, Dimakopoulou K, Marcon A, Vartiainen E, Lanki T, Yli-Tuomi T, Oftedal B, Schwarze PE, Nafstad P, De Faire U, Pedersen NL, Östenson C-G, Fratiglioni L, Penell J, Korek M, Pershagen G, Eriksen KT, Overvad K, Sørensen M, Eeftens M, Peeters PH, Meliefste K, Wang M, Bueno-de-Mesquita HB, Sugiri D, Krämer U, Heinrich J, de Hoogh K, Key T, Peters A, Hampel R, Concin H, Nagel G, Jaensch A, Ineichen A, Tsai M-Y, Schaffner E, Probst-Hensch NM, Schindler C, Ragettli MS, Vilier A, Clavel-Chapelon F, Declercq C, Ricceri F, Sacerdote C, Galassi C, Migliore E, Ranzi A, Cesaroni G, Badaloni C, Forastiere F, Katsoulis M, Trichopoulou A, Keuken M, Jedynska A, Kooter IM, Kukkonen J, Sokhi RS, Vineis P, Brunekreef B (2015) Natural-cause mortality and long-term exposure to particle components: an analysis of 19 European cohorts within the Multi-Center ESCAPE project. Environ Health Perspect 123(6):525–533. https://doi.org/10.1289/ehp.1408095

Blair AZT, Linos A, Stewart PA, Zhang YW, Cantor KP (2001) Occupation and leukemia: a population-based case-control study in Iowa and Minnesota. Am J Ind Med 40(1):3–14

Bodin M, Hartig T (2003) Does the outdoor environment matter for psychological restoration gained through running? Psychol Sport Exerc 4:141–153. https://doi.org/10.1016/s1469-0292(01)00038-3

Boers D, Zeegers M, Swaen G, Kant I, van den Brandt PA (2005) The influence of occupational exposure to pesticides, polycyclic aromatic hydrocarbons, diesel exhaust, metal dust, metal fumes, and mineral oil on prostate cancer: a prospective cohort study. Occup Environ Med 62 (8):531–537. https://doi.org/10.1136/oem.2004.018622

Bowler DE, Buyung-Ali LM, Knight TM, Pullin AS (2010) A systematic review of evidence for the added benefits to health of exposure to natural environments. BMC Public Health 10(1):456. https://doi.org/10.1186/1471-2458-10-456

Breitner S, Liu L, Cyrys J, Bruske I, Franck U, Schlink U, Leitte AM, Herbarth O, Wiedensohler A, Wehner B, Hu M, Pan XC, Wichmann HE, Peters A (2011) Sub-micrometer particulate air pollution and cardiovascular mortality in Beijing, China. Sci Total Environ 409:5196–5204. https://doi.org/10.1016/j.scitotenv.2011.08.023

Brook RD, Rajagopalan S, Pope CA, Brook JR, Bhatnagar A, Diez-Roux AV, Holguin F, Hong Y, Luepker RV, Mittleman MA, Peters A, Siscovick D, Smith SC, Whitsel L, Kaufman JD (2010) Particulate matter air pollution and cardiovascular disease: an update to the scientific statement from the American heart association. Circulation 121:2331–2378. https://doi.org/10.1161/CIR.0b013e3181dbece1

Brown T, Young C, Rushton L, British Occupational Cancer Burden Study Group (2012) Occupational cancer in Britain: remaining cancer sites: brain, bone, soft tissue sarcoma and thyroid. Br J Cancer 107(Suppl 1):S85–S91. https://doi.org/10.1038/bjc.2012.124

Brunekreef B (2013) Exposure science, the exposome, and public health. Environ Mol Mutagen 54:596–598. https://doi.org/10.1002/em.21767

Buck Louis GM, Sundaram R (2012) Exposome: time for transformative research. Stat Med 31 (22):2569–2575. https://doi.org/10.1002/sim.5496

Canadian Centre for Occupational Health and Safety (2017) Cancer sites associated with occupational exposures. https://www.ccohs.ca/oshanswers/diseases/carcinogen_site.html. Accessed 14 Mar 2017

Cao J, Yang C, Li J, Chen R, Chen B, Gu D, Kan H (2011) Association between long-term exposure to outdoor air pollution and mortality in China: a cohort study. J Hazard Mater 186:1594–1600. https://doi.org/10.1016/j.jhazmat.2010.12.036

Case BW, Abraham JL, Meeker G, Pooley FD, Pinkerton KE (2011) Applying definitions of "asbestos" to environmental and "low-dose" exposure levels and health effects, particularly malignant mesothelioma. J Toxicol Environ Health B Crit Rev 14(1–4):3–39. https://doi.org/10.1080/10937404.2011.556045

Centers for Disease Control and Prevention (2017) NCHHSTP social determinants of health. https://www.cdc.gov/nchhstp/socialdeterminants/definitions.html. Accessed 14 Mar 2017

Chen L, Mengersen K, Tong S (2007) Spatiotemporal relationship between particle air pollution and respiratory emergency hospital admissions in Brisbane, Australia. Sci Total Environ 373:57–67. https://doi.org/10.1016/j.scitotenv.2006.10.050

Chuang C-S, Ho S-C, Lin C-L, Lin M-C, Kao C-H (2016) Risk of cerebrovascular events in pneumoconiosis patients: a population-based study, 1996–2011. Medicine 95(9):e2944. https://doi.org/10.1097/MD.0000000000002944

Cohen-Cline H, Turkheimer E, Duncan GE (2015) Access to green space, physical activity and mental health: a twin study. J Epidemiol Community Health 69(6):523

Cost Council on State Taxation (2015) Study: total state and local business taxes (fy15). http://www.cost.org/page.aspx?id=69654. Accessed 16 Mar 2017

Coughlin SS, Smith SA (2015) The impact of the natural, social, built, and policy environments on breast cancer. J Environ Health Sci 1(3). https://doi.org/10.15436/2378-6841.15.020

Crimmins EM, Hayward MD, Seeman TE (2004) Race/ethnicity, socioeconomic status, and health. In: Anderson NB, Bulatao RA, Cohen B (eds) National Research Council (US) panel on race ethnicity, and health in later life. National Academies Press, Washington DC

Crouse DL, Peters PA, van Donkelaar A, Goldberg MS, Villeneuve PJ, Brion O, Khan S, Atari DO, Jerrett M, Pope CA, Brauer M, Brook JR, Martin RV, Stieb D, Burnett RT (2012) Risk of nonaccidental and cardiovascular mortality in relation to long-term exposure to low concentrations of fine particulate matter: a Canadian national-level cohort study. Environ Health Perspect 120:708–714. https://doi.org/10.1289/ehp.1104049

Dadvand P, Nieuwenhuijsen MJ, Esnaola M, Forns J, Basagaña X, Alvarez-Pedrerol M, Rivas I, López-Vicente M, De Castro Pascual M, Su J, Jerrett M, Querol X, Sunyer J (2015) Green spaces and cognitive development in primary schoolchildren. Proc Natl Acad Sci 112 (26):7937–7942. https://doi.org/10.1073/pnas.1503402112

Dannemiller KC, Murphy JS, Dixon SL, Pennell KG, Suuberg EM, Jacobs DE, Sandel M (2013) Formaldehyde concentrations in household air of asthma patients determined using colorimetric detector tubes. Indoor Air 23(4):285–294. https://doi.org/10.1111/ina.12024

Dehbi H-M, Blangiardo M, Gulliver J, Fecht D, de Hoogh K, Al-Kanaani Z, Tillin T, Hardy R, Chaturvedi N, Hansell AL (2017) Air pollution and cardiovascular mortality with over 25 years follow-up: a combined analysis of two British cohorts. Environ Int 99:275–281. https://doi.org/10.1016/j.envint.2016.12.004

Dishaw L, Macaulay L, Roberts SC, Stapleton HM (2014) Exposures, mechanisms, and impacts of endocrine-active flame retardants. Curr Opin Pharmacol 19:125–133. https://doi.org/10.1016/j.coph.2014.09.018

Dong GH, Zhang P, Sun B, Zhang L, Chen X, Ma N, Yu F, Guo H, Huang H, Lee YL, Tang N, Chen J (2012) Long-term exposure to ambient air pollution and respiratory disease mortality in Shenyang, China: a 12-year population-based retrospective cohort study. Respiration 84:360–368. https://doi.org/10.1159/000332930

Elwell Bostrom H, Shulaker B, Rippon J, Wood R (2017) Strategic and integrated planning for healthy, connected cities: Chattanooga case study. Prev Med 95(Suppl):S115–S119. https://doi.org/10.1016/j.ypmed.2016.11.002

Evans GW (2003) The built environment and mental health. J Urban Health 80(4):536–555. https://doi.org/10.1093/jurban/jtg063

Fay JWJ, Ashford JR (1961) A survey of the methods developed in the national coal board's pneumoconiosis field research for correlating environmental exposure with medical condition. Br J Ind Med 18(3):175–196

GBD 2013 Risk Factors Collaborators, Forouzanfar MH, Alexander L, Anderson HR, Bachman VF, Biryukov S et al (2015) Global, regional, and national comparative risk assessment of 79 behavioural, environmental and occupational, and metabolic risks or clusters of risks in 188 countries, 1990–2013: a systematic analysis for the Global Burden of Disease Study 2013. Lancet 386(10010):2287–2323. https://doi.org/10.1016/S0140-6736(15)00128-2

Golden R (2011) Identifying an indoor air exposure limit for formaldehyde considering both irritation and cancer hazards. Crit Rev Toxicol 41(8):672–721. https://doi.org/10.3109/10408444.2011.573467

Goodson WH, Lowe L, Carpenter DO, Gilbertson M, Manaf Ali A, Lopez de Cerain Salsamendi A, Lasfar A, Carnero A, Azqueta A, Amedei A, Charles AK, Collins AR, Ward A, Salzberg AC, Colacci AM, Olsen A-K, Berg A, Barclay BJ, Zhou BP, Blanco-Aparicio C, Baglole CJ, Dong C, Mondello C, Hsu C-W, Naus CC, Yedjou C, Curran CS, Laird DW, Koch DC, Carlin DJ, Felsher DW, Roy D, Brown DG, Ratovitski E, Ryan EP, Corsini E, Rojas E, Moon E-Y, Laconi E, Marongiu F, Al-Mulla F, Chiaradonna F, Darroudi F, Martin FL, Van Schooten FJ, Goldberg GS, Wagemaker G, Nangami GN, Calaf GM, Williams GP, Wolf GT, Koppen G, Brunborg G, Lyerly HK, Krishnan H, Ab Hamid H, Yasaei H, Sone H, Kondoh H, Salem HK, Hsu H-Y, Park HH, Koturbash I, Miousse IR, Scovassi A, Klaunig JE, Vondráček J, Raju J, Roman J, Wise JP, Whitfield JR, Woodrick J, Christopher JA, Ochieng J, Martinez-Leal JF, Weisz J, Kravchenko J, Sun J, Prudhomme KR, Narayanan KB, Cohen-Solal KA, Moorwood K, Gonzalez L, Soucek L, Jian L, D'Abronzo LS, Lin L-T, Li L, Gulliver L, McCawley LJ, Memeo L, Vermeulen L, Leyns L, Zhang L, Valverde M, Khatami M, Romano MF, Chapellier M, Williams MA, Wade M, Manjili MH, Lleonart ME, Xia M, Gonzalez Guzman MJ, Karamouzis MV, Kirsch-Volders M, Vaccari M, Kuemmerle NB, Singh N, Cruickshanks N, Kleinstreuer N, van Larebeke N, Ahmed N, Ogunkua O, Krishnakumar PK, Vadgama P, Marignani PA, Ghosh PM, Ostrosky-Wegman P, Thompson PA, Dent P,

Heneberg P, Darbre P, Leung PS, Nangia-Makker P, Cheng Q, Robey R, Al-Temaimi R, Roy R, Andrade-Vieira R, Sinha RK, Mehta R, Vento R, Di Fiore R, Ponce-Cusi R, Dornetshuber-Fleiss R, Nahta R, Castellino RC, Palorini R, Hamid RA, Langie SAS, Eltom SE, Brooks SA, Ryeom S, Wise SS, Bay SN, Harris SA, Papagerakis S, Romano S, Pavanello S, Eriksson S, Forte S, Casey SC, Luanpitpong S, Lee T-J, Otsuki T, Chen T, Massfelder T, Sanderson T, Guarnieri T, Hultman T, Dormoy V, Odero-Marah V, Sabbisetti V, Maguer-Satta V, Rathmell W, Engström W, Decker WK, Bisson WH, Rojanasakul Y, Luqmani Y, Chen Z, Hu Z (2015) Assessing the carcinogenic potential of low-dose exposures to chemical mixtures in the environment: the challenge ahead. Carcinogenesis 36(Suppl 1):S254–S296. https://doi.org/10.1093/carcin/bgv039

Goswami E, Craven V, Dahlstrom DL, Alexander D, Mowat F (2013) Domestic asbestos exposure: a review of epidemiologic and exposure data. Int J Environ Res Public Health 10 (11):5629–5670. https://doi.org/10.3390/ijerph10115629

GreenFacts Scientific Board (2017) Pesticides: occupational exposure and associated health effects. http://www.greenfacts.org/en/pesticides-occupational-risks/index.htm. Accessed 14 Mar 2017

Greiser EM, Greiser KH, Ahrens W, Hagen R, Lazszig R, Maier H, Schick B, Zenner HP (2012) Risk factors for nasal malignancies in German men: the South-German Nasal cancer study. BMC Cancer 12:506–506. https://doi.org/10.1186/1471-2407-12-506

Hamula C, Wang Z, Zhang H, Kwon E, Li X-F, Gabos S, Le XC (2006) Chromium on the hands of children after playing in playgrounds built from chromated copper arsenate (CCA)–treated wood. Environ Health Perspect 114(3):460–465. https://doi.org/10.1289/ehp.8521

Hancock DG, Langley ME, Chia KL, Woodman RJ, Shanahan EM (2015) Wood dust exposure and lung cancer risk: a meta-analysis. Occup Environ Med 72(12):889–898. https://doi.org/10.1136/oemed-2014-102722

Handy DE, Castro R, Loscalzo J (2011) Epigenetic modifications: basic mechanisms and role in cardiovascular disease. Circulation 123(19):2145–2156. https://doi.org/10.1161/CIRCULATIONAHA.110.956839

Hunter RF, Christian H, Veitch J, Astell-Burt T, Hipp JA, Schipperijn J (2015) The impact of interventions to promote physical activity in urban green space: a systematic review and recommendations for future research. Soc Sci Med 124:246–256. https://doi.org/10.1016/j.socscimed.2014.11.051

Hystad P, Davies HW, Frank L, Van Loon J, Gehring U, Tamburic L, Brauer M (2014) Residential greenness and birth outcomes: evaluating the influence of spatially correlated built-environment factors. Environ Health Perspect 122(10):1095–1102. https://doi.org/10.1289/ehp.1308049

Institute of Medicine (US) Committee on Assessing Interactions Among Social B, and Genetic Factors in Health (2006) Genes, behavior, and the social environment: moving beyond the nature/nurture debate. National Academies Press, Washington, DC

International Human Genome Sequencing Consortium (2004) Finishing the euchromatic sequence of the human genome. Nature 431:931–945

Jones DP (2016) Sequencing the exposome: a call to action. Toxicol Rep 3:29–45. https://doi.org/10.1016/j.toxrep.2015.11.009

Juarez P (2013) Sequencing the public health genome. J Health Care Poor Underserved 24 (1 Suppl):114–120. https://doi.org/10.1353/hpu.2013.0035

Juarez PD, Matthews-Juarez P, Hood DB, Im W, Levine RS, Kilbourne BJ, Langston MA, Al-Hamdan MZ, Crosson WL, Estes MG, Estes SM, Agboto VK, Robinson P, Wilson S, Lichtveld MY (2014) The public health exposome: a population-based, exposure science approach to health disparities research. Int J Environ Res Public Health 11(12):12866–12895. https://doi.org/10.3390/ijerph111212866

Kibble A, Harrison R (2005) Point sources of air pollution. Occup Med 55(6):425–431. https://doi.org/10.1093/occmed/kqi138

Kim I, Ha J, Lee J-H, Yoo K-M, Rho J (2014) The relationship between the occupational exposure of trichloroethylene and kidney cancer. Ann Occup Environ Med 26:12–12. https://doi.org/10.1186/2052-4374-26-12

Kornartit C, Sokhi RS, Burton MA, Ravindra K (2010) Activity pattern and personal exposure to nitrogen dioxide in indoor and outdoor microenvironments. Environ Int 36(1):36–45. https://doi.org/10.1016/j.envint.2009.09.004

Korpi A, Pasanen A-L, Pasanen P (1998) Volatile s. Appl Environ Microbiol 64(8):2914–2919

Lacourt A, Pintos J, Lavoué J, Richardson L, Siemiatycki J (2015) Lung cancer risk among workers in the construction industry: results from two case–control studies in Montreal. BMC Public Health 15:941. https://doi.org/10.1186/s12889-015-2237-9

Langston MA, Levine RS, Kilbourne BJ, Rogers GL, Kershenbaum AD, Baktash SH, Coughlin SS, Saxton AM, Agboto VK, Hood DB, Litchveld MY, Oyana TJ, Matthews-Juarez P, Juarez PD (2014a) Scalable combinatorial tools for health disparities research. Int J Environ Res Public Health 11(10):10419–10443. https://doi.org/10.3390/ijerph111010419

Langston MA, Levine RS, Kilbourne BJ, Rogers GL, Kershenbaum AD, Baktash SH, Coughlin SS, Saxton AM, Agboto VK, Hood DB, Litchveld MY, Oyana TJ, Matthews-Juarez P, Juarez PD (2014b) Scalable combinatorial tools for health disparities research. Int J Environ Res Public Health 11(10):10410–10413

Latifovic L, Villeneuve PJ, Parent MÉ, Johnson KC, Kachuri L, Canadian Cancer Registries Epidemiology Group, Harris SA (2015) Bladder cancer and occupational exposure to diesel and gasoline engine emissions among Canadian men. Cancer Med 4(12):1948–1962. https://doi.org/10.1002/cam4.544

Lee ACK, Jordan HC, Horsley J (2015) Value of urban green spaces in promoting healthy living and wellbeing: prospects for planning. Risk Manag Healthc Policy 8:131–137. https://doi.org/10.2147/RMHP.S61654

Lee D, Shaddick G (2010) Spatial modeling of air pollution in studies of its short-term health effects. Biometrics 66:1238–1246. https://doi.org/10.1111/j.1541-0420.2009.01376.x

Lentz TJ, Dotson GS, Williams PR, Maier A, Gadagbui B, Pandalai SP, Lamba A, Hearl F, Mumtaz M (2015) Aggregate exposure and cumulative risk assessment—integrating occupational and non-occupational risk factors. J Occup Environ Hyg 12(sup1):S112–S126. https://doi.org/10.1080/15459624.2015.1060326

Lepeule J, Laden F, Dockery D, Schwartz J (2012) Chronic exposure to fine particles and mortality: an extended follow-up of the Harvard six cities study from 1974 to 2009. Environ Health Perspect 120:965–970. https://doi.org/10.1289/ehp.1104660

Levy JI, Hanna SR (2011) Spatial and temporal variability in urban fine particulate matter concentrations. Environ Pollut 159:2009–2015. https://doi.org/10.1016/j.envpol.2010.11.013

Liu H, Li F, Li J, Zhang Y (2017) The relationships between urban parks, residents' physical activity, and mental health benefits: a case study from Beijing, China. J Environ Manag 190:223–230. https://doi.org/10.1016/j.jenvman.2016.12.058

Lobdell DT, Jagai JS, Rappazzo K, Messer LC (2011) Data sources for an environmental quality index: availability, quality, and utility. Am J Public Health 101(S1):S277–S285. https://doi.org/10.2105/AJPH.2011.300184

Lovekamp-Swan T, Davis BJ (2003) Mechanisms of phthalate ester toxicity in the female reproductive system. Environ Health Perspect 111(2):139–145

Malambo P, Kengne AP, De Villiers A, Lambert EV, Puoane T (2016) Built environment, selected risk factors and major cardiovascular disease outcomes: a systematic review. PLoS One 11(11):e0166846. https://doi.org/10.1371/journal.pone.0166846

Malhotra J, Sartori S, Brennan P, Zaridze D, Szeszenia-Dabrowska N, Świątkowska B, Rudnai P, Lissowska J, Fabianova E, Mates D, Bencko V, Gaborieau V, Stücker I, Foretova L, Janout V, Boffetta P (2015) Effect of occupational exposures on lung cancer susceptibility: a study of gene–environment interaction analysis. Cancer Epidemiol Biomark Prev 24(3):570–579. https://doi.org/10.1158/1055-9965.epi-14-1143-t

Marmot M (2005) Social determinants of health inequalities. Lancet 365(9464):1099–1104. https://doi.org/10.1016/S0140-6736(05)71146-6

Marmot M, Ryff CD, Bumpass LL, Shipley M, Marks NF (1997) Social inequalities in health: next questions and converging evidence. Soc Sci Med 44(6):901–910. https://doi.org/10.1016/S0277-9536(96)00194-3

Mendell MJ, Mirer AG, Cheung K, Tong M, Douwes J (2011) Respiratory and allergic health effects of dampness, mold, and dampness-related agents: a review of the epidemiologic evidence. Environ Health Perspect 119(6):748–756. https://doi.org/10.1289/ehp.1002410

Miller KA, Siscovick DS, Sheppard L, Shepherd K, Sullivan JH, Anderson GL, Kaufman JD (2007) Long-term exposure to air pollution and incidence of cardiovascular events in women. N Engl J Med 356:447–458. https://doi.org/10.1056/NEJMoa054409

National Health Policy Forum (2010) Governmental public health: an overview of state and local public health agencies. BACKGROUND PAPER No. 77. George Washington University

National Research Council (1983) Risk assessment in the federal government: managing the process. The National Academies Press, Washington, DC

National Research Council (2004) Air quality management in the United States. The National Academies Press, Washington DC. https://doi.org/10.17226/10728

National Research Council (2012) Exposure science in the 21st century: a vision and a strategy. The National Academies Press, Washington, DC

National Research Council (US) Committee on Applications of Toxicogenomic Technologies to Predictive Toxicology (2007) Applications of toxicogenomic technologies to predictive toxicology and risk assessment. National Academies Press, Washington DC

Nielsen LS, Bælum J, Rasmussen J, Dahl S, Olsen KE, Albin M, Hansen NC, Sherson D (2014) Occupational asbestos exposure and lung cancer—a systematic review of the literature. Arch Environ Occup Health 69(4):191–206. https://doi.org/10.1080/19338244.2013.863752

Noonan AS, Velasco-Mondragon HE, Wagner FA (2016) Improving the health of African Americans in the USA: an overdue opportunity for social justice. Public Health Rev 37(1):12. https://doi.org/10.1186/s40985-016-0025-4

Occupational Safety and Health Administration (2017) US DOL. https://www.osha.gov/SLTC/lead/. Accessed 14 Mar 2017

Oddone E, Modonesi C, Gatta G (2014) Occupational exposures and colorectal cancers: a quantitative overview of epidemiological evidence. World J Gastroenterol: WJG 20 (35):12431–12444. https://doi.org/10.3748/wjg.v20.i35.12431

Olsen J, Sabroe S (1984) Occupational causes of laryngeal cancer. J Epidemiol Community Health 38(2):117–121

Pardo-Crespo MR, Narla NP, Williams AR, Beebe TJ, Sloan J, Yawn BP, Wheeler PH, Juhn YJ (2013) Comparison of individual-level versus area-level socioeconomic measures in assessing health outcomes of children in Olmsted County, Minnesota. J Epidemiol Community Health 67 (4):305–310. https://doi.org/10.1136/jech-2012-201742

Parent M, Siemiatycki J, Fritschi L (2000) Workplace exposures and oesophageal cancer. Occup Environ Med 57(5):325–334. https://doi.org/10.1136/oem.57.5.325

Pećina-Šlaus N, Pećina M (2015) Only one health, and so many omics. Cancer Cell Int 15:64. https://doi.org/10.1186/s12935-015-0212-2

Pesch B, Pierl CB, Gebel M, Gross I, Becker D, Johnen G, Rihs HP, Donhuijsen K, Lepentsiotis V, Meier M, Schulze J, Brüning T (2008) Occupational risks for adenocarcinoma of the nasal cavity and paranasal sinuses in the German wood industry. Occup Environ Med 65(3):191

Ruiz JDC, Quackenboss JJ, Tulve NS (2016) Contributions of a child's built, natural, and social environments to their general cognitive ability: a systematic scoping review. PLoS One 11(2): e0147741. https://doi.org/10.1371/journal.pone.0147741

Samet JM, Zeger SL, Dominici F, Curriero F, Coursac I, Dockery DW, Schwartz J, Zanobetti A (2000) The national morbidity, mortality, and air pollution study. Part II: morbidity and mortality from air pollution in the United States. Res Rep Health Eff Inst 94(Pt 2):5–70

Sexton K, Linder SH (2011) Cumulative risk assessment for combined health effects from chemical and nonchemical stressors. Am J Public Health 101(Suppl 1):S81–S88. https://doi.org/10.2105/AJPH.2011.300118

Smith TD, DeJoy DM (2012) Occupational Injury in America: an analysis of risk factors using data from the General Social Survey (GSS). J Saf Res 43(1):67–74. https://doi.org/10.1016/j.jsr.2011.12.002
Sobel E, Davanipour Z, Sulkava R, Erkinjuntti T, Wikstrom J, Henderson VW, Buckwalter G, Bowman JD, Lee P-J (1995) Occupations with exposure to electromagnetic fields: a possible risk factor for Alzheimer's disease. Am J Epidemiol 142(5):515–524. https://doi.org/10.1093/oxfordjournals.aje.a117669
Song W, Cao Y, Wang D, Hou G, Shen Z, Zhang S (2015) An investigation on formaldehyde emission characteristics of wood building materials in chinese standard tests: product emission levels, measurement uncertainties, and data correlations between various tests. PLoS One 10(12):e0144374. https://doi.org/10.1371/journal.pone.0144374
Stapleton HM, Klosterhaus S, Keller A, Ferguson PL, van Bergen S, Cooper E, Webster TF, Blum A (2011) Identification of flame retardants in polyurethane foam collected from baby products. Environ Sci Technol 45(12):5323–5331. https://doi.org/10.1021/es2007462
Sun E, Shi Y (2015) MicroRNAs: small molecules with big roles in neurodevelopment and diseases. Exp Neurol 268:46–53. https://doi.org/10.1016/j.expneurol.2014.08.005
Surdu SFE, Bloom MS, Boscoe FP, Carpenter DO, Haase RF et al (2013) Occupational exposure to ultraviolet radiation and risk of non-melanoma skin cancer in a multinational European study. PLoS One 8(4):e62359. https://doi.org/10.1371/journal.pone.0062359
Syed ST, Gerber BS, Sharp LK (2013) Traveling towards disease: transportation barriers to health care access. J Community Health 38(5):976–993. https://doi.org/10.1007/s10900-013-9681-1
Tarvainen L, Kyyrönen P, Kauppinen T, Pukkala E (2008) Cancer of the mouth and pharynx, occupation and exposure to chemical agents in Finland [in 1971–95]. Int J Cancer 123:653–659
U.S. Census Bureau (2014) American community survey design and methodology. https://www.census.gov/programssurveys/acs/methodology/design-and-methodology.html. Accesed 14 Mar 2017
U.S. Census Bureau (2017) My congressional district. https://www.census.gov/mycd/. Accessed 14 Mar 2017
UC Berkeley Safe Transportation Research and Education Center (2017) Berkeley SafeTREC. https://safetrec.berkeley.edu/. Accessed 14 Mar 2017
United States Comptroller General (1977) National water quality goals cannot be attained without attention to pollution from diffused or "nonpoint" sources. vol Rep. CED-78-6 USGO, Washington, DC
Urban Institute and Brookings Institution (2017) State (and Local) Taxes. http://www.taxpolicycenter.org/briefing-book/what-are-sources-revenue-local-governments. Accessed 14 Mar 2017
USDHHS (2017) Health information privacy. https://www.hhs.gov/hipaa/index.html. Accessed 14 Mar 2017
USEPA (2003) Framework for cumulative risk assessmen. EPA, Washington DC
USEPA (2017a) Carbon monoxide's impact on indoor air quality. USEPA. https://www.epa.gov/indoor-air-quality-iaq/carbon-monoxides-impact-indoor-air-quality. Accessed 14 Mar 2017
USEPA (2017b) Exposure assessment tools by media—water and sediment. https://www.epa.gov/expobox/exposure-assessment-tools-media-water-and-sediment. Accessed 14 Mar 2017
USEPA (2017c) The inside story: a guide to indoor air quality. https://www.epa.gov/indoor-air-quality-iaq/inside-story-guide-indoor-air-quality. Accessed 14 Mar 2017
USEPA (2017d) Particulate matter (PM) pollution. https://www.epa.gov/pm-pollution. Accessed 14 Mar 2017
USEPA (2017e) Sulfur dioxide (SO2) pollution. https://www.epa.gov/so2-pollution. Accessed 14 Mar 2017
USEPA (2017f) Types of pesticide ingredients. USEPA. https://www.epa.gov/ingredients-used-pesticide-products/types-pesticide-ingredients. Accessed 14 Mar 2017
USGPO (2017) Electronic code of federal regulations. http://www.ecfr.gov/cgi-bin/ECFR?page=browse. Accessed 14 Mar 2017

USGS (2017) NLCD 92 land cover class definitions. https://landcover.usgs.gov/classes.php. Accessed 14 Mar 2017

Volchek K, Thouin G, Kuang W, Li K, Tezel FH, Brown CE (2014) The release of lindane from contaminated building materials. Environ Sci Pollut Res Int 21(20):11844–11855. https://doi.org/10.1007/s11356-014-2742-x

Wang Y, Eliot MN, Wellenius GA (2014) Short-term changes in ambient particulate matter and risk of stroke: a systematic review and meta-analysis. J Am Heart Assoc 3(4):e000983. https://doi.org/10.1161/JAHA.114.000983

Wells NM (2017) How natural and built environments impact human health. http://www.human.cornell.edu/outreach/upload/CHE_DEA_NaturalEnvironments.pdf. Accessed 14 Mar 2017

Wild CP (2005) Complementing the genome with an "exposome": the outstanding challenge of environmental exposure measurement in molecular epidemiology. Cancer Epidemiol Biomark Prev 14(8):1847–1850

Wild CP (2012) The exposome: from concept to utility. Int J Epidemiol 41(1):24–32. https://doi.org/10.1093/ije/dyr236

Williams D (1990) Socioeconomic differentials in health: a review and redirection. Soc Psychol Q 53(2):81–99

Williams PRD, Dotson GS, Maier A (2012) Cumulative risk assessment (CRA): transforming the way we assess health risks. Environ Sci Technol 46(20):10868–10874. https://doi.org/10.1021/es3025353

Wimalawansa SA, Wimalawansa SJ (2016) Environmentally induced, occupational diseases with emphasis on chronic kidney disease of multifactorial origin affecting tropical countries. Ann Occup Environ Med 28:33. https://doi.org/10.1186/s40557-016-0119-y

World Health Organization (2017) Social determinants of health. http://www.who.int/social_determinants/en/. Accessed 14 Mar 2017

Yang S-N, Hsieh C-C, Kuo H-F, Lee M-S, Huang M-Y, Kuo C-H, Hung C-H (2014) The effects of environmental toxins on allergic inflammation. Allergy, Asthma Immunol Res 6(6):478–484. https://doi.org/10.4168/aair.2014.6.6.478

Zhang LJ, Zheng JY, Yin SS, Peng K, Zhong LJ (2010) Development of non-road mobile source emission inventory for the Pearl River Delta region. Huan Jing Ke Xue 31(4):886–891

Zhao X, Wang H, Li J, Shan Z, Teng W, Teng X (2015) The correlation between polybrominated diphenyl ethers (PBDEs) and thyroid hormones in the general population: a meta-analysis. PLoS One 10(5):e0126989. https://doi.org/10.1371/journal.pone.0126989

Zimmer KE, Gutleb AC, Ravnum S, Krayer von Krauss M, Murk AJ, Ropstad E, Skaare JU, Eriksen GS, Lyche JL, Koppe JG, Magnanti BL, Yang A, Bartonova A, Keune H (2012) Policy relevant results from an expert elicitation on the health risks of phthalates. Environ Health 11 (Suppl 1):S6–S6. https://doi.org/10.1186/1476-069X-11-S1-S6

Chapter 3
The U.S. Military and the Exposome

Christopher Bradburne and John A. Lewis

Abstract The US military has great potential to study and utilize the exposome, as well as great need. By their very nature, military Service members serve in diverse environments, under a variety of stressors, with pressure to perform and execute tasks under any conditions. The US military is unique in the range of environments they deploy to, as well as the kinds of industrial chemicals and materials they are exposed to, making them one of the most dynamic occupational health populations in the world. Unique environmental exposures include documented incidents such as burn-pit exposure, sulfur fires at the Mishraq sulfur mine, and water quality at Camp Lejeune, versus subtle exposures such as lead from munitions training, diesel exhaust, or JP-8 jet fuel exposure. Balancing the environmental health concerns of individuals with the operational needs of a unit can be challenging and dynamic. This has led the military to prioritize efforts for exposure surveillance, mitigation strategies, and leading-edge research initiatives. In this chapter, we will discuss the unique operational environments and exposures warfighters encounter, as well as the biomonitoring, military records for exposures, and how this relates to individual exposomes. The military has unique assets for exposome monitoring, such as uniform electronic health records (EHR), individual longitudinal exposure records (ILER), serum collection pre- and post-deployment for biobanking and surveillance, and a more normalized population relative to nutrition and demographics. This makes the military exposome unique, and provides important avenues for study and application.

Keywords U.S. Military · Warfighters · Electronic health records · Exposure Surveillance

C. Bradburne (✉)
Applied Physics Laboratory, Johns Hopkins University, Laurel, MD, USA

McKusick-Nathans Institute of Genetic Medicine, Johns Hopkins University, Baltimore, MD, USA
e-mail: Chris.Bradburne@jhuapl.edu

J. A. Lewis
US Army Center for Environmental Health Research, Ft Detrick, MD, USA

S. Dagnino, A. Macherone (eds.), *Unraveling the Exposome*,
https://doi.org/10.1007/978-3-319-89321-1_3

Introduction

The research approach of occupational health involves trying to determine the environmental causes of acute and chronic health issues, and how they interact with internal systems to produce adverse health effects over time. Understanding these dynamics can provide utility in prevention, mitigation, treatment, etc. New technologies such as genomics are revolutionizing our understanding of human health; however, an individual's genome (or other 'omics, such as transcriptome, proteome, etc.) represents only part of the story. An individual represents a complex system, containing a biochemical "record" of most of the exposures that have ever occurred (the exposome), as well as the background (genetic, psychological, social, etc.) on which those exposures occurred. This individual exposome can cloud the search in populations for causative agents, biomarkers, and individual disease trajectories.

Linking environmental causes to human health effects requires clear health endpoints, supporting data, and environmental information. A good example, discussed elsewhere in this book, is the association of aflatoxin from contaminated food storage to increased risk for hepatocellular carcinoma (Beasley et al. 1981). This linkage involved a straightforward application of basic and applied sciences, from epidemiology, to laboratory investigation, to population health approaches. But what about linking environmental causes to less obvious endpoints of human health? The WHO has defined health as "a state of complete physical, mental, and social well-being and not merely the absence of disease or infirmity (World Health Organization 1949)." Tools that can be used to discover more mechanistic causes such as chemical adducts may fail to link environmental factors with more vague effects such as malaise, depression, etc. In occupational health, there is a simple need for varied types of data: Data which describes the genomic and 'omic background, data which provides good information about the environmental, social, and psychological environment, and lots of longitudinal collection for everything.

The US military offers unique opportunities for addressing the occupational health research issues mentioned above. Throughout its history, the military has been faced with many situations in which chronic health effects appear in active-duty Service members or veterans, and they must employ methods to "discover" the source and mitigate the situation. Examples range from "smoking-gun" environmental agents such as aflotoxins to health effects that are still not linked to causes to this day, such as Gulf-War syndrome. Therefore, there has been an ethical, but also political, need for the military to establish good means for individual and environmental surveillance. Today, the military maintains an advanced systems of occupational health surveillance for its Service members and veterans.

Historical Need for Accurate Military Exposure Monitoring

Throughout the history of the US military, there has been a desire to protect Service members from potentially toxic exposures. Much of this effort has been aimed at chemical weapons due to the rapid devastation that they can inflict and the need for immediate countermeasures. In more recent years, the importance to protect Service members from all toxicants has become more apparent. Despite the nature of the toxicant, there are two important areas that need to be addressed (1) acute injury/incapacitation and (2) chronic health effects. In regard to the exposome and this chapter, we will focus on exposures that are more likely to result in chronic health effects and are typically more difficult to document.

Accurate and timely assessment and documentation of exposures is critical for developing intervention strategies and treatments, as well as for providing proper compensation. Despite significant efforts to perform environmental testing and surveillance, linking environmental concentration to actual exposure to the Service member is challenging. In practice, the DoD and VA have often relied on retrospective analysis of troop locations to identify and track potentially exposed individuals. This has been the case for some of the more well-known historical exposures that have affected US Service members.

Chemical Weapons

The concern for chemical weapons dates back to World War I when Germany first used chlorine gas against French and Canadian troops. Fortunately, despite significant chemical weapon production and stockpiling by both the USA and her adversaries, use of chemical weapons against US forces has been minimal since WWI. However, the great potential for injury and the ability to weaponize some commonly available toxic industrial chemicals or materials (TICs or TIMs) makes them an ongoing threat.

Over the years, there have been numerous lower dose exposures to chemical agents that have the potential for chronic effects. During World War II, Service members were involved in human subject testing, primarily for mustard agents, to determine the effectiveness of protective clothing and skin ointments. Between 1963 and 1969, approximately 5900 Service members were exposed during a series of tests, term Project 112/SHAD (Shipboard Hazard and Defense), to determine the effectiveness of shipboard defenses (National Academies of Sciences, Engineering, and Medicine 2016a). While little data exist for the actual level of exposures for the individuals who were involved in these incidents, the DoD maintains databases of those involved and an additional database of Service members exposed during the other Cold War activities. In total, there are nearly 44,000 individuals listed in these three databases.

Pre-Vietnam Exposures

The era during the Vietnam War marks increasing public concern with the effects of toxic exposure. Prior to the Vietnam era, the military, as well as many commercial entities, used toxic chemicals such as polychlorinated biphenyls (PCBs), industrial solvents, asbestos, and radiation. Service members in occupations using these chemicals and materials were at highest risk, but others may have also been exposed. Many service members were exposed to ionizing radiation through atmospheric testing of nuclear weapons, deployment to regions of Japan postwar, and occupational hazards such as working in a reactor plant. Both asbestos and ionizing radiation are recognized by the VA as compensatable exposures, though many other potential TIC and TIM exposure of the era go unrecognized.

Agent Orange

Perhaps one of the most politicized military exposures is Agent Orange during the Vietnam War. Agent Orange is a mixture of two herbicides 2,4,5-trichlorophenoxyacetic acid (2,4,5-T) and 2,4-dichlorophenoxyacetic acid (2,4-D), that was formulated to be dispersed from aircraft to defoliate jungle vegetation. Unfortunately, 2,3,7,8-tetrachlorodibenzo-*p*-dioxin (TCDD) was a by-product of the manufacturing process. TCDD is a potent teratogen and carcinogen. Despite the long half-life of TCDD in humans and the potential to estimate exposure levels even years later, little data exist to indicate which Service members were exposed to this toxicant while in Vietnam. However, the widespread use of Agent Orange (at least 17–19 million gallons) during the campaign led to Public Law 102–4, passed in 1991, which created a presumptive determination for any Service member contracting one of the 14 diseases likely to be caused by dioxin and who was in Vietnam during the periods of Agent Orange use. Estimates of those affected vary, but 2.6–4.3 million US Service members served in Vietnam (National Academies of Sciences, Engineering, and Medicine 2016b).

Gulf War Illness

Gulf War Veterans' Medically Unexplained Illnesses, more commonly called Gulf War Illness, is a collection of chronic medical symptoms that have been reported by returning veterans from the First Gulf War. While there is a strong belief that Gulf War Illness is likely the result of some combination of exposures occurring during the war, no linkage has been discernable. Known exposures included oil-well fire smoke, dust, high ambient temperatures, depleted uranium, pesticides, and pyridostigmine bromide (PB). After over 25 years and millions of dollars of research, it is unlikely that the causative factor will ever be determined. Findings from the National Academy Science, Institute of Medicine review conclude that exposure data must be collected before, during, and after deployment to permit an accurate assessment (National Academies of Sciences, Engineering, and Medicine 2016c). In response to concerns of adverse environmental exposure in Southwest Asia, a Gulf War Registry has been created by the VA and includes over 152,000 Service members across Operation Desert Shield, Operation Desert Storm, Operation Iraqi Freedom, and Operation New Dawn.

Pulmonary Exposures in Southwest Asia

Over the course of the US involvement in Southwest Asia, there have been numerous reports of pulmonary symptoms or injuries, ranging from increased rates of asthma to constrictive bronchiolitis. Exposures to high levels of particulate matter, burn pits, and a sulfur fire at the Mishraq State Sulfur mine have been recognized as potential causes. However, epidemiological studies have failed to definitively link any to the observed symptoms. In response to early concerns, the US Army Center Health Promotion and Preventive Medicine (USACHPPM) conducted the Enhanced Particulate Matter Surveillance Program (EPMSP). Despite this more rigorous environmental surveillance effort, the challenges with linking exposures to Service members and outcomes have persisted. The VA's Airborne Hazards Open Burn Pit Registry currently has over 100,000 Service members enrolled.

Response to Effective Exposure Monitoring, and Overall Military Exposure Life Cycle

Through the efforts to identify the etiology of Gulf War Illness, it became abundantly clear that the DoD was failing in its effort to protect the Service member from the complex set of exposures that are inherent in the operational environment. In response to this gap, President Clinton issued, through the National Science and Technology Council, Presidential Review Directive-5, which is aimed at minimizing or eliminating post-conflict health concerns. An important requirement in this directive is for the DoD and VA to establish a longitudinal record of the Service member's occupational and environmental exposures and events. While this goal has not been realized, the DoD has advanced exposure monitoring in many ways and is working toward this end goal. DoD Directive 6490.02E, Comprehensive Health Surveillance and DoD Instruction 6490.03, Deployment Health provide the DoD guidance which requires the implementation of exposure monitoring.

The DoD has developed two resources to increase documentation of environmental surveillance and occupational exposure (Table 3.1). The Periodic

Table 3.1 DoD efforts to maintain exposure records for use in risk assessment. ILER has been prototyped but is not yet fully implemented

Geographical site specific	Unit specific	Individual
POEMS: Periodic Occupational and Environmental Monitoring Summary	DOEHRS: Defense Occupational and Environmental Health Readiness System	ILER: Individual Longitudinal Exposure Record DoDSR: Department of Defense Serum Repository

Occupational and Environmental Monitoring Summary (POEMS) provides an assessment of occupational and environmental health (OEH) risk for a specific location and is based on extensive environmental surveillance sampling and monitoring data, field investigation, and health assessment reports. While these documents do not provide an individualized exposure record, they do provide a publically available record of potential exposures that can be used to guide health care decisions. The Defense Occupational and Environmental Health Readiness System (DOEHRS) Resources is a Defense Health Agency (DHA) managed system to document occupational and environmental exposure events for both deployed and garrison environments. This system is the first step in capturing exposure events that can be linked to a Service member's exposome.

While an Individual Longitudinal Exposure Record (ILER) has not been implemented, efforts are underway to establish one. In January 2013, a two-year pilot project to develop the ILER was initiated, and the DOEHRS is envisioned as the foundation for the fully implemented ILER. The primary challenges in implementing an ILER for the US military have been related to the unique diversity of operational environments and potential exposures that Service members can encounter. Throughout even a single deployment, a Service member can move through a diverse array of locations, each with its own potential set of exposures. Environmental surveillance efforts are targeted at known or potential risks, but there are currently no technologies that allow the full range of exposures to be quantified and document. In addition, the specific location of each Service member is unique, and the environmental surveillance only provides a point estimate for the area. This is further complicated when locations of specific military missions (and the participating Service member) must be classified for national security reasons. For an ILER to be fully realized, some form of wearable exposure monitoring will likely be required.

The US military encompasses both unique challenges and unique opportunities for exposure monitoring. Service members can be deployed to virtually anywhere in the world and, as such, have the potential to be exposed to nearly anything. As discussed above, this necessitates a comprehensive set of technologies to be able to identify and quantify this broad range of exposures. However, Service members must be able to accomplish missions, which are often life or death. This austere nature of the operational environment can make it impossible to commit even the bare minimum of time or resources to collecting exposure data. However, the structure and scope of the military provides unique resources that may not be available in the public sector. For example, as part of the bi-annual HIV testing requirement, the DoD has established a serum repository, the DoDSR, that contains longitudinally collected samples for all Service members. These samples have the potential to serve as an exposure record and provide pre-exposure baselines for yet undiscovered assays. Unfortunately, due to the nature of the collection and cold-chain requirements for their primary use (i.e., immunological assays), the serum samples are likely to have variations in levels of preservation for many biomolecules of interest. Nonetheless, their existence is a unique asset that is unlikely to be available in the civilian sector.

When considering potential exposures for Service members, the focus is often the deployed environment. This is clearly important due to the austere nature of the environment and the inability to provide effective controls to eliminate unanticipated exposures. However, when considering the entire exposome of military personnel, one must consider the life cycle of military employment. Upon entry into service, preexisting exposure or genetic conditions may predispose individuals to adverse effects from exposures during service. While on active duty, typical Service members will deploy to an operational environment multiple times throughout their career, each of which may have unique exposure risks. However, they may also experience adverse exposures within garrison settings, and as with all individuals, they may be exposed during their personal life (e.g., spraying a pesticide in their garden). After separation, Service members are under the care of the VA, which requires transition of exposure and health data, but also may be complicated with post-service occupational or personal exposures. The DoD and VA must consider many routes (e.g., military, occupational, and personal) when evaluating the exposome of military personnel.

Exposomics: Models and Approaches

As described by Wild (2012), the exposome constitutes "every exposure to which an individual is subjected from conception to death". In the military, the overall exposome will be cumulative over time: literally, from birth, through preservice-adulthood, through service (and deployment exposures), and then in the VA-care stage of post-service (Fig. 3.1). The latter stage is the most likely to present with chronic health effects from a lifetime of exposures (and age), while during service or deployments, acute effects from a specific exposure may be more common. This means that acute events may still be reflected in biological markers in later years.

The exposome itself can be separated into several domains: the general external, specific external, and the internal exposome (Wild 2012). The *general external exposome* represents the individual's social, financial, psychological, and geographic background. Military resources to provide some information on this element would include general service records and individual demographic data. The *specific external exposome* is the combination of specific events, such as chemical

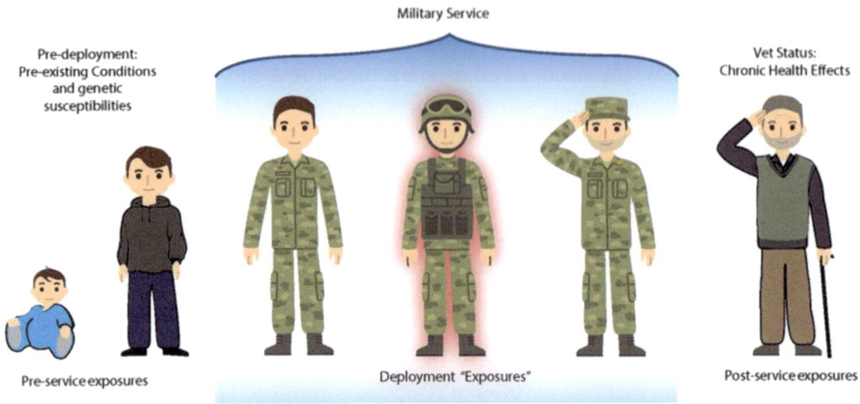

Fig. 3.1 Exposome of a Service member across their lifetime. The internal exposome will consist of general and specific exposome contributions over a lifetime

contaminants, environmental pollutants, diet, radiation, lifestyle factors such as smoking, infections, medical interventions and drugs, etc. Some lifestyle information can be found in medical records. Tracking more specific exposures falls to the DOEHRS, and the developing ILER described earlier. However, there is need for more passive, cumulative detectors of pollutants and chemicals, similar to the dosimeters used in ionizing radiation work. Finally, the *internal exposome* consists of the body systems and components affected by the external exposomes. Examples include gut microflora, inflammation, peroxidation, adducts, and oxidative stress.

An important quest of military occupation health research is to discover and exploit actionable biomarkers of the internal exposome that can be linked to military occupational exposures. This requires extensive biological sample collection and curation to be of any use for *both* population research, *and* individual exposure tracking. To this end, the military has collected and curated serum in the Department of Defense Serum Repository (DoDSR) since 1985, as a biological record of service and deployment exposures. The DoDSR is extensive: >54,000,000 samples as of 2015, making it one of the world's largest biobanks (Perdue et al. 2015). It has been called a national treasure by many in the military research community, but only a small proportion of its entire sample collection (0.42%) have ever been thawed for research or individual tracking (which amounts to only 226,800 samples) (Perdue et al. 2015). In addition, the samples are serum only. While serum is informative for antibodies (previous infections) and metabolites or xenobiotics from pollutants, many effects of exposures such as oxidative DNA damage, epigenetic changes, and changes in immune function may be missed. Currently, plans are being made to possibly expand its collection from blood to include DNA, RNA, protein fractions, and peripheral white blood cells (Lindler 2015). This would allow a more full range of 'omics-wide studies to be carried out to examine the internal exposome (Bradburne et al. 2015).

An additional component to the internal exposome would be *preexisting susceptibility*. This would include such things as variants in the individual's genome that could confer increased or decreased risk, nutritional differences, and preexisting diseases (Bradburne et al. 2015). The emerging field of genomic medicine fits into this category. It attempts to use a person's genomic information to infer risks for various chronic conditions over their lifetime, and it is likely that eventually every soldier will have their genome sequenced, incorporated into their medical record, and genetic susceptibilities determined. Policy and ethical issues in the military are being addressed and it is envisioned that as the information becomes actionable for screening and health care, large-scale genomic information will be implemented in the military (De Castro et al. 2016). However, an individual's genome contains only a small part of the risk for the majority of chronic diseases (Rappaport 2016). Other sources must be explored. An additional component of susceptibility on top of the genome is the microbiome: microbial populations that are literally the interface between the human organism and its environment (i.e., on the skin, in the gut, at mucosal interfaces, etc.). Individual microbiomes can vary greatly, and change the dynamics of xenobiotic interactions. Ultimately, an individual's microbiome may 1 day be measured for susceptibility similar to their individual genome (Bradburne and Hamosh 2016).

So far, we have described an exposome model and the military resources that exist or are being developed to support susceptibility determination, population health, and individual tracking… so how could useful biomarkers be discovered for actionable information in a specific exposure event? Figure 3.2 shows the "Toxicological paradigm," in which an external exposure results in an internal dose, which then becomes a biologically effective dose. This leads to early effects, then altered structure and function, and finally clinical disease. Genetic

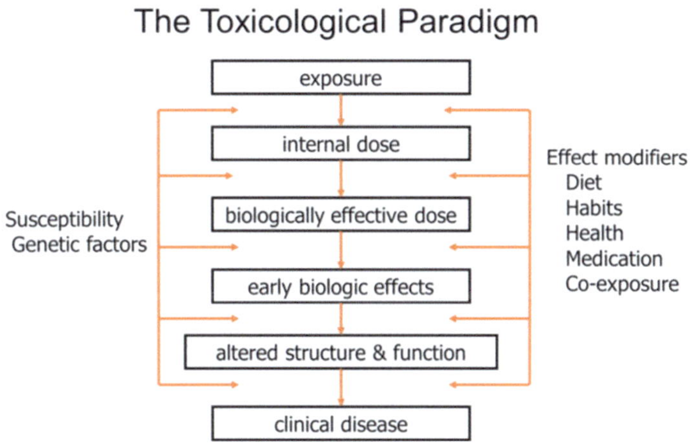

Fig. 3.2 The Toxicological Paradigm. Genetic susceptibility factors could interact across the pathway from exposure to clinical effect. Effect modifiers such as diet, habits, medication, etc. can also influence the toxicological process (Courtesy of Dr. John Groopman, JHU-BSPH)

Table 3.2 Factors that may contribute to susceptibility, and effect modifiers that may influence the trajectory of an exposure manifesting itself into clinical disease

Preexisting or concurrent nonmilitary homelife	Training and unit environment and characteristics	Exposure effect modifiers
Genomic and 'omic markers	Occupational (administrative vs active, etc.)	Diet
Social capital	Social and leadership	Tobacco, alcohol, drugs
Education	Unit cohesiveness	Other chemical exposures (TICs, TIMs) and specific xenobiotics
Financial	Career pressures and performance	Infections
Psychological stress	Guilt/self-blame	Medical interventions
Demographic stratification	Unit vs individual morale	Environment (radiation, altitude, cold, heat)
Self-perception	Ethics	Lifestyle
Urban or rural environment		

susceptibilities could be found throughout the process, as well as effect modifiers from the general and specific external exposome.

A good epidemiologic model using military exposomes would consist of establishing a baseline of preexisting susceptibilities such as a sequenced genome, and other items that can be assessed as contributing factors for risk assessment (education, psychological stress, etc.), and then tracking stressors, effect modifiers, and exposures throughout military service (Table 3.2). The overall picture provided may shed light on various factors involved in influencing the toxicological paradigm at any given point. From this point then, there are several types of physiological biomarkers that can be sought (1) biomarkers of susceptibility, (2) biomarkers of exposure, and (3) biomarkers of effect.

Biomarkers of susceptibility would include existing individual genomic variants described earlier. These interact across the spectrum of the toxicological paradigm. An example would be men who carry a susceptibility gene to pesticide exposure. Studies of 55,747 licensed, male pesticide applicators in Iowa and North Carolina showed 3.7X higher risk of prostate cancer after extended occupational exposure to petroleum products and 3X higher risk for men exposed to Terbufos who had the minor allele SNP rs1883633 in the oxidative stress gene glutamate–cysteine ligase (GCLC) (Koutros et al. 2011). As whole genomes are sequenced, military service members with this risk allele may warrant preventive measures given specific duties that put them at risk for these exposures such as extended duty in the motor pool.

Biomarkers of exposure would consist of blood or clinical sample components indicative of an exposure event. In the toxicological paradigm, these would typically be found in the exposure, and internal dose stages. Examples include levels of the offending compound in a clinical sample, such as finding arsenic in urine. Others might include adducts such as AFB-1 guanine found in urine, indicative of very recent or "acute" exposure to aflatoxin over the past 24 h, vs. AFB-1 albumin found in blood, indicative of "chronic" aflatoxin exposure over the previous weeks or even months. A chronically exposed service member returning from deployment would provide a post-deployment serum sample to the DoDSR. This could preserve AFB-1 albumin as a discoverable record of aflatoxin exposure.

Biomarkers of effect are markers that result from an exposure. An example would be bladder cell micronuclei that result from arsenic exposure through drinking contaminated water, and are correlated directly with urinary arsenic (Biggs et al. 1997). In this case, evaluating a urine sample of a service member with symptoms using a urinary micronucleus assay (MNu) could elucidate the effects of arsenic poisoning.

Determining System Links, Biomarkers, and Actionable Information

An overall aim of predictive toxicology is to develop risk assessments based on mechanistic information. One approach that has gained widespread utility is the grouping of similar mechanisms for various toxicants into "Adverse Outcome Pathways" (AOPs). An AOP is a conceptual framework that allows existing knowledge about a molecular toxicological event to be tied to an adverse outcome to inform a risk assessment (Ankley et al. 2010). Defining an AOP allows you to leverage knowledge about cellular pathways that are engaged in a particular toxicological mechanism, and apply it to others that may use the same pathway. A more extensive review of AOPs can be found here (Vinken 2013).

Taken together, AOPs can be modular: fitting together to describe an overall toxicological process for different exposures. Organization of information in an AOP consists of (1) the molecular initiating event (MIE), (2) intermediate steps, and (3) the adverse outcome (Table 3.3). AOPs can be proposed and developed in the community, and they are currently being developed for military use-cases.

As AOPs attempt to modularize information, they can be used to establish broad chemical categories with similar properties or mechanisms of action, inform test guidelines, and integrate testing and assessment approaches. In military exposure science, AOPs are beginning to gain traction and are in the process of being defined. Developing an AOP requires a starting knowledge of an adverse health effect, and some recognition of the key steps leading up to the adverse outcome (OECD 2012). This information can, in turn, be used to inform the search for an MIE if the MIE itself is not known. In this way, AOPs may become useful in providing clues for a toxicant in individual exposures, and they may also be useful for elucidating causes for broad, population-level exposure effects such as gulf war syndrome.

Table 3.3 An example of an AOP for chemical-induced skin sensitization

Molecular Initiating Event (MIE)	Intermediate steps			Adverse Outcome (AO)
Molecular event	Initial interaction	Cellular effects	Organ effects	Organism effects
Exposure of chemical; covalent modification of protein nucleophilic residues	Skin keratinocytes	Inflammatory cytokine induction	Histocompatibility complexes presented by dendritic cells; T-cell activation; T-cell proliferation	Skin inflammation
Penetration of compound	Dendritic cells	Inflammatory cytokine induction; dendritic cell mobilization		

Information adapted from Vinken (2013) and MacKay et al. (2013)

Exemplars in the Military for Exposome Surveillance, Research, Mitigation, and Actionable Information Development

Case Study #1: Hexavalent Chromium Exposure at Qarmat Ali Water Treatment Facility

In 2003, Service members, including active duty Soldiers, National Guard (NG) members and Reservists, assigned to guard the Qarmat Ali Water Treatment Plant in Iraq were potentially exposed to sodium dichromate, a form of hexavalent chromium, which is a known carcinogen. The water from this facility was being produced to inject into the Basra oil fields to increase oil production and was not for human consumption. The sodium dichromate was added to the water to limit corrosion. It is likely that the bags were purposely open and the contents dispersed by previous occupants to contaminate the site.

The first visits to the QA WTP began in late April 2003 and regular work at the site began in late May. The first notification to the on-site contractor of the use of sodium dichromate at the plant was in June, and during this period there were efforts by the contractor working on-site to cover discolored soil with clean soil to minimize aerosolization and exposure. In early August, an environmental assessment team from the civilian contractor working on-site revealed a possible contamination, and the contractor officially notified the US Army Corps of Engineers (USACE) contracting officer. Later in the month, the contractor began an effort, which lasted until October, to encapsulate the soil in gravel and asphalt. In September, site access restrictions were implemented and a request was sent by the Combined Forces Land Component Command Surgeon office to the US Army Center for Health Promotion and Preventative Medicine (USACHPPM) for occupational and environmental health personnel support. A Special Medical Augmentation Response Team,

Preventive Medicine (SMART-PM) arrived at the end of September and conducted a health risk assessment beginning in October.

The SMART-PM performed both environmental and health assessments. The results of the environmental testing showed elevated levels of hexavalent chromium in the soil, but none in breathing zones or general areas. It should be noted that this assessment occurred nearly a month after encapsulation efforts had begun, so actual concentrations may have been greater earlier and dependent on activities being conducted. The health assessments of civilians and Service members still on-site included measuring total chromium concentrations in whole blood, physical examinations, pulmonary function tests, and chest x-rays. The blood chromium levels were conducted approximately 1 month after the potential exposure, but none showed elevated levels that would be consistent with an occupational exposure. No medical findings were observed that would indicate acute hexavalent chromium exposures, though there were earlier reports of bronchio-nasal irritation that were assumed to result from the dry dusty conditions. Overall, it was concluded that there is only a low risk for long-term health effects from this potential exposure, but individuals with higher, undocumented exposures may be at higher risk. Following this initial response, efforts have been aimed at health communication and notifying others who served in the area.

In 2008, the Surgeon General of the Army requested a review of the CHPPM response to the incident. The Defense Health Board (DHB) review concluded that the CHPPM had met or exceeded the standard of practice for occupational medicine. It also recommended the creation of a registry of individuals involved, which led to broadly inclusive effort to contact any individual who may have been in the region. A review by the DoD Inspector General from 2009 to 2010 looked specifically at the efforts to identify and contact personnel who may have been exposed and showed that greater than 95% of the 977 civilian and Service members who were in the area had been notified. It was also noted in this report that three NG members had died of interstitial lung disease, lung cancer, and leukemia, respectively, and two civilians had died of heart attacks since the incident. There is currently no definitive link between these deaths and the exposure, but a court case in 2012 found the contractor negligent for the delay in identifying and notifying the exposure risk. While health risks are likely to be low, there are still unresolved concerns related to this incident.

Unanswered Questions:

- Can the identification of the exposure risk be accelerated?
- Can biological samples be captured prior to the arrival an OEH team that can be used for biological assessment?
- Can biological markers be developed that would last >30 days to allow assessment of exposure and effects postexposure?
- Are there methods to definitively link the deaths to the hexavelent chromium exposure?

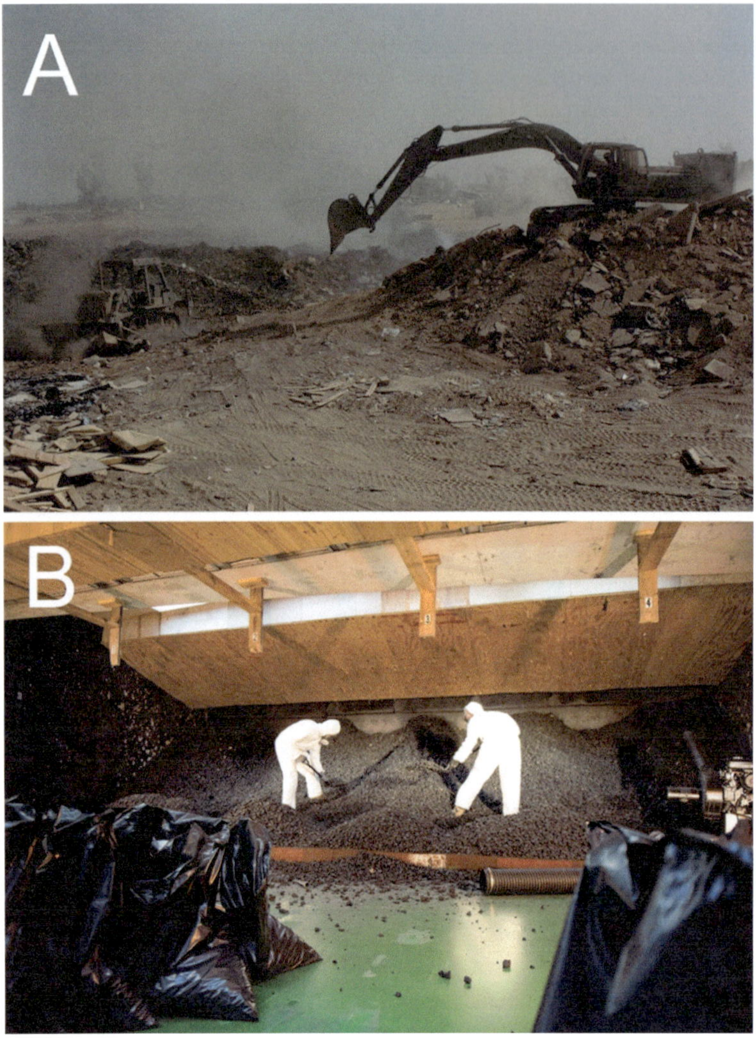

Fig. 3.3 (**a**) Soldiers from the 84th Combat Engineer Battalion use a bulldozer and excavator to maneuver trash and other burnable items around in the JBB burn pit. (Courtesy Photo, 28th Public Affairs Detachment) (**b**) Two contractors shoveling in order to remove lead bullets from behind the targets at an indoor firing range. (US Army photo by Visual Information Specialist Pierre-Etienne Courtejoie/Released)

Case Study #2: Occupational Lead Exposures in the US Military

The nature of modern warfare means that Service members will be exposed to lead from small arms munitions through vaporization of lead from the bullet as well as from combustion of lead styphnate, used as an explosive in primers. While all Service members are at risk, those working at firing ranges, particularly indoors, are at greater risk due to the increased exposure times and concentrations (Fig. 3.3b).

In the USA, occupational exposure limits for lead were set by the Occupational Safety and Health Administration (OSHA) lead standard (29 CFR 1910.1025) in 1978. This standard both sets permissible exposure level (PEL) and requires biomonitoring of blood lead levels (BLLs) and erythrocyte protoporphyrin (EP). Policies within the DoD (DoD 6055.05-M) and its branches are in place to meet these standards. The OSHA standard sets the PEL as an 8-h time-weighted average (TWA) of 50 μg/m^3 in air and a BLL of 30 μg/dL as the level requiring medical surveillance and 60 μg/dL (or three consecutive BLLs averaging 50 μg/dL or higher) as the level which requires mandatory removal for exposure level. Two consecutive BLLs of 40 μg/dL or lower after removal from exposure would allow an individual to return to work.

As these criteria were established at a level assumed to be safe, in setting this standard, one can conclude that OSHA had determined that 40 μg/dL is a BLL concentration that protects the worker.

Recent evidence suggests that the OSHA standard may not be protective and that chronic BLL as low as 10 μg/dL may produce adverse health effects in adults and 5 μg/dL may affect prenatal and neonatal development (NTP monograph and for review, see Environ Health Perspect. 2007 Mar; 115(3): 463–471). In response to these concerns, the DoD requested an evaluation of potential health risks related to recurrent lead exposure to firing-range personnel by the National Research Council. The NRC agreed with recent literature that the OSHA standards do not adequately protect the Service member. The NRC report highlighted concerns with using BLL as it generally represents recent exposures to lead and does not adequately account for total body burden, but it can also be affected by perturbations of the equilibrium with lead stored in bone from chronic exposures. The NRC found that current DoD policies do not adequately protect the Service members and recommended the review and revision of DoD policies based on currently available lead exposure data and the lowering of trigger values for BLL. In response, the US Army Public Health Center in collaboration with subject matter experts across the DoD develop new provisional BLL guidelines for the Office of the Deputy Under Secretary of Defense (Installations and Environment) (US Army Public Health Command 2014). Because the report only provided BLL and did not address occupational exposure levels, additional work is underway to develop suitable exposure concentration using pharmacokinetic models (Sweeney 2015). It is anticipated that these findings will lead to updates in the DOD 6055.05M. The great range of targets affected with varying BLL makes this a very complex issue that is well suited to exposome science.

Unanswered Questions:

- Can biomarkers of effect be developed that are more personalized than BLL in assessing risk?
- Can the US military implement proactive monitoring strategies that would identify at-risk exposure levels?
- Are suitable methods available to evaluate total body burden for chronic exposure that will be better predictors of risk than BLL?

Case Study #3: Burn-Pit Exposures in Southwest Asia
During operations in Southwest Asia (SWA), the burning of trash and other waste products was a common practice for disposal. The decision to burn trash was made, at least in part, for operational security reasons to prevent adversaries from utilizing discarded material and to eliminate the risk incurred by transportation of waste material outside of secured locations. While high temperature incinerators were implemented in later stages of operations, the early methodology used open air "burn pits," where waste material was placed in piles and burned with an accelerant (e.g., JP8 fuel). This resulted in incomplete combustion and the generation of many potential harmful combustion products. Further, the large breadth of material that was burned included some items that containing toxic material that was aerosolized during the combustion (e.g., heavy metals). Exposure to these toxic combustion products and materials may result in adverse health effects if exposure concentrations are sufficiently high. However, there are insufficient data available to know the actual exposure level for any individual Service member and whether he or she is at risk for injury or disease.

One of the largest burn pits in operations in Southwest Asia was at Joint Base Balad (JBB) near Baghdad, Iraq (Fig. 3.3a). This burn pit was several acres large and burned an estimated 100–200 tons of waste per day in 2007 (Institute of Medicine 2011). The use of burn pits began in Iraq in 2003 and environmental surveillance and health studies around the JBB burn pit began in 2004, but it was a memorandum by the acting Bioenvironmental Engineer Flight Commander, Lt Col Curtis, that brought attention to the health risk resulting from the burn pit at JBB. In 2007 and 2009, the USACHPPM and the Air Force Institute for Operational Health conducted ambient-air sampling and screening health-risk assessments of burn-pit exposures at JBB. Findings included elevated ambient airborne particulate matter (PM), polyaromatic hydrocarbons (PAH), and volatile organic compounds (VOCs), and detection of low levels of polychlorinated dibenzo-p-dioxins and dibenzo-p-furans (PCDDs/Fs). However, the PM, PAH, and VOC levels were likely background due geogenic dust and urban pollution, respectively. Only the PCDDs/Fs were significantly derived from the burn pit. By 2009, the use of burn pits had been restricted by the DoD and the burn pit at JBB had been closed and replaced by high temperature incinerators.

In the same time frame, numerous anecdotal reports began to emerge from Southwest Asia about the potential for multiple lung injuries in returning Service members. Notable, a VA studied identified an increased risk for asthma in Soldiers returning from SWA (Szema et al. 2010), and a cohort of 38 Soldiers were diagnosed with constrictive bronchiolitis (King et al. 2011). In a prospective population-based study using data collected from questionnaires completed by the Millennium Cohort Study participants, an increase in new onset respiratory symptoms was observed in Service members who were deployed, but no increases were detected for asthma or constrictive bronchiolitis (Smith et al. 2009). In 2010, a working group of academic, DoD, and VA physicians and exposure scientists reviewed the available data and was unable to link adverse pulmonary outcomes with specific exposure events in SWA, and cited the need for increased and standardized pre- and post-deployment medical surveillance (Rose et al. 2012). The IOM, upon the request of the VA, reviewed the potential for long-term health effects due to exposure to burn pits in Iraq and Afghanistan and concluded that the air pollution and high PM levels represented a greater risk than the burn pit (Institute of Medicine 2011). However, the study does recognize that individuals with an existing susceptibility and those with increased exposure levels (e.g., working at the burn pit) are at higher risk for long-term health consequences from exposure to the burn pit.

Research within the DoD is ongoing to help elucidate health risks associated with burn-pit exposure and other pulmonary exposures in SWA. The Defense Health Program funds pulmonary health research through a working group chaired at the US Army Center for Environmental Health research. This work has included evaluation of the toxicity of the PM from Iraq (Porter et al. 2015), and a Study of active duty military for pulmonary disease related to environmental deployment exposures (STAMPEDE) (Morris et al. 2014). Overall findings are suggestive that the high ambient levels of PM are the primary contributor to pulmonary symptoms observed in Service members deployed to SWA. Additional work by the Naval Medical Research Unit-Dayton (NAMRU-D) is underway to assessing the potential health risks using a model burn pit based on the JBB waste stream.

Despite limited data, the concern for adverse health effects resulting from exposure to burn pits remains in the public press (Foxnews 2016; FoxNews 2017), and activist groups, such as BurnPit360, remain committed. While it is well know that exposure to combustion products is a health risk and some of the material burned in the DoD burn pits could increase this toxicity, there is currently no retrospective way to identify exposure doses for individual Service members. Further, for individual Service members who have developed a disease and have been exposed to burn pits, the disease cannot be definitively linked to the burn-pit exposure versus some other undocumented cause. New methodologies are required to document exposure, assess exposure risks, and identify at-risk individuals for targeted intervention and treatment strategies.

Unanswered Questions:

- Can biomarkers of susceptibility be developed that could help to limit exposures to susceptible populations for operationally required risk, such as burn pits?
- Would wearable dosimeters for a small number of surrogate chemicals be sufficient to assess health risks for individual Soldiers?
- Can biomarkers of effect be developed that would allow earlier diagnosis, intervention, and treatment of health effects from environmental exposures?
- Are methods available that would allow causative linkages to be determined for individual diseases or injuries?

Potential Impact of New Technologies

New technologies will continue to emerge to enhance capabilities to determine susceptibilities, determine biomarkers, and monitor population and individual exposures. The military has an extensive process to select, evaluate (with T&E), procure, and deploy new technologies. This process is rigorous, but also typically much slower than adoption seen in the civilian world (it can take several years following initial technology downselection). In addition, individuals may choose to adopt or benefit from technologies that emerge in their civilian lives before widespread adoption by the military (e.g., the adoption of wearable physical activity trackers and geolocational sensors). Therefore, the impact of new tech may be seen in some soldiers before widespread adoption. In addition to new tech, Service members will benefit in the wider implementation of exposures tracking resources such as the ILER discussed earlier. The expansion of the DoDSR to cover extended 'omics capabilities will increase its utility for both research, and individual exposure discovery. In addition, genomic information associated with military medical records will increasingly inform an individual soldier's susceptibility.

New technologies on the horizon could include wearable, personal dosimeters that can absorb airborne toxicants, and be desorbed later for various mass-spectrometry-based analysis techniques such as selective reaction monitoring (SRM, discussed more extensively in (Bradburne et al. 2015)). Such an approach can be combined with sensitive, targeted-metabolomics approaches when known toxicants are suspected (such as benzene or toluene), or expanded to less sensitive, but broader untargeted-metabolomics approaches. The DoDSR in its current form should already be able to support this. Other technologies on the horizon include nanopore sequencing for biomarker discovery or detection. An example could be discovery of transcriptomic biomarkers that could provide presymptomatic actionable information for preventive medicine. Nanopore sequencing (which is already a hand-held, flash-drive-sized device) could even eventually be deployed with soldiers and used in the field.

Other data sources already available for soldiers include an Android-based platform called "NetWarrior," which is allows multiple apps to be used for warfighters (Army.mil 2013). One app in development by the US Army Center for Environmental Health Research (USACEHR) and the JHU Applied Physics Laboratory (JHU/APL) is the "eHARM" (Environmental Health And Response Management) app. This Android app will be deployed on the NetWarrior platform, and provide geolocational tracking for exposure purposes, information resources, such as all chemical toxicological information contained in the NIOSH (National Institute of Occupational Safety and Health) handbook, picture capture and curation of environmental threats, protocol tracking, and information entry for exposure documentation.

Finally, social media and health care informatics will continue to play a role, as symptom and health care utilization surveillance systems continue to develop. An example is the DoD's ESSENCE system: "Electronic Surveillance System for the

Early Notification of Community-based Epidemics." As not every health care anomaly is infectious disease, systems like ESSENCE could eventually be utilized for passive longitudinal monitoring of specific, acute exposure events.

Future Outlook

Those who cannot remember the past are condemned to repeat it—George Santayana

The US military has an exceptional history of success in achieving its mission; however, it unfortunately has a less stellar record in protecting its Service members from exposures. In nearly every major conflict, troops have returned from deployment with exposure-related injuries and diseases. Limitation of resources results in a prioritization that places the focus on the mission and protecting the Warfighter from immediate, catastrophic outcomes. A new paradigm is required that adequately protects the Service member from exposures while minimizing the burdens placed on him or her to achieve this protection.

Identification and Documentation of Exposures

Accurate identification and documentation of exposures is an initial step in being able to address exposure-related health outcomes. Knowing what the Service member has been exposed to allows proper medical interventions and is the first step in health risk assessment for long-term health consequences. Current practices which utilize environmental sampling based on geographical locations are not capable of providing accurate exposure levels for individual Service members.

Broad-range, wearable dosimeters are the next stepping stone in collecting accurate exposure data. Due to limitations in current technology and the diversity of potentially toxic chemicals, it is not realistic to assume a single detection platform will quantify all exposures. A more realistic approach may be to develop dosimeters based on class of chemicals and assign different detectors to individuals within a squad. This minimizes the burden on each Service member while providing an exposure assessment for a co-localized group.

For a more personalized approach, biomarkers of exposure will need to be developed. For some compounds (e.g., hexavalent chromium) these biomarkers exist, so technologies, such as blood spot cards, that are able to collect and preserve biological specimens at the appropriate time points post exposure will allow accurate documentation of these exposures. For other toxicants, development of host response biomarkers through the use of system biology methodologies is the most feasible approach.

A true ILER will need to be implemented to capture this exposure data, and an automatic data process will be required to ensure it is reliably transitioned to the database. This can be built off existing infrastructure of the DOERS, but conceptually it will be much grander. The current process is designed to capture known toxicants. The concept of the exposome requires all chemicals to be included, as their interplay contributes to the effects on human health. Early phases of the ILER will focus on documenting known high-priority toxicants, but ultimately it will include the full exposome of each individual.

The identification and documentation of specific exposures is an intermediate strategy. It assumes that all of the hundreds of thousands of chemical can be assessed at concentrations of importance, that there is a known understanding of the level of toxicity for each of these chemicals, and that knowledge required for dealing with mixtures is understood. This methodology is suitable for prioritized high-risk chemicals and is also satisfying to individuals because they want to know what they were exposed to. However, the future of personalized medicine is more likely to lie in treating the individual response and will not require knowledge of the causative chemical or mixture of exposures that is perturbing the health of the individual.

Evaluate Personalized Health Risks

It is well understood that individuals respond differently to the same toxicant, so even with accurate exposure data, the prediction of personalized health risk will be limited. Health risk assessment can be enhanced through the utilization of biomarkers of susceptibility and biomarkers of effect.

Biomarkers of Susceptibility

Biomarkers of susceptibility can be utilized in a variety of ways: (1) they allow decisions makers to better understand the exposure risk to individual Service members which can be used to guide pre-exposure to prevent highly susceptible populations from being exposed; (2) they can be used during operations to make determinations on when to pull Warfighters out of the fight because they have reached critical thresholds; and (3) they can be used postexposure to make individualized assessments for long-term health consequences.

The development of biomarkers of susceptibility will be extremely challenging. The best understood susceptibility markers relate to genetic variations that effect bioavailability of toxicants (e.g., detoxification enzymes). These enzymes are well suited to in vitro studies and can be studied for toxicants of interest. Other genetic susceptibility markers are less well understood and are often identified through genome-wide association studies (GWAS), which are not available for a broad range of toxic exposure. Perhaps the most challenging concepts for susceptibility are transient changes that affect an individual. For example, changes in the microbiome are known to affect bioavailability of compounds, and the microbiome in its nature is transient. A full breadth of biomarkers will be required to fully understand an individual's susceptibility, but initial steps can be focused on factors that have the largest impact on susceptibility.

Biomarkers of Effect

The actual response to a chemical exposure will be the basis for biomarkers of effect. By understanding the mechanistic pathways underlying adverse outcome pathways related to an exposure, biomarkers can be developed to assess steps along the pathway. An accurate measurement of AOPs triggered by an exposure will be the basis for determining health risks. These mechanistic based biomarkers are likely to be evolutionary conserved and can be developed using animal models. The mechanistic understanding gained through biomarkers of effect will set the groundwork for personalized treatment strategies.

Personalized Intervention and Treatment Strategies

With history as our guide, exposures are going to occur. The question is how to maximize the treatment to minimize the adverse outcome for the Service member. Personalized intervention and treatment strategies are the ultimate goal to protect the Service member. To reach this end assume a full understanding of the mechanistic nature of the toxicant (e.g., AOP and biomarkers of effect), and it also requires specific knowledge on how the individual will respond to a given treatment. By fully understanding the role of the toxicant and the role of the treatment, physicians of the not so distant future will be able to tune their treatment plans to maximize their effectiveness.

Conclusion

The US military operates in a complex environment with many documented and undocumented exposure. To fully protect the Service member, new methodologies must be developed and incorporated in existing practices that enhance the DoD and VA's ability to (1) identify and document exposure; (2) evaluate personalized health risks associated with exposure; and (3) provide personalized intervention and treatment strategies for injuries and diseases. With the advent of personalized medicine, and the availability of inexpensive genomics and wearable sensors, this vision is within reach. When viewed in the context of past exposure concerns, such as Agent Orange, oil-well fires, and burn pits, it is easy to envision the benefit of such practices. However, the implementation of this new exposome-based medicine will require a concerted across research communities and unwavering commitment by senior leadership to see it through.

Conflict of Interest The views, opinions, and/or findings contained in this report are those of the authors and should not be construed as official Department of the Army position, policy, or decision, unless so designated by other official documentation.

References

Ankley GT, Bennett RS, Erickson RJ, Hoff DJ, Hornung MW, Johnson RD, Mount DR, Nichols JW, Russom CL, Schmieder PK, Serrano JA, Tietge JE, Villeneuve DL (2010) Adverse outcome pathways: a conceptual framework to support ecotoxicology research and risk assessment. Environ Toxicol Chem 29:730–741. https://doi.org/10.1002/etc.34

Army.mil (2013) Nett Warrior gets new end-user device. [Online] [Cited: 5 6, 2016] https://www.army.mil/article/107811/

Beasley RP, Lin CC, Hwang LY, Chien CS (1981) Hepatocellular carcinoma and hepatitis B virus: a prospective study of 22 707 men in Taiwan. Lancet 2(8256):1129–1133

Biggs ML, Kalman DA, Moore LE, Hopenhayn-Rich C, Smith MT, Smith AH (1997) Relationship of urinary arsenic to intake estimates and a biomarker of effect, bladder cell micronuclei. Mutat Res 386:185–195

Bradburne C, Hamosh A (2016) Integrating the microbiome into precision medicine. Expert Rev Precis Med Drug Dev 1:475–477

Bradburne C, Graham D, Kingston HM, Brenner R, Pamuku M, Carruth L (2015) Overview of omics technologies for military occupational health surveillance and medicine. Mil Med 180:34–48

De Castro M, Biesecker L, Turner C, Brenner R, Witkop C, Mehlman M, Bradburne C, Green R (2016) Genomic medicine in the military. NPJ Genom Med 1:15008

Foxnews (2016) Thousands-iraq-afghan-war-vets-sickened-after-working-at-burn-pits. News Report

FoxNews (2017) Bipartisan-bill-to-provide-assistance-to-burn-pit-vets-introduced-in-senate. News Report

Institute of Medicine (2011) Long-term health consequences of exposure to burn pits in Iraq and Afghanistan. National Academies Press, Washington, DC

King MS, Eisenberg R, Newman JH et al (2011) Constrictive bronchiolitis in soldiers returning from Iraq and Afghanistan. N Engl J Med 365:222–230

Koutros S, Andreotti G, Berndt SI, Hughes Barry K, Lubin JH, Hoppin JA, Kamel F, Sandler DP, Burdette LA, Yuenger J, Yeager M, Alavanja MC, Freeman LE (2011) Xenobiotic metabolizing gene variants, pesticide use, and risk of prostate cancer. Pharmacogenet Genomics 10:615–623

Lindler LE (2015) Enhancing the department of defense's capability to identify environmental exposures into the 21st century. Mil Med 180:5–9

MacKay C, Davies M, Summerfield V, Maxwell G (2013) From pathways to people: applying the adverse outcome pathway (AOP) for skin sensitization to risk assessment. ALTEX 30:473–486

Morris MJ, Dodson DW, Lucero PF, Haislip GD, Gallup RA, Nicholson KL, Zacher LL (2014) Study of active duty military for pulomary disease related to environmental deployment resources. Am J Respir Crit Care Med 190:77–84. https://doi.org/10.1164/rccm.201402-0372OC

National Academies of Sciences, Engineering, and Medicine (2016a) Assessing health outcomes among veterans of project SHAD. National Academy of Sciences, Washington, DC

National Academies of Sciences, Engineering, and Medicine (2016b) Veterans and agent orange: update 2014. National Academy of Science, Washington, DC

National Academies of Sciences, Engineering, and Medicine (2016c) Gulf war and health: volume 10: update of health effects of serving in the gulf war. National Academy of Sceince, Washington, DC

OECD (2012) Proposal for a template, and guidance on developing and assessing the completeness of adverse outcome pathways, pp 1–17

Perdue CL, Eick-Cost AA, Rubertone MV (2015) A brief description of the operation of the DoD serum repository. Mil Med 180:10–12

Porter KL, Green FH, Harley RA, Vallyathan V, Castranova V, Waldron NR, Leonard SS, Nelson DE, Lewis JA, Jackson DA (2015) Evaluation of the pulmonary toxicity of ambient particulate matter from camp victory, Iraq. J Toxicol Environ Health 78:1385–1408. https://doi.org/10.1080/15287394.2015.1072611

Rappaport SM (2016) Genetic factors are not the major causes of chronic diseases. PLoS One 11: e0154387

Rose C, Abraham J, Harkins D, Miller R, Morris M, Zacher L, Meehan R, Szema A, Tolle J, King M, Jackson D, Lewis J, Stahl A, Lyles MB, Hodgson M, Teichman R, Salihi W, Matwiyoff G, Meeker G, Mormon S, Bird K, Baird C (2012) Overview and recommendations for medical screening and diagnostic evaluation for postdeployment lung disease in returning US warfighters. J Occup Environ Med 54:746–751. https://doi.org/10.1097/JOM.0b013e31825297ba

Smith B, Wong CA, Smith TC, Boyko EJ, Gackstetter GD (2009) Newly reported respiratory symptoms and conditions among military personnel deployed to Iraq and Afghanistan: a prospective population-based study. Am J Epidemiol 170:1433–1442

Sweeney LM (2015) Evaluation of pharmacokinetic models for the disposition of lead (Pb) in humans, in support of application to occupational exposure limit derivation. Naval Medical Research Unit Dayton, US Navy, NAMRU-D-16-11

Szema AM, Peters MC, Weissinger KM, Gagliano CA, Chen JJ (2010) New-onset asthma among soldiers serving in Iraq and Afghanistan. Allergy Asthma Proc 31:67–71

US Army Public Health Command (2014) Provisional blood lead guidelines for occupational monitoring of lead exposure in the DoD. US Army, Aberdeen Proving Ground, MD

Vinken M (2013) The adverse outcome pathway: a pragmatic tool in toxicology. Toxicology 312:158–165

Wild CP (2012) The exposome: from concept to utility. Int J Epidemiol 41:24–32

World Health Organization (1949) The preamble of the constitution of the World Health Organization. Bull World Health Organ 80(12):982

Chapter 4
Framing Fetal and Early Life Exposome Within Epidemiology

Jessica E. Laine and Oliver Robinson

Abstract The time periods that influence fetal and early life development are identified in this chapter as key windows of susceptibility to exposures and critical developmental stages of preconception, and the prenatal, perinatal, and postnatal periods. We highlight in this chapter these key developmental windows that characterize the fetal and early life exposome, and present a review of studies that have identified fetal and early life external and internal domains of the exposome. We also present a discussion of issues in exposome study design, including choice of biological samples and statistical complexities, specific to the key developmental times of fetal and early life. While notable studies and consortia have been established to investigate the exposome during the times of fetal development and early life, we argue that future exposome research must expand to incorporate the preconception period, build upon the existing and large body of knowledge of reproductive and peri/pre-natal epidemiological methods and study design, and utilize methods of causal inference. Collectively, this will aid in strengthening both the internal and external validity of our studies, and in the identification of potential causal mechanisms underlying many preventable diseases. Such advancements will lead to better risk assessments and potential policy and medical interventions.

Keywords Early-life exposome · Pre-conception monitoring

Introduction

The exposome, in its original form, was proposed to encompass the totality of exposures across the life course from conception onward (Wild 2005), and the definition has since been further modified. The exposome paradigm provides a framework within epidemiological research to identify new environmental risk

J. E. Laine (✉) · O. Robinson
Department of Epidemiology and Biostatistics, MRC-PHE Centre for Environment and Health, School of Public Health, Imperial College, London, UK
e-mail: j.laine@imperial.ac.uk

S. Dagnino, A. Macherone (eds.), *Unraveling the Exposome*,
https://doi.org/10.1007/978-3-319-89321-1_4

factors, elucidate associations between multiple exposures and diseases, determine putative mechanisms that act as intermediates in pathways between exposures and diseases, and extrapolate subsequent causal effects of diseases for disease treatment and prevention. Since its conception, the exposome has been divided into external and internal domains, generally understood to correspond to exogenous and endogenous exposures. The external exposome domain includes both general factors such as socioeconomic position and urbanization and specific factors such as environmental toxicants (e.g. air and water contaminants) and diet, among others. The internal exposome may be assessed via the analysis of biological materials, that may be suitable biomarkers of exposures or disease states or act to support mechanisms that may underlie exposure-disease relationships. These include a wide range of molecules, with a strong emphasis on 'omics, represented by metabolites (metabolomics), proteins (proteomics), miRNA and mRNA (transcriptomics), DNA ((epi)genomics) and adducts in various biological tissues. Of course, this categorization is not straightforward: Environmental contaminants of external exposome origin are often assessed via biomarker measurements within biological samples, representative of the internal (i.e., within body) environment. Furthermore, this conceptual framework is stretched in areas of reproductive and perinatal epidemiology since one may consider that the *in-utero* environment represents the external exposome domain of the developing fetus and the internal domain of the mother.

A further key complexity of the exposome, in contrast to the static genome, is its dynamic nature. Such complexity makes estimating the totality of the exposome a challenge; however, methods to address this have been put forth, where researchers either combine variations of the exposome over time from individuals in a prospective manner or measure a series of snapshots at specific times over the life span of an individual, allowing for an approximation of the full exposome (Wild 2012). Subsequently, as it is more feasible, many studies of the exposome have targeted specific stages of the exposome as snapshots. Rappaport and Smith (2010) proposed key life stages for theses snapshots where the exposome could be measured cross-sectionally, including gestation, early childhood, puberty, and reproductive years. There has been a concerted effort to characterize the exposome at these sensitive and/or critical stages of life that may be of most importance for both exposure and disease susceptibility.

Pregnancy and early life periods have been a focus of many studies, as it is well established that these periods are comprised of a series of developmentally sensitive times for many exposures. Naturally, research aimed at describing and quantifying the exposome during fetal development and early life has been conducted mostly within pregnancy and birth cohorts. The focus on these stages of life is based on the mounting evidence that early life exposures (Vrijheid et al. 2016), particularly during fetal developmental windows, have implications for health at birth, across the life span, and into further generations. This has been described as the fetal origins of adult disease' hypothesis (FOAD) (Barker 2004), where one's health status (healthy or diseased) is based on biological programming from exposures that are experienced *in utero*. Notable consortia have been established to investigate the exposome in early life, and some have expanded into investigations across the life cycle with extensive data. These include the Human Exposome Project in the USA, and in Europe, the Human Early-Life Exposome (HELIX) (Vrijheid et al. 2014), Health and Environment-Wide Associations based on Large Population Surveys (HEALS) (http://www.heals-eu.eu/index.php/project/), and the Enhanced Exposure Assessment and Omic Profiling for High Priority Environmental Exposures in Europe (EXPOsOMICS) (Vineis et al. 2016) projects. Furthermore, there are a number of individual and collective research projects targeting specific aspects of the exposome.

Because the original description of the exposome was intended to capture life exposures beginning at the start of conception, most early life exposome research has been carried out using snapshots throughout pregnancy. However, there are other windows of the exposome to consider that may influence fetal development, such as the preconception period. Additionally, while the snapshot approach is appropriate for some reasearch questions, it does not detail the collective spectrum of fetal development within the context of the exposome. Taken together, there are areas of the early life exposome that have yet to be fully characterized, resulting in a fragmented view of early life exposures. Furthermore, exposome research of early life must build on the existing and large body of knowledge of reproductive and peri/pre-natal epidemiological methodology and study design, to improve our assessments of health during these time periods. Ultimately, there is a need to fully address and improve upon methods of exposome research that underlie fetal development and early life.

The concepts of the exposome have recently developed into an exciting field of research, with a particular focus on characterising the early life exposome. As such, there are two notable extensive reviews describing the pregnancy exposome (Robinson and Vrijheid 2015; Robinson et al. 2015), and one on the maternal exposome (Wright et al. 2016), and a review article highlighting opportunities for exposome research for reproductive and perinatal epidemiology (Buck Louis et al. 2013). We will expand on some of the concepts covered in these previous reviews in this chapter. We will first discuss the key developmental windows of the fetal and early life exposome. We will then review available studies characterizing the periods of fetal development and early life external and internal domains of the exposome. Despite the overlapping nature of these domains, we consider as external exposome studies, those studies that have attempted to provide a more comprehensive assessment of the external environment for exposures, which may be considered a "bottom-up" approach moving from multiple exposures to health. We review as internal exposome studies those that have taken a more "top-down" approach, going from say birth and/or child health outcomes to biomarkers and/or multiple molecular features as measured by the post-genomic 'omics platforms. We then consider those studies that have linked the domains through the examination of the relationships between specific external exposome factors and biomarkers and/or 'omics profiles. The studies discussed in this chapter demonstrate the complexities in measuring and analyzing the fetal and early life exposome; although none has achieved "complete" coverage of the exposome (or is ever likely to), they share a similar holistic, rather than single exposure approach that defines them as exposome studies. Finally, we present issues and new approaches for fetal and early life exposome study design and methodology, including choice of biological samples and statistical complexities. By addressing some of these overarching fetal and early life exposome research principles, we can determine mechanisms that underlie adverse reproductive and pregnancy outcomes, such as birthweight and fetal growth and those of FOAD, such as obesity, cancers, cardiovascular disease, Collectively, this will aid in reducing the burden of diseases experienced from early life exposures and improve development, health, and quality of life.

Developmental Windows for Fetal and Early Life Exposome Studies

Many studies of the exposome that target fetal and neonatal exposures and disease risks have focused on the time period of pregnancy, and often within narrow windows of the prenatal period. However, given that these snapshots of sensitive and/or critical time periods are an important theory for study design and methods of exposome research, there is a need to better characterize and clarify the terms "sensitive" and "critical." This distinction is necessary as differences between the two influence the times targeted to characterize the exposome and subsequently, methods and approaches of study design and analyses of data. It is common to use the terms "critical" and "sensitive" interchangeably without regard of their meaning. To distinguish between the two, we highlight theories from life course epidemiology literature, reviewed by Mirshra et al. (2010). A critical time period is a limited window in which an exposure can have either adverse or protective effects on disease risks, and outside this window, there is no excess disease risk associated with the exposure, whereas a sensitive window is a time period where an exposure has a stronger effect on development and disease risk than it would at other time periods. A critical time period may be that of the fetal period where there may be permanent and irreversible implications, such as the example of prenatal exposure to thalidomide and birth defects. However, for many life course and chronic disease models there may be later life effect modifiers of disease risks from critical exposures. For example, shifts in metabolic and hormonal systems from exposures during fetal development may be modified by several early life and/or later in life adult exposures. Therefore, if there is a later in life effect modifier of an exposure at a critical time period, the critical period may only be critical for those individuals who experience some other exposure. In contrast, factors that raise disease risk or promote good health may accumulate gradually over the life course, although there may be developmental periods when their effects have greater impact on later health than factors operating at other times. This idea is complementary to the notion of allostatic load, which aims to measure biological "wear and tear" due to cumulative physiological adaptation to adverse environmental conditions (Ben-Shlomo and Kuh 2002). Not only do these two different models of disease development emphasize the need to characterize the exposome over the course of fetal development and early life, but also the need to carefully examine the chosen periods, and how they relate to one another that represent fetal and early life development. Currently, there are defined time periods that influence fetal development and early life specified within the fields of reproductive and perinatal epidemiology that include key windows of susceptibility to exposures and critical developmental stages namely, preconception, and the prenatal, perinatal, and postnatal periods.

Preconception

Preconception is biologically defined as the period before a woman becomes pregnant, although this term has also been used to describe the time before a woman knows that she is pregnant. The preconception period is a sensitive and/or critical window because exposures during this time can directly impact gametes and subsequent successful reproduction and fetal development. According to the Agency for Toxic Substances and Disease Registry (ATSDR 2017), parental preconception exposures to environmental toxicants can affect, individually or collectively, fertility, conception, pregnancy, and/or delivery (ATSDR 2017). Overall, there is very little research of the exposome during the preconception time period, especially surrounding both male and female fecundity. However, some studies have investigated single and/or a handful of exposures during the preconception period, and results from these studies bring to light the importance of measuring exposures during this time, as they may influence birth and later in life health outcomes.

There are several maternal-related preconception exposures and behaviors that have been established as risk factors for poorer birth and later in life health outcomes, including BMI (both being underweight and overweight/obese) (Moussa et al. 2016), excessive caffeine and alcohol intake and smoking (Lassi et al. 2014), increased levels of stress (Baird et al. 2009), and poor dietary habits (King 2016). There have also been toxicant-specific studies and/or classes of toxic compounds surrounding preconception, where, for example, maternal preconception exposure to several different persistent organic pollutants is associated with lower birthweights (Robledo et al. 2015). Furthermore, there is evidence that there are paternal contributions that influence reproductive and birth outcomes. Broadly, paternal preconception external exposures such as stress, diet, and exposure to environmental toxicants are associated with adverse health effects in offspring, such as hormone dysregulation, birth defects, childhood cancers, metabolic diseases such as growth, obesity, and cardiometabolic risk markers (as reviewed by Braun et al. 2017). However, despite these known associations of paternal health during the preconception period on adverse early life health, a recent review on preconception health behaviors highlights that only 11% of studies at the time included male participants (Toivonen et al. 2017).

There is great potential for the extension of the fetal and early life exposome to expand into the preconception period. As a guide, we can look to concepts put forth from Braun et al. (2017) of five potential study frameworks for the preconception period that could be used to advance the understanding of the impact preconception exposures have on offspring health. These include developing new population-based prospective cohorts, clinic-based prospective cohorts, and case–control studies, recruiting subsequent siblings from existing cohorts, and constructing retrospective exposure assessments within existing cohorts. Many of these study designs may provide a way to capture the embryonic internal exposome in relation to implantation success, among other conception based outcomes. Such populations may include and target those seeking assisted reproductive therapy (ART). A unique set of biospecimens could be collected to examine the external and/or internal exposome, such as follicular fluid, blastocyst cells, or the spent IVF culture media dish or during pregnancy (Chason et al. 2011). This could represent a large growing population, with over four million babies born from ART worldwide since 1978 (Braun et al. 2017). Given these potentials and needs for better understanding this important time period, it is evident that future exposome studies will need to broaden their study designs to include the preconception period.

Prenatal

A majority of studies that estimate the exposome experienced *in utero* are framed around the prenatal/antenatal time periods. The time period of pregnancy starts approximately 266 days post conception or roughly 280 days past the last menstrual period. This period includes the embryonic stage of development of the first 8 weeks (56 days post conception) and a fetal development period of 9–40 weeks (approximately 30 weeks post conception). Thus, in total, the period of gestation is typically 40 weeks, though there is some variation in this. Given that this time period is the most common window of fetal and early life development explored in exposome studies, specific studies of the internal and external exposome are discussed further throughout this chapter.

Perinatal and Postnatal

After delivery, in the perinatal and postnatal period, the infant undergoes biological changes that are influenced by exposures and biological mechanisms that occurred prior to delivery. The perinatal period is usually used to describe the period directly after birth. However, it is a term that varies, due in part to legal differences across countries in the time periods that define perinatal outcomes, such as stillbirth (Wilcox 2010). However, the Internal Classification of Disease Code (ICD) ten defines “perinatal” as the time period starting at 22 completed weeks (154 days) gestation and lasting through 7 days after birth (Wilcox 2010). The postnatal period is followed by the neonatal period (birth to 1 month). Most studies of the perinatal and postnatal exposome are typically aimed at reconstructing exposures experienced prenatally and/or focus on outcomes experienced during this time period.

During these periods of development outlined in this section there are various outcomes measured to determine and characterise health experienced in early life. Many of these outcomes are actual health endpoints and are detrimental, such as mortality at birth, while others may serve as intermediates for later in life health outcomes, such as shifts in birthweight and later in life body mass. Several of these conditions and outcomes have been widely investigated in fetal and early life studies, and a thorough discussion of these is beyond the scope of this chapter and has been more adequately discussed by others (Wilcox 2010). However, in the studies highlighted in this chapter there is a particular focus on fetal growth patterns, birthweight, and preterm birth (PTB), therefore a brief description of these is warranted. Parameters that represent fetal growth (i.e., birthweight) are phenotypes that are readily discussed in this review, as it is widely studied in perinatal epidemiology. Being born at a suboptimal birthweight has many health consequences, starting at birth, through childhood, and later into adulthood. To note, while a majority of perinatal studies focus on the association between low birthweight (LBW), defined as a baby weighing $\leq$2500 g regardless of gestational age, and small for gestational age (SGA), those with normal birthweights and large for gestational age (LGA) may still be at risk for adverse health outcomes in early life and for later in life diseases. Given that up to 40% of fetal growth is influenced by genetic factors (Valero De Bernabé et al. 2004), and the remaining is thought to be due to environmental exposures, there is a need to characterize the exposome in relation to this phenotypic condition. Other outcomes and birth characteristics discussed here are gestational age, PTB, and preeclampsia as they are also important indicators of neonatal health and are unfortunately common pregnancy complications. PTB deliveries are those that occur at less than 37 weeks' gestational age; however, the low-gestational age cutoff, used to determine PTB from spontaneous abortion, varies by location specific definitions. PTB affects between 5 and 13% of pregnancies globally, and is rising in industrialized countries (Goldenberg et al. 2008). Preeclampsia is a condition defined by hypertension and proteinuria. The estimated incidence of preeclampsia is that it occurs between 3 and 10% of all pregnancies (Jeyabalan 2013). We will not elaborate further as to the implications for many of these outcomes for public health disease burdens, whether experienced immediately at birth, such as mortality, or later in life such as the effects of being born at a suboptimal birthweight and the impacts of this on later in life obesity, however, the detrimental impacts of these on individuals and society are not to be dismissed.

Fetal and Early Life Exposome

Collectively, the time periods discussed in this section are important for characterizing the fetal and early life exposome, and a more comprehensive assessment of the impacts of the timing of exposures is warranted. It is still not fully understood whether exposures are more impactful to the developing fetus during one or more of these specific isolated periods, or due to a combination of exposures over these specific critical periods, or if the cumulative effects of exposures across the full life cycle experienced from these time periods and beyond lead to the development of diseases. It is likely that the answers will depend on specific exposures and outcomes. Therefore, future exposome studies will need to look collectively and individually at these specific time periods. Of course, the complexity of measuring the exposome during these time periods is not to be dismissed, where traditional epidemiological methodological challenges surrounding recruitment and follow-up, avoiding selection bias, and addressing the difficulties of generalizability from population specific studies are present. These challenges, along with others, and potential ways to overcome them are presented later in this chapter. Additionally, it is important to note that there are critical and/or sensitive time periods of early life that follow the postnatal period but are beyond the scope of this chapter.

There are a few larger consortiums established in Europe that were designed specifically to investigate the influence of pre- and postnatal exposures (not necessarily the exposome) on the health of the offspring, including the Developmental ORIgins of healthy and unhealthy AgeiNg (DORIAN) (Iozzo et al. 2014), the Human Early-Life Exposome (HELIX) (Vrijheid et al. 2014), and most recently the LifeCycle a EU Child Cohort Network (https://lifecycle-project.eu/) has been established to investigate socioeconomic status, migration, urban environment, and lifestyle during the preconception as a targeted research area. Additionally, there are individual cohorts that have been established focusing on this time period. For example, the INMA (INfancia y Medio Ambiente) is a birth cohort study in seven regions of Spain that examined the role of environmental pollutants during pregnancy and early childhood in relation to child growth and development (Guxens et al. 2012).

Characterizing the Domains of Fetal and Early Life Exposome Studies

Many of the domains discussed in this chapter surround health endpoints and outcomes of pregnancy complications, birth, and later in life health effects that may be associated with exposures experienced prior to and/or during pregnancy. Measuring these outcomes is advantageous for a top-down approach of exposome research. Not only are data readily available, especially for parameters such as birthweight, across the many studies, but are also important predictors of early and later in life disease and are, unfortunately, commonly experienced. For example, pregnancy complications represent disorders that effect up to 20% of all gestations (Sõber et al. 2015). Exposome studies that investigate these relationships can greatly contribute in the ability to reduce many of these preventable disease burdens, as the etiologies of many of these adverse health outcomes are not readily understood.

External Exposome

The characterization of the external exposome has been a historical undertaking, even before the development of the concept of the exposome, where multiple studies examined single exposures. Here we provide an overview of external exposome studies with the aims of capturing multiple exposures that influence fetal development and later in life health. While no current studies have been able to capture the full extent of the external exposome, the literature shows an increasing trend toward a more holistic assessment of environmental exposure in early life.

The "public health exposome" framework, proposed by Juarez (discussed in Chap. 2), has been applied to understanding the determinants of PTB (Oyana et al. 2015) and low birthweight (Kershenbaum et al. 2014) at an ecological level of analysis of counties in the USA. The studies applied geographic information systems (GIS) and spatiotemporal modeling to hundreds of variables from national databases covering the natural, built, social, and political environments. Through this place-based approach they identified "hotspots," counties with unusual rates of these outcomes, which may be important in designing targeted interventions among pregnant women and aide in measures to reduce health disparities. Interestingly, there were multiple predictors, including teen birth rate, adult obesity, uninsured adults, physically unhealthy days, percent of adults who smoke, ozone, and fine particulate matter that were important contributors to rates of LBW in multivariate models, while race, obesity and diabetes, sexually transmitted infection rates, mother's age, income, marriage rates, pollution, and temperature were among the predictors of PTB rate.

Air pollution research is increasingly moving toward multi-pollutant analyses (Billionnet et al. 2012) due to both the high collinearity between air pollutants and the existence of national routine monitoring networks that provide simultaneous measurements of multiple pollutants. Within fetal and early life exposome research, linkage of these monitoring data to state birth registries has led to multi-pollutant research of hundreds of thousands of observations of air pollutants. This has also led to a better understanding of the temporal effects of exposure to air pollutants across pregnancy. For example, Warren et al. (2012) applied a hierarchical Bayesian framework to examine the critical time windows of effect of both $PM_{2.5}$ and ozone on the incidence of PTB in Texas, USA. Their findings illustrate the importance of understanding the temporal nature of exposures during pregnancy since $PM_{2.5}$ had a stronger effect overall, and in particular in weeks 4 to 22 of pregnancy, while ozone was found to have the greatest impact on PTB in weeks 1 to 5 of pregnancy. Swartz et al. (2015) also examined data from the Texan birth registry to identify risk factors of the birth defect spina bifida, for 25 different hazardous air pollutants. Using a novel Bayesian stochastic search variable selection, which is effective in the context of correlated exposures, they identified quinoline and trichloroethylene as significant contributors to spina bifida risk. Le et al. (2012) examined the effects of four air pollutants on PTB and SGA births in Detroit, USA. In the study area, correlations between exposures were modest due to the diverse pollutant sources, which allowed the use of multi-pollutant regression models to separate the effects of each pollutant individually. Significant effects at various critical windows were found for SGA births for PM_{10} (first month and first trimester), CO (first and second trimester), SO_2 (all trimesters), and NO_2 (first month and all trimesters). For PTB births, odds ratios were significantly elevated during only the first month of SO2 and NO2 exposures. They also explored interactive effects with social context, finding greater effects of air pollution on PTB among mothers with lower educational level and black ethnicity. Furthermore, there was heterogeneity of the effects of pollutants on PTB between smokers and nonsmokers.

Population-based prospective birth cohorts, which often collect many hundreds of individual-level exposure variables, are beginning to apply exposome approaches in the analysis of their databases. North et al. (2017) describes the Kingston Allergy Birth Cohort, a cohort that has recently begun recruiting in Canada to study the developmental origins of allergic disease. They employed exposome globes (Patel and Manrai 2015) (also referred to as circos plots) to visualize the interconnectivity between multiple factors between and within the general external (e.g., socioeconomic status, rural or urban residence), specific external (e.g., cigarette smoke, breastfeeding, mold or dampness), and internal (e.g., respiratory health, gestational age) realms of the exposome. Based on these analyses, they constructed a multivariable exposome predictive model of respiratory symptoms of children at 2 years, finding that all three domains of the exposome had effects on respiratory health. Exposome-wide association study approaches (EWAS) have been applied to the thousands of preconception and prenatal environmental risk factors collected in the large and long-established Avon Longitudinal Study of Parents and Children (ALSPAC) in the UK. Golding et al. (2014) conducted multiple univariate analyses on 1755 variables to identify associations with gross motor skills at 8 years and then incorporated the most significant variables into a final regression model. They identified the novel association between an unhappy childhood of the mother and motor skills in the child. However, with only minimal adjustment for the generation of false positives, it is difficult to draw conclusions from this study. Steer et al. (2015) then scanned 3855 variables, representing prenatal exposures, in association with communication impairments in 9-year-old children and identified 615 significant associations, after correction for false discovery rate. These variables were then subjected to a series of stepwise regression analyses to build a final model containing 19 variables. The analysis suggested six possible causal pathways for children's communication impairments: socioeconomic disadvantage; parental personality that affects future parenting skills; aspects of the home environment; poor maternal health with a novel finding concerning maternal hearing loss; maternal education which was partially mediated by the child's IQ; and the consumption of fatty or processed foods during pregnancy.

Within the INMA birth cohort in Spain, Robinson et al. (2015) examined the correlation structure of multiple environmental exposures of pregnant women. The work followed analysis by Patel and Ioannidis (2014) of the US National Health and Nutrition Examination Survey (NHANES) biomonitoring data, where they proposed that transparent knowledge of the correlation structure of exposome data is required to interpret the significance of results using those data. Among 81 exposures, assessed through GIS, biomarker measurements, and questionnaire in the INMA women, effectively all the variance could be explained by 41 principal components due to strong correlation structure, particularly within exposure families. Correlations between exposure families measured by biomarkers, for instance, between brominated and fluorinated compounds, however, were weak suggesting little potential for confounding by other unreported exposures in this study area. However, between-family correlations for those exposures encountered in the outdoor urban environment measured by GIS methods, including noise, air pollutants, the built environment, and temperature, were moderate to strong, indicating that these exposures should be considered jointly in relation to health. This approach has been extended to other birth cohorts participating in the HELIX project (discussed in Chap. 15), where the "urban exposome," a set of over 30 environmental indicators assessed through GIS, remote sensing, and spatiotemporal modeling, has been characterized for almost 30,000 pregnant women in nine cities from across Europe. In a subsequent analysis (Robinson et al. 2018), it was found that almost a third of the variance in the urban exposome could be described by a component describing areas that were less polluted with higher surrounding greenness, which can be interpreted as capturing the relative rurality of different urban environments. Other important variance patterns included high traffic and less populous areas and areas that were walkable with good access to natural space. The relationship between these types of the urban exposome and determinants such as socioeconomic status varied considerably between cities, indicating that the relationship between the general and specific domains of the external exposome depends greatly upon the local context. Dadvand et al. (2014) performed a similar comprehensive exposome assessment in their analysis of term LBW in Barcelona, Spain. They incorporated measures including the built environment (proximity to major roads), temperature, noise, air pollutants, and the buffering effects of roadside tree planting. Living within 200 m of a major road while pregnant was associated with a 46% increased risk of LBW, with measured air pollutants and heat together explaining around one third of this association. Furthermore, there was some evidence for attenuation of the effect size with increasing tertiles of tree cover.

Multiple biomarker measurements are leading to the specific external exposome to be increasingly well characterized. The NHANES biomonitoring dataset has been used by many researchers to identify prenatal exposures and subsequent health outcomes. For example, Rosofsky et al. (2017) detected on average 92 biomarkers per women with the most commonly detected chemical classes including phthalates, metals, phytoestrogens, and polycyclic aromatic hydrocarbons. They observed there was high between-class correlation among urinary biomarkers, while there was high within-class correlation for biomarkers measured in blood. Although for most biomarkers, levels were less than observed in comparable women in the USA in NHANES data, their work demonstrated that chemical contamination is ubiquitous in Denmark. The correlations the authors observed suggest common exposure sources, specific lifestyle factors, or shared metabolism pathways. Woodruff et al. (2011) conducted an analysis of the exposome of pregnant women using NHANES data and they found that analytes including polychlorinated biphenyls, organochlorine pesticides, PFCs, phenols, PBDEs, phthalates, polycyclic aromatic hydrocarbons, and perchlorate were detected in 99–100% of pregnant women, with levels were found to be similar between pregnant and nonpregnant women. Vafeaidi et al. (2014) combined measurements of multiple correlated exposures, including pesticides, polychlorinated diphenyls, and polybrominated diphenyl ether (PBDE)-47, into a single exposure score among pregnant women during the first trimester in Crete, Greece. They observed a 42-g decrease in birthweight with a unit increase in exposure score, which was robust to adjustment for social and maternal factors. Lenters et al. (2016) evaluated the effects of multiple correlated exposures, including phthalates, perfluoroalkyl acids, and organochlorines, on fetal growth among three cohorts from Greenland, Poland, and Ukraine. Using elastic net regression, a penalized variable selection technique, they observed independent effects of four compounds with MEHHP, PFOA, and p,p´-DDE associated with lower birthweight while MOiNP was associated with higher birthweight. Agay Shay et al. (2015) examined associations between 27 endocrine disrupting chemicals and child weight status at 7 years in the INMA cohort in Spain. Chemicals including phthalates, bisphenol A, metals, organochlorines, and polybrominated diphenyl ethers were measured in multiple biological tissues reflecting potential fetal exposure, and organochlorine factor was positively associated with BMI z-scores and with odds of children being overweight. These associations remained after adjustment for the other components and lifestyle factors.

The effects of multiple chemical exposures on child neurodevelopment are of concern and have been prospectively examined in a handful of birth cohort studies. Forns et al. (2016) analyzed measurements of 25 different persistent organic pollutants in breast milk samples from new mothers collected on average a month after delivery. The authors examined the relationship with these exposures and behavioral problems in the child at 12 and 24 months of age, using three different multivariate analytical approaches including principal component analysis, elastic net regression, and Bayesian model averaging. All three methods resulted in the legacy pesticide DDT being statistically significant in association with behavioral problems at 12 months. However, these effects were not found in the analysis of similar outcomes at 24 months. Braun et al. (2014) measured levels of 52 endocrine-disrupting chemicals among pregnant women in the USA and applied a semi-hierarchical Bayesian analysis to examine the relationship with autistic behaviors in children at 4 and 5 years of age. They found that increasing concentrations of PBDE-28 and the pesticide trans-nonachlor were associated with more autistic behaviors; however, the small study size prevented the dismissal of an association with the chemicals for which null associations were found. The study of Yorifuji et al. (2011) only examined two exposures, cord blood lead concentration and methylmercury, but demonstrated the importance of considering combined effects of exposures. They observed clear associations between lead and cognitive function at 7 and 14 years of age in a Faroese birth cohort, but this was only apparent in the lowest methylmercury exposure group. The results suggest a less-than additive or antagonistic interaction, which may be explained by competition of toxicants for protein-binding sites in similar biological cascades.

Internal Exposome

The exploration of internal markers of the exposome related to phenotypic outcomes of reproduction and pregnancy is just beginning to emerge. Despite much research on exposures–outcome associations we still do not fully understand many of the underlying causes and mechanisms of many health outcomes related to fetal development and early life. One approach to address this in the realms of exposome research is to investigate molecular mechanisms underlying diseases using a top-down approach. This approach starts with an outcome and identifies the association with internal biomarkers and/or molecular mechanisms, either measured cross-sectionally or before the outcome. Here we focus our discussion on the internal exposome in terms of 'omics, including transcriptomic, epigenomic, metabolomic, and proteomic studies. A majority of exposome studies have focused on epigenetics, as this was one of the first 'omics techniques to be used widely in epidemiological studies; however, recent advances in the field of metabolomics have led to an increase in studies analyzing the metabolome in relation to adverse outcomes. Therefore, our review of the internal exposome is heavily weighed toward epigenetics and other commonly used 'omics techniques such as metabolomics. However, as we continue to develop multi-'omics studies, further discussion of the proteome and other 'omics will need to be expanded on. Additionally, other realms of the internal exposome, such as biomarkers of nutrition and other clinical markers, are not discussed here but are significant for exposome research.

Epigenetics (discussed in Chap. 6) refers to mitotically heritable changes to DNA that do not alter the DNA sequence but can potentially modify its function. Epigenetic changes include acetylation, methylation, and histone modification of DNA. DNA methylation is the most widely studied epigenetic alteration in population studies mainly due to advances in assays and technologies. For example, commercial platforms, such as the Illumina HumanMethylation arrays, now allow whole genome measurement of patterns of DNA methylation to cytosine-phosphate-guanine (CpG) sites. Studying DNA methylation may prove an important avenue for describing and understanding the fetal and early life exposome for several reasons. DNA methylation is a dynamic process known to be influenced by both genetic and environmental factors (Marsit 2015) and may therefore be a link between the genome and the exposome with subsequent effects on health. Early development encompasses series of profound changes in DNA methylation patterns and thus is likely to be a crucial period for exposure-induced DNA methylation changes. These changes can persist over several years and may provide an important mechanistic link underlying the developmental origins of health and disease, with DNA methylation mediating the effects of the fetal and early life exposures on organ structure and function in adult life. Finally, methylation profiles may serve as a powerful biomarker of exposures, even without linking to a causal role for the outcome. These may prove particularly useful for capturing chronic long-term exposures that are more difficult to accurately measure by traditional methods. Through the study of epigenetic changes associated with birth outcomes, information is potentially provided on the in-utero exposures and the specific biological processes that are relevant to the outcome.

Numerous studies have examined methylation of DNA from leukocytes collected from umbilical cord blood in association with birthweight. The largest study to date of 1046 births was conducted in the Norwegian MoBa cohort (Engel et al. 2014). The study detected 19 epigenome-wide significant differentially methylated CpG sites, 13 of which were within genes and 6 were within intergenic regions. Identified genes had important developmental roles such as the gene *ARID5B*, involved in adipogenesis, and the gene *XRCC3*, involved in chromosome maintenance and DNA repair. More recently in the UK ALSPAC cohort, birthweight was associated with 23 CpG sites, within 14 genes. Two of these, including *ARID5B*, were previously detected in the MoBa study. The ALSPAC study (Simpkin et al. 2015) also examined the persistence of methylation in the same children later in life, finding that most birthweight-related methylation differences resolved during childhood. Despite the non-persistence of these methylation patterns, there was evidence for mediating roles of methylation of the developmentally important *NFIX* and *LTA* genes in cord blood and the observed associations between birthweight and developmental outcomes in later childhood. The Project VIVA cohort in the USA also noted an inverse relationship between methylation of CpGs in the *ARID5B* gene and birthweight (Agha et al. 2016). The study detected 34 differentially methylated CpG loci including four within the *PBX1* gene involved in regulation of embryonic regulation. Associations were also observed between birthweight and methylation of these four CpG sites in later childhood.

Targeted analysis of methylation of specific genes or regions also allows for testing of specific hypotheses regarding mechanisms underlying fetal growth. Differentially methylated regions (DMRs) are genomic regions thought to be important in embryonic development as methylation patterns are established before gastrulation, and are maintained in somatic tissues. Lack of fidelity in preserving these marks through epigenetic reprogramming may initiate changes that should be detectable in all tissues and may inform changes in multiple genes. McCullough et al. (2015) measured methylation levels at four DMRs known to be associated with birthweight. They investigated the role of these DMRs in linking physical activity of the mother to birthweight, finding that methylation at the *PLAGL1* DMR was related to total non-sedentary time and may account for the link between sedentary behavior and risk of large-birthweight babies. The aryl hydrocarbon receptor is an important "xenosensor" that regulates the expression of metabolizing enzymes in response to exposure to chemicals such as dioxins. Smoking during pregnancy, a risk factor of fetal growth restriction, is associated with altered methylation of the *aryl-hydrocarbon receptor repressor (AHRR)* gene in cord blood leukocytes. Burris et al. (2015) reasoned that methylation of this gene may also be relevant to fetal growth, independent of tobacco smoke exposure. They examined a range of predictors in a population of nonsmokers, finding that methylation of *AHRR* in cord blood was positively associated with maternal BMI and negatively associated with both gestational age and birthweight-for-gestational age, suggesting the importance of this gene in fetal development. Volberg et al. (2017) examined methylation at multiple sites within the *proliferator-activated receptor γ (PPARγ)* gene, which is crucial in adipogenesis and regulation of metabolism-related hormones, in blood collected at birth and child age 9 years. They found methylation patterns within this gene and its promotor regions to be conserved over time and methylation to be inversely correlated with both birthweight and BMI at age 9 years. Rangel et al. (2014) investigated the role of the *angiotensin-converting enzyme (ACE)* gene, which has been studied extensively as a key gene for cardiovascular disease. The authors compared methylation in the promoter region of the gene in blood collected from low- and normal-birthweight children. They found that low-birthweight children had significantly lower levels of methylation of *ACE* and higher levels of systolic blood pressure, independent of *ACE* polymorphisms, suggesting a potential mechanism linking the in-utero environment with cardiovascular disease risk in later life. Simpkin et al. (2015) 5 performed an epigenome-wide scan for differentially methylated CpG sites in cord blood associated with gestational age, identifying associations with 224 different probes annotated to 155 genes. These associations did not persist in later childhood, and there was little evidence for associations between gestational age and methylation at later time points. Nor was there evidence for a role for methylation in linking gestational age to developmental phenotypes in childhood. Over half of the probes identified replicated previous findings and 72 were also detected in a previous case–control study of preterm birth (Cruickshank et al. 2013). Many of the replicated probes were in parturition-related genes. Since methylation differences associated with PTB were found to be no longer detectable in adults (Cruickshank et al. 2013), it suggests that differences at delivery (when not at term) merely reflect methylation changes during a normal developmental trajectory. This has been explored further by Knight et al. (2016) who developed a highly accurate predictor of gestational age based on 148 CpG sites selected through elastic net regression. Interestingly, having a more advanced DNA methylation-predicted gestational age than actual gestational age was associated with higher birthweight, reflecting greater developmental maturity. Furthermore, those mothers with private medical insurance were more likely to deliver babies with advanced DNA methylation gestational age, than mothers receiving state medical assistance, indicating the importance of the socioeconomic environment in fetal development.

A recent review of transcriptome studies in relation to PTB (Eidem et al. 2015) included 134 studies that investigated genome-wide association studies of mRNA, miRNA, or DNA methylation. They found that of the 10,993 unique genetic elements reported to be transcriptionally active, only 23 elements were recovered ten or more times. Meta-analysis of the 93 gene expression studies across nine distinct gestational tissues (e.g., placenta, maternal blood, fetal membranes, umbilical cord blood) and 29 differently defined clinical phenotypes showed a limited overlap of differentially expressed genes. They found that genes involved in hormone regulation (i.e., *CGB*, *CRH*, *INHA*, and *GH2*), key in the maintenance of pregnancy, show substantial overlap in studies relating PTB to preeclampsia. Genes involved in inflammation such as *IL8* are differentially expressed in multiple studies. The review highlighted the need for larger systematic studies due to the large heterogeneity in expression profiles observed both between and within clinical phenotypes related to PTB and tissues analyses. However, it was recently demonstrated that the placental transcriptome is highly variable both between and within individuals (Hughes et al. 2015) which may account for the lack of consistency in profiles related to PTB. Analysis of gene expression has also been used to investigate SGA, where lower expression of genes from the growth hormone locus and reduced expression of the insulin-like growth factor-2 (IGF-2) gene associated with SGA (Sõber et al. 2015).

Metabolomics is increasingly used in maternal-fetal medicine (Fanos et al. 2013) to identify biological changes associated with adverse birth outcomes, such as fetal growth. Metabolomics (discussed in Chap. 6) is considered a particularly promising approach in exposome research as it can potentially assess both exogenous and endogenous chemicals, allowing both external and internal domains to be profiled in a single analysis. Horgan et al. (2011) examined ultrahigh-performance liquid chromatography–mass spectrometry (UHPLC-MS) profiles of first-trimester maternal plasma samples to identify metabolites predictive of small for gestational age babies. Maitre et al. used nuclear magnetic resonance (NMR) spectroscopy on maternal pregnancy urine samples to identify metabolites predictive of preterm birth, small for gestational age, and fetal growth restriction (Maitre et al. 2014) and birthweight (Maitre et al. 2016). Dessi et al. (2011) identified four metabolites associated with fetal growth restriction in urine samples. However, only a handful of studies have investigated metabolic changes in cord blood, which is a particularly relevant tissue as it contains the essential nutrients, hormones, and immunological factors, and potentially harmful xenobiotic metabolites, to which the developing fetus is directly exposed. Horgan et al. (2011) compared six small for gestational age babies with controls observing differences in levels of sphingolipids, phospholipids, and carnitines. Ivorra et al. (2012) and Tea et al. (2012) both used nuclear magnetic resonance spectroscopy to compare a small number of low-birthweight and very-low-birthweight newborns, respectively, with controls and detected some differences in metabolite levels. Recently, Hellmuth et al. (2017) applied a targeted mass spectrometry-based analysis and observed a positive association between lysophosphatidylcholines (lysoPCs) and birthweight. While initial studies have so far been based on small samples or limited subsets of molecules, they demonstrate the potential of metabolic profiling to detect biological pathways related to fetal development. Within the EXPOsOMICS project, we have employed untargeted UHPLC-MS-based metabolomics to identify metabolic features associated with birthweight in 500 cord blood samples collected from four European birth cohorts. We identified 68 metabolome-wide significant metabolites, which together suggested the importance of specific processes including mitochondrial function, hormonal signaling, fatty acid metabolism, and nutrient availability in fetal growth. In addition to identifying new metabolites associated with birthweight, we identified the novel mechanism of vitamin A depletion in linking maternal smoking with lower birthweight (Robinson et al. 2018). Interestingly, metabolomic profiling may be used as a screening tool for adverse birth outcomes. For example, metabolomics scans of amniotic fluid have distinguished patients who delivered at term from patients who delivered preterm, where a decrease in carbohydrates was associated with preterm delivery in the presence or absence of inflammation, whereas an increase in amino acid metabolites was a unique feature of preterm labor with inflammation (Romero et al. 2010b).

There have been few studies that have addressed proteomic realms of the fetal and early life exposome from a top-down approach. Studies of the proteome may take the approach to analyze actual proteins and peptides or proteolytic peptides that surrogate the proteins of interest, and these techniques can be further delimited as either discovery-based, untargeted approaches or targeted proteomics. Wang et al. (2013) investigated a comparative proteome analysis of placental tissues between normal pregnancies and preeclamptic pregnancies. The authors found that 171 proteins were differentially expressed between normal pregnancies and preeclampsia, where 147 proteins were down-regulated and 24 proteins were up-regulated in preeclampsia. In a cross-sectional study, amniotic fluid samples from 75 patients with an episode of uterine contractions at preterm gestations and intact membranes were used to investigate the proteome of PTB (Romero et al. 2010a). Overall, 77 proteins were up-regulated in PTB, whereas six proteins were down-regulated. Many of these proteins had been previously associated with infection/inflammation related to PTB and play a role in host defence, mobility, localization, and targeting; anti-apoptosis; and metabolism/catabolism. In a small nested study of the maternal–infant research on environmental chemicals research study (MIREC), protein interaction networks and maternal biological pathways that may be related to LBW were identified from targeted and global plasma proteomic markers (Kumarathasan et al. 2014).

Linking the External and Internal Domains of the Exposome

The last of the domains discussed of fetal and early life exposome research is the linkage of the external exposome to the internal exposome. Several studies have taken a single-pollutant/exposure approach to understand the influence of exposures on alternations of 'omics. However, there have been a few studies that have looked at multiple exposures and even fewer that have addressed multiple 'omics associated with external exposures.

While there has been great success in understanding the epigenetic effects of maternal smoking during pregnancy (Joubert et al. 2012), the influence of other environmental factors such as air pollution, known to associated with intrauterine growth restriction at a lower magnitude than the effects of smoking is likely to exert much smaller effects on the epigenome. To perform epigenome-wide association scans, which implies a large multiple testing burden, for exposures that are expected to have smaller effect sizes requires increasingly large sample sizes. To address this issue and others, the Pregnancy And Childhood Epigenetics (PACE) Consortium was established compiling 39 studies from around the world with over 29,000 samples and DNA methylation data in pregnant women, newborns, and/or children (Felix et al. 2017). The first analysis from the consortium addressed the effects of maternal smoking and there are now plans to examine a range of other environmental exposures. Most recently, they examined the effects of prenatal air pollution exposure in over 1500 participants, analyzing the relationship between NO_2 exposure during pregnancy at the residential addresses and cord blood DNA methylation (Gruzieva et al. 2017). They identified three CpG sites that were epigenome-wide significant, located in the mitochondria function-related genes, *LONP1*, *HIBADH*, and *SLC25A28*. Mitochondria are known to be involved in many pathways related to cellular responses to environmental stressors (Shaughnessy et al. 2014), and recent evidence has supported their important role in fetal growth (Robinson et al. 2018). They additionally adopted a more hypothesis-driven analysis where they looked up methylation patterns in a reduced set of antioxidant and anti-inflammatory related genes, chosen based on existing literature on the biological response to air pollutants. The identified two significantly methylated genes in this analysis, *CAT* and *TPO*, both involved in defense against oxidative stress. The investigation of the effects of air pollution on the early life methylome is an area of active research with analyses underway in other consortia such as the HELIX and EXPOsOMICs projects, with further smaller studies underway around the world (Goodrich et al. 2016; Rossnerova et al. 2013).

In the US Child Heath Study, they focused on the association between air pollution and global assessment of methylation through analysis of methylation at repetitive transposable elements distributed throughout the genome, namely, long interspersed nuclear elements (LINEs) and AluYb8 (Breton et al. 2016). They assessed exposure to multiple air pollutants during each of the pregnancy trimesters and assessed DNA methylation in 459 newborn dried blood spots, which are routinely collected in many American states. These repetitive elements generally remain in a hyper-methylated state following a global re-methylation early in embryogenesis. Altered DNA methylation during this stage may therefore have wide-ranging downstream effects on the developing fetus. Indeed, they observed that O_3 exposure during the first trimester, independently of other measured air pollutants, was associated with lower methylation of LINE1 elements. Furthermore, there was an interaction with genotypes of the *DNA methyltransferase 1* gene, which plays a pivotal role in methylation during embryogenesis, with a greater magnitude of association for certain genotypes of this gene. Kingsley et al. (2016) also noted lower methylation of LINE1 elements in placental tissue in association with mothers living near to major roads, a proxy of air pollutant exposure. The ENVIRONAGE study in Belgium also examined global DNA methylation of placental tissue in association with air pollution, assessed at various time periods in pregnancy. Global DNA methylation was assessed by measuring levels of methylated deoxycytidine in hydrolyzed DNA extracted from placental tissue. They observed 2.19% decrease in global DNA methylation for every 5 μg/m^3 increase in exposure to $PM_{2.5}$. Mutual adjustment by time period indicated that only exposure during the first trimester was significantly associated with global methylation (Janssen et al. 2013).

The EVIRONAGE study has also associated air pollution during pregnancy to numerous other markers of the internal exposome to explore the mechanistic pathways through which air pollution exerts its effects. Saenen et al. (2017) examined methylation levels of CpG sites within the leptin gene, which encodes a key energy-regulating hormone involved in fetal growth. They reported that both second-trimester $PM_{2.5}$ exposure and placental levels of the oxidative stress biomarker 3-nitrotyrosine were associated with methylation levels at this gene, identifying a potential link between air pollution and fetal growth. Grevendonk et al. (2016) assessed mitochondrial DNA damage through measurement of 8-hydroxy-2-′-deoxyguanosine (8-OHdG) in mitochondrial DNA which is formed through oxidation of guanine following DNA damage by reactive oxidative species. They found that PM10 throughout pregnancy was associated with levels 8-OHdG in mitochondrial DNA extracted from maternal blood after delivery, while exposure to PM10 earlier in pregnancy was associated with levels of 8-OHdG in mitochondrial DNA extracted from cord blood. The work demonstrated the role of particulate air pollution exposure in early life in increasing systemic oxidative stress at the level of mitochondria. Mitochondria play a vital role in embryogenesis, mainly due to their role in cellular energy provision and it is hypothesized that oxidative stress induced by air pollutants may be particularly damaging to mitochondrial DNA as they lack many of the protective structures and mechanisms found in nuclear DNA. Janssen et al. (2013) reported the effects of air pollution on mitochondrial DNA content in both placenta and cord blood. Mitochondrial DNA copy number (i.e., content) is an established marker of mitochondrial damage and dysfunction. The authors reported that placental mitochondrial DNA content was negatively associated with the PM_{10} exposure in the third trimester of pregnancy. No associations were observed with cord blood mitochondrial DNA content. Janssen et al. (2015) further examined mechanisms underlying the effects of particulate matter on placental mitochondrial DNA content. They measured methylation at two important regulatory regions in mitochondrial DNA, the DNA D-loop control region and *MT-RNR1* that encodes 12S rRNA, required for proper translation of mitochondrial RNA. They hypothesized that methylation may be altered in response to air pollution and impact on the replication or transcription of mitochondrial DNA. They reported that methylation of both these regions was associated with both exposure during pregnancy to $PM_{2.5}$ and mitochondrial DNA content. Furthermore, they estimated that mtDNA methylation of D-loop and *MT-RNR1* was estimated to mediate 27% and 54%, respectively, of the inverse association between $PM_{2.5}$ exposure and mitochondrial DNA content.

Prenatal inorganic arsenic (iAs) exposure and metabolism of iAs has also been linked to multiple 'omics features, including the fetal epigenome (5-methyl cytosine), transcriptome (mRNA expression), and/or proteome (protein levels), as reviewed by Laine and Fry (2016). Many of these studies only investigated single 'omics. However, in the biomarkers of exposure to arsenic (BEAR) pregnancy cohort, a genome-wide approach of cord leukocytes of 40 mother–baby pairs identified that 54 genes were altered both at the gene expression level and in CpG methylation status (Rojas et al. 2015). Within this cohort, there were 334 differentially expressed mRNAs from a genome-wide analysis in cord blood leukocytes in 40 mother–baby pairs (Rager et al. 2014). Additionally, using the largest proteome assessment (at the time the study was published) related to prenatal iAs exposure, a panel of 507 proteins was assessed in newborns exposed to iAs prenatally, where a total of 111 proteins were found to be associated with prenatal iAs exposure, which are known to be regulated by tumor necrosis factor (TNF) and are enriched in functionality related to immune/inflammatory response and cellular development/proliferation (Bailey et al. 2014). Furthermore, maternal biotransformation of iAs and neonatal levels of iAs and its metabolites were associated with differences in neonate cord metabolomics (NMR) profiles (Laine et al. 2017).

Epidemiological Study Design and Methods for Fetal and Early Life Exposome Research

As we have moved from the concept of the exposome, to proof of practice in the original studies of the exposome, and now into more robust studies, there are several factors that must be considered to continue the advancement of the field. This is especially true for characterizing the exposome during sensitive and critical life stages, such as fetal development and early life, as there are several characteristics that are unique to these time periods. Specifically, these life stages are comprised of complex biological and pathophysiological properties with many rapid changes over the course of pregnancy for the developing fetus, such as rapidly growing cells and immature repair processes. Additionally, while many of these properties influence immediate health they also are important determinants for later in life health. Based on these complexities, among others, there is a need for special attention to the types of study designs employed and statistical approaches used, the timing of when exposures are experienced, and the types and measurements of biomarkers and tissue matrices selected when characterizing the external and internal exposome of fetal and early life. Providentially, strengths of the field of epidemiology are in the consistency of attention to study design and methods, and the continual dedication to the improvement of both. This has increasingly lead to more robust causal thinking, the development and application of stronger methodologies, and an overall improvement of the internal validity of our work (Galea 2017). Similarly, close attention to study design and the need for methods improvement is a strong principle underlying the exposome paradigm, as discussed in Chap. 1. Many challenges and recommendations to improve studies of the exposome have been put forth for future epidemiological studies in the field, and was an area of emphasis in the recent report on *Using twenty-first Century Science to Improve Risk-Related Evaluations*

(National Academies of Sciences, Engineering, and Medicine et al. 2017). It is fortunate that the development of epidemiological exposome studies has coincided with a focus on utilizing the most appropriate study design and research methods, and such a focus will remain imperative as we continue to advance the field.

Strengthening Study Design

Ultimately, the study design and subsequent methods employed for any study should reflect the research question of interest. However, research of the exposome is convoluted by having a milieu of several research questions that span multiple dimensions, well not in the metaphysical term of the word (yet). Other study design challenges are presented by one of the major underlying principles of the exposome, to capture one's exposures and disease risks over their life course. These elements necessitate advances in study design for exposome studies, beyond traditional epidemiological methods, where improved study design will largely benefit the field. Such efforts will need to focus on minimizing information and selection bias, and on increasing the precision of our estimates and external validity to allow for more generalizability of our findings. Additionally, the application of formal methods rooted in causal inference will aid in improving our causal conclusions from exposome studies. Together, this will result in better estimates of associations and in identifying potential causal mechanisms of exposures and diseases that can then be translated into risk assessment and both policy and medical interventions.

Thus far, exposome studies of the fetal and early life have been particularly plagued by issues of reverse causation and misclassification bias, among other issues. Of course, to minimize these biases, it would be ideal to develop longitudinal birth cohort studies, capturing both the external and internal exposome during prepregnancy, throughout pregnancy, and continuous follow-up of offspring into later life. This study design would provide a thorough assessment of the key developmental windows mentioned in previous sections of this chapter, but is problematic due to the enormous expense and time required. However, the development of several large prospective longitudinal designs may not be necessary (depending on the research question) to determine associations of exposures and disease over the life course, as proposed in the theoretical development of the exposome. This theory has been supported and recently expanded on in a proposed method to improve risk assessment for children's health to include snapshots of exposures, captured by focusing on key windows of susceptibility for target organs, in what is termed life stage exposome snapshots (LEnS) (Shaffer et al. 2017).

Many of the preliminary exposome studies have been structured by using a variety and combination of study designs. For example, in both the EXPOsOMICS and HELIX projects, data from historical (legacy) cohorts were used, along with further follow-up and resampling nested within these cohorts. There were also a range of epidemiological study designs employed, including prospective cohort studies, case–control studies, and randomized clinical trials, as well as the design of new short-term exposure studies. It is clearly advantageous to use existing cohorts to reduce costs, increase sample size by pooling studies, reduce follow-up time and lag time to disease, and to utilize existing banked biospecimens. Based on preliminary exposome studies, it is clear that much can be gained by utilizing existing cohorts, including the successful use of stored samples for external and internal exposome classification and the improved statistical power obtained from pooling datasets, as exemplified in our review of previous studies in this chapter and others. However, there are some challenges presented by using historical cohorts. For example, the differences in the types of and timing of exposures between cohorts from combining existing/historical and contemporary cohorts may present issues for pooling data.

The development of future prospective study designs in addition to other study designs may be needed to address certain research gaps in exposome studies. For example, new prospective cohorts may be needed to capture the time period of preconception. This could be achieved by recruiting couples that are intending to get pregnant with follow-up throughout the prenatal and into the postnatal periods and possibly further. Additionally, new prospective designs may be needed to ensure temporality of exposures to diseases, and to reduce misclassification of exposures, to improve our assessments of causality. Furthermore, the external exposome has grown since its conception, as have the number and types of contaminants; therefore, historical cohorts may be limited in the types of exposure data available. We may also need to collect more biospecimens prospectively, as our techniques to measure the internal exposome evolve, particularly in the realms of 'omics.

Cross-sectional designs can be implemented for specific exposure windows to describe prevalence of diseases; however, may be limited in the interpretation of 'omics, mainly due to the potential for reverse causation. Several studies have used a case–control approach to examine the exposome in relation to specific diseases. Case–control studies can be particularly useful as a first-level exposome approach, for generating new hypotheses. Rappaport and Smith (2010) proposed to use case–control studies as an initial discovery phase, where targeted and nontargeted 'omics approaches could be used to compare exposures of cases that have a specific disease with exposures of controls that do not. After the initial discovery phase, the replication phase within a prospective framework can then be used to improve analytical (and other) methods. This framework would likely work best in case–control studies that are nested within prospective cohort studies and those that have banked biospecimens. Also, for case–control studies, there is an opportunity unique to pregnancy cohorts for women to contribute to a study more than one pregnancy, which may allow for them to act as their own controls. Lastly, there is great opportunity for future exposome studies to use retrospective designs, to reconstruct the *in utero* and prepregnancy external and internal exposome. This design may be particularly useful for case–control studies of rare health outcomes, for example, childhood leukemia.

Perhaps one of the most important factors to consider in determining causality within the context of fetal and early exposome research is temporality. In the context of life course studies of the exposome, issues of temporality can arise from simply the variability and dynamic evolution experienced in individuals. Additionally, since exposome studies evaluate multiple exposures, 'omics as intermediate markers, and outcomes that may be measured on different temporal scales, we need to pay close attention to this in our estimations of life course risks. Furthermore, time-varying confounders will need to be carefully measured, identified, and appropriately modeled and controlled for. This is ever more important during the fetal developmental period, as the nature of traditional birth cohorts is characterized by the long-time gap between exposures and outcomes, increasing the likelihood of confounding bias. Additionally, given the high correlation between maternal exposures and behaviors during pregnancy with those postnatally, it is often challenging to tease apart intrauterine from postnatal effects, necessitating a decomposition of these effects. Another important consideration for many studies that measure the external and/or internal exposome from tissues collected at birth in association with birth outcomes is reverse causality (as mentioned previoulsy as a limitation in cross-sectional studies). For example, methylation markers of birthweight from cord blood CpGs may be due to external exposures experienced during pregnancy leading to altered outcomes such as altered fetal growth and/or timing of delivery or, it may act as a snapshot of the developmental processes at delivery (Engel et al. 2014). Furthermore, for internal measurements of 'omics, we will need to test hypothesis of cross-omic processes beyond the central biological dogma of a linear relationship, that is, genome (DNA) → transcriptome (RNA) → proteome → metabolome, as there is extensive cross talk among them (Rappaport 2012).

Lastly, as we begin to advance our studies and methods, there will need to be a primary focus on addressing causal questions and developing studies that are appropriate to answer these questions. A limited discussion for these approaches and potential methods are detailed below; however, it is important to note that such efforts will need to be addressed first in the study design phase, not as an afterthought in the application of statistical methods or in deriving causal conclusions from other approaches (e.g. triangulation). Importantly, at the study design phase, we will need to consider the internal exposome as potential causal intermediates and/or effect modifiers, and therefore direct close attention as to how these data are collected (e.g., temporality and potential confounders) that could influence our ability to come to causal conclusions. Using directed acyclical graphs at the study design phase could serve as conceptual models to aide in our methods, although this may be more difficult (though not impossible) in EWAS studies where many exposures are addressed.

Advancing Methodological Approaches

The discussion of methods for exposome research covers a wide range of topics, from tools and accuracy of exposure assessments in characterizing the external exposome, to laboratory methods of quantifying the internal exposome, to the challenges of statistical analyses and approaches of exposome data. Fortunately, many of the challenges of exposure assessment have been addressed and continue to be expanded upon as the field grows. One of the biggest challenges of exposome work is the dimensionality of the data, spanning from the identification of multiple external exposures and several different 'omics to the combination of these with particular phenotypes and multiple timepoints, as the combinations possible quickly becomes exponential (Patel 2017). Many of these challenges have been thoroughly discussed in the literature and are discussed more thoroughly in Chap. 12. However, there are some methods specific to the windows of susceptibility that are the focus of this chapter that necessitate further discussion. Additionally, many methods are discussed at length in the reproductive/perinatal epidemiological literature, but not within the context of the exposome. As we continue to refine our methodology in fetal and early life exposome research, we will need to pay close attention to causal intermediates, effect measure modification, confounding and other biases, and the accuracy and validity of biomarkers and 'omics in our modeling methods and biological interpretations.

There has been a thorough discussion concerning issues of conditioning and stratifying on causal intermediates in the field of perinatal epidemiology, warranting emphasis and attention to these for early life exposome research. For example, there is a long debate, that still necessitates further attention, on whether to adjust for pregnancy factors such as gestational age and other pregnancy/birth characteristics in assessments of risks for adverse birth and later in life outcomes, as many of these can lie on biological pathways from exposures to disease (Wilcox et al. 2011). Many traditional approaches treat factors such as gestational age as something to simply adjust for or as a determinant, where *in utero* and postnatal factors are treated separately; however, there are alternative approaches. For example, the fetuses at risk (FAR) approach, or most recently suggested an extended FAR approach, is a proposed causal model where gestational age is treated as survival time, and therefore incidence rates may be estimated as a continuous factors from the time period of fetal to postnatal windows (Joseph 2016). Additionally, it is important to note efforts to stratify on intermediates are not recommended, particularly if one is interested in assessing the overall effect of an exposure on an outcome, as this can create collider stratification bias. To avoid such biases, one may instead condition on the predicted risk of the intermediate, condition on the intermediate itself with sensitivity analysis, and/or condition on the principal stratum (Rothman et al. 2008). Additionally, we must take into account in our models the intermediates that serve as important mediating factors on the causal path of exposure and disease within the exposome framework. This is particularly important as ignoring intermediates may lead to miss-specified models and potential false interpretations of causality, and missing (i.e., not seeing associations) important exposure/disease responses. Specific to studies of the fetal and neonatal periods, many birth outcomes share common pathways, or multiple risk factors. For example, PTB is associated with other pregnancy conditions such as preeclampsia or eclampsia, and intrauterine growth restriction, thus this may necessitate more advance models of the multiple causes of birth outcomes and later in life health. Furthermore, alterations to specific 'omics and pathways are biological intermediates and must be treated as such. It will also be imperative for mechanistic plausibility to assess multiple mediators and their interaction with one another and exposures.

Addressing causal intermediates may be possible either through traditional mediation approaches (where appropriate), but more so via the application of methods common in the causal inference field, such as the counterfactual/potential outcomes framework. The use of model free estimands as put forth by Robins and Greenland (1992) and Pearl (2001) who used potential outcomes for the direct and indirect effects will need to be incorporated into statistical mediation language of the exposome. Using counterfactual mediation methods are particularly attractive as they allow for nonlinear models and interactive effects, while addressing the potential bias arising from both incorrect statistical analysis and suboptimal study design (Liu et al. 2016). An example of such applications include G-estimation, and multilevel models, as put forth by VanderWeele (2015), among others. Furthermore, causal estimands can be used to support lifecourse approaches that are foundational for the field of the exposome. For example, critical and sensitive time periods could be identified, as described by De Stavola and Daniel (2017), whereby estimating the controlled direct effects under different scenarios (e.g., single or multiple mediators at fixed values) could be used to disentangle sensitive periods, cumulative exposure effects, and/or a critical times for exposures. Lastly, by using causal mediation analyses efforts in mitigation/intervention on the exposure and/or causal intermediates can be addressed, signifying potential policy recommendations. In addition to mediation, other biological factors that can influence effect estimates will need to be assessed via effect measure modification analyses, as there may be heterogeneous effects not currently explored in exposome studies. This is particularly important because summary effect estimates may be incomplete or misleading if the effect of exposure varies by another factor. There are a limited number of fetal and early life exposome studies that have investigated effect measure modification. This may need to be further explored as, for example, there may be sex-specific 'omics responses that may modify the association between the external and internal exposome. For example, there are sex-specific differences in genome-wide methylation (Liu et al. 2010). The incorporation of mediation and modification into early life and fetal exposome studies will greatly aide our abilities to postulate biological mechanisms and potential causal relationships.

Improving Biological Sampling

The fetal and early life exposome is unique in the types of biomarkers assessed and tissue matrices available to measure external exposures and the internal responses to exposures. However, there are several considerations to take into account for these assessments, including determining what are the most appropriate tissues to use, paying careful attention to the biological interpretations of differences of the internal exposome based on exposures and/or disease states, addressing quality control and reproducibility of assays and techniques, and determining the best timing for sample collection. Ultimately, we want to move the field towards stronger standardization and replication of techniques, to reduce measurement error and exposure misclassification. The potential for such issues should be minimized in the study design; however, it may be possible to address these issues post data collection. For example, in the HELIX study, authors suggest to use traditional regression calibration and other techniques such as structural equation models and Bayesian profile regression models to account for exposure measurement error and uncertainty (Vrijheid et al. 2014).

The question of the most applicable tissues to use for measurements of both the external and internal exposome, whether for the use of biomarkers of exposure or for the characterization of internal 'omics, in exposome research is not specific to studies of fetal development and early life. In many studies, biological matrices are selected based on the least-invasive tissue and samples are used from existing clinical databases, commonly comprising of blood, saliva, and urine. However, there are several reasons to go beyond the most commonly used tissues, and there are many examples where they may better represent exposure, and/or the biological properties of the internal exposome. For a full review of biological matrices available for collection for different life stages, refer to Barr et al. (2005). Specific to the sensitive and/or critical time periods defined in this chapter, there are several tissues that may be used for exposome research and can be classified as maternal and/or fetal, and include gestational tissues such as the placenta, decidua, myometrium, maternal blood (serum and plasma), cervix, fetal membranes (chorion and amnion), amnionic fluid, umbilical cord, fetal blood, and basal plate or postnatal samples that may represent the pregnancy period such as dried blood spots, meconium, breast milk, uterine fluids, neonatal hair, and paternal contributions, such as sperm. Traditionally, many pregnancy studies use maternal biomarkers to represent fetal exposure, as they are less invasive and may serve as important clinical markers of the health status of the mom and/or developing fetus. Indeed, for some exposures, these maternal biomarkers of exposure may represent fetal exposure; however, it is important to acknowledge that these biological samples may not provide accurate measures of fetal exposure for all chemicals (Andra et al. 2016). Importantly, by relying solely on maternal biomarkers of fetal exposure, we may not be capturing the variability in placental transport and metabolism, and therefore we may be missing important maternal-fetal biological interactions (Yoon et al. 2009).

There are limitations to using the routinely collected biological tissues, such as blood, serum, and urine in quantifying the exposome, as for some analytes, the accumulation may occur only in particular tissues. Additionally, while umbilical cord blood has been successfully collected at birth in epidemiologic studies and has provided valuable exposure information for certain compounds with a short half-life in blood, cord blood levels cannot provide information about earlier pregnancy/prepregnancy exposures (Andra et al. 2016). Still, matrixes, and in particular those from maternal fluids, are important for clinical practice and may serve as important biomarkers for exposome studies. Both cord and maternal blood samples have also been widely used for measuring 'omics, as discussed previously in this chapter.

While the question of the specificity of these circulating (i.e., non-tissue-specific) 'omics is still being debated, their use may serve as important noninvasive biological intermediates of a particular response or pathway between an exposures and disease. For example, circulating mRNAs may provide information on potential fetal-maternal communication via the delivery and/or exchange of extracellular vesicles and mobile RNAs (Zhang et al. 2017).

Outside of maternal biological tissues, there is a great need to look into paternal contributions via paternal tissues. For example, tissues such as sperm may be important for the exposome as they also carry epigenetic information consisting of methylated DNA, noncoding RNAs, protamines, and histones that are critical to fertilization and programming early embryonic development (Carrell and Hammoud 2010). As reviewed by Day et al. (2016), there is also evidence that paternal epigenetics, such as DNA methylation, histone modification, and miRNA expression, may play an important role in birth defects and later in life developments of health outcomes, including obesity and metabolic disorders.

Additionally, the uses and development of biomarkers and the detection of 'omics in samples that are collected postpartum could aide in addressing many of the limitations of prospective study designs for periods that influence fetal and early life. For example, the placenta may serve as an important tissue that could address issues of temporality and specificity, as circulation is established approximately 17 days after fertilization, and the placenta acts as an important route for exposures and biological features such as key regulation of 'omics. The concept of the placental exposome has recently been put forth by Lewis et al. (2013) where they argue that the overall effects of the maternal environment on the placenta are the product of its exposures throughout gestation. The use of placental tissue, rather than circulating or umbilical cord blood for methylation markers, for example, may be useful for the high tissue specificity of epigenetic mechanisms. Placental tissues may also be relevant for pregnancy complications, as dysfunctional placental development is known to be involved in most pregnancy complications, for example, preeclampsia. Another postnatal tissue for exposome studies is the use of dried blood spots (DBS) to characterize the external and/or internal exposome. This technique may be particularly useful for historical data as DBS have been traditionally and are still widely collected from >98% of the newborns in the United States and in countries worldwide. The use of DBS to measure the internal and external exposome is also appealing as they are composed of whole blood, providing access to potential biomarkers from both serum and red and white blood cells. Recently, the use of DBS for metabolomics has been supported where 1000 prevalent small molecules were identified in samples, and in lipidomics where more than 1200 lipid species were identified and quantified (Gao et al. 2017). Another potential matrix for retrospective exposome studies is the use of baby, children, and even adult teeth (Andra et al. 2016). For example, teeth have been used to estimate the long-term cumulative levels of metals that children were exposed to (Andra et al. 2016). Furthermore, histological and chemical analysis have been recently established to precisely sample tooth layers that correspond to specific life stages (Andra et al.

2016). Together many of these tissues could create opportunities to reconstruct the exposome representing fetal development and early life.

There is also a need to increase the validity of 'omics markers. While 'omics measurements usually reflect changes within cells, we still need validation of many of 'omics biological endpoints. Many 'omics are measured as a surrogate tissue such as maternal blood and/or urine; however, we know that this may not reflect the target organ as some markers may be tissue-specific (such as epigenetic markers). Additionally, 'omics may reflect one or a few points in time. Combining different 'omics tools, and implementing cross-omics analyses, may increase the understanding of how different external exposures interact with internal molecules, for example, by inducing mutations (genomics), causing epigenetic changes (epigenomics), or modifying the internal cell environment in more complex ways. While selecting a tissue-specific matrix may be preferable for 'omics analyses (part of the internal exposome), using many of these tissue types for exposure assessment may present new challenges beyond the practical and ethical limitations. For example, many chemicals have different pharmacodynamics and biotransformation properties, thus it may ultimately be essential that multiple sample types are collected from each individual to fully define the exposome (Dennis et al. 2016). As molecular biology continues to advance, there will likely be the opportunity to address current gaps of these most widely used biological samples, including the investigation of tissue specific and cell specific 'omics, reflecting a more complete biological picture of cellular and systems-wide mechanisms. In the meantime, we will need to be cautious in our interpretations of internal 'omics as they relate to health outcomes.

Given that many of the results from larger exposome studies are just emerging, we are in the formative stages for creating gold standards for many exposome methods and techniques. However, there have been several initiatives put forth to address many of the previously mentioned gaps and challenges that exist in current studies. For example, the CHEAR initiative, led by the National Institutes of Health Sciences (NIEHS), may provide an opportunity for several studies to have access to standardized laboratory tools for targeted and untargeted analyses of biospecimens, as is described in Chap. 13. Additionally, many of the challenges of exposures and 'omics assessments were recently discussed when experts in exposome research met to deliberate on the current state and potential advancement of the field for the NIEHS Exposome Workshop (Dennis et al. 2016). From this workshop, recommendations for researchers for future methods to advance biomonitoring were broadly as follows: encourage secondary analysis for samples, use a standardized measurement platform, utilize existing databases, integrate interdisciplinary knowledge for targeted and non-targeted analysis, develop methods to detect low-abundance and differentiation of endogenous and exogenous molecules, enrich bioinformatics tools and methods, and incorporate and develop pharmacokinetic models (Dennis et al. 2016). An important conclusion for future studies is that exposome researchers may need to use a hybrid approach for biological monitoring and mechanistic investigations. For example, in the HELIX study hybrid approaches for data collection to investigate the prenatal and early life exposome are employed where investigators use personal external exposure monitoring, traditional biomonitoring techniques, and untargeted 'omics analyses of metabolomics, proteomics, transcriptomics, epigenomics, including repeat sampling to capture nonpersistent biomarkers (Vrijheid et al. 2014).

Conclusions

In a short time span, we have moved from conceptual frameworks of the exposome to the implementation of large-scale exposome studies, with many of the first exposome studies serving as a proof of concept for the application of the theory of the exposome. These studies have undergone an impressive feat of establishing large-scale multi-exposure, multi-internal biomarkers, and multi-'omics characterizations of the exposome. Findings from these studies serve to improve exposure assessment, inform us of potential risks of a variety of disease states and poor health from a multitude of exposures, and provide corroboration for the use of 'omics and pathway perturbation as evidence of mechanisms that underlie adverse health outcomes. In this review, we have highlighted several successful studies that characterized the exposome for sensitive and/or critical time periods that contribute to fetal and early life development.

Ultimately, future frameworks for research of the exposome will require multiple approaches and more interdisciplinary collaborations. As Wild (2012) suggested, exposome researchers will need to use different paradigms, tools, and languages to embrace the juxtapositions of molecular mechanisms, biotechnology, bioinformatics, biostatistics, epidemiology, social sciences, and clinical research. Furthermore, he suggested that we will need to encompass the challenges of differences in socioeconomic and global responses for causality and prevention of disease. Thus, future exposome studies will benefit from incorporating new tools from within the field and other fields of research, such as mathematical and/or pharmacokinetic modeling, machine learning, causal inference, and an incorporation of the social sciences. Furthermore, many studies will need replication, hence the need to facilitate more collaboration across exposome studies.

The research field of the exposome is essentially twenty-first century epidemiology, and holds promise to make more precise and causal effect estimates for fetal and early life epidemiological assessments through the incorporation of mechanistic information and improved exposure assessment. While the exposome was first proposed in the context of public health, it may also improve the provision of personalized maternal and antenatal care.

References

Agay-Shay K, Martinez D, Valvi D, Garcia-Esteban R, Basagana X, Robinson O, Casas M, Sunyer J, Vrijheid M (2015) Exposure to endocrine-disrupting chemicals during pregnancy and weight at 7 years of age: a multi-pollutant approach. Environ Health Perspect 123 (10):1030–1037. https://doi.org/10.1289/ehp.1409049

Agha G, Hajj H, Rifas-Shiman SL, Just AC, Hivert MF, Burris HH, Lin X, Litonjua AA, Oken E, DeMeo DL, Gillman MW, Baccarelli AA (2016) Birth weight-for-gestational age is associated with DNA methylation at birth and in childhood. Clin Epigenetics 8:118. https://doi.org/10.1186/s13148-016-0285-3
Andra SS, Austin C, Arora M (2016) The tooth exposome in children's health research. Curr Opin Pediatr 28(2):221–227. https://doi.org/10.1097/MOP.0000000000000327
ATSDR (2017) Agency for toxic substances and disease registry. Accessed Oct 2017. http://www.atsdr.cdc.gov
Bailey KA, Laine J, Rager JE, Sebastian E, Olshan A, Smeester L, Drobná Z, Styblo M, Rubio-Andrade M, García-Vargas G, Fry RC (2014) Prenatal arsenic exposure and shifts in the newborn proteome: interindividual differences in tumor necrosis factor (TNF)-responsive signaling. Toxicol Sci 139(2):328–337. https://doi.org/10.1093/toxsci/kfu053
Baird J, Hill CM, Kendrick T, Inskip HM, SWS Study Group (2009) Infant sleep disturbance is associated with preconceptional psychological distress: findings from the Southampton Women's Survey. Sleep 32(4):566–568
Barker DJ (2004) The developmental origins of adult disease. J Am Coll Nutr 23(6 Suppl):588S–595S
Barr DB, Wang RY, Needham LL (2005) Biologic monitoring of exposure to environmental chemicals throughout the life stages: requirements and issues for consideration for the National Children's Study. Environ Health Perspect 113(8):1083–1091
Ben-Shlomo Y, Kuh D (2002) A life course approach to chronic disease epidemiology: conceptual models, empirical challenges and interdisciplinary perspectives. Int J Epidemiol 31(2):285–293
Billionnet C, Sherrill D, Annesi-Maesano I (2012) Estimating the health effects of exposure to multi-pollutant mixture. Ann Epidemiol 22(2):126–141. https://doi.org/10.1016/j.annepidem.2011.11.004
Braun JM, Kalkbrenner AE, Just AC, Yolton K, Calafat AM, Sjodin A, Hauser R, Webster GM, Chen A, Lanphear BP (2014) Gestational exposure to endocrine-disrupting chemicals and reciprocal social, repetitive, and stereotypic behaviors in 4- and 5-year-old children: the HOME study. Environ Health Perspect 122(5):513–520. https://doi.org/10.1289/ehp.1307261
Braun JM, Messerlian C, Hauser R (2017) Fathers matter: why it's time to consider the impact of paternal environmental exposures on children's health. Curr Epidemiol Rep 4(1):46–55. https://doi.org/10.1007/s40471-017-0098-8
Breton CV, Yao J, Millstein J, Gao L, Siegmund KD, Mack W, Whitfield-Maxwell L, Lurmann F, Hodis H, Avol E, Gilliland FD (2016) Prenatal air pollution exposures, DNA methyl transferase genotypes, and associations with newborn LINE1 and alu methylation and childhood blood pressure and carotid intima-media thickness in the Children's Health Study. Environ Health Perspect 124(12):1905–1912. https://doi.org/10.1289/ehp181
Buck Louis GM, Yeung E, Sundaram R, Laughon SK, Zhang C (2013) The exposome—exciting opportunities for discoveries in reproductive and perinatal epidemiology. Paediatr Perinat Epidemiol 27(3):229–236. https://doi.org/10.1111/ppe.12040
Burris HH, Baccarelli AA, Byun HM, Cantoral A, Just AC, Pantic I, Solano-Gonzalez M, Svensson K, Tamayo y Ortiz M, Zhao Y, Wright RO, Tellez-Rojo MM (2015) Offspring DNA methylation of the aryl-hydrocarbon receptor repressor gene is associated with maternal BMI, gestational age, and birth weight. Epigenetics 10(10):913–921. https://doi.org/10.1080/15592294.2015.1078963
Carrell DT, Hammoud SS (2010) The human sperm epigenome and its potential role in embryonic development. Mol Hum Reprod 16(1):37–47. https://doi.org/10.1093/molehr/gap090
Chason RJ, Csokmay J, Segars JH, DeCherney AH, Armant DR (2011) Environmental and epigenetic effects upon preimplantation embryo metabolism and development. Trends Endocrinol Metab 22(10):412–420. https://doi.org/10.1016/j.tem.2011.05.005

Cruickshank MN, Oshlack A, Theda C, Davis PG, Martino D, Sheehan P, Dai Y, Saffery R, Doyle LW, Craig JM (2013) Analysis of epigenetic changes in survivors of preterm birth reveals the effect of gestational age and evidence for a long term legacy. Genome Med 5(10):96. https://doi.org/10.1186/gm500

Dadvand P, Ostro B, Figueras F, Foraster M, Basagana X, Valentin A, Martinez D, Beelen R, Cirach M, Hoek G, Jerrett M, Brunekreef B, Nieuwenhuijsen MJ (2014) Residential proximity to major roads and term low birth weight: the roles of air pollution, heat, noise, and road-adjacent trees. Epidemiology 25(4):518–525. https://doi.org/10.1097/ede.0000000000000107

Day J, Savani S, Krempley BD, Nguyen M, Kitlinska JB (2016) Influence of paternal preconception exposures on their offspring: through epigenetics to phenotype. Am J Stem Cells 5(1):11–18

Dennis KK, Auerbach SS, Balshaw DM, Cui Y, Fallin MD, Smith MT, Spira A, Sumner S, Miller GW (2016) The importance of the biological impact of exposure to the concept of the exposome. Environ Health Perspect 124(10):1504–1510. https://doi.org/10.1289/EHP140

Dessì A, Atzori L, Noto A, Adriaan Visser GH, Gazzolo D, Zanardo V, Barberini L, Puddu M, Ottonello G, Atzei A, Magistris AD, Lussu M, Murgia F, Fanos V (2011) Metabolomics in newborns with intrauterine growth retardation (IUGR): urine reveals markers of metabolic syndrome. J Matern Fetal Neonatal Med 24(Suppl 2):35–39. https://doi.org/10.3109/14767058.2011.605868

De Stavola BL, Daniel RM (2017) Commentary: Incorporating concepts and methods from causal inference into life course epidemiology. Int J Epidemiol 46(2):771. https://doi.org/10.1093/ije/dyw367

Eidem HR, Ackerman WE, McGary KL, Abbot P, Rokas A (2015) Gestational tissue transcriptomics in term and preterm human pregnancies: a systematic review and meta-analysis. BMC Med Genet 8:27. https://doi.org/10.1186/s12920-015-0099-8

Engel SM, Joubert BR, Wu MC, Olshan AF, Håberg SE, Ueland PM, Nystad W, Nilsen RM, Vollset SE, Peddada SD, London SJ (2014) Neonatal genome-wide methylation patterns in relation to birth weight in the Norwegian Mother and Child Cohort. Am J Epidemiol 179 (7):834–842. https://doi.org/10.1093/aje/kwt433

Fanos V, Atzori L, Makarenko K, Melis GB, Ferrazzi E (2013) Metabolomics application in maternal-fetal medicine. Biomed Res Int 2013:720514. https://doi.org/10.1155/2013/720514

Felix JF, Joubert BR, Baccarelli AA, Sharp GC, Almqvist C, Annesi-Maesano I, Arshad H, Baiz N, Bakermans-Kranenburg MJ, Bakulski KM, Binder EB, Bouchard L, Breton CV, Brunekreef B, Brunst KJ, Burchard EG, Bustamante M, Chatzi L, Cheng Munthe-Kaas M, Corpeleijn E, Czamara D, Dabelea D, Davey Smith G, De Boever P, Duijts L, Dwyer T, Eng C, Eskenazi B, Everson TM, Falahi F, Fallin MD, Farchi S, Fernandez MF, Gao L, Gaunt TR, Ghantous A, Gillman MW, Gonseth S, Grote V, Gruzieva O, Haberg SE, Herceg Z, Hivert MF, Holland N, Holloway JW, Hoyo C, Hu D, Huang RC, Huen K, Jarvelin MR, Jima DD, Just AC, Karagas MR, Karlsson R, Karmaus W, Kechris KJ, Kere J, Kogevinas M, Koletzko B, Koppelman GH, Kupers LK, Ladd-Acosta C, Lahti J, Lambrechts N, Langie SAS, Lie RT, Liu AH, Magnus MC, Magnus P, Maguire RL, Marsit CJ, McArdle W, Melen E, Melton P, Murphy SK, Nawrot TS, Nistico L, Nohr EA, Nordlund B, Nystad W, Oh SS, Oken E, Page CM, Perron P, Pershagen G, Pizzi C, Plusquin M, Raikkonen K, Reese SE, Reischl E, Richiardi L, Ring S, Roy RP, Rzehak P, Schoeters G, Schwartz DA, Sebert S, Snieder H, Sorensen TIA, Starling AP, Sunyer J, Taylor JA, Tiemeier H, Ullemar V, Vafeiadi M, Van Ijzendoorn MH, Vonk JM, Vriens A, Vrijheid M, Wang P, Wiemels JL, Wilcox AJ, Wright RJ, Xu CJ, Xu Z, Yang IV, Yousefi P, Zhang H, Zhang W, Zhao S, Agha G, Relton CL, Jaddoe VWV, London SJ (2017) Cohort profile: pregnancy and childhood epigenetics (PACE) consortium. Int J Epidemiol 47 (1):22–23u. https://doi.org/10.1093/ije/dyx190

Forns J, Mandal S, Iszatt N, Polder A, Thomsen C, Lyche JL, Stigum H, Vermeulen R, Eggesbo M (2016) Novel application of statistical methods for analysis of multiple toxicants identifies DDT as a risk factor for early child behavioral problems. Environ Res 151:91–100. https://doi.org/10.1016/j.envres.2016.07.014

Galea S (2017) Making epidemiology matter. Int J Epidemiol 46(4):1083–1085. https://doi.org/10.1093/ije/dyx154
Gao F, McDaniel J, Chen EY, Rockwell HE, Drolet J, Vishnudas VK, Tolstikov V, Sarangarajan R, Narain NR, Kiebish MA (2017) Dynamic and temporal assessment of human dried blood spot MS/MS(ALL) shotgun lipidomics analysis. Nutr Metab (Lond) 14:28. https://doi.org/10.1186/s12986-017-0182-6
Goldenberg RL, Culhane JF, Iams JD, Romero R (2008) Epidemiology and causes of preterm birth. Lancet 371(9606):75–84. https://doi.org/10.1016/S0140-6736(08)60074-4
Golding J, Gregory S, Iles-Caven Y, Lingam R, Davis JM, Emmett P, Steer CD, Hibbeln JR (2014) Parental, prenatal, and neonatal associations with ball skills at age 8 using an exposome approach. J Child Neurol 29(10):1390–1398. https://doi.org/10.1177/0883073814530501
Goodrich JM, Reddy P, Naidoo RN, Asharam K, Batterman S, Dolinoy DC (2016) Prenatal exposures and DNA methylation in newborns: a pilot study in Durban, South Africa. Environ Sci Process Impacts 18(7):908–917. https://doi.org/10.1039/c6em00074f
Grevendonk L, Janssen BG, Vanpoucke C, Lefebvre W, Hoxha M, Bollati V, Nawrot TS (2016) Mitochondrial oxidative DNA damage and exposure to particulate air pollution in mother-newborn pairs. Environ Health 15:10. https://doi.org/10.1186/s12940-016-0095-2
Gruzieva O, Xu CJ, Breton CV, Annesi-Maesano I, Anto JM, Auffray C, Ballereau S, Bellander T, Bousquet J, Bustamante M, Charles MA, de Kluizenaar Y, den Dekker HT, Duijts L, Felix JF, Gehring U, Guxens M, Jaddoe VV, Jankipersadsing SA, Merid SK, Kere J, Kumar A, Lemonnier N, Lepeule J, Nystad W, Page CM, Panasevich S, Postma D, Slama R, Sunyer J, Soderhall C, Yao J, London SJ, Pershagen G, Koppelman GH, Melen E (2017) Epigenome-wide meta-analysis of methylation in children related to prenatal NO2 air pollution exposure. Environ Health Perspect 125(1):104–110. https://doi.org/10.1289/ehp36
Guxens M, Ballester F, Espada M, Fernandez MF, Grimalt JO, Ibarluzea J, Olea N, Rebagliato M, Tardon A, Torrent M, Vioque J, Vrijheid M, Sunyer J (2012) Cohort profile: the INMA—INfancia y medio ambiente—(environment and childhood) project. Int J Epidemiol 41(4):930–940. https://doi.org/10.1093/ije/dyr054
Hellmuth C, Uhl O, Standl M, Demmelmair H, Heinrich J, Koletzko B, Thiering E (2017) Cord blood metabolome is highly associated with birth weight, but less predictive for later weight development. Obes Facts 10(2):85–100
Horgan RP, Broadhurst DI, Walsh SK, Dunn WB, Brown M, Roberts CT, North RA, McCowan LM, Kell DB, Baker PN, Kenny LC (2011) Metabolic profiling uncovers a phenotypic signature of small for gestational age in early pregnancy. J Proteome Res 10(8):3660–3673. https://doi.org/10.1021/pr2002897
Hughes DA, Kircher M, He Z, Guo S, Fairbrother GL, Moreno CS, Khaitovich P, Stoneking M (2015) Evaluating intra- and inter-individual variation in the human placental transcriptome. Genome Biol 16(1):54. https://doi.org/10.1186/s13059-015-0627-z
Iozzo P, Holmes M, Schmidt MV, Cirulli F, Guzzardi MA, Berry A, Balsevich G, Andreassi MG, Wesselink JJ, Liistro T, Gómez-Puertas P, Eriksson JG, Seckl J (2014) Developmental ORIgins of Healthy and Unhealthy AgeiNg: the role of maternal obesity—introduction to DORIAN. Obes Facts 7(2):130–151. https://doi.org/10.1159/000362656
Ivorra C, García-Vicent C, Chaves FJ, Monleón D, Morales JM, Lurbe E (2012) Metabolomic profiling in blood from umbilical cords of low birth weight newborns. J Transl Med 10:142. https://doi.org/10.1186/1479-5876-10-142
Janssen BG, Godderis L, Pieters N, Poels K, Kicinski M, Cuypers A, Fierens F, Penders J, Plusquin M, Gyselaers W, Nawrot TS (2013) Placental DNA hypomethylation in association with particulate air pollution in early life. Part Fibre Toxicol 10:22. https://doi.org/10.1186/1743-8977-10-22
Janssen BG, Byun HM, Gyselaers W, Lefebvre W, Baccarelli AA, Nawrot TS (2015) Placental mitochondrial methylation and exposure to airborne particulate matter in the early life environment: an ENVIRONAGE birth cohort study. Epigenetics 10(6):536–544. https://doi.org/10.1080/15592294.2015.1048412

Jeyabalan A (2013) Epidemiology of preeclampsia: impact of obesity. Nutr Rev 71(Suppl 1):S18–S25. https://doi.org/10.1111/nure.12055

Joseph KS (2016) A consilience of inductions supports the extended fetuses-at-risk model. Paediatr Perinat Epidemiol 30(1):11–17. https://doi.org/10.1111/ppe.12260

Joubert BR, Haberg SE, Nilsen RM, Wang X, Vollset SE, Murphy SK, Huang Z, Hoyo C, Midttun O, Cupul-Uicab LA, Ueland PM, Wu MC, Nystad W, Bell DA, Peddada SD, London SJ (2012) 450K epigenome-wide scan identifies differential DNA methylation in newborns related to maternal smoking during pregnancy. Environ Health Perspect 120 (10):1425–1431. https://doi.org/10.1289/ehp.1205412

Kershenbaum AD, Langston MA, Levine RS, Saxton AM, Oyana TJ, Kilbourne BJ, Rogers GL, Gittner LS, Baktash SH, Matthews-Juarez P, Juarez PD (2014) Exploration of preterm birth rates using the public health exposome database and computational analysis methods. Int J Environ Res Public Health 11(12):12346–12366. https://doi.org/10.3390/ijerph111212346

King JC (2016) A summary of pathways or mechanisms linking preconception maternal nutrition with birth outcomes. J Nutr 146(7):1437S–1444S. https://doi.org/10.3945/jn.115.223479

Kingsley SL, Eliot MN, Whitsel EA, Huang YT, Kelsey KT, Marsit CJ, Wellenius GA (2016) Maternal residential proximity to major roadways, birth weight, and placental DNA methylation. Environ Int 92-93:43–49. https://doi.org/10.1016/j.envint.2016.03.020

Knight AK, Craig JM, Theda C, Baekvad-Hansen M, Bybjerg-Grauholm J, Hansen CS, Hollegaard MV, Hougaard DM, Mortensen PB, Weinsheimer SM, Werge TM, Brennan PA, Cubells JF, Newport DJ, Stowe ZN, Cheong JL, Dalach P, Doyle LW, Loke YJ, Baccarelli AA, Just AC, Wright RO, Tellez-Rojo MM, Svensson K, Trevisi L, Kennedy EM, Binder EB, Iurato S, Czamara D, Raikkonen K, Lahti JM, Pesonen AK, Kajantie E, Villa PM, Laivuori H, Hamalainen E, Park HJ, Bailey LB, Parets SE, Kilaru V, Menon R, Horvath S, Bush NR, LeWinn KZ, Tylavsky FA, Conneely KN, Smith AK (2016) An epigenetic clock for gestational age at birth based on blood methylation data. Genome Biol 17(1):206. https://doi.org/10.1186/s13059-016-1068-z

Kumarathasan P, Vincent R, Das D, Mohottalage S, Blais E, Blank K, Karthikeyan S, Vuong NQ, Arbuckle TE, Fraser WD (2014) Applicability of a high-throughput shotgun plasma protein screening approach in understanding maternal biological pathways relevant to infant birth weight outcome. J Proteome 100:136–146. https://doi.org/10.1016/j.jprot.2013.12.003

Laine JE, Fry RC (2016) A systems toxicology-based approach reveals biological pathways dysregulated by prenatal arsenic exposure. Ann Glob Health 82(1):189–196. https://doi.org/10.1016/j.aogh.2016.01.015

Laine JE, Bailey KA, Olshan AF, Smeester L, Drobná Z, Stýblo M, Douillet C, García-Vargas G, Rubio-Andrade M, Pathmasiri W, McRitchie S, Sumner SJ, Fry RC (2017) Neonatal metabolomic profiles related to prenatal arsenic exposure. Environ Sci Technol 51 (1):625–633. https://doi.org/10.1021/acs.est.6b04374

Lassi ZS, Imam AM, Dean SV, Bhutta ZA (2014) Preconception care: caffeine, smoking, alcohol, drugs and other environmental chemical/radiation exposure. Reprod Health 11(Suppl 3):S6. https://doi.org/10.1186/1742-4755-11-S3-S6

Le HQ, Batterman SA, Wirth JJ, Wahl RL, Hoggatt KJ, Sadeghnejad A, Hultin ML, Depa M (2012) Air pollutant exposure and preterm and term small-for-gestational-age births in Detroit, Michigan: long-term trends and associations. Environ Int 44:7–17. https://doi.org/10.1016/j.envint.2012.01.003

Lenters V, Portengen L, Rignell-Hydbom A, Jonsson BA, Lindh CH, Piersma AH, Toft G, Bonde JP, Heederik D, Rylander L, Vermeulen R (2016) Prenatal phthalate, perfluoroalkyl acid, and organochlorine exposures and term birth weight in three birth cohorts: multi-pollutant models based on elastic net regression. Environ Health Perspect 124(3):365–372. https://doi.org/10.1289/ehp.1408933

Lewis RM, Demmelmair H, Gaillard R, Godfrey KM, Hauguel-de Mouzon S, Huppertz B, Larque E, Saffery R, Symonds ME, Desoye G (2013) The placental exposome: placental

determinants of fetal adiposity and postnatal body composition. Ann Nutr Metab 63 (3):208–215. https://doi.org/10.1159/000355222
Liu J, Morgan M, Hutchison K, Calhoun VD (2010) A study of the influence of sex on genome wide methylation. PLoS One 5(4):e10028. https://doi.org/10.1371/journal.pone.0010028
Liu SH, Ulbricht CM, Chrysanthopoulou SA, Lapane KL (2016) Implementation and reporting of causal mediation analysis in 2015: a systematic review in epidemiological studies. BMC Res Notes 9:354. https://doi.org/10.1186/s13104-016-2163-7
Maitre L, Fthenou E, Athersuch T, Coen M, Toledano MB, Holmes E, Kogevinas M, Chatzi L, Keun HC (2014) Urinary metabolic profiles in early pregnancy are associated with preterm birth and fetal growth restriction in the Rhea mother-child cohort study. BMC Med 12:110. https://doi.org/10.1186/1741-7015-12-110
Maitre L, Villanueva CM, Lewis MR, Ibarluzea J, Santa-Marina L, Vrijheid M, Sunyer J, Coen M, Toledano MB (2016) Maternal urinary metabolic signatures of fetal growth and associated clinical and environmental factors in the INMA study. BMC Med 14(1):177. https://doi.org/10.1186/s12916-016-0706-3
Marsit CJ (2015) Influence of environmental exposure on human epigenetic regulation. J Exp Biol 218(Pt 1):71–79. https://doi.org/10.1242/jeb.106971
McCullough LE, Mendez MA, Miller EE, Murtha AP, Murphy SK, Hoyo C (2015) Associations between prenatal physical activity, birth weight, and DNA methylation at genomically imprinted domains in a multiethnic newborn cohort. Epigenetics 10(7):597–606. https://doi.org/10.1080/15592294.2015.1045181
Mishra GD, Cooper R, Kuh D (2010) A life course approach to reproductive health: theory and methods. Maturitas 65(2):92–97. https://doi.org/10.1016/j.maturitas.2009.12.009
Moussa HN, Alrais MA, Leon MG, Abbas EL, Sibai BM (2016) Obesity epidemic: impact from preconception to postpartum. Future Sci OA 2(3):FSO137. https://doi.org/10.4155/fsoa-2016-0035
National Academies of Sciences, Engineering, and Medicine, Division on Earth and Life Studies, Board on Environmental Studies and Toxicology, Committee on Incorporating 21st Century Science into Risk-Based Evaluations (2017) Using 21st century science to improve risk-related evaluations. National Academies Press (US), Washington, DC. https://doi.org/10.17226/24635
North ML, Brook JR, Lee EY, Omana V, Daniel NM, Steacy LM, Evans GJ, Diamond ML, Ellis AK (2017) The Kingston Allergy Birth Cohort: exploring parentally reported respiratory outcomes through the lens of the exposome. Ann Allergy Asthma Immunol 118(4):465–473. https://doi.org/10.1016/j.anai.2017.01.002
Oyana TJ, Matthews-Juarez P, Cormier SA, Xu X, Juarez PD (2015) Using an external exposome framework to examine pregnancy-related morbidities and mortalities: implications for health disparities research. Int J Environ Res Public Health 13(1):ijerph13010013. https://doi.org/10.3390/ijerph13010013
Patel CJ (2017) Analytic complexity and challenges in identifying mixtures of exposures associated with phenotypes in the exposome era. Curr Epidemiol Rep 4(1):22–30. https://doi.org/10.1007/s40471-017-0100-5
Patel CJ, Ioannidis JP (2014) Placing epidemiological results in the context of multiplicity and typical correlations of exposures. J Epidemiol Community Health 68(11):1096–1100. https://doi.org/10.1136/jech-2014-204195
Patel CJ, Manrai AK (2015) Development of exposome correlation globes to map out environment-wide associations. Pac Symp Biocomput 2015:231–242
Pearl J (2001) Direct and indirect effects. In: Proceedings of the 17th conference in uncertainty in artificial intelligence. Morgan Kaufmann Publishers Inc, San Francisco, pp 411–420
Rager JE, Bailey KA, Smeester L, Miller SK, Parker JS, Laine JE, Drobná Z, Currier J, Douillet C, Olshan AF, Rubio-Andrade M, Stýblo M, García-Vargas G, Fry RC (2014) Prenatal arsenic exposure and the epigenome: altered microRNAs associated with innate and adaptive immune signaling in newborn cord blood. Environ Mol Mutagen 55(3):196–208. https://doi.org/10.1002/em.21842

Rangel M, dos Santos JC, Ortiz PH, Hirata M, Jasiulionis MG, Araujo RC, Ierardi DF, Franco Mdo C (2014) Modification of epigenetic patterns in low birth weight children: importance of hypomethylation of the ACE gene promoter. PLoS One 9(8):e106138. https://doi.org/10.1371/journal.pone.0106138

Rappaport SM (2012) Biomarkers intersect with the exposome. Biomarkers 17(6):483–489. https://doi.org/10.3109/1354750X.2012.691553

Rappaport SM, Smith MT (2010) Epidemiology. Environment and disease risks. Science 330 (6003):460–461. https://doi.org/10.1126/science.1192603

Robins JM, Greenland S (1992) Identifiability and exchangeability for direct and indirect effects. Epidemiology 3:143–155

Robinson O, Vrijheid M (2015) The pregnancy exposome. Curr Environ Health Rep 2(2):204–213. https://doi.org/10.1007/s40572-015-0043-2

Robinson O, Basagana X, Agier L, de Castro M, Hernandez-Ferrer C, Gonzalez JR, Grimalt JO, Nieuwenhuijsen M, Sunyer J, Slama R, Vrijheid M (2015) The pregnancy exposome: multiple environmental exposures in the INMA-Sabadell Birth Cohort. Environ Sci Technol 49 (17):10632–10641. https://doi.org/10.1021/acs.est.5b01782

Robinson O, Tamayo O, de Castro M, Valentin A, Giorgis-Allemand L, Hjertager Krog N, et al (2018) The urban exposome during pregnancy and its socio-economic determinants. Environ Health Perspect (in press)

Robledo CA, Yeung E, Mendola P, Sundaram R, Maisog J, Sweeney AM, Barr DB, Louis GM (2015) Preconception maternal and paternal exposure to persistent organic pollutants and birth size: the LIFE study. Environ Health Perspect 123(1):88–94. https://doi.org/10.1289/ehp.1308016

Rojas D, Rager JE, Smeester L, Bailey KA, Drobná Z, Rubio-Andrade M, Stýblo M, García-Vargas G, Fry RC (2015) Prenatal arsenic exposure and the epigenome: identifying sites of 5-methylcytosine alterations that predict functional changes in gene expression in newborn cord blood and subsequent birth outcomes. Toxicol Sci 143(1):97–106. https://doi.org/10.1093/toxsci/kfu210

Romero R, Kusanovic JP, Gotsch F, Erez O, Vaisbuch E, Mazaki-Tovi S, Moser A, Tam S, Leszyk J, Master SR, Juhasz P, Pacora P, Ogge G, Gomez R, Yoon BH, Yeo L, Hassan SS, Rogers WT (2010a) Isobaric labeling and tandem mass spectrometry: a novel approach for profiling and quantifying proteins differentially expressed in amniotic fluid in preterm labor with and without intra-amniotic infection/inflammation. J Matern Fetal Neonatal Med 23 (4):261–280. https://doi.org/10.3109/14767050903067386

Romero R, Mazaki-Tovi S, Vaisbuch E, Kusanovic JP, Chaiworapongsa T, Gomez R, Nien JK, Yoon BH, Mazor M, Luo J, Banks D, Ryals J, Beecher C (2010b) Metabolomics in premature labor: a novel approach to identify patients at risk for preterm delivery. J Matern Fetal Neonatal Med 23(12):1344–1359. https://doi.org/10.3109/14767058.2010.482618

Rosofsky A, Janulewicz P, Thayer KA, McClean M, Wise LA, Calafat AM, Mikkelsen EM, Taylor KW, Hatch EE (2017) Exposure to multiple chemicals in a cohort of reproductive-aged Danish women. Environ Res 154:73–85. https://doi.org/10.1016/j.envres.2016.12.011

Rossnerova A, Tulupova E, Tabashidze N, Schmuczerova J, Dostal M, Rossner P Jr, Gmuender H, Sram RJ (2013) Factors affecting the 27K DNA methylation pattern in asthmatic and healthy children from locations with various environments. Mutat Res 741-742:18–26. https://doi.org/10.1016/j.mrfmmm.2013.02.003

Rothman KJ, Greenland S, Lash TL (2008) Modern epidemiology, 3rd edn. Lippincott, Williams & Wilkins, Philadelphia, PA

Saenen ND, Vrijens K, Janssen BG, Roels HA, Neven KY, Vanden Berghe W, Gyselaers W, Vanpoucke C, Lefebvre W, De Boever P, Nawrot TS (2017) Lower placental leptin promoter methylation in association with fine particulate matter air pollution during pregnancy and placental nitrosative stress at birth in the ENVIRONAGE cohort. Environ Health Perspect 125(2):262–268. https://doi.org/10.1289/ehp38

Shaffer RM, Smith MN, Faustman EM (2017) Developing the regulatory utility of the exposome: mapping exposures for risk assessment through lifestage exposome snapshots (LEnS). Environ Health Perspect 125(8):085003. https://doi.org/10.1289/EHP1250

Shaughnessy DT, McAllister K, Worth L, Haugen AC, Meyer JN, Domann FE, Van Houten B, Mostoslavsky R, Bultman SJ, Baccarelli AA, Begley TJ, Sobol RW, Hirschey MD, Ideker T, Santos JH, Copeland WC, Tice RR, Balshaw DM, Tyson FL (2014) Mitochondria, energetics, epigenetics, and cellular responses to stress. Environ Health Perspect 122(12):1271–1278. https://doi.org/10.1289/ehp.1408418

Simpkin AJ, Suderman M, Gaunt TR, Lyttleton O, McArdle WL, Ring SM, Tilling K, Davey Smith G, Relton CL (2015) Longitudinal analysis of DNA methylation associated with birth weight and gestational age. Hum Mol Genet 24(13):3752–3763. https://doi.org/10.1093/hmg/ddv119

Sõber S, Reiman M, Kikas T, Rull K, Inno R, Vaas P, Teesalu P, Marti JM, Mattila P, Laan M (2015) Extensive shift in placental transcriptome profile in preeclampsia and placental origin of adverse pregnancy outcomes. Sci Rep 5:13336. https://doi.org/10.1038/srep13336

Steer CD, Bolton P, Golding J (2015) Preconception and prenatal environmental factors associated with communication impairments in 9 year old children using an exposome-wide approach. PLoS One 10(3):e0118701. https://doi.org/10.1371/journal.pone.0118701

Swartz MD, Cai Y, Chan W, Symanski E, Mitchell LE, Danysh HE, Langlois PH, Lupo PJ (2015) Air toxics and birth defects: a Bayesian hierarchical approach to evaluate multiple pollutants and spina bifida. Environ Health 14:16. https://doi.org/10.1186/1476-069x-14-16

Tea I, Gall GL, Küster A, Guignard N, Alexandre-Gouabau MC, Darmaun D, Robins RJ (2012) 1H-NMR-based metabolic profiling of maternal and umbilical cord blood indicates altered materno-foetal nutrient exchange in preterm infants. PLoS One 7:6–9. https://doi.org/10.1371/journal.pone.0029947

Toivonen KI, Oinonen KA, Duchene KM (2017) Preconception health behaviours: a scoping review. Prev Med 96:1–15. https://doi.org/10.1016/j.ypmed.2016.11.022

Vafeiadi M, Vrijheid M, Fthenou E, Chalkiadaki G, Rantakokko P, Kiviranta H, Kyrtopoulos SA, Chatzi L, Kogevinas M (2014) Persistent organic pollutants exposure during pregnancy, maternal gestational weight gain, and birth outcomes in the mother-child cohort in Crete, Greece (RHEA study). Environ Int 64:116–123. https://doi.org/10.1016/j.envint.2013.12.015

Valero De Bernabé J, Soriano T, Albaladejo R, Juarranz M, Calle ME, Martínez D, Domínguez-Rojas V (2004) Risk factors for low birth weight: a review. Eur J Obstet Gynecol Reprod Biol 116(1):3–15. https://doi.org/10.1016/j.ejogrb.2004.03.007

VanderWeele T (2015) Explanation in causal inference: methods for mediation and interaction, 1st edn. Oxford University Press, New York, NY

Vineis P, Chadeau-Hyam M, Gmuender H, Gulliver J, Herceg Z, Kleinjans J, Kogevinas M, Kyrtopoulos S, Nieuwenhuijsen M, Phillips DH, Probst-Hensch N, Scalbert A, Vermeulen R, Wild CP (2016) The exposome in practice: design of the EXPOsOMICS project. Int J Hyg Environ Health 220(2 Pt A):142–151. https://doi.org/10.1016/j.ijheh.2016.08.001

Volberg V, Yousefi P, Huen K, Harley K, Eskenazi B, Holland N (2017) CpG methylation across the adipogenic PPARgamma gene and its relationship with birthweight and child BMI at 9 years. BMC Med Genet 18(1):7. https://doi.org/10.1186/s12881-016-0365-4

Vrijheid M, Slama R, Robinson O, Chatzi L, Coen M, van den Hazel P, Thomsen C, Wright J, Athersuch TJ, Avellana N, Basagana X, Brochot C, Bucchini L, Bustamante M, Carracedo A, Casas M, Estivill X, Fairley L, van Gent D, Gonzalez JR, Granum B, Grazuleviciene R, Gutzkow KB, Julvez J, Keun HC, Kogevinas M, McEachan RR, Meltzer HM, Sabido E, Schwarze PE, Siroux V, Sunyer J, Want EJ, Zeman F, Nieuwenhuijsen MJ (2014) The human early-life exposome (HELIX): project rationale and design. Environ Health Perspect 122(6):535–544. https://doi.org/10.1289/ehp.1307204

Vrijheid M, Casas M, Gascon M, Valvi D, Nieuwenhuijsen M (2016) Environmental pollutants and child health-a review of recent concerns. Int J Hyg Environ Health 219(4–5):331–342. https://doi.org/10.1016/j.ijheh.2016.05.001

Wang F, Shi Z, Wang P, You W, Liang G (2013) Comparative proteome profile of human placenta from normal and preeclamptic pregnancies. PLoS One 8(10):e78025. https://doi.org/10.1371/journal.pone.0078025

Warren J, Fuentes M, Herring A, Langlois P (2012) Spatial-temporal modeling of the association between air pollution exposure and preterm birth: identifying critical windows of exposure. Biometrics 68(4):1157–1167. https://doi.org/10.1111/j.1541-0420.2012.01774.x

Wilcox AJ (2010) Fertility and pregnancy: an epidemiologic perspective. Oxford University Press, Oxford

Wilcox AJ, Weinberg CR, Basso O (2011) On the pitfalls of adjusting for gestational age at birth. Am J Epidemiol 174(9):1062–1068. https://doi.org/10.1093/aje/kwr230

Wild CP (2005) Complementing the genome with an "exposome": the outstanding challenge of environmental exposure measurement in molecular epidemiology. Cancer Epidemiol Biomark Prev 14(8):1847–1850. https://doi.org/10.1158/1055-9965.epi-05-0456

Wild CP (2012) The exposome: from concept to utility. Int J Epidemiol 41(1):24–32. https://doi.org/10.1093/ije/dyr236

Woodruff TJ, Zota AR, Schwartz JM (2011) Environmental chemicals in pregnant women in the United States: NHANES 2003–2004. Environ Health Perspect 119(6):878–885. https://doi.org/10.1289/ehp.1002727

Wright ML, Starkweather AR, York TP (2016) Mechanisms of the maternal exposome and implications for health outcomes. ANS Adv Nurs Sci 39(2):E17–E30. https://doi.org/10.1097/ANS.0000000000000110

Yoon M, Nong A, Clewell HJ, Taylor MD, Dorman DC, Andersen ME (2009) Evaluating placental transfer and tissue concentrations of manganese in the pregnant rat and fetuses after inhalation exposures with a PBPK model. Toxicol Sci 112(1):44–58. https://doi.org/10.1093/toxsci/kfp198

Yorifuji T, Debes F, Weihe P, Grandjean P (2011) Prenatal exposure to lead and cognitive deficit in 7- and 14-year-old children in the presence of concomitant exposure to similar molar concentration of methylmercury. Neurotoxicol Teratol 33(2):205–211. https://doi.org/10.1016/j.ntt.2010.09.004

Zhang Y, Wang Q, Wang H, Duan E (2017) Uterine fluid in pregnancy: a biological and clinical outlook. Trends Mol Med 23(7):604–614. https://doi.org/10.1016/j.molmed.2017.05.002

Part II
Measuring the Internal Exposome

Chapter 5
Epigenetics and the Exposome

Michelle Plusquin, Nelly D. Saenen, and Tim S. Nawrot

Abstract Epigenetic regulation is hereditable but can be influenced by environmental stimuli, *in utero* circumstances, and aging. Different layers of epigenetic remodeling including DNA methylation, modifications of histone tails, and noncoding RNAs control the spatial and temporal transcriptomic activity. In addition, the epigenome is involved in sustaining chromosome stability. Genomic DNA isolated from blood cells or other pertinent tissues is being expansively exploited for the discovery of biomarkers of effect and exposure. Technology to measure epigenetic marks on a genomic scale complemented with novel tools for data-analysis have recently been developed and continue to be enhanced. Here, we describe common techniques that are applied for untargeted approaches; and to measure regional modifications and gene-specific aberrations. Alterations in epigenetic marks have been associated with various exposures such as tobacco smoke, air pollution, and metal exposures in population-based studies. On the other hand, deviant DNA methylation is a major epigenetic mechanism of epigenetic silencing in a wide range of human diseases including cancers. Epigenetic modifications might play a prominent role in explaining biological mediation of exposures and their effect on health. This is of particular significance in early life exposures where epigenetic alterations can explain how diseases linked to *in utero* or childhood conditions occur later in life. We discuss relevant examples of how epigenetic remodeling by environmental stimuli affects several health outcomes in adults and in early life.

Keywords Epigenetics · Genomic DNA · Biomarkers of exposure

M. Plusquin (✉) · N. D. Saenen · T. S. Nawrot
Centre for Environmental Sciences, Hasselt University, Hasselt, Belgium
e-mail: michelle.plusquin@uhasselt.be; nelly.saenen@uhasselt.be; tim.nawrot@uhasselt.be

S. Dagnino, A. Macherone (eds.), *Unraveling the Exposome*,
https://doi.org/10.1007/978-3-319-89321-1_5

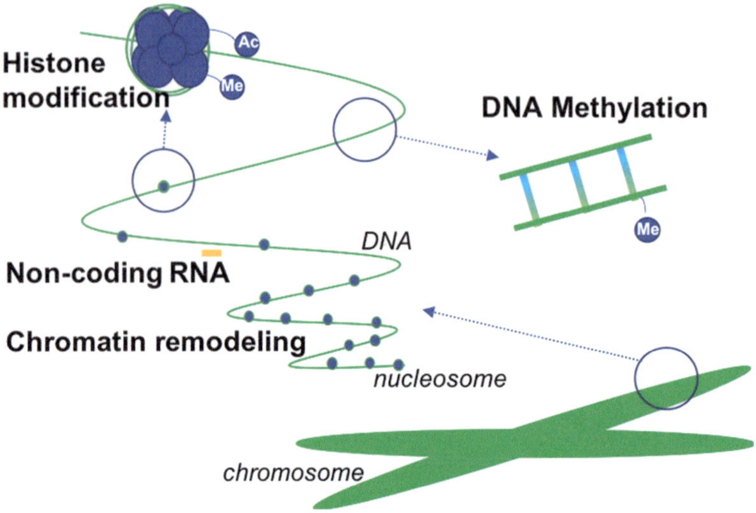

Fig. 5.1 The four interacting systems of epigenetics. *Me* methylation, *Ac* acetylation

Introduction

Epigenetics is the science of nongenetic mitotically heritable variation in the gene potential (Waterland and Michels 2007; Rakyan et al. 2011). Although we are only just beginning to understand the multitude of effects of the environment, lifestyle, and life course on epigenetics, growing evidence suggests that the exposome paradigm may provide a better understanding of the etiology of diseases and the role of epigenetically regulated gene-expression herein. In this chapter, we discuss different levels of human epigenetic control in relation to the exposome and present current challenges to study epigenetic alterations in this field.

Epigenetic Regulation

Four interacting systems ensure epigenetic control: DNA methylation, histone modification, noncoding RNAs, and chromatin remodeling (Fig. 5.1). These communicating mechanisms ensure the somatically heritable states of gene expression (Jones and Liang 2012).

DNA Methylation

The most extensively studied epigenetic mechanism is DNA methylation. It predominantly occurs as a covalent addition of methyl groups (CH_3) to cytosine that precedes a guanine in the DNA sequence (the CpG nucleotide) and determines the process by which genes are converted to mRNA (Yoder et al. 1997). Less is known about the function of the relatively rarer cytosine modifications including 5-hydroxymethylcytosine (Pang et al. 2016). The pattern of methylation is strongly associated to gene expression. While methylated gene promotors usually lead to transcriptional silencing, methylated gene bodies may lead to transcriptional activation (Jones 2012). Methylation patterns can be clustered in differentially methylated regions (DMRs), contiguous genomic regions that respond to regulatory signals. DMRs are involved in imprinting, aging, and diseases such as cancer (Rakyan et al. 2011). Cancer studies have shown critical links between cancer development and an overall genomic reduction of methylation. Global DNA hypomethylation induces genomic instability, for instance, through chromatin structure modeling (You and Jones 2012), loss of imprinting, and increased activation of oncogenes. DNA methylation patterns are largely erased after conception, first during gametogenesis, and again in early embryogenesis, with demethylation and remethylation occurring each time (Messerschmidt et al. 2014). Also during aging DNA methylation undergoes a profound remodeling: (1) global hypomethylation of the genome, (2) hypermethylation at specific loci, and (3) an increase of interindividual variation and in stochastic changes of DNA methylation (Gensous et al. 2017). In this regard, several epigenetic clocks have been proposed to estimate the biological age of a tissue, cell type, or organ (Horvath 2013; Hannum et al. 2013).

Besides the nuclear genome, the mitochondrial genome can also undergo epigenetic modifications. DNA methylation in specific regions of the mitochondrial genome has been shown to substantially mediate cellular processes (Janssen et al. 2015; Ghosh et al. 2014).

A major part of the exposome includes detrimental effects of environmental exposure on health. Here, we describe influences of arsenic, air pollution, and smoking on DNA methylation. Arsenic is a potent environmental pollutant and a class I human carcinogen (Howe and Gamble 2016). Both DNA methylation and arsenic metabolism require *S*-adenosylmethionine (SAM) as the methyl donor. Competitive demand between arsenic metabolism and DNA methylation for SAM could affect DNA methylation status throughout the genome (Lee et al. 2009). *In vitro* studies showed a decrease of methylation after arsenic exposure (Intarasunanont et al. 2012; Zhao et al. 1997; Reichard et al. 2007). Several population-based studies have identified an inverse association between arsenic exposure and global methylation status (Bandyopadhyay et al. 2016; Broberg et al. 2014; Hossain et al. 2017; Niedzwiecki et al. 2013; Pilsner et al. 2009; Tajuddin et al. 2013; Tellez-Plaza et al. 2014; Wilhelm et al. 2010), albeit not all (Hossain et al. 2012; Kile et al. 2012; Lambrou et al. 2012; Majumdar et al. 2010; Niedzwiecki et al. 2015; Pilsner et al. 2007, 2012; Intarasunanont et al. 2012). The findings from these observational studies are not consistent possibly because they include a number of differences in exposure measures and doses across studies.

A few recent epigenome-wide studies using the Illumina Human Methylation 450K array identified CpG loci significantly associated with increased arsenic concentrations (Argos et al. 2015), several studies investigated newborns (Kile et al. 2014; Koestler et al. 2013; Rojas et al. 2015; Broberg et al. 2014; Cardenas et al. 2015) and two studies did not identify significant CpG sites (Liu et al. 2014; Seow et al. 2014). The study by Cardenas et al. (2015) included different surrogate tissues and identified differential CPG methylation in placenta and umbilical artery tissue, but not in human umbilical vein endothelial cells (HUVEC). Smeester et al. (2011) examined epigenome-wide promoter DNA methylation in peripheral blood leukocytes among 16 arsenic-exposed women from Mexico in relation to skin lesion status, and constructed an arsenic-induced genetic network, identifying a complex of 17 tumor suppressors known to be silenced in human cancers (Smeester et al. 2011). Bailey et al. (2013) observed in this study population associations with arsenic metabolism reflected by the concentration of several arsenic metabolites in urine and differential promoter DNA methylation (Bailey et al. 2013). Targeted studies have evaluated arsenic in relation to gene-specific DNA methylation of which have found associations on *p16* and *p53* promoter methylation (Chanda et al. 2006; Chen et al. 2007; Engstrom et al. 2013; Hossain et al. 2012; Intarasunanont et al. 2012). In conclusion, growing data have linked DNA methylation alterations to arsenic exposure, but a large-scale study including multiple cohorts of epigenome-wide analyses is still outstanding.

Ambient air pollution includes gaseous components such as nitrogen oxides, benzene, and sulfur dioxide as well as particulate matter (PM), which consists of acids, organic chemicals, metals, and soil or dust particles of varying aerodynamic diameters. Because of their small size, these particles can be inhaled deeply into the lungs and deposited in the alveoli, and the smallest particles (<100 nm) can directly reach the blood stream (Brook et al. 2010). Recently, Vaiserman (2015) summarized clinical and epidemiological evidence in support of epigenetic factors that may mediate the link between early-life exposures and long-term health outcomes (Vaiserman 2015). Epigenetic studies report an inverse association between global methylation in leukocytes and long-term exposure to ambient air pollution (De Prins et al. 2013; Janssen et al. 2013; Sanchez-Guerra et al. 2015; Tao et al. 2014; Baccarelli et al. 2009; Madrigano et al. 2011), especially affecting 5-hydroxymethylcytosine methylation (Sanchez-Guerra et al. 2015). In newborns of the ENVIR*ON*AGE (ENVIRonmental influence *ON* AGEing in early life) birth cohort (Janssen et al. 2017), an increase of 10 μg/m^3 of particles smaller than 10 μM ($PM_{2.5}$) exposure in the third trimester of the pregnancy was associated with a significant relative decrease of 17% in global placental DNA methylation (Janssen et al. 2015).

Recent studies using Illumina Infinium Human Methylation 450 K technology (1) identified CpG sites whose methylation levels were affected by short- and mid-term exposure to particulate matter and its components (Panni et al. 2016; Dai et al. 2017); (2) investigated the association between long-term ambient air pollution exposure and DNA methylation (candidate sites and global) in monocytes of adults (Chi et al. 2016); (3) identified differential offspring DNA methylation in mitochondria-related genes in association with NO_2 exposure during pregnancy (Gruzieva et al. 2016). Also, long-term exposure to air pollution has been associated with epigenetic aging measures (Ward-Caviness et al. 2016). Methylation levels at specific loci in genes including tissue factor F3, interferon gamma (*IFN-γ*), interleukin 6 (*IL-6*), toll-like receptor 2 (*TLR-2*), intercellular adhesion molecule 1 (*ICAM-1*), and ten-eleven translocation (*TET1*) have been reported to be affected by exposure to air pollution (Bind et al. 2014, 2015; Lepeule et al. 2014; Somineni et al. 2015). A study by Hou et al. (2011) reported differential DNA methylation levels of tumor suppressor genes of *APC*, *p16, p53*, and *RASSF1A* in healthy steel workers (Hou et al. 2011). Recent studies are deciphering whether DNA methylation mediates health outcomes, for example, (1) a study by Wang et al. (2016) investigated short-term exposure to $PM_{2.5}$, blood pressure, and angiotensin converting enzyme, showing that methylation of the ACE gene plays a role in elevating blood pressure (Wang et al. 2016); (2) Chen et al. (2016) showed via an intervention study that *sCD40L* methylation mediates its protein levels (Chen et al. 2016); (3) Peng et al. (2016) showed that *ICAM-1* promoter methylation in part mediated short- and medium-term $PM_{2.5}$ exposure and higher fasting blood glucose levels (Peng et al. 2016). Mitochondrial DNA (mtDNA) methylation levels in blood of boiler makers were negatively associated with $PM_{2.5}$ exposure and modified the adverse relationships between $PM_{2.5}$ exposure and heart rate variability outcomes (Byun et al. 2016). MtDNA methylation substantially mediates the association between $PM_{2.5}$ exposure during gestation and placental mtDNA content which could reflect signs of mitophagy and mitochondrial senescence (Janssen et al. 2015). A recent exciting result in preventative measures was demonstrated via a human intervention trial, in which it was shown that acute ambient $PM_{2.5}$-induced DNA methylation changes can be reduced by B-vitamin supplementation (Zhong et al. 2017). Further exploration of air pollution exposure effects on gene expression regulation by DNA methylation is of paramount importance to unravel these potentially important relationships.

Cigarette smoke is a well-established toxicant associated with many health effects (Richmond and Joubert 2017). Even decades after stopping, former smokers are at long-term risk of developing diseases including some cancers, chronic obstructive pulmonary disease, and stroke. Smoking exposure has a profound effect on DNA methylation and CpG sites have been associated with more than 7000 genes in blood of adults (Joehanes et al. 2016), and over 6000 in cord blood (Joubert et al. 2016). For people who stopped smoking, the majority of DNA methylation sites returned to levels seen in never smokers within 5 years of quitting smoking. However, smoking does leave its "footprint" on the human genome in the form of DNA methylation (Guida et al. 2015; Joehanes et al. 2016). Methylation could therefore serve as sensitive and stable biomarkers of lifetime exposure to tobacco smoke. DNA methylation signatures should be further elaborated in the context of the exposome where individuals exposed to environmental pollutants, stress, or other lifestyle factors might provide biosensors or molecular archives of one's past.

Histone Modifications and Chromatin Recombination

In humans, protection and packaging of genetic material is largely realized by nucleosomes which are organized via eight histones. Modifications on tails of histones such as methylation or acetylation of particular lysine residues (Shahbazian and Grunstein 2007) are associated with relaxation of euchromatin, granting access to transcription factors and initiation of gene transcription. The silencing of transcription may be reversible or irreversible, depending on further changes.

Huang et al. state that the nucleosome and linker region of DNA coil into different hierarchical levels of organization, depending upon the histone variants of the region and the modifications to the DNA and the histone components of the nucleosomes. Together, these components of chromatin function as integral components of the genome or more precisely, when contemplated relative to their coordinated roles in gene expression (Adkins et al. 2013; Anderson et al. 2001; Barth and Imhof 2010; Huang et al. 2015).

There are several distinct classes of posttranslational histone modifications. The most common are histone acetylation, histone phosphorylation, and histone methylation (lysine methylation, histone demethylases, arginine methylation). Other modifications are deimination, *β-N*-acetylglucosamine, ADP ribosylation, ubiquitylation and sumoylation, histone tail clipping, and histone proline isomerization (Bannister and Kouzarides 2011). With regard to environmental exposures, most studies identified global levels of histone modifications using enzyme-linked immunoassays. For example, a repeated-measures study in 60 truck drivers and 60 office workers investigated the association of traffic-derived PM exposure with four types of global histone H3 modifications: H3 lysine 9 acetylation (H3K9ac), H3 lysine 9 tri-methylation (H3K9me3), H3 lysine 27 tri-methylation (H3K27me3), and H3 lysine 36 tri-methylation (H3K36me3). Results suggested a possible role of the global histone H3 modifications in the effects of traffic-derived PM exposures, particularly black carbon exposure (Zheng et al. 2017b). Another study in China examined global levels of H3K18ac, H3K9me2, and H3K36 in 138 arsenic-exposed and arsenicosis subjects. Results showed that altered global levels of all three histone modifications correlated with both urinary and hair-arsenic levels of the subjects. Furthermore, modifications of H3K18ac and H3K36me3 were also associated with urinary 8-OHdG, suggesting a critical role of histone modification in response to arsenic-induced oxidative damage (Ma et al. 2016).

Noncoding RNA

Noncoding RNAs (ncRNAs) are transcriptional regulators involved in numerous key cellular processes, such as metabolism, cell cycle control and differentiation (Kornfeld and Bruning 2014). Mechanistically, ncRNAs can influence transcription by recruiting or blocking factors to or from genomic loci as well as altering the epigenetic state of DNA. ncRNAs are transcripts that contain little or no open reading frame and are, in general, divided into 2 main categories: small or long ncRNAs (Hon et al. 2017). miRNAs are endogenous, single-stranded, short non-coding RNA sequences (~22 nucleotides) that regulate gene expression at the posttranscriptional level. Diverse disease states, such as cancer and heart failure, are associated with distinct miRNA signatures, suggesting that specific miRNA programs are activated in pathophysiological processes (Calin et al. 2005). Recent advances in molecular biology opened the opportunity for new approaches in population-based studies, in which exposures to a broad spectrum of environmental pollutants are evaluated in concert with biological systems. In this context, genome-wide miRNA analyses have already been used to identify altered plasma miRNA profiles of former smokers compared to nonsmokers. Levänen et al. (2013) observed distinct miRNA expression profiles in patients with asthma compared with controls after subway exposure. Another approach for miRNA identification is candidate-based microRNA expression. In this context, Bollati et al. showed in a cohort of steel plant workers a positive correlation between miR-21 and urinary 8-OHdG, a response that might reflect oxidative stress (Bollati et al. 2010).

In the previous decade, long ncRNAs have emerged from obscurity to being defined as a category of genetic elements. In cancer biology, the recent application of next-generation sequencing revealed a growing number of long ncRNAs whose expression is associated with different cancer types. In addition, the field is moving from annotation of long ncRNAs in various tumors toward understanding their importance in key cancer signaling networks (Huarte 2015; Bartonicek et al. 2016).

Only a few studies have investigated long ncRNAs with environmental pollutants. Gao et al. (2016) explored whether alterations of long ncRNA expression were correlated with PAH exposure. They found that expression of *HOTAIR*, *MALAT1*, and *TUG1* was positively correlated with external PAH exposure, but only *HOTAIR* and *MALAT1* were also significantly associated with internal PAH exposure levels (Gao et al. 2016). Although long ncRNAs are not extensively studied in the field of environmental exposures, it is promising for future developments in the field of epigenetics and the exposome.

The Life Course and Epigenetics

The development of diseases can find its origin in every stage of human life. Because of their dynamic and heritable properties, epigenetic mechanisms have emerged as a promising biological explanation linking events and exposures across life to long-term health. Recently, several studies have reported epigenetic signatures linked to former lifestyle characteristics.

With regard to smoking exposure, Guida et al. (2015) conducted a series of epigenome-wide association studies in blood samples from 745 women from two European populations to capture the dynamics of smoking-induced epigenetic changes with time since smoking cessation. They identified CpG sites that remained differentially methylated even more than 35 years after smoking cessation. These findings highlight the persistence of epigenetic changes due to smoking, which can potentially be detected decades after cessation (Guida et al. 2015).

Low socioeconomic status (SES) across the life course is suggested to be associated with a dampened immune response. Stringhini et al. (2015) found that indicators of socioeconomic status over the life course were inversely associated with DNA methylation of genes involved in inflammation, such as *NFATC1*, *IL1A*, *GPR132*, and genes belonging to the *MAPK* family. In addition, *CXCL2* and *PTGS2* were consistently more methylated among low SES individuals (Stringhini et al. 2015). Needham et al. (2015) observed that low childhood SES was associated with increased DNA methylation in three stress-related genes (*AVP, FKBP5, OXTR*) and two inflammation-related genes (*CD1D, CCL1*), while low adult SES was associated with one stress-related gene (AVP) and five inflammation-related genes (*CD1D, F8, KLRG1, NLRP12, TLR3*) (Needham et al. 2015). Both studies support the hypothesis that the social environment leaves an epigenetic signature but underscore the need to investigate how these epigenetic changes may contribute to the social patterning of disease.

It has been suggested that the etiology of diseases in adulthood may have a fetal origin and may be attributed to the effects of adverse environmental exposures *in utero*. This causality concept is known as the Barker or Developmental Origins of Health and Disease (DOHaD) hypothesis. Professor David Barker was the first to recognize this potential link when he became concerned about the association between malnutrition during pregnancy and the development of coronary heart disease in adult life (Barker 1995). Since then, many implications of this hypothesis have been reported. Early embryonic development is of special interest in this respect, because this is a crucial period for establishing and maintaining epigenetic marks (Reik et al. 2001). Here, we discuss three examples that support the DOHad theory.

The Dutch Hunger Winter study was the first study that supported the hypothesis that early-life environmental conditions can cause epigenetic changes in humans that persist throughout life. Here, individuals prenatally exposed to famine during the Dutch Hunger Winter in 1944–1945 had, six decades later, less DNA methylation of the imprinted insulin-like growth factor 2 (*IGF2*) gene and increases in the methylation of *IL-10*, *leptin*, *ATP-binding cassette A1*, and *guanine nucleotide-binding protein* genes compared with their unexposed, same-sex siblings (Heijmans et al. 2008). These results suggest that maternal nutritional constraint induces long-term epigenetic changes in key metabolic regulatory genes and point to a mechanism through which fetal famine exposure has an effect on adult metabolism.

In the Cincinnati Childhood Allergy and Air Pollution Study (CCAAPS) birth cohort, saliva *FOXP3* methylation was found to be associated with diesel exhaust exposure during the first year of life and persistent wheezing and asthma diagnosis at age 7 (Brunst et al. 2013), suggesting that epigenetics may contribute to the impact of early traffic-related air pollution exposure on childhood asthma risk. Within the ENVIR*ON*AGE birth cohort, maternal pre-pregnancy BMI was an important determinant of newborns telomere length (Martens et al. 2016). The authors suggest potential epigenetic pathways of which maternal BMI might influence molecular longevity of the next generation.

Conradt et al. (2013) showed that infants whose mothers reported depression during pregnancy had higher methylation of placental *NR3C1*, poorer self-regulation, more hypotonia, and more lethargy than infants whose mothers did not report depression. Furthermore, infants whose mothers reported anxiety during pregnancy showed higher methylation of placental *11β-HSD-2* and were more hypotonic compared with infants of mothers who did not report anxiety during pregnancy (Conradt et al. 2013). An *intra-uterine* environment that could be characterized by increased exposure to maternal cortisol may lead to alterations of neurodevelopmental outcomes.

Challenges and Opportunities

To further explore the exposome paradigm in epigenetics, it is essential to incorporate identified biomarkers to biological responses and determine dose–response relationships, in order to identify biological pathways underlying diseases. To further study the role of epigenetics in lifetime exposure and disease, important methodological and conceptual challenges remain; however, opportunities may arise for preventative strategies.

Assays to Measure Epigenetics

The technology to measure epigenetic marks on a genomic scale has recently been developed and continues to be enhanced. Until now the vast majority of exposure-related studies and epigenetics have focused on DNA methylation because of (1) interpretability of DNA methylation (it is located on or in between genes); (2) the stability of the methylation marks; (3) the smaller amounts of DNA that are needed for laboratory analyses; and (4) the availability of high-throughput techniques. Sodium bisulfite conversion of genomic DNA to differentiate and detect un-methylated versus methylated cytosines is the gold standard for DNA methylation analysis. This procedure can then be followed by PCR (polymerase chain reaction) amplification, sequencing, microarray, etc. to reveal the methylation status of every cytosine in gene-specific amplification or whole genome amplification.

A wide variety of techniques is available to test the methylation status of selected genes or regions of interest: beads array, PCR and (pyro)sequencing, methylation-specific PCR, PCR with high-resolution melting, and COLD-PCR (Kurdyukov and Bullock 2016). To measure region-specific DNA methylation regions, the mass spectrometry-based EpiTYPER method can be used.

Methylation microarray after bisulfide treatment is currently the most used genome-wide application. Available examples of such arrays include the Infinium human methylation EPIC and its preceding 450K Kit or Agilent's Human DNA Methylation microarrays. Bisulfide-treated DNA is amplified and put into contact with probes recognizing CpG sites. This method is well liked because of its price and "ease of use." Drawbacks are that the targets are limited to the companies' selection and it does not identify other DNA modifications such as 5-hydroxymethylcytosine methylation. Bisulfite sequencing allows a comprehensive study across the genome (Li and Tollefsbol 2011). DNA fragments are bisulfite-converted, amplified, and sequenced. Drawbacks are that this technique is labor-intensive and demands a difficult postsequencing alignment (Kurdyukov and Bullock 2016). Genome-wide assays will play an important role in the to further evaluate exposome effects on the epigenome, efforts such as platform development, reduction of costs, and importantly development of biostatistical tools are still needed.

To study the global methylation state of the genome, techniques are designed to either measure the 5-methylcytosine content or estimate the methylation status of representative parts of the DNA such as repetitive sequences. HPLC-UV (high-performance liquid chromatography–ultraviolet) is considered as gold standard to measure the global methylation. Although they generally under- or overestimate the amount of global DNA, other techniques also assess global methylation and have high specificity, sensitivity, and minimal variation: LC-MS/MS (LC coupled with mass spectrometry), ELISA-based (Enzyme-Linked Immunosorbent Assay) methods, repetitive sequences (LINE-Alu), AFLP (Amplification Length Polymorphism) and RFLP (restriction length polymorphism) or LUMA (luminometric methylation assay) (Kurdyukov and Bullock 2016).

Histone studies either focus on global or specific modifications on the proteins using ELISA (Tauheed et al. 2017). A complication of histone studies is the large amount of samples required for the laboratory analyses. MiRNAs are analyzed via miRNA-wide tools (microarray, or RNA sequencing) or targeted approaches (quantitative PCR).

Cell-Type Heterogeneity

Cell populations are heterogeneous, composed of multiple cellular states, each exhibiting stochastic noise in its gene expression and epigenetics (Singer et al. 2014). In contrast with genetics where genotype can be consistently determined from a diploid cell, epigenetics are tissue, cell, and context specific. The stochastic nature inherent to cellular processes such as gene regulation gives rise to cell-to-cell variation, enhancing survival under adverse conditions and stress (Elowitz et al. 2002; Lehner and Kaneko 2011; Raser and O'Shea 2005). Human studies that extensively characterize participants often store blood samples, are limited in tissue accessibility, and often do not have access to samples with individual cell types. As cellular composition explains much of the observed variability in DNA methylation, therefore, when the outcome of interest correlates with cell composition, as for example, exposures that trigger inflammation, failure to account for cellular heterogeneity may result in false positive results (Jaffe and Irizarry 2014).

Ultimately, purification of specific cells of interest might be ideal, but in practice this is not always feasible. Algorithms to overcome this problem have been designed and can be broadly classified as reference-based (if they use reference DNA methylation profiles of representative cell types) (Houseman et al. 2012) or reference-free (if they do not require such reference profiles) (Houseman et al. 2014; Bakulski et al. 2016). A frequently applied reference-based method in blood has been proposed by Houseman et al. (2012). They have developed a method to estimate relative proportions of 6 major blood cell types from whole blood 450K data (Houseman et al. 2012). These cell proportion estimates can then be used as adjustment factors in epigenetic association analyses. Recently, similar methods have been developed for cord blood (Bakulski et al. 2016; Gervin et al. 2016; Zheng et al. 2017a). Expanding these efforts to multiple tissues will improve the study of exposome-related effects on DNA methylation.

Underexplored Tissues

Tissue/cell type-specific DNA methylation profiles can provide a novel viewpoint for understanding normal and aberrant development, in terms of both differentiation and reproduction. Tissues that are noninvasive might serve as an excellent source for human samples and are currently underexplored. Recently exposure-related epigenetic alterations have been found in placenta tissue (Saenen et al. 2017; Green et al. 2016), saliva (Langie et al. 2016; Vriens et al. 2016), and blood spots (Breton et al. 2016).

Transgenerational Epigenetics

Evidence of transgenerational effects is mostly found in experimental studies such as in agouti viable yellow (A^{vy}) mice (Morgan et al. 1999), dietary changes before conception (Carone et al. 2010), and exposure to environmental toxins such as endocrine disruptors (Anway et al. 2005). The presumed responsible alleles that are variably expressed in genetically identical individuals due to epigenetic modifications established during early development and that are thought to be particularly vulnerable to environmental influences are metastable alleles. Also in human populations, transgenerational effects have been reported, for example, the study of Yehuda et al. (2016) showed that psychophysiological trauma can have intergenerational epigenetic effects by analyzing the DNA methylation status of *FKBP5* in the grandchildren of holocaust survivors. *FKBP5* methylation was associated with wake-up cortisol levels, indicating functional relevance of the methylation measures (Yehuda et al. 2016). Literature on maternal exposure and epigenetic changes is growing but also changes due to environmental exposures, such as diet, can occur in the male germ line and can be further passed on to the offspring (Soubry et al. 2014). A study by Pauwels et al. (2017) found an association between paternal methyl-group donor intake and offspring global and *IGF2* methylation, and prenatal growth was found.

A better understanding of the role of transgenerational epigenetics in the etiology of disorders may ultimately have implications for public health recommendations; a healthy environment and lifestyle for future mothers and fathers may be more important than realized before.

References

Adkins NL, Niu H, Sung P, Peterson CL (2013) Nucleosome dynamics regulates DNA processing. Nat Struct Mol Biol 20(7):836–842. https://doi.org/10.1038/nsmb.2585

Anderson JD, Lowary PT, Widom J (2001) Effects of histone acetylation on the equilibrium accessibility of nucleosomal DNA target sites. J Mol Biol 307(4):977–985. https://doi.org/10.1006/jmbi.2001.4528

Anway MD, Cupp AS, Uzumcu M, Skinner MK (2005) Epigenetic transgenerational actions of endocrine disruptors and male fertility. Science 308(5727):1466–1469. https://doi.org/10.1126/science.1108190

Argos M, Chen L, Jasmine F, Tong L, Pierce BL, Roy S, Paul-Brutus R, Gamble MV, Harper KN, Parvez F, Rahman M, Rakibuz-Zaman M, Slavkovich V, Baron JA, Graziano JH, Kibriya MG, Ahsan H (2015) Gene-specific differential DNA methylation and chronic arsenic exposure in an epigenome-wide association study of adults in Bangladesh. Environ Health Perspect 123 (1):64–71. https://doi.org/10.1289/ehp.1307884

Baccarelli A, Wright RO, Bollati V, Tarantini L, Litonjua AA, Suh HH, Zanobetti A, Sparrow D, Vokonas PS, Schwartz J (2009) Rapid DNA methylation changes after exposure to traffic particles. Am J Respir Crit Care Med 179(7):572–578. https://doi.org/10.1164/rccm.200807-1097OC

Bailey KA, Wu MC, Ward WO, Smeester L, Rager JE, Garcia-Vargas G, Del Razo LM, Drobna Z, Styblo M, Fry RC (2013) Arsenic and the epigenome: interindividual differences in arsenic metabolism related to distinct patterns of DNA methylation. J Biochem Mol Toxicol 27 (2):106–115. https://doi.org/10.1002/jbt.21462

Bakulski KM, Feinberg JI, Andrews SV, Yang J, Brown S, LM S, Witter F, Walston J, Feinberg AP, Fallin MD (2016) DNA methylation of cord blood cell types: applications for mixed cell birth studies. Epigenetics 11(5):354–362. https://doi.org/10.1080/15592294.2016.1161875

Bandyopadhyay AK, Paul S, Adak S, Giri AK (2016) Reduced LINE-1 methylation is associated with arsenic-induced genotoxic stress in children. Biometals 29(4):731–741. https://doi.org/10.1007/s10534-016-9950-4

Bannister AJ, Kouzarides T (2011) Regulation of chromatin by histone modifications. Cell Res 21 (3):381–395. https://doi.org/10.1038/cr.2011.22

Barker DJ (1995) Fetal origins of coronary heart disease. BMJ 311(6998):171–174

Barth TK, Imhof A (2010) Fast signals and slow marks: the dynamics of histone modifications. Trends Biochem Sci 35(11):618–626. https://doi.org/10.1016/j.tibs.2010.05.006

Bartonicek N, Maag JL, Dinger ME (2016) Long noncoding RNAs in cancer: mechanisms of action and technological advancements. Mol Cancer 15(1):43. https://doi.org/10.1186/s12943-016-0530-6

Bind MA, Lepeule J, Zanobetti A, Gasparrini A, Baccarelli A, Coull BA, Tarantini L, Vokonas PS, Koutrakis P, Schwartz J (2014) Air pollution and gene-specific methylation in the Normative Aging Study: association, effect modification, and mediation analysis. Epigenetics 9 (3):448–458. https://doi.org/10.4161/epi.27584

Bind MA, Coull BA, Peters A, Baccarelli AA, Tarantini L, Cantone L, Vokonas PS, Koutrakis P, Schwartz JD (2015) Beyond the mean: quantile regression to explore the association of air pollution with gene-specific methylation in the normative aging study. Environ Health Perspect 123(8):759–765. https://doi.org/10.1289/ehp.1307824

Bollati V, Marinelli B, Apostoli P, Bonzini M, Nordio F, Hoxha M, Pegoraro V, Motta V, Tarantini L, Cantone L, Schwartz J, Bertazzi PA, Baccarelli A (2010) Exposure to metal-rich particulate matter modifies the expression of candidate microRNAs in peripheral blood leukocytes. Environ Health Perspect 118(6):763–768. https://doi.org/10.1289/ehp.0901300

Breton CV, Yao J, Millstein J, Gao L, Siegmund KD, Mack W, Whitfield-Maxwell L, Lurmann F, Hodis H, Avol E, Gilliland FD (2016) Prenatal air pollution exposures, DNA methyl transferase genotypes, and associations with newborn LINE1 and alu methylation and childhood blood pressure and carotid intima-media thickness in the Children's Health Study. Environ Health Perspect 124(12):1905–1912. https://doi.org/10.1289/EHP181

Broberg K, Ahmed S, Engstrom K, Hossain MB, Jurkovic Mlakar S, Bottai M, Grander M, Raqib R, Vahter M (2014) Arsenic exposure in early pregnancy alters genome-wide DNA methylation in cord blood, particularly in boys. J Dev Orig Health Dis 5(4):288–298. https://doi.org/10.1017/S2040174414000221

Brook RD, Rajagopalan S, Pope CA 3rd, Brook JR, Bhatnagar A, Diez-Roux AV, Holguin F, Hong Y, Luepker RV, Mittleman MA, Peters A, Siscovick D, Smith SC Jr, Whitsel L, Kaufman JD, American Heart Association Council on Epidemiology and Prevention, Council on the Kidney in Cardiovascular Disease, and Council on Nutrition, Physical Activity and Metabolism (2010) Particulate matter air pollution and cardiovascular disease: an update to the scientific statement from the American Heart Association. Circulation 121(21):2331–2378. https://doi.org/10.1161/CIR.0b013e3181dbece1

Brunst KJ, Leung YK, Ryan PH, Khurana Hershey GK, Levin L, Ji H, Lemasters GK, Ho SM (2013) Forkhead box protein 3 (FOXP3) hypermethylation is associated with diesel exhaust exposure and risk for childhood asthma. J Allergy Clin Immunol 131(2):592–594.e3. https://doi.org/10.1016/j.jaci.2012.10.042

Byun HM, Colicino E, Trevisi L, Fan T, Christiani DC, Baccarelli AA (2016) Effects of air pollution and blood mitochondrial DNA methylation on markers of heart rate variability. J Am Heart Assoc 5(4):pii: e003218. https://doi.org/10.1161/JAHA.116.003218

Calin GA, Ferracin M, Cimmino A, Di Leva G, Shimizu M, Wojcik SE, Iorio MV, Visone R, Sever NI, Fabbri M, Iuliano R, Palumbo T, Pichiorri F, Roldo C, Garzon R, Sevignani C, Rassenti L, Alder H, Volinia S, Liu CG, Kipps TJ, Negrini M, Croce CM (2005) A MicroRNA signature associated with prognosis and progression in chronic lymphocytic leukemia. N Engl J Med 353 (17):1793–1801. https://doi.org/10.1056/NEJMoa050995

Cardenas A, Houseman EA, Baccarelli AA, Quamruzzaman Q, Rahman M, Mostofa G, Wright RO, Christiani DC, Kile ML (2015) In utero arsenic exposure and epigenome-wide associations in placenta, umbilical artery, and human umbilical vein endothelial cells. Epigenetics 10 (11):1054–1063. https://doi.org/10.1080/15592294.2015.1105424

Carone BR, Fauquier L, Habib N, Shea JM, Hart CE, Li R, Bock C, Li C, Gu H, Zamore PD, Meissner A, Weng Z, Hofmann HA, Friedman N, Rando OJ (2010) Paternally induced transgenerational environmental reprogramming of metabolic gene expression in mammals. Cell 143(7):1084–1096. https://doi.org/10.1016/j.cell.2010.12.008

Chanda S, Dasgupta UB, Guhamazumder D, Gupta M, Chaudhuri U, Lahiri S, Das S, Ghosh N, Chatterjee D (2006) DNA hypermethylation of promoter of gene p53 and p16 in arsenic-exposed people with and without malignancy. Toxicol Sci 89(2):431–437. https://doi.org/10.1093/toxsci/kfj030

Chen R, Meng X, Zhao A, Wang C, Yang C, Li H, Cai J, Zhao Z, Kan H (2016) DNA hypomethylation and its mediation in the effects of fine particulate air pollution on cardiovascular biomarkers: a randomized crossover trial. Environ Int 94:614–619. https://doi.org/10.1016/j.envint.2016.06.026

Chen WT, Hung WC, Kang WY, Huang YC, Chai CY (2007) Urothelial carcinomas arising in arsenic-contaminated areas are associated with hypermethylation of the gene promoter of the death-associated protein kinase. Histopathology 51(6):785–792. https://doi.org/10.1111/j.1365-2559.2007.02871.x

Chi GC, Liu Y, MacDonald JW, Barr RG, Donohue KM, Hensley MD, Hou L, McCall CE, Reynolds LM, Siscovick DS, Kaufman JD (2016) Long-term outdoor air pollution and DNA methylation in circulating monocytes: results from the Multi-Ethnic Study of Atherosclerosis (MESA). Environ Health 15(1):119. https://doi.org/10.1186/s12940-016-0202-4

Conradt E, Lester BM, Appleton AA, Armstrong DA, Marsit CJ (2013) The roles of DNA methylation of NR3C1 and 11beta-HSD2 and exposure to maternal mood disorder in utero on newborn neurobehavior. Epigenetics 8(12):1321–1329. https://doi.org/10.4161/epi.26634

Dai L, Mehta A, Mordukhovich I, Just AC, Shen J, Hou L, Koutrakis P, Sparrow D, Vokonas PS, Baccarelli AA, Schwartz JD (2017) Differential DNA methylation and PM2.5 species in a 450K epigenome-wide association study. Epigenetics 12(2):139–148. https://doi.org/10.1080/15592294.2016.1271853

De Prins S, Koppen G, Jacobs G, Dons E, Van de Mieroop E, Nelen V, Fierens F, Int Panis L, De Boever P, Cox B, Nawrot TS, Schoeters G (2013) Influence of ambient air pollution on global DNA methylation in healthy adults: a seasonal follow-up. Environ Int 59:418–424. https://doi.org/10.1016/j.envint.2013.07.007

Elowitz MB, Levine AJ, Siggia ED, Swain PS (2002) Stochastic gene expression in a single cell. Science 297(5584):1183–1186. https://doi.org/10.1126/science.1070919

Engstrom KS, Hossain MB, Lauss M, Ahmed S, Raqib R, Vahter M, Broberg K (2013) Efficient arsenic metabolism—the AS3MT haplotype is associated with DNA methylation and expression of multiple genes around AS3MT. PLoS One 8(1):e53732. https://doi.org/10.1371/journal.pone.0053732

Gao C, He Z, Li X, Bai Q, Zhang Z, Zhang X, Wang S, Xiao X, Wang F, Yan Y, Li D, Chen L, Zeng X, Xiao Y, Dong G, Zheng Y, Wang Q, Chen W (2016) Specific long non-coding RNAs response to occupational PAHs exposure in coke oven workers. Toxicol Rep 3:160–166

Gensous N, Bacalini MG, Pirazzini C, Marasco E, Giuliani C, Ravaioli F, Mengozzi G, Bertarelli C, Palmas MG, Franceschi C, Garagnani P (2017) The epigenetic landscape of age-related diseases: the geroscience perspective. Biogerontology 18(4):549–559. https://doi.org/10.1007/s10522-017-9695-7

Gervin K, Page CM, Aass HC, Jansen MA, Fjeldstad HE, Andreassen BK, Duijts L, van Meurs JB, van Zelm MC, Jaddoe VW, Nordeng H, Knudsen GP, Magnus P, Nystad W, Staff AC, Felix JF, Lyle R (2016) Cell type specific DNA methylation in cord blood: a 450K-reference data set and cell count-based validation of estimated cell type composition. Epigenetics 11(9):690–698. https://doi.org/10.1080/15592294.2016.1214782

Ghosh S, Sengupta S, Scaria V (2014) Comparative analysis of human mitochondrial methylomes shows distinct patterns of epigenetic regulation in mitochondria. Mitochondrion 18:58–62. https://doi.org/10.1016/j.mito.2014.07.007

Green BB, Karagas MR, Punshon T, Jackson BP, Robbins DJ, Houseman EA, Marsit CJ (2016) Epigenome-wide assessment of DNA methylation in the placenta and arsenic exposure in the new hampshire birth cohort study (USA). Environ Health Perspect 124(8):1253–1260. https://doi.org/10.1289/ehp.1510437

Gruzieva O, Xu CJ, Breton CV, Annesi-Maesano I, Anto JM, Auffray C, Ballereau S, Bellander T, Bousquet J, Bustamante M, Charles MA, de Kluizenaar Y, den Dekker HT, Duijts L, Felix JF, Gehring U, Guxens M, Jaddoe VV, Jankipersadsing SA, Merid SK, Kere J, Kumar A, Lemonnier N, Lepeule J, Nystad W, Page CM, Panasevich S, Postma D, Slama R, Sunyer J, Soderhall C, Yao J, London SJ, Pershagen G, Koppelman GH, Melen E (2016) Epigenome-wide meta-analysis of methylation in children related to prenatal NO2 air pollution exposure. Environ Health Perspect 125(1):104–110. https://doi.org/10.1289/EHP36

Guida F, Sandanger TM, Castagne R, Campanella G, Polidoro S, Palli D, Krogh V, Tumino R, Sacerdote C, Panico S, Severi G, Kyrtopoulos SA, Georgiadis P, Vermeulen RC, Lund E,

Vineis P, Chadeau-Hyam M (2015) Dynamics of smoking-induced genome-wide methylation changes with time since smoking cessation. Hum Mol Genet 24(8):2349–2359. https://doi.org/10.1093/hmg/ddu751

Hannum G, Guinney J, Zhao L, Zhang L, Hughes G, Sadda S, Klotzle B, Bibikova M, Fan JB, Gao Y, Deconde R, Chen M, Rajapakse I, Friend S, Ideker T, Zhang K (2013) Genome-wide methylation profiles reveal quantitative views of human aging rates. Mol Cell 49(2):359–367. https://doi.org/10.1016/j.molcel.2012.10.016

Heijmans BT, Tobi EW, Stein AD, Putter H, Blauw GJ, Susser ES, Slagboom PE, Lumey LH (2008) Persistent epigenetic differences associated with prenatal exposure to famine in humans. Proc Natl Acad Sci U S A 105(44):17046–17049. https://doi.org/10.1073/pnas.0806560105

Hon CC, Ramilowski JA, Harshbarger J, Bertin N, Rackham OJ, Gough J, Denisenko E, Schmeier S, Poulsen TM, Severin J, Lizio M, Kawaji H, Kasukawa T, Itoh M, Burroughs AM, Noma S, Djebali S, Alam T, Medvedeva YA, Testa AC, Lipovich L, Yip CW, Abugessaisa I, Mendez M, Hasegawa A, Tang D, Lassmann T, Heutink P, Babina M, Wells CA, Kojima S, Nakamura Y, Suzuki H, Daub CO, de Hoon MJ, Arner E, Hayashizaki Y, Carninci P, Forrest AR (2017) An atlas of human long non-coding RNAs with accurate 5′ ends. Nature 543(7644):199–204. https://doi.org/10.1038/nature21374

Horvath S (2013) DNA methylation age of human tissues and cell types. Genome Biol 14(10): R115. https://doi.org/10.1186/gb-2013-14-10-r115

Hossain K, Suzuki T, Hasibuzzaman MM, Islam MS, Rahman A, Paul SK, Tanu T, Hossain S, Saud ZA, Rahman M, Nikkon F, Miyataka H, Himeno S, Nohara K (2017) Chronic exposure to arsenic, LINE-1 hypomethylation, and blood pressure: a cross-sectional study in Bangladesh. Environ Health 16(1):20. https://doi.org/10.1186/s12940-017-0231-7

Hossain MB, Vahter M, Concha G, Broberg K (2012) Environmental arsenic exposure and DNA methylation of the tumor suppressor gene p16 and the DNA repair gene MLH1: effect of arsenic metabolism and genotype. Metallomics 4(11):1167–1175. https://doi.org/10.1039/c2mt20120h

Hou L, Zhang X, Tarantini L, Nordio F, Bonzini M, Angelici L, Marinelli B, Rizzo G, Cantone L, Apostoli P, Bertazzi PA, Baccarelli A (2011) Ambient PM exposure and DNA methylation in tumor suppressor genes: a cross-sectional study. Part Fibre Toxicol 8:25. https://doi.org/10.1186/1743-8977-8-25

Houseman EA, Accomando WP, Koestler DC, Christensen BC, Marsit CJ, Nelson HH, Wiencke JK, Kelsey KT (2012) DNA methylation arrays as surrogate measures of cell mixture distribution. BMC Bioinformatics 13:86. https://doi.org/10.1186/1471-2105-13-86

Houseman EA, Molitor J, Marsit CJ (2014) Reference-free cell mixture adjustments in analysis of DNA methylation data. Bioinformatics 30(10):1431–1439. https://doi.org/10.1093/bioinformatics/btu029

Howe CG, Gamble MV (2016) Influence of arsenic on global levels of histone posttranslational modifications: a review of the literature and challenges in the field. Curr Environ Health Rep 3(3):225–237. https://doi.org/10.1007/s40572-016-0104-1

Huang S, Litt M, Blakey A (2015) Chap. 2: Epigenetic gene expression and regulation. In: Histone modifications—models and mechanisms. Elsevier, pp 21–42

Huarte M (2015) The emerging role of lncRNAs in cancer. Nat Med 21(11):1253–1261. https://doi.org/10.1038/nm.3981

Intarasunanont P, Navasumrit P, Waraprasit S, Chaisatra K, Suk WA, Mahidol C, Ruchirawat M (2012) Effects of arsenic exposure on DNA methylation in cord blood samples from newborn babies and in a human lymphoblast cell line. Environ Health 11:31. https://doi.org/10.1186/1476-069X-11-31

Jaffe AE, Irizarry RA (2014) Accounting for cellular heterogeneity is critical in epigenome-wide association studies. Genome Biol 15(2):R31. https://doi.org/10.1186/gb-2014-15-2-r31

Janssen BG, Godderis L, Pieters N, Poels K, Kicinski M, Cuypers A, Fierens F, Penders J, Plusquin M, Gyselaers W, Nawrot TS (2013) Placental DNA hypomethylation in association with particulate air pollution in early life. Part Fibre Toxicol 10:22. https://doi.org/10.1186/1743-8977-10-22

Janssen BG, Byun HM, Gyselaers W, Lefebvre W, Baccarelli AA, Nawrot TS (2015) Placental mitochondrial methylation and exposure to airborne particulate matter in the early life environment: an ENVIRONAGE birth cohort study. Epigenetics 10(6):536–544. https://doi.org/10.1080/15592294.2015.1048412

Janssen BG, Madlhoum N, Gyselaers W, Bijnens E, Clemente DB, Cox B, Hogervorst J, Luyten L, Martens DS, Peusens M, Plusquin M, Provost EB, Roels HA, Saenen ND, Tsamou M, Vriens A, Winckelmans E, Vrijens K, Nawrot TS (2017) Cohort profile: the ENVIRonmental influence ON early AGEing (ENVIRONAGE): a birth cohort study. Int J Epidemiol 46(5):1386–1387m. https://doi.org/10.1093/ije/dyw269

Joehanes R, Just AC, Marioni RE, Pilling LC, Reynolds LM, Mandaviya PR, Guan W, Xu T, Elks CE, Aslibekyan S, Moreno-Macias H, Smith JA, Brody JA, Dhingra R, Yousefi P, Pankow JS, Kunze S, Shah SH, McRae AF, Lohman K, Sha J, Absher DM, Ferrucci L, Zhao W, Demerath EW, Bressler J, Grove ML, Huan T, Liu C, Mendelson MM, Yao C, Kiel DP, Peters A, Wang-Sattler R, Visscher PM, Wray NR, Starr JM, Ding J, Rodriguez CJ, Wareham NJ, Irvin MR, Zhi D, Barrdahl M, Vineis P, Ambatipudi S, Uitterlinden AG, Hofman A, Schwartz J, Colicino E, Hou L, Vokonas PS, Hernandez DG, Singleton AB, Bandinelli S, Turner ST, Ware EB, Smith AK, Klengel T, Binder EB, Psaty BM, Taylor KD, Gharib SA, Swenson BR, Liang L, DeMeo DL, O'Connor GT, Herceg Z, Ressler KJ, Conneely KN, Sotoodehnia N, Kardia SL, Melzer D, Baccarelli AA, van Meurs JB, Romieu I, Arnett DK, Ong KK, Liu Y, Waldenberger M, Deary IJ, Fornage M, Levy D, London SJ (2016) Epigenetic signatures of cigarette smoking. Circ Cardiovasc Genet 9(5):436–447. https://doi.org/10.1161/CIRCGENETICS.116.001506

Jones PA (2012) Functions of DNA methylation: islands, start sites, gene bodies and beyond. Nat Rev Genet 13(7):484–492. https://doi.org/10.1038/nrg3230

Jones PA, Liang G (2012) The human epigenome. In: Michels KB (ed) Epigentic epidemiology, 1st edn. Springer, New York, pp 5–20. https://doi.org/10.1007/978-94-007-2495-2

Joubert BR, Felix JF, Yousefi P, Bakulski KM, Just AC, Breton C, Reese SE, Markunas CA, Richmond RC, Xu CJ, Kupers LK, Oh SS, Hoyo C, Gruzieva O, Soderhall C, Salas LA, Baiz N, Zhang H, Lepeule J, Ruiz C, Ligthart S, Wang T, Taylor JA, Duijts L, Sharp GC, Jankipersadsing SA, Nilsen RM, Vaez A, Fallin MD, Hu D, Litonjua AA, Fuemmeler BF, Huen K, Kere J, Kull I, Munthe-Kaas MC, Gehring U, Bustamante M, Saurel-Coubizolles MJ, Quraishi BM, Ren J, Tost J, Gonzalez JR, Peters MJ, Haberg SE, Xu Z, van Meurs JB, Gaunt TR, Kerkhof M, Corpeleijn E, Feinberg AP, Eng C, Baccarelli AA, Benjamin Neelon SE, Bradman A, Merid SK, Bergstrom A, Herceg Z, Hernandez-Vargas H, Brunekreef B, Pinart M, Heude B, Ewart S, Yao J, Lemonnier N, Franco OH, Wu MC, Hofman A, McArdle W, Van der Vlies P, Falahi F, Gillman MW, Barcellos LF, Kumar A, Wickman M, Guerra S, Charles MA, Holloway J, Auffray C, Tiemeier HW, Smith GD, Postma D, Hivert MF, Eskenazi B, Vrijheid M, Arshad H, Anto JM, Dehghan A, Karmaus W, Annesi-Maesano I, Sunyer J, Ghantous A, Pershagen G, Holland N, Murphy SK, DeMeo DL, Burchard EG, Ladd-Acosta C, Snieder H, Nystad W, Koppelman GH, Relton CL, Jaddoe VW, Wilcox A, Melen E, London SJ (2016) DNA methylation in newborns and maternal smoking in pregnancy: genome-wide consortium meta-analysis. Am J Hum Genet 98(4):680–696. https://doi.org/10.1016/j.ajhg.2016.02.019

Kile ML, Baccarelli A, Hoffman E, Tarantini L, Quamruzzaman Q, Rahman M, Mahiuddin G, Mostofa G, Hsueh YM, Wright RO, Christiani DC (2012) Prenatal arsenic exposure and DNA methylation in maternal and umbilical cord blood leukocytes. Environ Health Perspect 120 (7):1061–1066. https://doi.org/10.1289/ehp.1104173

Kile ML, Houseman EA, Baccarelli AA, Quamruzzaman Q, Rahman M, Mostofa G, Cardenas A, Wright RO, Christiani DC (2014) Effect of prenatal arsenic exposure on DNA methylation and leukocyte subpopulations in cord blood. Epigenetics 9(5):774–782. https://doi.org/10.4161/epi.28153

Koestler DC, Avissar-Whiting M, Houseman EA, Karagas MR, Marsit CJ (2013) Differential DNA methylation in umbilical cord blood of infants exposed to low levels of arsenic in utero. Environ Health Perspect 121(8):971–977. https://doi.org/10.1289/ehp.1205925

Kornfeld JW, Bruning JC (2014) Regulation of metabolism by long, non-coding RNAs. Front Genet 5:57. https://doi.org/10.3389/fgene.2014.00057

Kurdyukov S, Bullock M (2016) DNA methylation analysis: choosing the right method. Biology (Basel) 5(1):pii: E3. https://doi.org/10.3390/biology5010003

Lambrou A, Baccarelli A, Wright RO, Weisskopf M, Bollati V, Amarasiriwardena C, Vokonas P, Schwartz J (2012) Arsenic exposure and DNA methylation among elderly men. Epidemiology 23(5):668–676. https://doi.org/10.1097/EDE.0b013e31825afb0b

Langie SA, Szarc Vel Szic K, Declerck K, Traen S, Koppen G, Van Camp G, Schoeters G, Vanden Berghe W, De Boever P (2016) Whole-genome saliva and blood DNA methylation profiling in individuals with a respiratory allergy. PLoS One 11(3):e0151109. https://doi.org/10.1371/journal.pone.0151109

Lee DH, Jacobs DR Jr, Porta M (2009) Hypothesis: a unifying mechanism for nutrition and chemicals as lifelong modulators of DNA hypomethylation. Environ Health Perspect 117 (12):1799–1802. https://doi.org/10.1289/ehp.0900741

Lehner B, Kaneko K (2011) Fluctuation and response in biology. Cell Mol Life Sci 68 (6):1005–1010. https://doi.org/10.1007/s00018-010-0589-y

Lepeule J, Bind MA, Baccarelli AA, Koutrakis P, Tarantini L, Litonjua A, Sparrow D, Vokonas P, Schwartz JD (2014) Epigenetic influences on associations between air pollutants and lung function in elderly men: the normative aging study. Environ Health Perspect 122(6):566–572. https://doi.org/10.1289/ehp.1206458

Levänen B, Bhakta NR, Torregrosa Paredes P, Barbeau R, Hiltbrunner S, Pollack JL, Skold CM, Svartengren M, Grunewald J, Gabrielsson S, Eklund A, Larsson BM, Woodruff PG, Erle DJ, Wheelock AM (2013) Altered microRNA profiles in bronchoalveolar lavage fluid exosomes in asthmatic patients. J Allergy Clin Immunol 131(3):894–903. https://doi.org/10.1016/j.jaci.2012.11.039

Li Y, Tollefsbol TO (2011) DNA methylation detection: bisulfite genomic sequencing analysis. Methods Mol Biol 791:11–21. https://doi.org/10.1007/978-1-61779-316-5_2

Liu X, Zheng Y, Zhang W, Zhang X, Lioyd-Jones DM, Baccarelli AA, Ning H, Fornage M, He K, Liu K, Hou L (2014) Blood methylomics in response to arsenic exposure in a low-exposed US population. J Expo Sci Environ Epidemiol 24(2):145–149. https://doi.org/10.1038/jes.2013.89

Ma L, Li J, Zhan Z, Chen L, Li D, Bai Q, Gao C, Li J, Zeng X, He Z, Wang S, Xiao Y, Chen W, Zhang A (2016) Specific histone modification responds to arsenic-induced oxidative stress. Toxicol Appl Pharmacol 302:52–61. https://doi.org/10.1016/j.taap.2016.03.015

Madrigano J, Baccarelli A, Mittleman MA, Wright RO, Sparrow D, Vokonas PS, Tarantini L, Schwartz J (2011) Prolonged exposure to particulate pollution, genes associated with glutathione pathways, and DNA methylation in a cohort of older men. Environ Health Perspect 119 (7):977–982. https://doi.org/10.1289/ehp.1002773

Majumdar S, Chanda S, Ganguli B, Mazumder DN, Lahiri S, Dasgupta UB (2010) Arsenic exposure induces genomic hypermethylation. Environ Toxicol 25(3):315–318. https://doi.org/10.1002/tox.20497

Martens DS, Plusquin M, Gyselaers W, De Vivo I, Nawrot TS (2016) Maternal pre-pregnancy body mass index and newborn telomere length. BMC Med 14(1):148. https://doi.org/10.1186/s12916-016-0689-0

Messerschmidt DM, Knowles BB, Solter D (2014) DNA methylation dynamics during epigenetic reprogramming in the germline and preimplantation embryos. Genes Dev 28(8):812–828. https://doi.org/10.1101/gad.234294.113

Morgan HD, Sutherland HG, Martin DI, Whitelaw E (1999) Epigenetic inheritance at the agouti locus in the mouse. Nat Genet 23(3):314–318. https://doi.org/10.1038/15490

Needham BL, Smith JA, Zhao W, Wang X, Mukherjee B, Kardia SL, Shively CA, Seeman TE, Liu Y, Diez Roux AV (2015) Life course socioeconomic status and DNA methylation in genes related to stress reactivity and inflammation: the multi-ethnic study of atherosclerosis. Epigenetics 10(10):958–969. https://doi.org/10.1080/15592294.2015.1085139

Niedzwiecki MM, Hall MN, Liu X, Oka J, Harper KN, Slavkovich V, Ilievski V, Levy D, van Geen A, Mey JL, Alam S, Siddique AB, Parvez F, Graziano JH, Gamble MV (2013) A dose-response study of arsenic exposure and global methylation of peripheral blood mononuclear cell DNA in Bangladeshi adults. Environ Health Perspect 121(11–12):1306–1312. https://doi.org/10.1289/ehp.1206421

Niedzwiecki MM, Liu X, Hall MN, Thomas T, Slavkovich V, Ilievski V, Levy D, Alam S, Siddique AB, Parvez F, Graziano JH, Gamble MV (2015) Sex-specific associations of arsenic exposure with global DNA methylation and hydroxymethylation in leukocytes: results from two studies in Bangladesh. Cancer Epidemiol Biomark Prev 24(11):1748–1757. https://doi.org/10.1158/1055-9965.EPI-15-0432

Pang AP, Sugai C, Maunakea AK (2016) High-throughput sequencing offers new insights into 5-hydroxymethylcytosine. Biomol Concepts 7(3):169–178. https://doi.org/10.1515/bmc-2016-0011

Panni T, Mehta AJ, Schwartz JD, Baccarelli AA, Just AC, Wolf K, Wahl S, Cyrys J, Kunze S, Strauch K, Waldenberger M, Peters A (2016) A genome-wide analysis of dna methylation and fine particulate matter air pollution in three study populations: KORA F3, KORA F4, and the normative aging study. Environ Health Perspect 124:983–990. https://doi.org/10.1289/ehp.1509966

Pauwels S, Truijen I, Ghosh M, Duca RC, Langie SA, Bekaert B, Freson K, Huybrechts I, Koppen G, Devlieger R, Godderis L (2017) The effect of paternal methyl-group donor intake on offspring DNA methylation and birth weight. J Dev Orig Health Dis 8(3):1–11. https://doi.org/10.1017/S2040174417000046

Peng C, Bind MC, Colicino E, Kloog I, Byun HM, Cantone L, Trevisi L, Zhong J, Brennan K, Dereix AE, Vokonas PS, Coull BA, Schwartz JD, Baccarelli AA (2016) Particulate air pollution and fasting blood glucose in nondiabetic individuals: associations and epigenetic mediation in the normative aging study, 2000-2011. Environ Health Perspect 124(11):1715–1721. https://doi.org/10.1289/EHP183

Pilsner JR, Liu X, Ahsan H, Ilievski V, Slavkovich V, Levy D, Factor-Litvak P, Graziano JH, Gamble MV (2007) Genomic methylation of peripheral blood leukocyte DNA: influences of arsenic and folate in Bangladeshi adults. Am J Clin Nutr 86(4):1179–1186

Pilsner JR, Liu X, Ahsan H, Ilievski V, Slavkovich V, Levy D, Factor-Litvak P, Graziano JH, Gamble MV (2009) Folate deficiency, hyperhomocysteinemia, low urinary creatinine, and hypomethylation of leukocyte DNA are risk factors for arsenic-induced skin lesions. Environ Health Perspect 117(2):254–260. https://doi.org/10.1289/ehp.11872

Pilsner JR, Hall MN, Liu X, Ilievski V, Slavkovich V, Levy D, Factor-Litvak P, Yunus M, Rahman M, Graziano JH, Gamble MV (2012) Influence of prenatal arsenic exposure and newborn sex on global methylation of cord blood DNA. PLoS One 7(5):e37147. https://doi.org/10.1371/journal.pone.0037147

Rakyan VK, Down TA, Balding DJ, Beck S (2011) Epigenome-wide association studies for common human diseases. Nat Rev Genet 12(8):529–541. https://doi.org/10.1038/nrg3000

Raser JM, O'Shea EK (2005) Noise in gene expression: origins, consequences, and control. Science 309(5743):2010–2013. https://doi.org/10.1126/science.1105891

Reichard JF, Schnekenburger M, Puga A (2007) Long term low-dose arsenic exposure induces loss of DNA methylation. Biochem Biophys Res Commun 352(1):188–192. https://doi.org/10.1016/j.bbrc.2006.11.001

Reik W, Dean W, Walter J (2001) Epigenetic reprogramming in mammalian development. Science 293(5532):1089–1093. https://doi.org/10.1126/science.1063443

Richmond RC, Joubert BR (2017) Contrasting the effects of intra-uterine smoking and one-carbon micronutrient exposures on offspring DNA methylation. Epigenomics 9(3):351–367. https://doi.org/10.2217/epi-2016-0135

Rojas D, Rager JE, Smeester L, Bailey KA, Drobna Z, Rubio-Andrade M, Styblo M, Garcia-Vargas G, Fry RC (2015) Prenatal arsenic exposure and the epigenome: identifying sites of 5-methylcytosine alterations that predict functional changes in gene expression in newborn cord blood and subsequent birth outcomes. Toxicol Sci 143(1):97–106. https://doi.org/10.1093/toxsci/kfu210

Saenen ND, Vrijens K, Janssen BG, Roels HA, Neven KY, Vanden Berghe W, Gyselaers W, Vanpoucke C, Lefebvre W, De Boever P, Nawrot TS (2017) Lower placental leptin promoter methylation in association with fine particulate matter air pollution during pregnancy and placental nitrosative stress at birth in the environage cohort. Environ Health Perspect 125 (2):262–268. https://doi.org/10.1289/EHP38

Sanchez-Guerra M, Zheng Y, Osorio-Yanez C, Zhong J, Chervona Y, Wang S, Chang D, McCracken JP, Diaz A, Bertazzi PA, Koutrakis P, Kang CM, Zhang X, Zhang W, Byun HM, Schwartz J, Hou L, Baccarelli AA (2015) Effects of particulate matter exposure on blood 5-hydroxymethylation: results from the Beijing truck driver air pollution study. Epigenetics 10(7):633–642. https://doi.org/10.1080/15592294.2015.1050174

Seow WJ, Kile ML, Baccarelli AA, Pan WC, Byun HM, Mostofa G, Quamruzzaman Q, Rahman M, Lin X, Christiani DC (2014) Epigenome-wide DNA methylation changes with development of arsenic-induced skin lesions in Bangladesh: a case-control follow-up study. Environ Mol Mutagen 55(6):449–456. https://doi.org/10.1002/em.21860

Shahbazian MD, Grunstein M (2007) Functions of site-specific histone acetylation and deacetylation. Annu Rev Biochem 76:75–100. https://doi.org/10.1146/annurev.biochem.76.052705.162114

Singer ZS, Yong J, Tischler J, Hackett JA, Altinok A, Surani MA, Cai L, Elowitz MB (2014) Dynamic heterogeneity and DNA methylation in embryonic stem cells. Mol Cell 55 (2):319–331. https://doi.org/10.1016/j.molcel.2014.06.029

Smeester L, Rager JE, Bailey KA, Guan X, Smith N, Garcia-Vargas G, Del Razo LM, Drobna Z, Kelkar H, Styblo M, Fry RC (2011) Epigenetic changes in individuals with arsenicosis. Chem Res Toxicol 24(2):165–167. https://doi.org/10.1021/tx1004419

Somineni HK, Zhang X, Biagini Myers JM, Kovacic MB, Ulm A, Jurcak N, Ryan PH, Khurana Hershey GK, Ji H (2015) Ten-eleven translocation 1 (TET1) methylation is associated with childhood asthma and traffic-related air pollution. J Allergy Clin Immunol 137(3):797–805.e5. https://doi.org/10.1016/j.jaci.2015.10.021

Soubry A, Hoyo C, Jirtle RL, Murphy SK (2014) A paternal environmental legacy: evidence for epigenetic inheritance through the male germ line. BioEssays 36(4):359–371. https://doi.org/10.1002/bies.201300113

Stringhini S, Polidoro S, Sacerdote C, Kelly RS, van Veldhoven K, Agnoli C, Grioni S, Tumino R, Giurdanella MC, Panico S, Mattiello A, Palli D, Masala G, Gallo V, Castagne R, Paccaud F, Campanella G, Chadeau-Hyam M, Vineis P (2015) Life-course socioeconomic status and DNA methylation of genes regulating inflammation. Int J Epidemiol 44(4):1320–1330. https://doi.org/10.1093/ije/dyv060

Tajuddin SM, Amaral AF, Fernandez AF, Rodriguez-Rodero S, Rodriguez RM, Moore LE, Tardon A, Carrato A, Garcia-Closas M, Silverman DT, Jackson BP, Garcia-Closas R, Cook AL, Cantor KP, Chanock S, Kogevinas M, Rothman N, Real FX, Fraga MF, Malats N, Spanish Bladder Cancer ESI (2013) Genetic and non-genetic predictors of LINE-1 methylation in leukocyte DNA. Environ Health Perspect 121(6):650–656. https://doi.org/10.1289/ehp.1206068

Tao MH, Zhou J, Rialdi AP, Martinez R, Dabek J, Scelo G, Lissowska J, Chen J, Boffetta P (2014) Indoor air pollution from solid fuels and peripheral blood DNA methylation: findings from a population study in Warsaw, Poland. Environ Res 134:325–330. https://doi.org/10.1016/j.envres.2014.08.017

Tauheed J, Sanchez-Guerra M, Lee JJ, Paul L, Ibne Hasan MO, Quamruzzaman Q, Selhub J, Wright RO, Christiani DC, Coull BA, Baccarelli AA, Mazumdar M (2017) Associations between post translational histone modifications, myelomeningocele risk, environmental arsenic

exposure, and folate deficiency among participants in a case control study in Bangladesh. Epigenetics 12(6):484–491. https://doi.org/10.1080/15592294.2017.1312238
Tellez-Plaza M, Tang WY, Shang Y, Umans JG, Francesconi KA, Goessler W, Ledesma M, Leon M, Laclaustra M, Pollak J, Guallar E, Cole SA, Fallin MD, Navas-Acien A (2014) Association of global DNA methylation and global DNA hydroxymethylation with metals and other exposures in human blood DNA samples. Environ Health Perspect 122(9):946–954. https://doi.org/10.1289/ehp.1306674
Vaiserman A (2015) Epidemiologic evidence for association between adverse environmental exposures in early life and epigenetic variation: a potential link to disease susceptibility? Clin Epigenetics 7:96. https://doi.org/10.1186/s13148-015-0130-0
Vriens A, Nawrot TS, Saenen ND, Provost EB, Kicinski M, Lefebvre W, Vanpoucke C, Van Deun J, De Wever O, Vrijens K, De Boever P, Plusquin M (2016) Recent exposure to ultrafine particles in school children alters miR-222 expression in the extracellular fraction of saliva. Environ Health 15(1):80. https://doi.org/10.1186/s12940-016-0162-8
Wang C, Chen R, Cai J, Shi J, Yang C, Tse LA, Li H, Lin Z, Meng X, Liu C, Niu Y, Xia Y, Zhao Z, Kan H (2016) Personal exposure to fine particulate matter and blood pressure: a role of angiotensin converting enzyme and its DNA methylation. Environ Int 94:661–666. https://doi.org/10.1016/j.envint.2016.07.001
Ward-Caviness CK, Nwanaji-Enwerem JC, Wolf K, Wahl S, Colicino E, Trevisi L, Kloog I, Just AC, Vokonas P, Cyrys J, Gieger C, Schwartz J, Baccarelli AA, Schneider A, Peters A (2016) Long-term exposure to air pollution is associated with biological aging. Oncotarget 7 (46):74510–74525. https://doi.org/10.18632/oncotarget.12903
Waterland RA, Michels KB (2007) Epigenetic epidemiology of the developmental origins hypothesis. Annu Rev Nutr 27:363–388. https://doi.org/10.1146/annurev.nutr.27.061406.093705
Wilhelm CS, Kelsey KT, Butler R, Plaza S, Gagne L, Zens MS, Andrew AS, Morris S, Nelson HH, Schned AR, Karagas MR, Marsit CJ (2010) Implications of LINE1 methylation for bladder cancer risk in women. Clin Cancer Res 16(5):1682–1689. https://doi.org/10.1158/1078-0432.CCR-09-2983
Yehuda R, Daskalakis NP, Bierer LM, Bader HN, Klengel T, Holsboer F, Binder EB (2016) Holocaust exposure induced intergenerational effects on FKBP5 methylation. Biol Psychiatry 80(5):372–380. https://doi.org/10.1016/j.biopsych.2015.08.005
Yoder JA, Walsh CP, Bestor TH (1997) Cytosine methylation and the ecology of intragenomic parasites. Trends Genet 13(8):335–340
You JS, Jones PA (2012) Cancer genetics and epigenetics: two sides of the same coin? Cancer Cell 22(1):9–20. https://doi.org/10.1016/j.ccr.2012.06.008
Zhao CQ, Young MR, Diwan BA, Coogan TP, Waalkes MP (1997) Association of arsenic-induced malignant transformation with DNA hypomethylation and aberrant gene expression. Proc Natl Acad Sci U S A 94(20):10907–10912
Zheng SC, Beck S, Jaffe AE, Koestler DC, Hansen KD, Houseman AE, Irizarry RA, Teschendorff AE (2017a) Correcting for cell-type heterogeneity in epigenome-wide association studies: revisiting previous analyses. Nat Methods 14(3):216–217. https://doi.org/10.1038/nmeth.4187
Zheng Y, Sanchez-Guerra M, Zhang Z, Joyce BT, Zhong J, Kresovich JK, Liu L, Zhang W, Gao T, Chang D, Osorio-Yanez C, Carmona JJ, Wang S, McCracken JP, Zhang X, Chervona Y, Diaz A, Bertazzi PA, Koutrakis P, Kang CM, Schwartz J, Baccarelli AA, Hou L (2017b) Traffic-derived particulate matter exposure and histone H3 modification: a repeated measures study. Environ Res 153:112–119. https://doi.org/10.1016/j.envres.2016.11.015
Zhong J, Karlsson O, Wang G, Li J, Guo Y, Lin X, Zemplenyi M, Sanchez-Guerra M, Trevisi L, Urch B, Speck M, Liang L, Coull BA, Koutrakis P, Silverman F, Gold DR, Wu T, Baccarelli AA (2017) B vitamins attenuate the epigenetic effects of ambient fine particles in a pilot human intervention trial. Proc Natl Acad Sci U S A 114(13):3503–3508. https://doi.org/10.1073/pnas.1618545114

Chapter 6
Metabonomics

Toby Athersuch

Abstract The exposome concept places substantial weight on the internal chemical milieu of individuals, as this is the primary integrator of the human genome and the wider external environment. Small molecule metabolites of both endogenous and exogenous origin are involved in a plethora of cellular and systemic functions, and collectively contribute to the mechanistic linkage of exposures, responses, and associated adverse outcomes. Temporal and spatial responses of metabolic phenotypes to various environmental stimuli provide a direct report on multiple interacting and conditional processes that are modulated by numerous factors including diet, lifestyle, pharmaceutical use, microbial activity, age, sex, and many others. Measuring and integrating information about the human metabolome represents a critical part of the path toward understanding the environmental determinants of chronic disease.

The size and diversity of the chemical space that metabolites occupy means that measuring the human metabolome via serum, urine, or other biofluids or tissues represents a huge analytical challenge, addressed by the application of high-resolution platforms, typically incorporating liquid- and/or gas-chromatography for separation, and nuclear magnetic resonance spectroscopy and/or mass spectrometry for detection. Advances in the performance of these platforms now permit the measurement of many hundreds or thousands of metabolites, in either targeted or untargeted assays. The focus of this chapter is on the utility of the different analytical platforms, their complementarity, and application to large-scale sample set analysis. Considerations for data analysis and integration with other omics, exposure, and outcome data are discussed, alongside approaches for interpreting findings in the context of the human exposome.

Keywords Metabonomics · Analytical platforms to measure the exposome · NMR · Mass spectrometry

---

T. Athersuch (✉)
Computational and Systems Medicine, Department of Surgery and Cancer, Faculty of Medicine, Imperial College London, London, UK
e-mail: toby.athersuch@imperial.ac.uk

S. Dagnino, A. Macherone (eds.), *Unraveling the Exposome*,
https://doi.org/10.1007/978-3-319-89321-1_6

Introduction

Since the formulation of the exposome concept, the idea that it is the *internal chemical environment* that is critically important in mediating gene–environment interactions has remained central (Wild 2005, 2012; Wild et al. 2013; Miller and Jones 2014). Therefore, it is appropriate that the identity of the constituents of this chemical milieu, how they are measured, and what they represent be considered in detail. In this chapter, the main research strategies for characterizing the human metabolome are presented, the main analytical platforms are discussed, and a description of some of the challenges that exist for metabolic phenotyping of large-scale sample sets in the context of human exposome studies is provided.

While a detailed treatment of data analysis methods for the integration of metabolome data with other measures is provided elsewhere in this book (e.g., parallel omics, exposure assessments, sample set metadata, etc.), some of the specialized chemometric tools commonly used to explore metabolome datasets, and how these approaches can also assist in the annotation of unknowns in the spectral outputs from metabolic phenotyping, are highlighted. Similarly, community standards for data reporting and analysis relevant to metabonomics are surveyed in brief, alongside specific resources for archiving and sharing metabolic profile data.

Characteristics of the Human Metabolome

With respect to the characterization of the internal chemical environment in the context of exposome studies, it is the methods and outputs of metabolic phenotyping (i.e., measuring the metabolome) which are most relevant. With respect to the wider exposome framework, it is the modeling and understanding of the variation in these phenotypes over space and time (i.e., interpreting the metabolome) that are of greatest interest. Before launching into how metabonomics can beneficially be integrated into studies of the human exposome, it is useful to take a step back and acquire some clarity on precisely what the various terms mean.

Terminology and Definitions

The term *metabolome* refers to the complement of low-molecular-weight chemical species present in a biological entity, whether this be a single cell, an organ, a system, or superorganism (see Fig. 6.1). An initial definition was provided by Oliver (1998): “The quantitative complement of all of the low molecular weight molecules present in cells in a particular physiological or developmental state”. The term *metabolomics* refers to the characterization of metabolomes and their components, defined as “A comprehensive and quantitative analysis of all metabolites… the

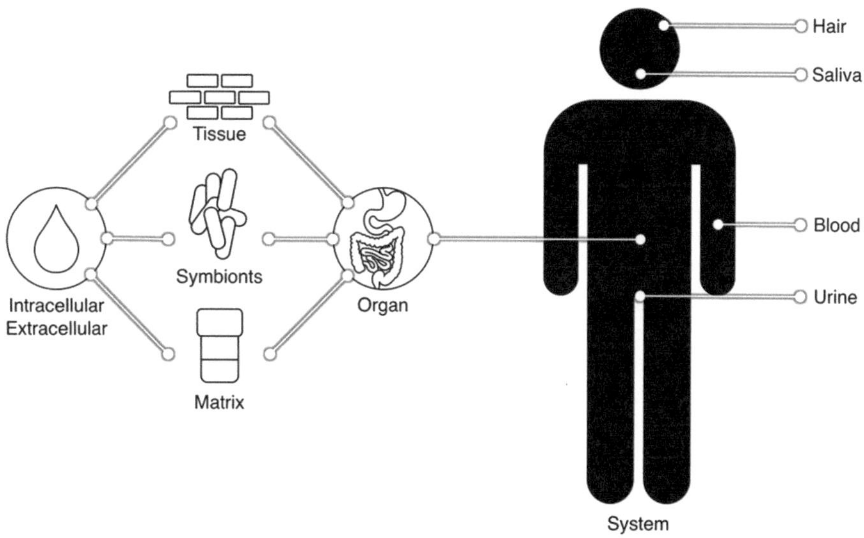

Fig. 6.1 The term metabolome is used to describe the complement of small molecule metabolites in a specified biological system, for example, a fluid, a cell, an organ, an organism, or a superorganism. It is therefore possible to define multiple metabolomes that describe aspects of complex systems such as humans that may include the contributions from symbiotic/commensal organisms such as enteral and epidermal microbiota, in addition to the primary endogenous function and metabolic response to xenobiotic agents

metabolome of the biological system under study" (Fiehn 2001). The related term metabonomics refers the metabolic responses that biological entities display, with a particular emphasis on system-level dynamism, originally defined by Nicholson et al. (1999): "the quantitative measurement of the dynamic multiparametric metabolic response of living systems to pathophysiological stimuli or genetic modification".

In short, the measurements we can make are a subset of the outwardly observable characteristics of the biological sample(s) from which they were derived, specifically a *metabolic phenotype*, and the dynamics this phenotype exhibits in response to various stimuli. Thus, the approach of *metabolic phenotyping* as a way of informing metabonomics (and other) studies is commonly cited (Holmes et al. 2008, 2010; Holmes and Nicholson 2008).

Components of the Human Metabolome

By most common definitions, small molecules in the metabolome have a molar mass of <1500 g/mol, and span a wide range of chemical classes including lipids, organic acids, sugars, nucleotides, and so on (Oliver 1998; Fiehn 2001; Nicholson et al. 1999). Attempts to derive a "metabolic atlas" for the human blood and urine metabolomes have been conducted, revealing each to contain many thousands of distinct, biologically-relevant small molecules. Bouatra et al. (2013) used a multiplatform approach to profile urine, providing positive identification and ability to quantify 209 metabolites, with a literature review enabling an online searchable database of over 2200 metabolites to be created. The blood metabolome received a similar treatment, with Psychogios et al. (2011) conducting a similar multiplatform analysis of human serum, reporting a total of 4229 compounds with associated concentration information and available literature relevant to reports in man. In other work that sampled from the UK population, the HUSERMET project (www.husermet.org) used a combination of GC-MS and UPLC-MS analyses to characterize the serum metabolome of 1200 healthy adult participants, and provided an elegant analysis of metabolite association and distributions in relation to anthropomorphic factors including sex, age, and BMI, as well as smoking status as a key environmental lifestyle variable (Dunn et al. 2014). Less accessible, although clinically relevant biofluids and tissues have also been subjected to such analysis, including the metabolome of human cerebrospinal fluid (Wishart et al. 2008). The authors of these papers note that while extensive in their coverage, such studies can never provide complete comprehensive coverage of all small molecule metabolites for these systems—something that would only be true for all but the simplest biological systems or matrices—because of their vast chemical complexity.

Why Is the Metabolome So Chemically Complex?

The observed complexity in the low-molecular-weight metabolite profile of biofluids results from several factors, discussed in turn below.

Environmental Exposures Are Chemically Diverse

Upon entering/interacting with the body, small molecules of exogenous origin become part of the chemical milieu for a period, and therefore a (temporary) part of the metabolome. The structural/chemical space from which these small molecules may be drawn is vast; it has been estimated that the possible number of biologically active chemicals (i.e., relevant to any discussion relating to the exposome) is approximately 10^{60}, although it is worthwhile to point out that of these possibilities, only a very small fraction exist in any appreciable quantity, and an even smaller fraction represents molecules that are relevant either as components of real chemical exposures, or of the human metabolome (Reymond et al. 2010, 2012; Ruddigkeit et al. 2012; Fink and Reymond 2007; Uppal et al. 2016). However, these exposures *are* possible and therefore relevant when considering the totality of lifetime exposures encapsulated by the exposome concept (Athersuch 2016).

Environmental Exposures Are Dynamic

Few environmental exposures remain constant, and although some may be highly controlled, there is considerable variation in dose, periodicity, and frequency that makes global exposure profiles inherently complex. Quantitatively, the diet represents the largest input to the metabolism of an individual. Dietary exposures represent a complex mixture of small molecules and macromolecules, with a complexity similar in magnitude to the human metabolome (comprising as it does other organisms' metabolomes, and co-exposures such as pesticides); people rarely eat the same combinations of foods at the same time for prolonged periods, and there is considerable variation in the volume and combination of these environmental inputs (Scalbert et al. 2014). Other high-level exposures include intake of pharmaceuticals (which exhibit a lower chemical complexity but include both the pharmacologically active agents and any excipients). Quantitatively lower exposures include deliberate and/or regular intake/contact with products in an occupational setting, cosmetics, as well as exposures such as components of ambient air or water.

Metabolism Increases Chemical Complexity

The human body is capable of metabolizing a wide variety of small molecules (and indeed catabolic products of macromolecules) to which it is exposed, and commonly does so by utilizing a range of enzyme systems to functionalize and/or conjugate molecules to expedite their excretion from the body or facilitate their incorporation (Nicholson and Wilson 2003). Consequently, when considering the xenobiotic exposures, the complex patterns of in vivo metabolites are also relevant; it is rare for exogenous chemicals to remain metabolically inert. For example, even a simplified metabolism scheme for paracetamol shows multiple metabolites (some of which exhibit pharmacological action or are responsible for toxicity) that are observed as an appreciable proportion to the dose (Fig. 6.2). In this instance, the main intermediate and terminal (excreted) metabolites of the parent compound are also accompanied by short-lived, highly reactive chemical intermediates that potentially influence/produce exposome measures such as the adductome profile of human serum albumin through covalent binding (Damsten et al. 2007).

Metabolic Activation of Protoxicants Is Common

Metabolic processes (activation) are necessary for many of the observed pharmacological and/or toxicological of chemical exposures. For example, benzo[a]pyrene (B [α]P) is a polycyclic aromatic hydrocarbon (PAH) that is a known human toxicant, and classed as a Group 1 carcinogen by the International Agency for Research on Cancer (IARC) (Humans IWGOTEOCRT 2012). It is formed during incomplete combustion processes, originating from multiple environmental sources including

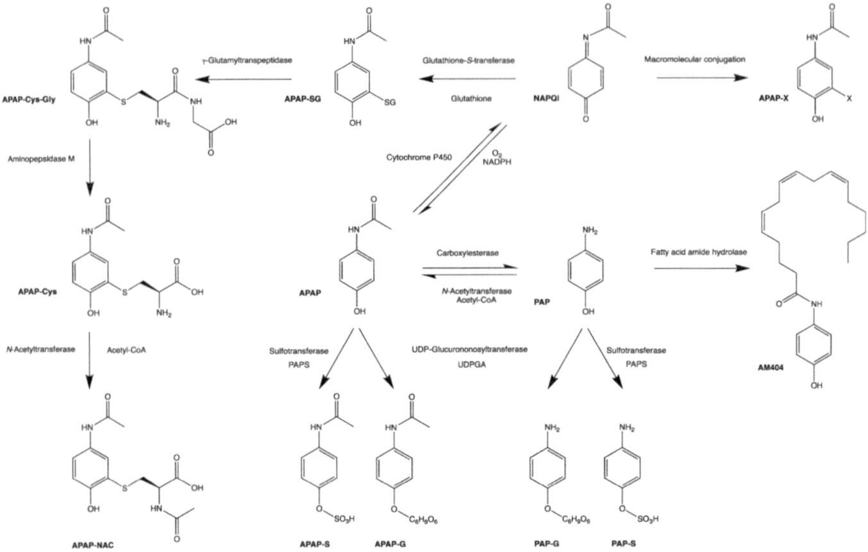

Fig. 6.2 Principal routes of metabolism of paracetamol (*N*-acetyl-*p*-aminophenol, APAP) in man. UDPGA uridine 5′-diphosphoglucuronic acid; PAPS 3′-phosphoadenosine-5′-phosphosulfate; AM404 *N*-arachidonoylaminophenol; NADPH nicotinamide adenine dinucleotide phosphate

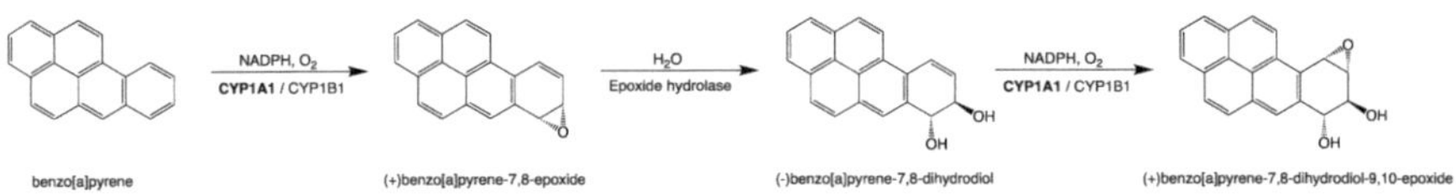

Fig. 6.3 Metabolism of the polycyclic aromatic hydrocarbon benzo[a]pyrene by cytochrome P450 isoforms CYP1A1 and/or CYP1B1. Initial oxidoreductase action results in the formation of an epoxide; dihydroxylation by epoxide hydrolase yields a diol; second epoxidation step forms (+)benzo[a]pyrene-7,8-dihydrodiol-9,10-epoxide, a metabolite capable of covalently binding to guanine bases in DNA

fuel smoke, tobacco smoke, and grilled foods, and is consequently a ubiquitous human environmental exposure. One of the most extensively studied mechanisms of B[a]P toxicity requires metabolism (Fig. 6.3) (Peter Guengerich and Shimada 1998; Shimada et al. 2001). In brief, the oxidoreductases CYP1A1 and CYP1B1 result in epoxidation of the parent compound, producing (+)benzo[a]pyrene-7,8-epoxide (alongside other metabolites). Epoxide hydrolase catalyses the formation of a diol, yielding (−)benzo[a]pyrene-7,8-dihydrodiol, which undergoes a second epoxidation to form (+)benzo[a]pyrene-7,8-dihydrodiol-9,10-epoxide, a metabolite has an affinity for DNA, covalently binding to guanine bases.

Here we have an example of how the true chemical exposure profile of a relatively simple molecule—and ubiquitous exposure—can become complicated very quickly; the mechanism of B[a]P genotoxicity and relationship with other cellular macromolecules is mediated by metabolism, with links to genetic variation that can modulate individual responses (discussed below). Furthermore, B[a]P induces expression of CYP1A1 through binding the aryl hydrocarbon receptor (AHR), and therefore is responsible in part for activating the mechanism for its own metabolism. *Note:* the absence of monooxygenase activity and failure to detoxify B[a]P via metabolism leads to alternative toxicological outcomes.

Not All Transformations Require Biological Catalysts

Although a minor contributor to the overall metabolite flux, it is possible for *metabonates* (products of nonenzymatic reactions) to form, which may confound analysis based solely on known enzyme-catalyzed biotransformations, such as metabolic network reconstructions (Beckett 2008).

Not Everyone Is the Same

Metabolic complexity arising from inter- and intraindividual variation in both endogenous and xenobiotic metabolism represents one of the most challenging aspects of metabonomics (Thévenot et al. 2015; Illig et al. 2010). For example, in the design of phenotyping studies, statistical power estimates have remained a persistent issue, largely because expected effect size is unknown in most cases, which makes sample size justification difficult, although recent attempts to improve estimates using simulation have been attempted (Blaise et al. 2016). Several studies have attempted to decompose the metabolic variation observed between and within individuals according to multiple factors including genetics, temporal/diurnal variation, fasting status, sleep, and age (Nicholson et al. 2011; Maitre et al. 2017; Sampson et al. 2013; Giskeødegård et al. 2015; Davies et al. 2014; McShane et al. 2013; Kim et al. 2014). With respect to the utility of metabolic phenotyping in exposome studies, there is a need to conduct further analyses that make use of longitudinally collected samples that can help deconvolve short- and long-term changes in the metabolome, as is being done at present in several large-scale studies (Vineis et al. 2017; Vrijheid et al. 2014).

How Do We Distinguish Exposures and Responses in the Human Metabolome?

Faced with such chemical complexity and a need to identify environmental (and therefore potentially modifiable) determinants of disease, a delineation of metabolome components according to their origin (i.e., endogenous/exogenous) seems to be pragmatic; it is the relationships between features that represent either exposures or responses, that we seek in exposome studies. Holmes et al. (2007)

previously defined a discrete subset of the metabolome as the *xenometabolome*: "The multivariate description of the xenobiotic (foreign compound) metabolite profile of an individual or sample from an individual that has been exposed through any route (either deliberately or accidentally) to drugs, environmental pollutants, or dietary components that cannot be completely catabolized by endogenous metabolic enzyme systems". While such xenobiotic exposures may be relatively easy to identify, delineating all metabolome components proves more difficult, largely on account of the continuum that exists between those things that are central to endogenous function, and those that are clearly xenobiotic (Loo et al. 2009). Nicholson and Wilson (2003) provide an enhanced lexicon with which we can more accurately describe the human metabolome, defining terms for processes and compounds that lie in this liminal zone between endogenous and xenobiotic extremes:

- *Sym*-endogenous: "... essential to host biological function and which can be metabolized or further utilized by host, but for which there is no biosynthetic capability in the host genome, for example, vitamins and essential amino acids."
- *Sym*-xenobiotic: "... involving co-metabolism by two or more organisms that are commensal or symbiotic (for example, bile acid metabolism). Not necessarily essential to the host, but can influence endogenous or other xenobiotic metabolic processes."
- *Trans*-xenobiotic: "... of extra-genomic or chemical origin but which are metabolically converted to endogenous species or metabolites that can be utilized directly in endogenous processes, for example, ethanol."

What Controls the Metabolome?

Metabolic phenotypes reflect the outwardly observable characteristics/end points of *gene–environment interactions* and therefore can help provide evidence for causal linkage of exposures with adverse outcomes, and attribution of disease risk (Vineis et al. 2013; Rappaport et al. 2014). Phenotypes indicative of aberrant metabolism of genetic origin have long been studied in the clinic to identify individuals with inborn errors of metabolism; newborn babies are routinely screened (Guthrie test) (Guthrie and Susi 1963; Pitt 2010). For example, phenylketonuria (PKU) is an inherited autosomal recessive disorder, characterized by polymorphisms in the gene encoding the phenylalanine hydroxylase enzyme, and consequent deficiency in the conversion of its substrate phenylalanine, to tyrosine. Without intervention, the concentration of phenylalanine in the blood becomes elevated, and tyrosine depleted, leading to toxicity. Metabolic profiling offers an efficient route to elucidate such clear phenotypic traits in individuals and across populations (Dénes et al. 2012).

Exposures to a toxicant may result in variable toxicity, the degree of which will be determined in part by genetic predisposition, mediated by the action of enzymatic gene products involved in detoxication reactions. The complexification of exposures through metabolism, introduced above in relation to paracetamol, is also an exemplar of the interplay between genes and the environment; variation in metabolic genes contributes to the observed metabolic fate and toxicological profile at the individual level (Zhao and Pickering 2011). There are similar examples that relate to environmental contaminants. For example, several studies have suggested that genetic variation relating to the metabolic competence of individuals to detoxify inorganic arsenic may be associated with the degree of observed toxicity (Pierce et al. 2013; Jansen et al. 2016; Karagas et al. 2015; Melkonian et al. 2013). Contributions may also arise from interactions with the microbiome. Metabolically derived toxicity dependent on the host's microflora composition and capacity is possible, as recently illustrated in relation to melamine-induced kidney injury, that was conditional on the ability of the gut microflora to metabolize melamine to the toxin cyanuric acid (Zheng et al. 2013).

A more holistic understanding of the influence of genetic variation on metabolic processes has been the subject of several recent studies (Suhre et al. 2016; Kastenmüller et al. 2015). The availability of parallel metabolite and genomics data in large cohorts has enabled relationships between genetic variation and metabolic regulation to be explored using genome-wide analysis methods; genome-wide association studies (GWAS) using metabolic phenotypic features in blood (Illig et al. 2010; Gieger et al. 2008; Hicks et al. 2009) and urine (Suhre et al. 2011) have been conducted. Relationships between epigenetic regulation of metabolism and metabolism have also been explored; additional control over metabolic processes is exerted via epigenetic mechanisms (most studies to date have primarily focused on CpG site methylation status of DNA via regulation of gene transcription) (Petersen et al. 2014). Furthermore, the metabolome–transcriptome interface has been studied by Bartel et al. (2015); cross-omics interactions are discussed elsewhere in this book.

Measuring the Human Metabolome

Having considered the complexity of the metabolome, we turn now to studying it. Characterizing the metabolome with both breadth and depth represents a huge analytical challenge, including requirements to provide quantitatively useful information about multiple chemicals using a minimal sample, and be amenable to high-throughput analysis.

Why Is the Metabolome So Challenging to Measure Comprehensively?

The chemical space that encapsulates even the most prevalent environmental exposures and metabolome components spans a wide variety of chemical classes, and thus presents a considerable challenge to the analyst. Furthermore, individual components of the metabolome cover a huge concentration range, placing demands on sample availability, analytical limit of detection, and instrumental dynamic range.

Rappaport et al. (2014) surveyed the available literature and compiled a list of 1561 compounds that have been measured and reported in studies of normal human populations. This analysis showed that these metabolome components spanned a concentration range of 11 orders of magnitude, with those relating to core endogenous, diet and pharmaceuticals typically around three orders of magnitude greater than those of xenobiotics commonly classified as environmental pollutants.

We are presented with some key challenges when approaching an extensive characterization of the human metabolome:

- *Analytes*. Reporting on multiple chemical species is complex as many compounds have very similar physicochemical properties.
- *Samples*. Exposome studies commonly aim to use multiple bioanalytical approaches in parallel (e.g., exposure biomarkers, proteomics, epigenomics, adductomics, metabonomics), and therefore sample volume requirements for precious samples must be kept to a minimum. Samples may also have been collected using different protocols, and present specific analytical challenges (presence of preservatives, anticoagulants, contaminants, etc.).
- *Throughput*. Analyses must be amenable to a high rate of sample throughput to enable large-scale studies to be completed efficiently.
- *Cost*. For metabolic phenotyping to represent a cost-effective use of resources and to be feasibly applied to large-scale sample sets, both the per-sample and per-analyte cost must be low.

These challenges have been well addressed in several of the omics sciences (specifically genomics, transcriptomics, and epigenomics), and the result has been a dramatic increase in the availability and utility of postgenomic-era information. The principal reason for this has been that they are based on characterizing *coded* molecules, that is, molecules which are distinguished both structurally and functionally based on the order of subunits within their structure. This order commonly determines interaction specificity with *very* high precision (e.g., the molecular recognition of complementary nucleotide sequences) based on the ordered sequence. Analysis of such macromolecules takes advantage of either the exquisite specificity of interactions (e.g., in a microarray) or from being able to determine the code rapidly, because of it being formed from a small (and known) number of possibilities (e.g., DNA sequencing). By comparison, such approaches are not possible for characterizing the metabolome as neither the specificity nor the ability to "sequence" small molecules in a similar way is possible. The analytical challenge for metabolic phenotyping is to unpack complex chemical mixtures, for which all the possibilities are not yet known. Note: one might suggest metabolites are also intrinsically coded, as they can be distinguished both structurally and functionally based on the arrangement of their constituent atoms, selected from a known and restricted set; while true, this fails to recognize that we are typically not dealing with one *type* of molecule, and there is considerable opportunity for physicochemically similar molecules to confound our analyses.

Which Sample Matrices Are Amenable to Metabonomics?

Phenotyping of biofluids and tissues is essentially a complex chemical mixture analysis problem, and therefore no sample matrix can be realistically excluded from a technical perspective; analytical chemists are both resourceful and tenacious problem-solvers. Indeed, metabolic phenotyping has been used extensively to explore a range of human biofluids including blood serum/plasma, urine, cerebrospinal fluid, saliva, semen, ascites fluid, as well as extracts from a plethora of tissue types. The most important consideration is not whether the analysis is possible (it always will be, by some method), but whether the type of sample and the conditions under which it was collected represent a useful phenotypic description for the purposes of the experiment at hand; different fluids and tissue will provide different information, and these are also modulated in time with respect to the factors of interest (e.g., relationship to exposures, or influence of disease processes).

Clearly, as a chemical mixture problem, the type of sample matrix will determine the *specific* analytical challenges related to producing a meaningful metabolic phenotype dataset; different fluids/tissues exhibit large differences in their overall composition because their physiological function (e.g., urine contains a high proportion of hydrophilic species, whereas blood contains a substantial lipid profile, etc.). From a practical point of view, it may be most efficient to adjust the preparation of a sample matrix to fit one or more existing phenotyping assays. This eliminates the requirement to develop and characterize the analytical step from scratch, facilitates rapid transfer of previously validated metabolite annotations of detected features, and retains the logistical benefits of existing workflows (time, handling, batching, and data processing). While there is considerable variation in the metabolic composition of various biofluids and tissues, many metabolites that are integral to core metabolism processes (e.g., TCA cycle intermediate) will likely be present. The alternative case is for specific assays to be developed for different sample matrices, which offer the possible advantage of enhanced optimization, at the expense of interoperability of output datasets.

As with other biological entities, metabolomes are susceptible to modulation by factors relating to sampling, handling, and storage. Objective measures of biofluid quality are hard to define from the standpoint of metabolite composition. While the presence or absence of specific metabolites can indicate poor sample collection or contamination, there is no absolute measure of quality that can be applied. Unlike loss of genomic information from the degradation of coded molecules (which can be assessed, e.g., RNA integrity number), degradation of small molecules results in the interconversion to another small molecule; even knowledge of this process is not indicative of how all small molecules—and therefore the overall metabolic profile—may be affected. In addition to degradative processes, volatile compounds may be lost and therefore escape detection. Assessing the impact of preanalytical factors has been the subject of several studies aimed at characterizing the variation in phenotypes that is attributable to procedures involved in sample collection, storage, and handling prior to analysis, and to provide guidance to minimize phenotypic deterioration from the point of sampling (Teahan et al. 2006; Kamlage et al. 2014; Anton et al. 2015). With respect to the use of archived samples in exposome studies, one key study was that of Hebels et al. (2013) who investigated preanalytical factors relevant to blood samples analyzed in the EU FP7 EnviroGenomarkers project (www.envirogenomarkers.net). This study systematically explored the effect of several factors on the profiles of multiple omics platforms to generate criteria for the selection of cohort samples. In the case of metabolome analyses, the key influences on derived blood plasma UPLC-MS profiles related to the choice of anticoagulants (an influence observed by others), and the bench time. Preanalytical factors for metabonomics have been the subject of several comprehensive reviews (Hernandes et al. 2017; Li et al. 2016; Budde et al. 2016; Yin et al. 2015).

It should be noted that the minimization of chemical contamination or a carryover from the point of sampling to that of data acquisition is particularly important in the analysis of metabolites; inadvertent additions to the chemical matrix can lead to inaccurate measurement (e.g., through ionization suppression effects or spectral overlap), conflation with a true exposure (e.g., polyethylene glycol), or inappropriate conclusions being drawn.

How Is the Metabolome Measured in Practice?

No single analytical platform can provide the necessary sensitivity, resolution, selectivity, and coverage across all metabolome components; more complete coverage can be achieved through parallel analysis by different analytical platforms and complementary assays (Bouatra et al. 2013; Psychogios et al. 2011; Wishart et al. 2008). Protocols for routine metabolic phenotyping of common biofluids/tissues have been published, including both targeted and untargeted analysis by NMR spectroscopy (Beckonert et al. 2007, 2010; Keun and Athersuch 2010; Dona et al. 2014), LC-MS (Gika et al. 2008; Want et al. 2010; Lewis et al. 2016; Gray et al. 2017; Xu et al. 2005; Plumb et al. 2010; Qi et al. 2014; Dunn et al. 2011), and GC-MS (Dunn et al. 2011; García and Barbas 2010; Chan et al. 2011).

Two main analytical platforms are used for the quantitative detection of small molecule metabolites by metabonomics researchers: nuclear magnetic resonance (NMR) spectroscopy and mass spectrometry (MS).

The fact that metabolite analysis represents a chemical analysis challenge with a multitude of feasible solutions has made harmonization of protocols difficult to achieve. This challenge is due to the combination of data-driven approaches, the vast chemical space under scrutiny, and unstandardized analytical platforms with varying selectivity and sensitivity.

Two complementary models for promoting sample/data comparability through harmonization of protocols exist.

1. *Distributed analysis*. Sharing of protocols via traditional publication routes and deployment by individual laboratories/researchers. This has the benefit of

enabling researchers to benefit from existing knowledge (e.g., annotations provided and shared in publically available databases). A limitation of this approach is that individual laboratories require their own equipment and facility staff. Additionally, without overall harmonization and oversight, independent implementation of protocols may introduce systematic differences based on fundamental aspects of the analytics (solvent quality/batch/QC, laboratory glassware quality, maintenance schedule, local handling procedures, etc.) as well as fine instrumental details relating to either hardware specification or implementation.

2. *Centralized analysis*. Creation of high-throughput facilities that cater to large-scale sample sets using standardized and well-characterized analytical methods (e.g., MRC-NIHR National Phenome Centre, http://phenomecentre.org; national research cores (https://commonfund.nih.gov/metabolomics/researchcores).

Nuclear Magnetic Resonance Spectroscopy

NMR spectroscopy has long been a key analytical platform for characterizing biological samples in metabonomics studies (Lindon et al. 2000). While a full explanation of the technical aspects is beyond the scope of this chapter, the main principles that underpin its function and how they make NMR spectroscopy particularly useful are provided below. As a type of spectroscopy, NMR relates to the interaction of electromagnetic radiation and matter, with isotopes of several elements capable of displaying this interaction. The NMR-active nuclei interact with magnetic and electromagnetic fields. When placed into a magnetic field, NMR-active nuclei precess about the direction of that field, at a frequency (the Larmor frequency) that is proportional to the strength of the field experienced by the nucleus, and the intrinsic magnetic properties of the nucleus itself. Nuclei in different chemical environments (e.g., in different chemical groups in a molecule) experience a different net magnetic field; electrons present have the effect of shielding the nuclei to a variable degree, depending on the location of an atom in its parent molecule. Therefore, nuclei in different chemical environments will have different Larmor frequencies; if these differences are sufficiently large, they can be differentiated in an NMR experiment. In modern spectrometers, pulsed NMR experiments apply a broadband radiofrequency pulse over the range of Larmor frequencies in the sample, and observe the resulting induction decay that results as the nuclei in the sample return to a thermal equilibrium. Fourier analysis is used to convert a recorded time-domain signal to a readily interpreted frequency-domain spectrum (displayed as a chemical shift). Nuclei in identical chemical environments will contribute equally to the resulting spectrum, and therefore the output is inherently quantitative; the higher the concentration of a chemical species or the number of chemically equivalent nuclei, the more intense its spectral peaks will be in the resulting spectrum. The resolution and sensitivity of an NMR-based platform is largely determined by the applied field strength; higher fields give improved spectral characteristics. While NMR magnets with exceptionally high fields exist, those with field strengths of 14.1 T (equivalent to a ^{1}H Larmor frequency of 600 MHz) are a commonplace for laboratories conducting metabonomics; these provide a good compromise between spectral resolution, sensitivity, and both capital and operational cost (which can be prohibitive).

For one-dimensional (one frequency domain plus intensity/abundance) ^{1}H NMR spectroscopic profiles, it is common practice for raw acquired data (free induction decays in the time domain) to be transformed to a frequency spectrum using common spectral processing parameters (e.g., apodization function/line broadening), spectral phasing and baseline correction, and chemical shift calibration to an internal standard (commonly 3-(trimethylsilyl)propionic-2,2,3,3-d_4 (TSP-d_4) or 3-(trimethylsilyl)-1-propanesulfonic acid-d_6 (DSS-d_6)), and optional spectral alignment of features. Resulting spectra can then be reported in a number of ways, with the data table typically reflecting a chemical shift region, and a corresponding spectral integral. Those less familiar with the origin of the reported features should remember that compounds are often represented in the spectra by multiple signals, and that chemical shift regions can be occupied by the signals from multiple metabolites, that is, signal overlap.

NMR spectroscopy is an attractive platform for metabolic profiling for several reasons:

- *Untargeted.* While it is possible to perform selective NMR experiments that report on specific chemical functionalities, and the physicochemical properties of some small molecules mean they are not/poorly detected, typical pulse sequences for metabolic phenotyping report on all NMR-visible compounds in a sample with no *a priori* analyte selection. This makes the platform ideal for agnostic analysis, identification of outliers, and highlighting sample contaminant issues.
- *Quantitative.* The inherently quantitative nature of NMR spectroscopy is attractive for analyzing multicomponent mixtures. Furthermore, the technique has a high and linear dynamic range, which although less than that of metabolite concentrations found in biofluids, reduces the need for sample concentration adjustment prior to analysis.
- *Nondestructive.* NMR spectroscopic analysis is nondestructive, and the sample remains unaffected by the NMR experiment itself; this is particularly important in the context of analyzing rare/precious samples from biobank archives. It also means that multiple experiments can be conducted on a single sample, without the requirements for additional aliquots (e.g., acquisition of multidimensional experiments for improved spectral resolution and/or *de novo* structure elucidation of unknowns).
- *Representative.* Samples can be handled in a way to minimize any deterioration (e.g., addition of preservatives such as sodium azide, or refrigerated prior to analysis), and when analysed are in a similar state to that found physiologically.
- *Simple.* Sample preparation for NMR-based metabonomics is minimal, which advantageously minimizes laboratory demands. It also means that the analyzed sample remains highly representative of that which was sampled, with minimal opportunity for bias to be introduced in preparation.

- *High-throughput.* Spectra can be obtained in an automated manner, at high-throughput. For profiling purposes, the acquisition of one-dimensional metabolic profiles (spectra with a single frequency domain) can be conducted in matter of minutes (common protocols have total per-sample analysis time of ~20 min for a standard suite of profiling acquisitions) (Dona et al. 2014).
- *Isolated.* NMR spectroscopic analysis of samples prepared in individual tubes (as opposed to flow-cell systems) benefits from the complete absence of a carryover between the samples because the samples remain physically isolated from the instrument throughout the experiment. This makes NMR spectroscopy an ideal screening platform for identifying samples suffering from gross contamination or other compositional properties that might adversely affect other analytical platforms (e.g., those with a chromatographic component) and therefore serve to protect instrumentation.
- *Informative.* The basis of the profiles generated from NMR experiments intrinsically provides a wealth of structural information about the species detected, and therefore allows a direct connection between features of interest at their chemical identity.

Some additional thoughts on the application of NMR spectroscopy have been recently provided (Wist 2017).

One of the first exposome applications explored metabolic signatures related to low-level heavy metal exposures using ^{1}H NMR spectroscopy (Avonmouth study; Thomas et al. 2007; Ellis et al. 2012). Urine samples from 178 volunteers living nearby to a decommissioned industrial smelter were analyzed by 600 Mhz ^{1}H NMR spectroscopy to identify correlates to existing urinary markers of chronic cadmium exposure (Cd content, *N*-acetyl-β-D-glucosaminidase). Associations with cadmium included mitochondrial metabolism, and one-carbon metabolism. The authors noted that significant associations to urinary citrate concentration with cadmium exposure and smoking status survived adjusting for age and sex, known factors in cadmium body burden.

Mass Spectrometry

Mass spectrometry (MS) is an analytical technique that is based on the detection of charged species (ions) in the gas phase. Many variants exist, but those most relevant to metabonomics applications are described below. The resolution (i.e., the ability to distinguish two spectral peaks with differing mass-to-charge ratio (*m*/*z*) values) and sensitivity (i.e., the systems response for a given amount of sample introduced) of instruments are the strongest determinants of their utility in metabolic profiling. Other factors include instrumental parameters such as the scan rate (which can influence ability to characterize chromatographic separations), ionization mode (which can influence the type of analytes detected), and the linear response range.

The most commonly used mass spectrometers in metabolic phenotyping are based on time-of-flight mass analyzers (Lewis et al. 2016; Wilson et al. 2005b).

The principle by which ions are differentiated according to their *m*/*z* values is that of the time taken to travel a given distance to a detector, after acceleration by an electrostatic force (a pusher); ions are accelerated at different rates, related to their inertial mass (relative to the charge they have), and arrive at the detector in ascending *m*/*z* order. The transit time is related to the *m*/*z* using a calibration curve, and typically permits low- or sub-ppm accurate mass measurements to be made. The benefit of these analyzers is that they can operate at very high scan rates, and therefore provide good characterization of chromatographic separations (i.e., high number of points per chromatographic peak) while still retaining wide spectral width and good sensitivity.

Quadrupole spectrometers use a different principle—electromagnetic fields are adjusted to selectively focus ions with specific *m*/*z* onto a detector. Commonly, this is done as a sweep through over an *m*/*z* range, or by preselection of specific *m*/*z* values/ranges. Consequently, quadrupole instruments can be operated at very high sensitivity, by tuning the conditions to optimally ionize specific molecules and detect their molecular/fragment/adduct ions, and are therefore commonly used for targeted, quantitative assays (Gray et al. 2017).

Fourier transform ion cyclotron resonance mass spectrometry (FT-ICR-MS) instruments offer a very high mass resolution, and can provide *m*/*z* measurements with sublime high accuracy and precision (commonly <1 ppm). These instruments measure *m*/*z* values of ions based on their cyclotron frequencies in relation to a fixed magnetic field; the collective, induced signal from these ions is recorded as an induction decay, and deconvolved using Fourier transformation to yield the corresponding mass spectrum. Orbitrap mass spectrometers also utilize the principle of recording the image current of ions circulating in a magnetic field, and are another member of the FTMS family (Hu et al. 2005; Jones et al. 2014; Zubarev and Makarov 2013). FT-ICR and Orbitrap instrumentation has been the subject of a comprehensive review (Ghaste et al. 2016). FT-MS instruments have utility in the characterization of unknown analytes; for small molecules, the ability to measure the *m*/*z* with great accuracy and precision often enables the elemental composition of an analyte or its fragments to be determined (or at least provide a short list), and coupled with fragmentation data can provide structural information for *de novo* characterization/identification of unknowns. FT-MS can also provide exquisite sensitivity (Ghaste et al. 2016). The major operational limitation of FT-ICR-MS instruments is that the scan rate to achieve such high resolution (commonly ~1 Hz) is unfavorable when attempting to characterize chromatographic separations such as those of UPLC-MS; the scan rate affords only a small number of spectral points to be obtained for each peak, and therefore can lead to subpar integration of the peak area. FT-ICR-MS instruments also have very high capital costs and overheads, alongside requiring highly specialized operator knowledge to deliver high-quality data; use in metabolic phenotyping is currently limited, but the analytical capabilities of these instruments may prove particularly useful in the screening, identification, and monitoring of low-abundance environmental exposures.

Outputs of GC-MS or LC-MS analyses include both an *m*/*z* domain and a chromatographic domain, in additional to an intensity/abundance domain. Reporting the entire dataset at full resolution is prohibitive in terms of size and complexity, so preprocessing typically reduces the data down to a feature set that has been picked from the data according to predefined criteria, and reported as the intensity of *m/z*-retention time pairs. These features are commonly filtered according to other criteria including their alignment and/or presence with similar features across a sample set, or their analytical characteristics (e.g., linearity of response, signal-to-noise ratio), which reduces the number of spurious features in the final data table. Those less familiar with the way in which these data are derived should remember that (without the application of additional processing routines) one metabolite can be represented by multiple features in the final data table because adducts, isotopes, and fragments may be present in addition to the parent molecular ion.

Collectively, mass spectrometry methods are an attractive proposition for metabolic profiling studies for several reasons:

- *Targeted and untargeted analyses are possible*. It is possible to generate spectra that offer great coverage.
- *Sensitivity*. While instrumental sensitivity relies on the ionization of chemical species, which may be poor in some cases, one of the key advantages of modern mass spectrometry is their exquisite sensitivity.
- *Cost*. Infrastructure and running costs are relatively low; instrument footprint can be small (not so for FT-ICR-MS).

Mass spectrometric analysis has recently been deployed in an exposome study of air pollution (Vlaanderen et al. 2017). In this work, the authors exposed healthy volunteers to ambient air for a prolonged period, measuring common health markers, multiple external exposure markers, and sampling blood for UPLC-MS-based metabolic phenotyping. Associations of metabolites with both air pollutants and health markers were found for tyrosine, guanosine, and hypothanthine, with a subsequent bioinformatics analysis indicating several enriched pathways.

Other Spectroscopic Methods

Attempts to use other spectroscopic methods such as IR, Raman, and UV/vis in metabolic phenotyping have found limited applicability. One drawback of such techniques is their limited ability to provide sufficient molecular specificity for individual components of complex mixtures; the overlap of signals arising from similar chemical groups is often considerable, and is insufficient to provide information for specific metabolome components. However, as these techniques are inexpensive and easy to deploy and can provide a multivariate profile based on the chemical composition of the sample, they may yet find utility in achieving widespread exposome coverage, and as an augmentation of mainstream profiling methods (Cherney et al. 2007; Ellis and Goodacre 2006; Xiao et al. 2016).

What Analytical Techniques/Platforms Are Used for Metabolite Separation?

The physical separation of sample components affords several advantages in the analysis of complex biofluids, and therefore detectors are often *hyphenated* with chromatographic systems to increase the usable information content that can be obtained from metabolic phenotyping. Appropriate selection of the separation type and specific conditions used allows components to be preferentially separated by several properties including hydrophobicity, ionization, size, and enantiospecificity. The two main methods for biofluid separation in metabolic phenotyping applications—liquid chromatography (LC) and gas chromatography (GC)—are summarized below; other methods including capillary electrophoresis (CE) (Maier and Schmitt-Kopplin 2016; Ramautar et al. 2017) and supercritical fluid chromatography (SFC) (Kassner et al. 2008; Bamba et al. 2008, 2012) are yet to become well established but can offer complementary analysis.

Liquid Chromatography

Most common in the field of metabolic phenotyping is liquid chromatography (LC), based on the retention of analytes on a stationary adsorbent solid phase eluted with a solvent mobile phase. The properties of the stationary phase and the solvent can be adjusted to enable the separation of sample components based on their differential interactions with the stationary phase, which may result from multiple physicochemical properties including dipole–dipole interactions (polarity), dimensions (size), and ionic affinity (electrical charge). In most research applications, the stationary phase is packed into a column, and the solvent eluted under high pressure (HPLC) either isocratically (i.e., composition is kept constant) or as gradient (i.e., the composition is altered according to a predefined scheme). Modifiers may be added to the solvent to tune the separation, or to improve the characteristics of the solvent for hyphenation to MS- or NMR-based detection. The flow rate and composition of the mobile phase and column temperature can be altered, and also influence the separation, as do the physical characteristics of the column (length, diameter) and the particle size of the stationary phase; these have a bearing on the resulting chromatographic performance. Sample components sequentially elute with the mobile phase from the chromatographic column as "peaks" and may be detected, collected, or discarded.

Two main types of LC are common for metabonomics applications, often in parallel (Contrepois et al. 2015) to provide complementary chromatographic separations:

- *Reversed-phase chromatography (RPC)*. A nonpolar stationary phase (commonly based on C_8 or C_{18} functionalized material) is eluted with a polar solvent

(typically a combination of water and methanol or acetonitrile). Hydrophobic metabolites are well retained (Want et al. 2010; Lewis et al. 2016).
- *Hydrophilic interaction chromatography (HILIC).* A polar stationary phase (commonly functionalized with diol or amide groups) is eluted with a largely nonpolar solvent. Highly polar metabolites are well retained (Spagou et al. 2010).

Liquid chromatographic separations of biofluids for metabonomics applications are typically straightforward, with "dilute-and-shoot" (urine) and simple protein precipitation (blood) protocols in common usage (Want et al. 2010; Lewis et al. 2016; Dunn et al. 2011).

Gas Chromatography

Gas chromatography is also a commonly used platform for separation of biofluid components in profiling applications for large-scale sample sets (Dunn et al. 2014). A gas chromatograph (GC) separates sample components in a similar manner to LC, but the mobile phase used is an inert gas (e.g., helium, nitrogen, or hydrogen), and the stationary phase is a narrowly packed or functionalized capillary (or a liquid). As with LC, flow rate and column temperature can be controlled too. Efficient GC separations typically require the derivatization of sample components to make them readily volatile and separable (Moros et al. 2017; Abbiss et al. 2015). Optimized protocols for using gas chromatography for separations in metabonomics applications have been established (Dunn et al. 2011), and retention time-based libraries for metabolites are available for specific protocols (e.g., Agilent Fiehn 2013 GC-MS Metabolomics RTL Library) (Kind et al. 2009).

How Are Metabolic Profiles Annotated?

One of the main advantages of using an untargeted analysis for metabolic phenotyping is the absence of the need to make *a priori* selections of analytes. This allows for efficient data-driven analysis to reveal features of interest, and may uncover unexpected phenomena that would not have been captured by a targeted analysis (e.g., novel metabolites/profile features). For those features identified by untargeted methods, there is a need to provide an annotation and move from profile representations to discussion of metabolites and their relevance within the study. For many commonly observed metabolites, annotation is straightforward; spectral features relating to metabolites with high abundance can be rapidly assigned. To positively confirm spectral annotations, evidence is gained through multiple sources:

- *Aggregation of relevant available spectral information.* Experienced spectroscopists will be able to interrogate raw spectral data to provide putative identities for the unknown. The complexity of biofluid spectra may confound this analysis, and therefore tools to filter and/or connect related spectral features are useful. Common approaches include the use of statistical spectroscopy (STOCSY and related techniques) to identify sets of features that are highly correlated across spectra (either within or between spectral sets). These techniques have been the subject of a recent review (Robinette et al. 2013). Statistical methods are efficient as they can be conducted on existing profile data and rapidly implemented in silico, as recently demonstrated in relation to dietary biomarker identification (Posma et al. 2017).

- *Generation of metabolite-specific spectral information*. Additional spectroscopic analyses can be conducted to provide structural information about metabolites, and/or improve overall spectral resolution; multidimensional NMR spectroscopic experiments such as correlation spectroscopy (COSY), J-resolved (JRES), diffusion ordered (DOSY), or heteronuclear multiple bond correlation (HMBC); targeted or data-dependent acquisitions of tandem mass spectrometry (MSMS) can reveal complementary structural clues through (Marchand et al. 2017; Keun et al. 2002; Beckonert et al. 2007).
- *Biochemical intervention*. Addition of chemical (e.g., chelating) or biochemical (e.g., enzymes) agents that will result in perturbation of the spectral features arising from specific classes of compounds as additional confirmation of their identity.
- *Spectral database matching*. The spectral features obtained from profiling and/or other experiments can be submitted as search criteria to metabolite/chemical databases. For NMR spectroscopic data, this can include chemical shift, relative peak intensity, and scalar coupling information; for MS, molecular ion/adduct/fragment (intensity) pattern and isotope distribution are common inputs.
- *Spike-in experiments*. Confirmation with an authenticated chemical standard by addition.
- *Additional radiolabelled or stable labelled compounds*. Less relevant to human population studies, but use of isotopologues to trace or provide distinct isotope distribution patterns is common in other metabonomics applications (Creek et al. 2012; Creek 2013).

What Advances in Analytical Techniques/Platforms Have Been Most Significant in Metabolic Phenotyping and Metabolite Annotation?

Three main analytical platforms/techniques have had a substantial impact on the applicability of metabonomics methods to exposome studies as detailed below.

Ultraperformance Liquid Chromatography

Advances in column, pump, and fluidic technology signaled a step change in the resolution and throughput of routine LC-MS separations (Wilson et al. 2005b, b). Central to this was the ability to use much smaller chromatographic particles, while

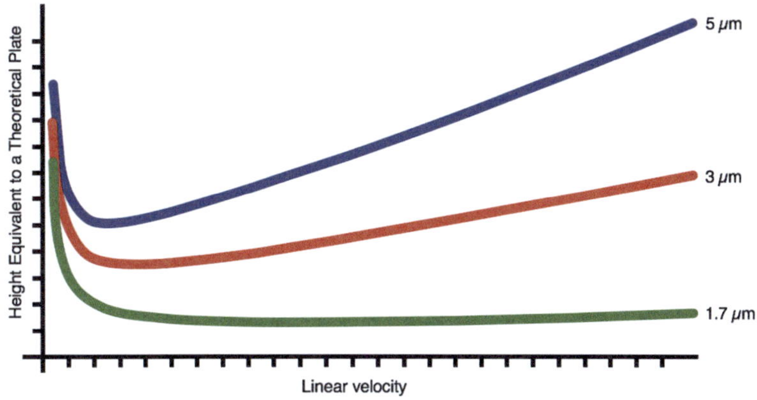

Fig. 6.4 Chromatographic resolving power as a function of the linear velocity for columns packed with particles of diameter 5 μm, 3 μm, and 1.7 μm. It is noted that the optimal linear velocity (lowest HETP) range increases as particle size is decreased

retaining high flow rates (the consequent increases in back pressure place demands on the column material and the solvent delivery system [valves, pumps, etc.]), termed ultraperformance or ultra-high-performance liquid chromatography UPLC/ UHPLC. While full treatment of chromatographic separations is beyond the scope of this chapter, we here consider the basic principles that govern column resolving power (height equivalent to a theoretical plate, HETP) according to the relation described by the van Deemter equation (Van Deemter et al. 1956).

The practical reality of this is that the use of smaller particles increases the range of linear velocities over which optimal resolving power (lowest HETP) can be achieved, and therefore separations with sub 2-μm particles can be performed at high flow rates without a loss of resolution (Fig. 6.4). Thus, instruments and columns capable of operating at such linear velocities, and the high column back pressures that accompany them, offer improvement in throughput while retaining or increasing peak capacity. An additional benefit of such separations is that chromatographic peak intensity increases, and therefore sensitivity of hyphenated detectors is improved.

Ion Mobility Separations

Ion-mobility (IM) separations offer an attractive addition to the separations detailed above. This technique separates ions in the gas phase according to their collision cross section (CCS) (Cumeras et al. 2015a, b). In addition to the orthogonal separation IM provides (with concomitant improvement in resolution and reduced spectral overlap), the CCS of molecular ions that can be derived from suitably calibrated mobility measurements provides an additional characteristic that can be used to identify metabolites; databases of such measurements offer potential for enhancing automated metabolite annotation approaches (Paglia et al. 2014a, b).

Targeted Analytical Assays

As above, targeted analyses provide information about individual, or multiple predefined metabolites. While this does not permit agnostic analysis and discovery of previously unknown entities that may be of interest (e.g., using data-driven analysis), the output of these assays can greatly enhance the ability of research teams to work collaboratively between disciplines. Targeted assays typically provide annotated data outputs that can be immediately inputted into informatics tools (e.g., pathway enrichment, network models, etc.) and the identification of features of interest is already known. This also reduces the bottleneck that is commonly presented by the need to positively identify features of interest, which may require substantial analytical resources. In several studies, kit-form targeted assays—which are typically focused on core metabolic intermediates—have helped integrate metabolite profiling with epidemiological analysis. At present, several options are available to researchers, with most being used to acquire targeted data in parallel with untargeted phenotyping (Suhre et al. 2010).

Reporting and Interpreting the Human Metabolome

Approaches to data analysis pipelines mirror those of in the laboratory:

1. *Distributed analysis*. Individuals/laboratories develop bespoke data pipelines that leverage in-house/proprietary knowledge, as well as open-access sources; computation often conducted at a relatively low rate on workstation or local server/cluster. Suited to highly flexible analyses that allow users full control on analysis resource and workflow; highly transparent to users; may result in processed data being locally useful but not directly compatible with other centers. Many tools (open-source and commercial) that perform similar analytical pipelines are available.
2. *Centralized analysis*. Online/remote server-based processing. Benefits from availability and transparency of existing protocols that maximize data interoperability. Discussed in a recent perspective on cloud computing of metabolite profile data (Warth et al. 2017).

How Are Metabolic Phenotypic Data Modelled?

A detailed description of data analysis tools for integrating omics measurements and their integration are provided elsewhere in this book. For metabolome-specific analyses, myriad options exist, but readers are directed to selected methods that include interactive tools for exploring transgenomic metabolic networks (Metabonetworks) (Posma et al. 2014), online processing pipelines PhenoMeNal (http://phenomenal-h2020.eu)/Galaxy-M (Davidson et al. 2016) XCMS-online (Huan et al. 2017; Tautenhahn et al. 2012), and MetaboAnalyst (Xia et al. 2015). One approach that amalgamates metabolite annotation and network analysis is the *mummichog* algorithm implementation. This approach has been developed to explore untargeted metabolic profile data, utilizing knowledge of documented metabolic pathways and networks reconstructions (Ma et al. 2007; Tanabe et al. 2017; Swainston et al. 2016; Duarte et al. 2007; Thiele et al. 2013) to predict functional activity from profile feature data prior to (or without reference to) positive identification. While this does not directly address the need for annotation, the computed network connectivity of a spectral feature may provide additional clues as to its identity.

How Are Metabolic Phenotypic Data Stored and Shared?

Metabolic profiles that combine the plurality of analytical methods used to generate them, has made sharing of data difficult to achieve. Raw spectral data generated from untargeted analysis is useful as it allows for complete reanalysis, although it is not easy for nonspecialists to use, and therefore is infrequently used as the primary format for data sharing. More common is for a limited set of preprocessing steps to be conducted to provide a more human-readable data table. What and how to report such data is not self-evident, and inconsistent reporting and formats have historically limited data interoperability.

Collaborative efforts between data generators and repositories are addressing these challenges, and facilitating a move toward more accessible data in metabonomics (Rocca-Serra et al. 2016). These include the Metabolomics Standards Initiative (MSI) and Coordination of Standards in MetabOlomicS (COSMOS) (Fiehn et al. 2007; Sansone et al. 2007; Salek et al. 2015), which built on other initiatives within the metabonomics research community (Lindon et al. 2005; Jenkins et al. 2004). To avoid the existence of multiple parallel repositories for metabolome data from causing data to be "siloed" based on the framework used for deposition, a community effort to enable efficient exchange of data/metadata between repositories has been developed by an EU-funded consortium http://metabolomexchange.org.

MetaboLights (http://www.ebi.ac.uk/metabolights) is a repository for metabolic phenotype datasets, providing a framework for deposition of experimental data (i.e., raw or processed biofluid spectra) and associated metadata in a systematized manner using the established ISA-Tab data format (Haug et al. 2013). The primary repository function enables deposited datasets to be identified (browse or search) according to available metadata (species, analytical platform, submission date, etc.). For those datasets with open-access agreements in place, researchers can download a local copy for reanalysis and integration. An additional function provided by MetaboLights is an efficient mechanism for retrieving metabolite-related information from many preexisting metabolome database resources (Table 6.1).

Table 6.1 Selected chemical, metabolite, and spectral databases/repositories for metabolome information

Name	Abbreviation	Curator	Website	Reference
Biological Magnetic Resonance Bank	BMRB/ BioMagResBank	University of Wisconsin	www.bmrb.wisc.edu	Ulrich et al. (2008)
Chemical Entities of Biological Interest	ChEBI	EMBL-EBI	www.ebi.ac.uk/chebi	Hastings et al. (2013)
ChemSpider		Royal Society of Chemistry	www.chemspider.com	
Human Metabolome Database	HMDB	The Metabolomics Innovation Centre	www.hmdb.ca	Wishart et al. (2012)
KNApSAcK		NAIST Comparative Genomics Laboratory	kanaya.naist.jp/KNApSAcK	Nakamura et al. (2014)
LIPID Metabolites and Pathways Strategy	LIPID MAPS®	LIPID MAPS® consortium	www.lipidmaps.org	Schmelzer et al. (2007)
Madison-Qingdao Metabolomics Consortium Database	MMCD	National Magnetic Resonance Facility at Madison	mmcd.nmrfam.wisc.edu/	Cui et al. (2008)
MassBank			www.massbank.jp	Horai et al. (2010)
MetaboLights		EMBL-EBI	www.ebi.ac.uk/metabolights	Haug et al. (2013)
MetaCyc		SRI International	metacyc.org	Caspi et al. (2014)
METLIN		The Scripps Research Institute	metlin.scripps.edu	Smith et al. (2005)
PubChem		NCBI	pubchem.ncbi.nlm.nih.gov	Kim et al. (2016)

Another comprehensive tool is the Metabolomics Workbench (www.metabolomicsworkbench.org) (Sud et al. 2016) which offers a similar functionality, alongside an established metabolite database. Further details on data sharing are discussed elsewhere in this book, and also in recent perspectives with a focus on metabolic profile data in particular (Haug et al. 2017).

Several community resources exist relating to the deposition and annotation of metabolic profile data, mostly with interfaces that allow searching using text, chemical structural information, or spectral feature data (m/z values, chemical shift, etc.). Selection is listed in Table 6.1.

How Is Quality Assessed in Metabolic Phenotype Data?

Exposome studies typically require many individuals (and consequently samples) to be analyzed, and therefore long-term assurance of analytical platform performance/stability is necessary to ensure the derived spectral data are both representative and comparable. While it is not possible to completely mitigate instrument failure or drift, it is possible to include sufficient measures to characterize performance, variability, and be able to confirm the analyses fit for purpose:

- *Split samples.* Common practice is for studies to include blinded, split samples that provide an opportunity for independent, objective measurement of analytical sufficiency.
- *Study pooled quality control samples.* A composite of all (or a specific subset of) study samples that commonly represent the geometric mean of all components across the set, typically achieved by mixing together equal volume aliquots of each study sample. This sample therefore contains all components from the entire sample set. Note: the concentration of metabolites detectable in individual samples may fall below the limit of detection in the pooled samples if only present in a small proportion of the overall set.
- *Long-term reference quality control sample.* Analyzed alongside regular samples, this provides an ability to compare platform performance between studies, and aids linkage by identifying changes to analytical performance that result in drift of observations (e.g., chromatographic retention time).
- *Dilution series.* A set of samples derived from the study pooled quality control sample diluted (e.g., with the final sample diluent) to enable an estimation of the linearity of response of individual sample components. This facilitates subsequent data analysis steps that can filter out spectral features that have insufficient analytical properties to be useful in analysis.

The nature of the analytical methods means that chemical contamination occurring before/during preparation and analysis can obfuscate the metabolic profiles obtained, and therefore it is recommended that blank samples be characterized using the analytical platform alongside the study samples:

- *Preparation blank.* A blank sample (e.g., final sample diluent) prepared alongside the study samples to enable assessment of contaminants introduced by the steps involved in sample preparation.
- *Analytical blank.* A sample of a representative solvent (e.g., final sample diluent) used to identify the presence and extent of carryover and spectral artifacts that are independent of the sample set or the preparative steps used prior to analysis.

Combination of phenotypic data across cohorts may help address the need to increase study power, and better characterize differential contributions from genetic and environmental factors in various populations. To do so, there is a need to establish that the platforms and protocols used to generate these data can be compared. To this end, several studies have investigated the performance characteristics of metabonomics analyses, including inter-day reproducibility and robustness in individual laboratory studies (Gika et al. 2012), as well as multicenter studies for both common untargeted and targeted platforms (Martin et al. 2015; Siskos et al. 2017; Benton et al. 2012).

Conclusions

This chapter has attempted to illustrate why metabolic measurements—that can provide a detailed phenotypic report on the internal chemical milieu—are important in exposome studies, with some examples of how this has been achieved to date. The factors that influence the makeup of the human metabolome have been discussed, painting a complex, dynamic picture that facilitates myriad gene–environment interactions. Given the metabolome includes the entire complement of small molecules present in each fluid, tissue, compartment, or system (i.e., defined biological entity), it is worth considering how this is related to an entire human. Compared with lower organisms that can be more completely characterized, that are less genotypically and phenotypically diverse, and that can be fed and maintained under strict dietary and environmental conditions, humans represent a much more heterogeneous bunch. Consequently, characterizing the composition of the human metabolome and the variability of its components within and between individuals is many orders of magnitude more complex, not least because it is very difficult (and possibly pointless) to define a "normal" human because combinations of genotypes and environmental exposures vary considerably.

Modern instrumentation and protocols for metabolome analysis can deliver wide coverage of metabolites, and can report on molecules from across the continuum of metabolism. Some researchers appear to restrict their focus of the metabolome to those species that are considered well-established endogenous molecules, and associated with human metabolic pathways/networks. This makes sense from the perspective of creating a tractable problem to tackle. While a perfectly reasonable approach, two issues surface when applying this thinking to the metabolome in the context of the human exposome. Firstly, even endogenous metabolism is far more complex than most model descriptions of the metabolome suggest. Secondly, only considering those entities that are considered "endogenous" seriously impacts on our ability to appreciate and more objectively characterize exposures; while it is true that—with good reason—the identity of many dietary exposures overlaps with endogenous metabolites, while most of the exposures we encounter do not. They *do*, however, both contribute to and (directly or indirectly) impact on the internal chemical milieu, which is a focus for our exposome studies. These aspects will be explored in ever greater detail as instrumentation becomes more sensitive, and we improve our ability of characterizing the in vivo chemical environment.

It is worth noting that our ability to cope with the impact of complex mixtures will become ever more crucial, to help understand their interactions: cooperative, inverse, concerted, additive, multiplicative, etc. While it is possible to detect quantitatively high xenobiotic exposures that occur, present routine analyses do not cover those chemicals that are present at very low abundance which may in combination be responsible for a substantial proportion of the overall lifetime exposure; specialist optimized methods that typically require significant sample volume are still required.

From a technical perspective, future exposome studies will likely benefit from the rapidly advancing field of ambient mass spectrometry methods such as desorption electrospray ionization (DESI) (Takáts et al. 2004, 2005; Dénes et al. 2009) and rapid evaporative ionization mass spectrometry (REIMS) (Balog et al. 2010; Damjanovich et al. 2013; Strittmatter et al. 2014), which have particularly attractive qualities for implementation at scale.

To service the need to push the boundaries of analytical performance, to characterize the feature set of the human metabolome, and to make progress in molecular epidemiological studies by sharing and combining datasets, it is likely that two complementary workflows will emerge featuring parallel agnostic and targeted analytical protocols, particularly those that can start accessing chemicals present in trace amounts (Jones 2016).

Similarly, the greatest benefit will likely result from applying parallel campaigns of laboratory studies which provide deep phenotyping of biofluids/tissues across multiple platforms, alongside in situ distributed analysis that can capture individual phenotypic information with very high temporal resolution.

Critical to the longer-term success (or at least the rate of advancement) of metabolic profiling in exposome studies will be consolidation of methods and sharing of data through structures such as those discussed above. Ultimately, no one group of researchers will be able to deliver on all aspects of exposome characterization, and therefore community efforts to provide sampling, analytical, and data sharing architectures will be key to realizing the potential of exposome science.

Conflict of Interest The author confirms no conflict of interest.

Acknowledgments TJA is supported by the EU FP7 EXPOsOMICS (grant agreement: 308610) and HELIX (grant agreement: 308333) projects.

References

Abbiss H, Rawlinson C, Maker GL, Trengove R (2015) Assessment of automated trimethylsilyl derivatization protocols for GC–MS-based untargeted metabolomic analysis of urine. Metabolomics 11:1908–1921

Anton G, Wilson R, Yu Z-H, Prehn C, Zukunft S, Adamski J et al (2015) Pre-analytical sample quality: metabolite ratios as an intrinsic marker for prolonged room temperature exposure of serum samples. Kim KH, editor. PLoS One e0121495:10

Athersuch T (2016) Metabolome analyses in exposome studies: profiling methods for a vast chemical space. Arch Biochem Biophys 589:177–186

Balog J, Szaniszlo T, Schaefer K-C, Dénes J, Lopata A, Godorhazy L et al (2010) Identification of biological tissues by rapid evaporative ionization mass spectrometry. Anal Chem 82:7343–7350

Bamba T, Shimonishi N, Matsubara A, Hirata K, Nakazawa Y, Kobayashi A et al (2008) High throughput and exhaustive analysis of diverse lipids by using supercritical fluid chromatography-mass spectrometry for metabolomics. J Biosci Bioeng 105:460–469

Bamba T, Lee JW, Matsubara A, Fukusaki E (2012) Metabolic profiling of lipids by supercritical fluid chromatography/mass spectrometry. J Chromatogr A 1250:212–219

Bartel J, Krumsiek J, Schramm K, Adamski J, Gieger C, Herder C et al (2015) The human blood metabolome-transcriptome interface. Inouye M, editor. PLoS Genet e1005274:11

Beckett AH (2008) Metabolic oxidation of aliphatic basic nitrogen atoms and their α-carbon atoms. Xenobiotica 1:365–384

Beckonert O, Keun HC, Ebbels TMD, Bundy J, Holmes E, Lindon JC et al (2007) Metabolic profiling, metabolomic and metabonomic procedures for NMR spectroscopy of urine, plasma, serum and tissue extracts. Nat Protoc 2:2692–2703

Beckonert O, Coen M, Keun HC, Wang Y (2010) High-resolution magic-angle-spinning NMR spectroscopy for metabolic profiling of intact tissues. Nat Protoc 5(6):1019–1032

Benton HP, Want E, Keun HC, Amberg A, Plumb RS, Goldfain-Blanc F et al (2012) Intra- and interlaboratory reproducibility of ultra performance liquid chromatography-time-of-flight mass spectrometry for urinary metabolic profiling. Anal Chem 84:2424–2432

Blaise BJ, Correia G, Tin A, Young JH, Vergnaud A-C, Lewis M et al (2016) Power analysis and sample size determination in metabolic phenotyping. Anal Chem 88:5179–5188

Bouatra S, Aziat F, Mandal R, Guo AC, Wilson MR, Knox C et al (2013) The human urine metabolome. Dzeja P, editor. PLoS One e73076:8

Budde K, Gök Ö-N, Pietzner M, Meisinger C, Leitzmann M, Nauck M et al (2016) Quality assurance in the pre-analytical phase of human urine samples by 1H NMR spectroscopy. Arch Biochem Biophys 589:10–17

Caspi R, Altman T, Billington R, Dreher K, Foerster H, Fulcher CA et al (2014) The MetaCyc database of metabolic pathways and enzymes and the BioCyc collection of Pathway/Genome Databases. Nucleic Acids Res 42:D459–D471

Chan ECY, Pasikanti KK, Nicholson JK (2011) Global urinary metabolic profiling procedures using gas chromatography-mass spectrometry. Nat Protoc 6:1483–1499

Cherney DP, Ekman DR, Dix DJ, Collette TW (2007) Raman spectroscopy-based metabolomics for differentiating exposures to triazole fungicides using rat urine. Anal Chem 79:7324–7332

Contrepois K, Jiang L, Snyder M (2015) Optimized analytical procedures for the untargeted metabolomic profiling of human urine and plasma by combining hydrophilic interaction (HILIC) and reverse-phase liquid chromatography (RPLC)-mass spectrometry. Mol Cell Proteomics 14:1684–1695

Creek DJ (2013) Stable isotope labeled metabolomics improves identification of novel metabolites and pathways. Bioanalysis 5:1807–1810. https://doi.org/10.4155/bio.13.131

Creek DJ, Chokkathukalam A, Jankevics A, Burgess KEV, Breitling R, Barrett MP (2012) Stable isotope-assisted metabolomics for network-wide metabolic pathway elucidation. Anal Chem 84:8442–8447

Cui Q, Lewis IA, Hegeman AD, Anderson ME, Li J, Schulte CF et al (2008) Metabolite identification via the Madison Metabolomics Consortium Database. Nat Biotechnol 26:162–164

Cumeras R, Figueras E, Davis CE, Baumbach JI, Gràcia I (2015a) Review on Ion Mobility Spectrometry. Part 1: current instrumentation. Analyst 140:1376–1390

Cumeras R, Figueras E, Davis CE, Baumbach JI, Gràcia I (2015b) Review on Ion Mobility Spectrometry. Part 2: hyphenated methods and effects of experimental parameters. Analyst 140:1391–1410

Damjanovich L, Darzi A, Nicholson JK (2013) Intraoperative tissue identification using rapid evaporative ionization mass spectrometry. Sci Transl Med 5(194):194ra93

Damsten MC, Commandeur JNM, Fidder A, Hulst AG, Touw D, Noort D et al (2007) Liquid chromatography/tandem mass spectrometry detection of covalent binding of acetaminophen to human serum albumin. Drug Metab Dispos 35:1408–1417

Davidson RL, Weber RJM, Liu H, Sharma-Oates A, Viant MR (2016) Galaxy-M: a Galaxy workflow for processing and analyzing direct infusion and liquid chromatography mass spectrometry-based metabolomics data. Gigascience 5:10

Davies SK, Ang JE, Revell VL, Holmes B, Mann A, Robertson FP et al (2014) Effect of sleep deprivation on the human metabolome. PNAS 111:10761–10766

Dénes J, Katona M, Hosszú Á, Czuczy N, Takáts Z (2009) Analysis of biological fluids by direct combination of solid phase extraction and desorption electrospray ionization mass spectrometry. Anal Chem 81:1669–1675

Dénes J, Szabó E, Robinette SL, Szatmári I, Szőnyi L, Kreuder JG et al (2012) Metabonomics of newborn screening dried blood spot samples: a novel approach in the screening and diagnostics of inborn errors of metabolism. Anal Chem 84:10113–10120

Dona AC, Jiménez B, Schäfer H, Humpfer E, Spraul M, Lewis MR et al (2014) Precision high-throughput proton NMR spectroscopy of human urine, serum, and plasma for large-scale metabolic phenotyping. Anal Chem 86:9887–9894

Duarte NC, Becker SA, Jamshidi N, Thiele I, Mo ML, Vo TD et al (2007) Global reconstruction of the human metabolic network based on genomic and bibliomic data. PNAS 104:1777–1782

Dunn WB, Broadhurst D, Begley P, Zelena E, Francis-McIntyre S, Anderson N et al (2011) Procedures for large-scale metabolic profiling of serum and plasma using gas chromatography and liquid chromatography coupled to mass spectrometry. Nat Protoc 6:1060–1083

Dunn WB, Lin W, Broadhurst D, Begley P, Brown M, Zelena E et al (2014) Molecular phenotyping of a UK population: defining the human serum metabolome. Metabolomics 11:9–26

Ellis DI, Goodacre R (2006) Metabolic fingerprinting in disease diagnosis: biomedical applications of infrared and Raman spectroscopy. Analyst 131:875–885

Ellis JK, Athersuch TJ, Thomas LD, Teichert F, Pérez-Trujillo M, Svendsen C et al (2012) Metabolic profiling detects early effects of environmental and lifestyle exposure to cadmium in a human population. BMC Med 10:61

Fiehn O (2001) Combining genomics, metabolome analysis, and biochemical modelling to understand metabolic networks. Comp Funct Genomics 2:155–168

Fiehn O, Robertson D, Griffin J, van der Werf M, Nikolau B, Morrison N et al (2007) The metabolomics standards initiative (MSI). Metabolomics 3:175–178

Fink T, Reymond J-L (2007) Virtual exploration of the chemical universe up to 11 atoms of C, N, O, F: assembly of 26.4 million structures (110.9 million stereoisomers) and analysis for new ring systems, stereochemistry, physicochemical properties, compound classes, and drug discovery. J Chem Inf Model 47:342–353

García A, Barbas C (2010) Gas chromatography-mass spectrometry (GC-MS)-based metabolomics. In: Fan J-B (ed) Metabolic profiling. Humana Press, Totowa, NJ, pp 191–204

Ghaste M, Mistrik R, Shulaev V (2016) Applications of fourier transform ion cyclotron resonance (FT-ICR) and orbitrap based high resolution mass spectrometry in metabolomics and lipidomics. Int J Mol Sci 17(6):pii: E816

Gieger C, Geistlinger L, Altmaier E, Hrabé de Angelis M, Kronenberg F, Meitinger T et al (2008) Genetics meets metabolomics: a genome-wide association study of metabolite profiles in human serum. PLoS Genet 4:e1000282

Gika HG, Theodoridis GA, Wilson ID (2008) Liquid chromatography and ultra-performance liquid chromatography–mass spectrometry fingerprinting of human urine. J Chromatogr A 1189:314–322

Gika HG, Theodoridis GA, Earll M, Wilson ID (2012) A QC approach to the determination of day-to-day reproducibility and robustness of LC–MS methods for global metabolite profiling in metabonomics/metabolomics. Bioanalysis 4:2239–2247. https://doi.org/10.4155/bio.12.212

Giskeødegård GF, Davies SK, Revell VL, Keun H, Skene DJ (2015) Diurnal rhythms in the human urine metabolome during sleep and total sleep deprivation. Sci Rep 5:14843

Gray N, Zia R, King A, Patel VC, Wendon J, McPhail MJW et al (2017) High-speed quantitative UPLC-MS analysis of multiple amines in human plasma and serum via precolumn

derivatization with 6-aminoquinolyl-N-hydroxysuccinimidyl carbamate: application to acetaminophen-induced liver failure. Anal Chem 89:2478–2487
Guthrie R, Susi A (1963) A simple phenylalanine method for detecting phenylketonuria in large populations of newborn infants. Pediatrics 32:338–343
Hastings J, de Matos P, Dekker A, Ennis M, Harsha B, Kale N et al (2013) The ChEBI reference database and ontology for biologically relevant chemistry: enhancements for 2013. Nucleic Acids Res 41:D456–D463
Haug K, Salek RM, Conesa P, Hastings J (2013) MetaboLights—an open-access general-purpose repository for metabolomics studies and associated meta-data. Nucleic Acids Res 41:D781–D786
Haug K, Salek RM, Steinbeck C (2017) Global open data management in metabolomics. Curr Opin Chem Biol 36:58–63
Hebels DGAJ, Georgiadis P, Keun HC, Athersuch TJ, Vineis P, Vermeulen R et al (2013) Performance in omics analyses of blood samples in long-term storage: opportunities for the exploitation of existing biobanks in environmental health research. Environ Health Perspect 121 (4):480–487
Hernandes VV, Barbas C, Dudzik D (2017) A review of blood sample handling and pre-processing for metabolomics studies. Electrophoresis 38(18):2232–2241
Hicks AA, Pramstaller PP, Johansson Å, Vitart V, Rudan I, Ugocsai P et al (2009) Genetic determinants of circulating sphingolipid concentrations in European populations. Gibson G, editor. PLoS Genet e1000672:5
Holmes E, Nicholson JK (2008) Human metabolic phenotyping and metabolome wide association studies. Oncogenes meet metabolism. Springer, Berlin, Heidelberg, pp 227–249
Holmes E, Loo RL, Cloarec O, Coen M, Tang H, Maibaum E et al (2007) Detection of urinary drug metabolite (xenometabolome) signatures in molecular epidemiology studies via statistical total correlation (NMR) spectroscopy. Anal Chem 79:2629–2640
Holmes E, Wilson ID, Nicholson JK (2008) Metabolic phenotyping in health and disease. Cell 134:714–717
Holmes E, Stamler J, Nicholson JK (2010) Opening up the "Black Box": metabolic phenotyping and metabolome-wide association studies in epidemiology. J Clin Epidemiol 63(9):970–979
Horai H, Arita M, Kanaya S, Nihei Y, Ikeda T, Suwa K et al (2010) MassBank: a public repository for sharing mass spectral data for life sciences. J Mass Spectrom 45:703–714
Hu Q, Noll RJ, Li H, Makarov A, Hardman M, Graham CR (2005) The Orbitrap: a new mass spectrometer. J Mass Spectrom 40:430–443
Huan T, Forsberg EM, Rinehart D, Johnson CH (2017) Systems biology guided by XCMS Online metabolomics. Nat Methods 14(5):461–462
IARC (2012) Chemical agents and related occupations. Benzo[α]pyrene. IARC Monogr Eval Carcinog Risk Chem Man 100F:111–144
Illig T, Gieger C, Zhai G, Römisch-Margl W, Wang Sattler R, Prehn C et al (2010) A genome-wide perspective of genetic variation in human metabolism. Nat Genet 42:137–141
Jansen RJ, Argos M, Tong L, Li J, Rakibuz-Zaman M, Islam MT et al (2016) Determinants and consequences of arsenic metabolism efficiency among 4,794 individuals: demographics, lifestyle, genetics, and toxicity. Cancer Epidemiol Biomark Prev 25(2):381–390
Jenkins H, Hardy N, Beckmann M, Draper J, Smith AR, Taylor J et al (2004) A proposed framework for the description of plant metabolomics experiments and their results. Nat Biotechnol 22:1601–1606
Jones DP (2016) Sequencing the exposome: a call to action. Toxicol Rep 3:29–45
Jones DR, Wu Z, Chauhan D, Anderson KC, Peng J (2014) A nano ultra-performance liquid chromatography–high resolution mass spectrometry approach for global metabolomic profiling and case study on drug-resistant multiple myeloma. Anal Chem 86:3667–3675
Kamlage B, Maldonado SG, Bethan B, Peter E, Schmitz O, Liebenberg V et al (2014) Quality markers addressing preanalytical variations of blood and plasma processing identified by broad and targeted metabolite profiling. Clin Chem 60:399–412

Karagas MR, Gossai A, Pierce B, Ahsan H (2015) Drinking water arsenic contamination, skin lesions, and malignancies: a systematic review of the global evidence. Curr Envir Health Rep 2:52–68

Kassner MK, Charney R, Fernandez FM (2008) Novel approach to employing SFC for metabolomics research. In: Aiche annual meeting, conference proceeding

Kastenmüller G, Raffler J, Gieger C, Suhre K (2015) Genetics of human metabolism: an update. Hum Mol Genet 24(R1):R93–R101

Keun HC, Athersuch TJ (2010) Nuclear magnetic resonance (NMR)-based metabolomics. In: Fan J-B (ed) Metabolic profiling. Humana Press, Totowa, NJ, pp 321–334

Keun HC, Beckonert O, Griffin JL, Richter C, Moskau D, Lindon JC, Nicholson JK et al (2002) Cryogenic probe 13C NMR spectroscopy of urine for metabonomic studies. Anal Chem 74:4588–4593

Kim K, Mall C, Taylor SL, Hitchcock S, Zhang C, Wettersten HI et al (2014) Mealtime, temporal, and daily variability of the human urinary and plasma metabolomes in a tightly controlled environment. Brennan L, editor. PLoS One e86223:9

Kim S, Thiessen PA, Bolton EE, Chen J (2016) PubChem substance and compound databases. Nucleic Acids Res 44:D1202–D1213

Kind T, Wohlgemuth G, Lee DY, Lu Y, Palazoglu M, Shahbaz S et al (2009) FiehnLib: mass spectral and retention index libraries for metabolomics based on quadrupole and time-of-flight gas chromatography/mass spectrometry. Anal Chem 81:10038–10048

Lewis MR, Pearce JTM, Spagou K, Green M, Dona AC, Yuen AHY et al (2016) Development and application of ultra-performance liquid chromatography-TOF MS for precision large scale urinary metabolic phenotyping. Anal Chem 88:9004–9013

Li N, Song YP, Tang H, Wang Y (2016) Recent developments in sample preparation and data pre-treatment in metabonomics research. Arch Biochem Biophys 589:4–9

Lindon JC, Nicholson JK, Holmes E, Everett JR (2000) Metabonomics: metabolic processes studied by NMR spectroscopy of biofluids. Concepts Magn Reson 12:289–320

Lindon JC, Nicholson JK, Holmes E, Keun HC (2005) Summary recommendations for standardization and reporting of metabolic analyses. Nat Biotechnol 23(7):833–838

Loo RL, Coen M, Ebbels T, Cloarec O, Maibaum E, Bictash M et al (2009) Metabolic profiling and population screening of analgesic usage in nuclear magnetic resonance spectroscopy-based large-scale epidemiologic studies. Anal Chem 81:5119–5129

Ma H, Sorokin A, Mazein A, Selkov A, Selkov E, Demin O et al (2007) The Edinburgh human metabolic network reconstruction and its functional analysis. Molecular Systems Biology. EMBO Press 3:135

Maier TV, Schmitt-Kopplin P (2016) Capillary electrophoresis in metabolomics. Methods Mol Biol 1483:437–470

Maitre L, Lau C-HE, Vizcaino E, Robinson O, Casas M, Siskos AP et al (2017) Assessment of metabolic phenotypic variability in children's urine using (1)H NMR spectroscopy. Sci Rep 7:46082

Marchand J, Martineau E, Guitton Y, Dervilly-Pinel G, Giraudeau P (2017) Multidimensional NMR approaches towards highly resolved, sensitive and high-throughput quantitative metabolomics. Curr Opin Biotechnol 43:49–55

Martin J-C, Maillot M, Mazerolles G, Verdu A, Lyan B, Migné C et al (2015) Can we trust untargeted metabolomics? Results of the metabo-ring initiative, a large-scale, multi-instrument inter-laboratory study. Metabolomics 11:807–821

McShane LM, Cavenagh MM, Lively TG, Eberhard DA, Bigbee WL, Williams PM et al (2013) Criteria for the use of omics-based predictors in clinical trials. Nature 502:317–320

Melkonian S, Argos M, Hall MN, Chen Y, Parvez F, Pierce B et al (2013) Urinary and dietary analysis of 18,470 bangladeshis reveal a correlation of rice consumption with arsenic exposure and toxicity. States JC, editor. PLoS One e80691:8

Miller GW, Jones DP (2014) The nature of nurture: refining the definition of the exposome. Toxicol Sci 137(1):1–2

Moros G, Chatziioannou AC, Gika HG, Raikos N, Theodoridis G (2017) Investigation of the derivatization conditions for GC-MS metabolomics of biological samples. Bioanalysis 9:53–65
Nakamura Y, Afendi FM, Parvin AK, Ono N, Tanaka K, Hirai Morita A et al (2014) KNApSAcK Metabolite Activity Database for retrieving the relationships between metabolites and biological activities. Plant Cell Physiol 55:e7
Nicholson G, Rantalainen M, Maher AD, Li JV, Malmodin D, Ahmadi KR et al (2011) Human metabolic profiles are stably controlled by genetic and environmental variation. Mol Syst Biol 7:525–525
Nicholson JK, Wilson ID (2003) Understanding "global" systems biology: metabonomics and the continuum of metabolism. Nat Rev Drug Discov 2:668–676
Nicholson JK, Lindon JC, Holmes E (1999) "Metabonomics": understanding the metabolic responses of living systems to pathophysiological stimuli via multivariate statistical analysis of biological NMR spectroscopic data. Xenobiotica 29:1181–1189
Oliver S (1998) Systematic functional analysis of the yeast genome. Trends Biotechnol 16:373–378
Paglia G, Angel P, Williams JP, Richardson K, Olivos HJ, Thompson JW et al (2014a) Ion mobility-derived collision cross section as an additional measure for lipid fingerprinting and identification. Anal Chem 87:1137–1144
Paglia G, Williams JP, Menikarachchi L, Thompson JW, Tyldesley-Worster R, Halldórsson S et al (2014b) Ion mobility derived collision cross sections to support metabolomics applications. Anal Chem 86:3985–3993
Peter Guengerich F, Shimada T (1998) Activation of procarcinogens by human cytochrome P450 enzymes. Mutat Res 400:201–213
Petersen A-K, Zeilinger S, Kastenmüller G, Römisch-Margl W, Brugger M, Peters A et al (2014) Epigenetics meets metabolomics: an epigenome-wide association study with blood serum metabolic traits. Hum Mol Genet 23:534–545
Pierce BL, Tong L, Argos M, Gao J, Jasmine F (2013) Arsenic metabolism efficiency has a causal role in arsenic toxicity: Mendelian randomization and gene-environment interaction. Int J Epidemiol 42(6):1862–1871
Pitt JJ (2010) Newborn screening. Clin Biochem Rev 31(2):57–68
Plumb RS, Shockcor J, Holmes E, Nicholson JK (2010) Global metabolic profiling procedures for urine using UPLC–MS. Nat Protoc 5(6):1005–1018
Posma JM, Robinette SL, Holmes E, Nicholson JK (2014) MetaboNetworks, an interactive Matlab-based toolbox for creating, customizing and exploring sub-networks from KEGG. Bioinformatics 30:893–895
Posma JM, Garcia-Perez I, Heaton JC, Burdisso P, Mathers JC, Draper J et al (2017) Integrated analytical and statistical two-dimensional spectroscopy strategy for metabolite identification: application to dietary biomarkers. Anal Chem 89:3300–3309
Psychogios N, Hau DD, Peng J, Guo AC, Mandal R (2011) The human serum metabolome. PLoS One 6(2):e16957
Qi Y, Song Y, Gu H, Fan G, Chai Y (2014) Global metabolic profiling using ultra-performance liquid chromatography/quadrupole time-of-flight mass spectrometry. Methods Mol Biol 1198:15–27
Ramautar R, Somsen GW, de Jong GJ (2017) CE-MS for metabolomics: developments and applications in the period 2014-2016. Rassi El Z, editor. Electrophoresis 38:190–202
Rappaport SM, Barupal DK, Wishart D, Vineis P, Scalbert A (2014) The blood exposome and its role in discovering causes of disease. Environ Health Perspect 122:769
Reymond J-L, van Deursen R, Blum LC, Ruddigkeit L (2010) Chemical space as a source for new drugs. Med Chem Commun 1:30–38
Reymond J-L, Ruddigkeit L, Blum L, van Deursen R (2012) The enumeration of chemical space. Wiley Interdiscip Rev Comput Mol Sci 2:717–733
Robinette SL, Lindon JC, Nicholson JK (2013) Statistical spectroscopic tools for biomarker discovery and systems medicine. Anal Chem 85:5297–5303

Rocca-Serra P, Salek RM, Arita M, Correa E, Dayalan S, Gonzalez-Beltran A et al (2016) Data standards can boost metabolomics research, and if there is a will, there is a way. Metabolomics 12:14

Ruddigkeit L, van Deursen R, Blum LC, Reymond J-L (2012) Enumeration of 166 billion organic small molecules in the chemical universe database GDB-17. J Chem Inf Model 52:2864–2875

Salek RM, Neumann S, Schober D, Hummel J, Billiau K, Kopka J et al (2015) COordination of standards in MetabOlomicS (COSMOS): facilitating integrated metabolomics data access. Metabolomics 11:1587–1597

Sampson JN, Boca SM, Shu XO, Stolzenberg-Solomon RZ, Matthews CE, Hsing AW et al (2013) Metabolomics in epidemiology: sources of variability in metabolite measurements and implications. Cancer Epidemiol Biomark Prev 22:631–640

Sansone S-A, Fan T, Goodacre R, Griffin JL, Hardy NW, Kaddurah-Daouk R et al (2007) The metabolomics standards initiative. Nat Biotechnol 25:846–848

Scalbert A, Brennan L, Manach C, Andres-Lacueva C, Dragsted LO, Draper J et al (2014) The food metabolome: a window over dietary exposure. Am J Clin Nutr 99:1286–1308

Schmelzer K, Fahy E, Subramaniam S, Dennis EA (2007) The lipid maps initiative in lipidomics. Methods Enzymol 432:171–183

Shimada T, Oda Y, Gillam EMJ, Guengerich FP, Inoue K (2001) Metabolic activation of polycyclic aromatic hydrocarbons and other procarcinogens by cytochromes P450 1A1 and P450 1B1 allelic variants and other human cytochromes P450 in Salmonella typhimurium NM2009. Drug Metab Dispos 29:1176–1182

Siskos AP, Jain P, Römisch-Margl W, Bennett M, Achaintre D, Asad Y et al (2017) Interlaboratory reproducibility of a targeted metabolomics platform for analysis of human serum and plasma. Anal Chem 89:656–665

Smith CA, O'Maille G, Want EJ, Qin C, Trauger SA, Brandon TR et al (2005) METLIN: a metabolite mass spectral database. Ther Drug Monit 27:747–751

Spagou K, Wilson ID, Masson P, Theodoridis G, Raikos N, Coen M et al (2010) HILIC-UPLC-MS for exploratory urinary metabolic profiling in toxicological studies. Anal Chem 83:382–390

Strittmatter N, Rebec M, Jones EA, Golf O, Abdolrasouli A, Balog J et al (2014) Characterization and identification of clinically relevant microorganisms using rapid evaporative ionization mass spectrometry. Anal Chem 86:6555–6562

Sud M, Fahy E, Cotter D, Azam K, Vadivelu I, Burant C et al (2016) Metabolomics Workbench: an international repository for metabolomics data and metadata, metabolite standards, protocols, tutorials and training, and analysis tools. Nucleic Acids Res 44:D463–D470

Suhre K, Meisinger C, Döring A, Altmaier E, Belcredi P, Gieger C et al (2010) Metabolic footprint of diabetes: a multiplatform metabolomics study in an epidemiological setting. Breant B, editor. PLoS One e13953:5

Suhre K, Wallaschofski H, Raffler J, Friedrich N (2011) A genome-wide association study of metabolic traits in human urine. Nature 43(6):565–569

Suhre K, Raffler J, Kastenmüller G (2016) Biochemical insights from population studies with genetics and metabolomics. Arch Biochem Biophys 589:168–176

Swainston N, Smallbone K, Hefzi H, Dobson PD, Brewer J, Hanscho M et al (2016) Recon 2.2: from reconstruction to model of human metabolism. Metabolomics 12:109

Takáts Z, Wiseman JM, Gologan B, Cooks RG (2004) Mass spectrometry sampling under ambient conditions with desorption electrospray ionization. Science 306:471–473

Takáts Z, Wiseman JM, Cooks RG (2005) Ambient mass spectrometry using desorption electrospray ionization (DESI): instrumentation, mechanisms and applications in forensics, chemistry, and biology. J Mass Spectrom 40:1261–1275

Tanabe M, Sato Y, Morishima K (2017) KEGG: new perspectives on genomes, pathways, diseases and drugs. Nucleic Acids Res 45(D1):D353–D361

Tautenhahn R, Patti GJ, Rinehart D, Siuzdak G (2012) XCMS online: a web-based platform to process untargeted metabolomic data. Anal Chem 84:5035–5039

Teahan O, Gamble S, Holmes E, Waxman J, Nicholson JK, Bevan C et al (2006) Impact of analytical bias in metabonomic studies of human blood serum and plasma. Anal Chem 78:4307–4318

Thévenot EA, Roux A, Xu Y, Ezan E, Junot C (2015) Analysis of the human adult urinary metabolome variations with age, body mass index, and gender by implementing a comprehensive workflow for univariate and OPLS statistical analyses. J Proteome Res 14(8):3322–3335

Thiele I, Swainston N, Fleming RMT, Hoppe A, Sahoo S, Aurich MK et al (2013) A community-driven global reconstruction of human metabolism. Nat Biotechnol 31:419–425

Thomas L, Hodgson S, Nieuwenhuijsen M, Jarup L (2007) Early renal damage in a population environmentally exposed to cadmium—the Avonmouth Pilot Study. Epidemiology 18:S123

Ulrich EL, Akutsu H, Doreleijers JF, Harano Y, Ioannidis YE, Lin J et al (2008) BioMagResBank. Nucleic Acids Res 36:D402–D408

Uppal K, Walker DI, Liu K, Li S, Go Y-M, Jones DP (2016) Computational metabolomics: a framework for the million metabolome. Chem Res Toxicol 29:1956–1975

Van Deemter JJ, Zuiderweg FJ, Klinkenberg A (1956) Longitudinal diffusion and resistance to mass transfer as causes of nonideality in chromatography. Chem Eng Sci 5:271–289

Vineis P, Veldhoven K, Chadeau-Hyam M, Athersuch TJ (2013) Advancing the application of omics-based biomarkers in environmental epidemiology. Environ Mol Mutagen 54(7):461

Vineis P, Chadeau-Hyam M, Gmuender H, Gulliver J, Herceg Z, KLEINJANS J et al (2017) The exposome in practice: design of the EXPOsOMICS project. Int J Hyg Environ Health 220:142–151

Vlaanderen JJ, Janssen NA, Hoek G, Keski-Rahkonen P, Barupal DK, Cassee FR et al (2017) The impact of ambient air pollution on the human blood metabolome. Environ Res 156:341–348

Vrijheid M, Slama R, Robinson O, Chatzi L, Coen M, van den Hazel P et al (2014) The human early-life exposome (HELIX): project rationale and design. Environ Health Perspect 122 (6):535–544

Want EJ, Wilson ID, Gika H, Theodoridis G, Plumb RS, Shockcor J et al (2010) Global metabolic profiling procedures for urine using UPLC|[ndash]|MS. Nat Protoc 5:1005–1018

Warth B, Levin N, Rinehart D, Teijaro J, Benton HP, Siuzdak G (2017) Metabolizing data in the cloud. Trends Biotechnol 35:481–483

Wild CP (2005) Complementing the genome with an “exposome”: the outstanding challenge of environmental exposure measurement in molecular epidemiology. Cancer Epidemiol Biomark Prev 14:1847–1850

Wild CP (2012) The exposome: from concept to utility. Int J Epidemiol 41:24–32

Wild CP, Scalbert A, Herceg Z (2013) Measuring the exposome: a powerful basis for evaluating environmental exposures and cancer risk. Environ Mol Mutagen 54:480–499

WILSON I, Plumb R, Granger J, Major H, WILLIAMS R, LENZ E (2005a) HPLC-MS-based methods for the study of metabonomics. J Chromatogr B 817:67–76

Wilson ID, Nicholson JK, Castro-Perez J, Granger JH, Johnson KA, Smith BW et al (2005b) High resolution “ultra performance” liquid chromatography coupled to oa-tof mass spectrometry as a tool for differential metabolic pathway profiling in functional genomic studies. J Proteome Res 4:591–598

Wishart DS, Lewis MJ, Morrissey JA, Flegel MD, Jeroncic K, Xiong Y et al (2008) The human cerebrospinal fluid metabolome. J Chromatogr B 871:164–173

Wishart DS, Jewison T, Guo AC, Wilson M, Knox C, Liu Y et al (2012) HMDB 3.0—the human metabolome database in 2013. Nucleic Acids Res 41:D801–D807

Wist J (2017) Complex mixtures by NMR and complex NMR for mixtures: experimental and publication challenges. A special issue of the associate ed, editor. Magn Reson Chem 55:22–28

Xia J, Sinelnikov IV, Han B, Wishart DS (2015) MetaboAnalyst 3.0—making metabolomics more meaningful. Nucl Acids Res 43(W1):W251–W257

Xiao R, Zhang X, Rong Z, Xiu B, Yang X, Wang C et al (2016) Non-invasive detection of hepatocellular carcinoma serum metabolic profile through surface-enhanced Raman spectroscopy. Nanomedicine 12:2475–2484

Xu X, Veenstra TD, Fox SD, Roman JM, Issaq HJ, Falk R et al (2005) Measuring fifteen endogenous estrogens simultaneously in human urine by high-performance liquid chromatography-mass spectrometry. Anal Chem 77:6646–6654
Yin P, Lehmann R, Xu G (2015) Effects of pre-analytical processes on blood samples used in metabolomics studies. Anal Bioanal Chem 407:4879–4892
Zhao L, Pickering G (2011) Paracetamol metabolism and related genetic differences. Drug Metab Rev 43:41–52
Zheng X, Zhao A, Xie G, Chi Y, Zhao L (2013) Melamine-induced renal toxicity is mediated by the gut microbiota. Sci Transl Med 5(172):172ra22
Zubarev RA, Makarov A (2013) Orbitrap mass spectrometry. Anal Chem 85:5288–5296

Chapter 7
Transcriptomics within the Exposome Paradigm

D. A. Sarigiannis

Abstract The advent of omics technologies has enhanced significantly our capacity to interpret mechanistically the association between environmental exposure and disease. Although understanding these interactions requires capturing perturbations at different levels of biological organization, transcriptomics holds a key role. Modulation of gene expression represents the initial biological perturbations due to environmental exposure. This is of particular importance when assessing real-life exposure that involves multiple stressors in highly variable time regimes. This chapter aims at (a) demonstrating the place of transcriptomics in modern risk assessment and environmental health associations, highlighting the respective bio-informatics tools that are necessary for the interpretation and (b) demonstrating the feasibility of transcriptomics of understanding environmental risk associated to real-life ubiquitous mixtures. Although environmental exposures occur to mixtures of chemicals rather than to individual agents, most of the toxic effects of air pollutants are ascribed to single chemicals. There is a growing feeling in both the scientific and regulatory communities, however, that there is a need for more comprehensive approaches toward managing the potential impact of complex environmental chemical mixtures on human health. In this perspective, it is expected that toxicogenomics would be the appropriate screening method for assessing biological effects of complex chemical mixtures, allowing us to review the whole spectrum of potential biological response rather than focusing on a predefined number of endpoints as in classical toxicological analysis. In this chapter, beyond the overview of the analytical and computational aspects necessary for implementing toxicogenomics in the context of the exposome, a concrete example of such an application on a typical indoor air mixture as defined in the EU-wide review study INDEX and on a mixture

D. A. Sarigiannis (✉)
HERACLES Research Center on the Exposome and Health, Center for Interdisciplinary Research and Innovation, Thessaloniki, Greece

Environmental Engineering Laboratory, School of Chemical Engineering, Aristotle University of Thessaloniki, Thessaloniki, Greece

University School for Advanced Study (IUSS), Pavia, Italy
e-mail: denis@eng.auth.gr

S. Dagnino, A. Macherone (eds.), *Unraveling the Exposome*,
https://doi.org/10.1007/978-3-319-89321-1_7

of polyaromatic hydrocarbons (PAHs) isolated from urban air in the city of Milan is given with the aim to identify specific sets of biomarkers for each of the two types of exposure (indoor or outdoor). A human cell line derived from a bronco-pulmonary system (A549) was used as the appropriate in vitro model to support the investigation of the molecular basis for adverse outcomes that are attributed to indoor and/or outdoor air pollution based on epidemiological evidence. Applying a Total Gene Expression assay by Applied Biosystems Microarrays, large sets of genes modulated by single mixtures exposure were profiled. This process led us to identify common biochemical pathways and specific molecular responses. Indoor air mixtures induced a higher level of gene modulation than ambient air PAHs. A closer look at the differences in biological response confirmed major discrepancies in the mode of action of the two mixtures. Indoor air induced primarily modulation of genes associated to protein targeting and localization including in particular cytoskeletal organization; PAHs modulated mostly the expression of genes related to cell motility and gene networks regulating cell–cell signaling, as well as cell proliferation and differentiation. These results provide biological information useful for articulating mechanistic hypotheses linking exposure to xenobiotic mixtures and physiological responses. The evidence on the latter is supported by a large amount of epidemiological evidence, associating exposure to urban air pollution with respiratory allergies, chronic obstructive pulmonary disease, cardiovascular disease, and cancer. Lately, such evidence has been extended to include associations of exposure to polluted ambient and indoor air with kidney disease and even neurodegenerative disorders, and in particular dementia.

Keywords Transcriptomics · Genetic susceptibility · Integrated exposure biology · Systems Biology

Introduction and Current State of Play

Owing to large-scale improvements in omics technologies, changes in biological processes can now be tracked in unprecedented detail. Despite these achievements, many existing studies still treat the genome, transcriptome, proteome, and metabolome as isolated biological layers failing to take stock of their interconnections. This shortcoming is largely because of the limited availability of multi-omics data collected on the same group of individuals, as well as the limited availability of sufficiently powerful tools for high-dimensional analysis (Yan et al. 2017). Although such measurements span a diverse range of cellular activities, quantitative associations and meaningful biological contextualization are still lacking (Ebrahim et al. 2016). Recent efforts in collecting multi-omics data in the same group of individuals open numerous opportunities for more comprehensive analyses of complex diseases. Example projects include the Alzheimer's Disease Neuroimaging Initiative (ADNI) (Weiner et al. 2010), the Cancer Genome Atlas (TCGA) Research Network (http://cancergenome.nih.gov/), the International Cancer Genome Consortium (ICGC; http://icgc.org/), and the HEALS project (Health and Environmental-wide

Associations via Large population Surveys) (http://www.heals-eu.eu) in Europe. Instead of limiting their perspective to a single-omics layer, these data collections create a molecular landscape spanning the genome, transcriptome, proteome, and even metabolome (Saykin et al. 2015). On the other hand, TCGA data have helped to drive the progress of cancer research by revealing a large-scale integrative view of the molecular aberrations in various cancers (TCGA 2011, 2014; Ciriello et al. 2015). Finally, the link of exposure to carbon nanotubes of different degrees of impurities and their effect on immunity system (Vitkina et al. 2016) was investigated through coupled analysis of transcriptomics and lipidomics; from the study, it was found that modulation of genes activating the NFkB pathway is associated to inflammatory responses that may cause a perturbation in the IL-6 signaling pathway in order to regulate inflammatory processes and compensate for apoptotic changes.

Adverse Outcome Pathways (AOPs) (Ankley et al. 2010) offer a valuable framework not only for aggregating information in a biologically plausible manner for communication, but also as a tool for risk assessment. In order to maximize the AOP concept for application to regulatory situations, several improvements can be made. Although the exposome and the AOP concepts evolved separately in the fields of human toxicology and ecotoxicology, respectively, the understating of the biological perturbations resulted from environmental exposures creates synergies among AOPs and the exposome. One of the key points of convergence in the concepts is how long-term, cumulative exposure (i.e., the exposome) results in sustained biological perturbations, which do not result only in activated, but sustained AOPs. The identification of sustained key events would also result in the establishment of early effect markers, readily applicable in molecular epidemiology. Overall, AOPs could provide additional insights into the biological plausibility of the hypothesis of the toxic effects of the combined exposures, by integrating networks of AOPs activated by multiple stressors.

Current approaches for documenting and establishing scientific confidence in the events and pathways are credible and transparent but need to incorporate semi-quantitative or quantitative measures of confidence such as weight of evidence, multi-criteria decision making, or Bayesian modeling in order to facilitate the use of AOPs in risk-based decisions (Linkov et al. 2015). Moreover, development of an AOP should not rely on the investigation of a single key event but rather on data able to describe the dose–response relationships between events (Perkins et al. 2015). To successfully incorporate AOPs into risk assessment, multiple interacting pathways or networks of AOPs (Villeneuve et al. 2014; Garcia-Reyero 2015) should be accounted for. The current linear AOPs must evolve into AOP networks that can be monitored using key event assays that are sufficiently linked by quantitative relationships to other key events or an outcome in order to understand what chemicals do (general hazard identification), not just identify if a chemical activates a specific AOP (specific hazard identification or characterization). Thus, the use of AOPs becomes a key component of understanding health effects, under the condition that a sufficient number of pathways of toxicity (PoT) are integrated, aiming at capturing all potential hazards and adversities. Mapping jointly multi-omics results into Pathways of Toxicity (PoT) will allow the buildup of plausible hypotheses that can be linked mechanistically to AOPs.

The opportunities provided by investigating health and disease on the omics scale come with the need to implement a novel modus operandi to address data generation, analysis, and sharing. It is critical to recognize that (multi)omics data, that is, omics data generated within isolated and not yet integrated contexts, need to be analyzed and interpreted as a whole through effective and integrative data analysis pipelines. The combination of -omics techniques with biologically derived in silico models of cell function and physiological processes can significantly enhance our understanding of the biological mechanisms leading to onset or exacerbation of human pathologies linked to xenobiotic exposure, if appropriately anchored to phenotypic evidence. These results would also have to account for inter- and intraindividual variability of expression profiles, as well as being able to differentiate between adaptation and defense mechanisms and actual response that may lead to pathology. Integrated health impact assessment of environmental stressors and mixtures thereof would need to follow a so-called "full chain" approach to take into account all relevant health stressors and their interaction. This highlights the need for comprehensive data interpretation of the molecular, biochemical, and physiological processes that couple exposure to health outcome; hence, the need to forge a new paradigm for interdisciplinary scientific work in the area of environment and health. We shall call that "the connectivity paradigm for risk assessment," denoting an approach that builds on the exploration of the interconnections between the coexistence of multiple stressors and the different levels of biological organization that come to bear in order to produce the final adverse health outcome (Sarigiannis et al. 2009). The connectivity paradigm has been demonstrated in this context by Sarigiannis et al. (2009), where the effect of time–dose response in the identification of sustained network activation under various doses and time duration to the quaternary mixture of benzene, toluene, ethylbenzene, and xylenes (BTEX, ubiquitous in ambient and indoor air) has been comprehensively studied using transcriptomics. Results showed the time–dose dependence of important regulatory pathway activation, which were differentially expressed after treatment at different doses and at different time lengths such as the ones regarding inflammation mediated by chemokine and cytokine signaling, apoptosis, and oxidative stress.

Toward the development of a coherent biological context from multi-omics data, as well as a comprehensive understanding of the toxicity mechanisms associated to the adverse outcomes, computational toxicology methods and biological modeling help put together a systems approach toward the derivation of dose–time response functions and their effective application in a risk assessment context. This clearly requires the support of advanced bioinformatics and biostatistics. The needs for building big data infrastructure and joined pathway analysis for understanding multi-omics data in the context of the exposome have been recently highlighted by Manrai et al. (2017). At the moment, although there are several tools for developing molecular networks based on single-omics data, there is lack of comprehensive bioinformatics platforms for evaluating multiple-omics data (Yan et al. 2017). At the moment, the most suitable platform for multi-omics analysis is the GeneSpring platform by Agilent that has been effectively applied in the joint analysis of changes in selected transcripts and proteins related to the estrogen receptor following sulforaphane treatment in human breast epithelial MCF10A cells (Agyeman et al. 2012). However, in the frame of large-scale multicenter exposomics programs such as HEALS, joint pathway analysis of multi-omics data by Agilent GeneSpring has been applied in the study of combined exposure to heavy metals and endocrine disruptors in relation to neurodevelopmental disorders. Additional efforts include the use of Bayesian networks to combine clinical and omics data to predict the prognosis of breast cancer (Gevaert et al. 2006). Network-based integration of multilevel systems genetics data has been successfully applied to type 2 diabetes, by integrating co-segregating transcriptomes and metabolomes of rats (Dumas et al. 2016). To identify valuable biological information, graph measures (e.g., degree connectivity, centrality) and graph algorithm (e.g., subnetwork) have been used, such as the network-based concept performed for drug-target therapy recently published in France by the team of Inserm/UPD (Boezio et al. 2017). Additional insights have been provided by the molecular complex detection (MCODE) approach and the Markov Clustering algorithm (MCL) that have been applied in previous studies (Taboureau and Audouze 2017; Taboureau et al. 2013; Audouze et al. 2010). Toward a systems biology approach, functional annotation of gene sets based on the three Gene Ontology categories (molecular function, biological process, and cellular components) has been previously performed in a study of brown adipose tissue transcriptome profiling (Hao et al. 2015), while biological pathways have been also explored in proteomics analysis of adipocyte differentiation (Borkowski et al. 2014). With regard to proteins, it is important that protein lists have expanded by including their first-order protein–protein interaction partners using high-confidence human interactome (Szklarczyk et al. 2017; Li et al. 2017). For example, the InWeb interactome contains 428,429 unique protein–protein interactions based on refined experimental proteomics data. This approach has been proved to provide scored-based complexes that may be investigated for functional annotation, pathways characterization, and diseases as performed previously for a chemical mixture by Kongsbak et al. (2014).

The Role of the Transcriptome in the Frame of an Integrated Multi-Omics Paradigm for Elucidating the Exposome

An integrated multi-omics approach toward elucidating the internal exposome relies on evaluating the maximum available information from multiple-omics data. Toward this aim, a well-structured exposure biology workflow would have to be followed, as graphically illustrated in Fig. 7.1.

This workflow is specifically designed to decipher the causal links of genome, environment, and disease, entailing a stepwise process:

1. The overall assessment needs to be initiated from an agnostic point of view; no former hypothesis has to be formed unless driven by some evidence of perturbed pathways. Thus, the first step of the assessment is the investigation of untargeted

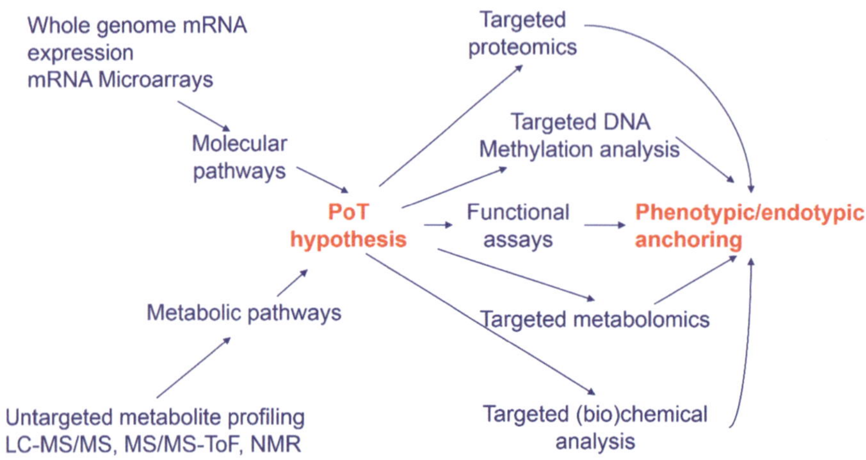

Fig. 7.1 From untargeted to targeted omics workflow

omics, including whole genome mRNA expression and untargeted metabolite profiling.

2. The results of untargeted metabolomics and transcriptomics will be jointly analyzed (molecular and metabolic pathway analysis) so as to construct putative hypotheses on Pathways of Toxicity (or of Adverse Health Outcome).
3. After building the putative hypotheses of the Pathways of Toxicity (thus limiting the number of potential pathways to be perturbed), additional evidence will be provided through targeted omics (including proteomics and metabolomics), DNA methylation, functional assays, and biochemical biomonitoring. Anchoring of the different endotypes will be further supported by in vitro assays for mechanistic confirmation, biology-inspired modeling, and bioinformatics for data analysis.

This results in

(a) Identification of endotypes (early biological event markers) that can be associated causally with phenotypes of adverse health outcomes.
(b) Determination of allostatic status of the analyzed individuals
(c) Association of allostatic status with individual risk of disease and eventually burden of disease on the population level

Technically, the different array of analytical exposure biology will include (a) transcriptomics, meaning the complete set of RNA transcripts that are produced by the genome, under specific circumstances or in a specific cell—using high-throughput methods, such as microarray analysis. (b) Metabolomics (untargeted and targeted) entails the global analysis of metabolites in easily accessible human body fluids. Established and robust sample preparation protocols (e.g., for plasma, urine) in combination with the latest generation of NMR and ultrahigh-pressure liquid chromatography (UPLC)–mass spectrometry (MS), high-pressure liquid chromatography–electrochemical detection (HPLC-ED), and LC-ECA (electrochemical coulometric arrays) have to be applied. These approaches should be applied to samples from selected human cohorts to determine the impact of exposure to chemicals (e.g., pesticides, industrial solvents, etc.) on the human exposome. (c) Adductomics deals with the measurement of adducts of electrophiles with DNA, blood proteins (hemoglobin, albumin), and glutathione. With respect to SNP profiling in toxicologically relevant genes, genotyping assays will be developed. Genome-wide DNA methylation profiling and miRNA expression are key to investigating the importance of epigenetic effects on environmental health. Array- and sequencing-based DNA methylation profiling technologies could be used to meet this goal. In particular, Agilent microarray miRNA profiling and genome-wide bisulfite sequencing could be used to analyze miRNA expression and DNA methylation, respectively, on the bio-samples of the cohorts to be analyzed in the frame of an exposome study:

1. Apply SNP analysis, miRNA analysis, and next-generation sequencing to define genetic susceptibility for chemicals at population level (e.g., DNA repair phenotypes, Phase II reaction genotypes)
2. Identify differences between epigenetically influenced and independent SNPs
3. Develop sample handling and metabolomics workflows
4. Generate metabolite expression data; identify biomarkers from cohorts and in vitro models (glutathione, s-adenosyl methionine, and bisphenol A)
5. Develop methodologies (FS-SRM) and generate DNA and protein adduct data from cohorts.
6. Determine methyl/hydroxymethylcytosine at specific genomic sequences (promoters, CpG islands, and repeated sequences)
7. Identify Illumina methylation at 20 genes, and apply Methylator phenotypic analysis for this set of genes.

The causal link between exposure and disease endpoint as revealed by the identification of—omics markers of effect requires mechanistic confirmation, which could be based on in vitro assays assessing complex toxic endpoints relevant to disease endpoints on cell models derived from liver (hepG2/HuH7, HepaRG), adipocytes (hMADS), or cells differentiating into neurons (C11) to establish for the different technologies omics compound specific hazard signatures. This will also support the system biology model of causation. Next to confirming the systems biology exposome model, these mechanistic studies provide kinetic constants and compound receptor-binding data useful to further develop the model. In this context, it is possible to use these data to design in vitro human cell-based assays and systems biology experiments, to mechanistically anchor omics observations from cohort studies in light of the linkage between exposure and health outcome data. The hypothesis is put forward that the correspondence of omics signatures in cohort biomaterials with those obtained from experimental in vitro models can demonstrate the validity of chosen biomarkers. In fact, this approach can be used to corroborate if (a) signatures are more closely related to the actual compound exposure (e.g., biotransformation) and thus represent a marker for exposure; (b) more closely related to mechanisms of disease, and consequently represent a marker for health impact. Overall, empirical perturbation detected from HBM data (including omics markers) can be integrated into in vitro testing to help verify system biology-level or "real-world" human toxicity of environmental stressors (Pleil 2012).

Bioinformatics and Joint Pathway Analysis

Statistics and Big Data Analytics for Exposome-Wide Association Studies

Introduction and Objectives

The scope of the currently available statistical methods is

(a) To understand biological functions comprising toxicity pathways and their interactions in relation to external/internal exposure
(b) To confirm the causative effect between exposure and disease endpoint through theoretical (computational) models
(c) To combine mixed data, resulted from various sources, through the utilization of advanced data mining analysis techniques
(d) To provide the methodological tools for integrating multiple biomarkers into a mechanistic description
(e) To derive the systems biology exposome model

Gene expression data can be a valuable tool for understanding gene regulation, biological networks, and cellular states. One goal in analyzing expression data is to try to determine how the expression of any particular gene might affect the expression of other genes; the genes involved in this case could belong to the same gene network. By a gene network, we mean a set of genes being expressed together in a nonrandom pattern. Another goal of expression data analysis is to try to determine what genes are expressed as a result of certain cellular conditions, for example, what genes are expressed in diseased cells that are not expressed in healthy cells. While early experiments using microarrays profiled only a few samples, more recent experiments profile on the order of dozens or even hundreds of samples, allowing for a more robust statistical analysis of the data. In the near future, datasets containing thousands of samples should become available. As gene expression datasets become larger and larger, spreadsheets will become less and less of an adequate tool for performing transcriptomics data analysis and data mining techniques using large databases should find more and more use in analyzing expression data. In the analysis of gene expression data, the items in an association rule can represent genes that are strongly expressed or repressed, as well as relevant facts describing the cellular environment of the genes (e.g., a diagnosis for a tumor sample that was profiled, or a drug treatment given to cells in the sample before profiling). An example of an association rule mined from expression data might be [cancer] $\Rightarrow$ gene A$\uparrow$, gene B$\downarrow$, gene C$\uparrow$}, meaning that, for the dataset that was mined, in most profile experiments where the cells used were cancerous, gene A was measured as being up (i.e., highly expressed), gene B was down (i.e., highly repressed), and gene C was up, altogether. Hence our scope is to interpret gene expression technology results via integration of gene expression profiles with corresponding biological knowledge (gene annotations, literature, etc.) extracted from biological databases. Consequently, the key task in the interpretation step is to detect the present co-expressed (sharing similar expression profiles) and co-annotated (sharing the same properties such as function, regulatory mechanism, etc.) gene group. Several approaches dealing with the interpretation problem have recently been reported. These approaches can be classified around three axes (Martinez and Collard 2007; Chang and Halgamuge 2002): expression-based approaches, knowledge-based approaches, and co-clustering approaches. The most currently used interpretation axis is the expression-based axis that gives more weight to gene expression profiles. However, it presents many well-known drawbacks. First, these approaches cluster genes by similarity in expression profiles across all biological conditions. However, gene groups involved in a biological process might be only co-expressed in a small subset of conditions (Altman and Raychaudhuri 2001). Second, many genes have different biological roles in the cell; they may be conditionally co-expressed with different groups of genes. Since almost all clustering methods used place each gene in a single cluster, that is, a single group of genes, relationships with different groups of conditionally regulated genes may remain undiscovered (Gasch and Eisen 2002). Third, discovering biological relationships among co-expressed genes is not a trivial task and requires a lot of additional work, even when similar gene expression profiles are related to similar biological roles (Shatkay et al. 2000).

In a wider exposomics analysis framework like the one heralded by the connectivity paradigm, bioinformatics techniques should be applied and enhanced in order to select the most relevant omics data and derive specific data profiles for given exposure/disease pathways. In particular, predictive biomarkers could be determined, based on heterogeneous datasets, resulting from human biomonitoring, omics and epigenetics analyses, and systems biology and/or physiology-based biokinetic (PBBK) modeling. This would require

(a) The preprocessing of the data produced
(b) The discovery of specific data patterns and/or clusters
(c) The creation of data models based on training sets and, finally
(d) The evaluation of the models with regard to their validity and prediction capacity on the basis of the test data

Results would be systematically assessed and the model that best describes exposome data would be employed on population survey datasets. Multiple-omics biomarkers could also be integrated to a mechanistic description of toxicity pathway interactions, in relation to external/internal exposure, achieved by developing systems biology pathway models and using the predictive bioinformatics approaches.

Problem Formulation

Data preprocessing is necessary before applying unsupervised learning algorithms to exposome data. This includes

- Technology-specific data preprocessing (e.g., spectra deconvolution)
- Noise removal, to ensure the consistency and high quality of our data from possible outliers or discrepancies in the measurements
- Data transformation, to normalize the values in our dataset and also increase their generalization
- Data reduction, to decrease both the apparent complexity in our data, through subset representations, and the dimensionality of the derived models
- Discretization, to scale the data and prepare them for further analysis by means of clustering and pattern extraction

After preprocessing, training and testing of the various datasets could follow the learning process. This is required in order to depict relations among the observed variables, to recognize complex patterns and to extract rules. Lastly, since the number of states required to be recognized may be much larger than the available dataset, a representative training dataset is selected to generalize to the entire dataset.

Data Mining

Data mining (Fayyad et al. 1996) systems combine techniques from many scientific disciplines, for example, statistics or computer science (database systems and machine learning). The systems can be categorized according to the task they solve. There are, for example, classification, regression, clustering, or descriptive systems. In addition, data mining systems can be also characterized by the type of knowledge they produce. In this regard, there are connectionist (artificial neural networks), statistical (Naïve Bayes classifier), or logical (decision trees or classification rules) systems. Some algorithms combine several methods such as Domingos' RISE algorithm (Dong et al. 1999) which integrates instance-based and rule-based learning. Lastly, data complexity criteria separate data mining systems into two groups: propositional and relational systems.

Although statistics or connectionist systems have achieved good results in text mining, that is, text classification and categorization, they may not be suitable for mining knowledge intended for further analysis. Nevertheless, it may be very difficult to process a complex dataset, such as data describing chemical molecules (structure of organic molecules) and biological (strings of DNA) data. Following the axiom that "a data mining algorithm is a well-defined procedure that takes data as input and produces output in the form of models or patterns," the data mining algorithms can be classified in accordance to Fig. 7.2. Accordingly, model structures are categorized according to prediction types (incl. Linear regression, piecewise linear, nonparametric regression, and classification), the probability distributions

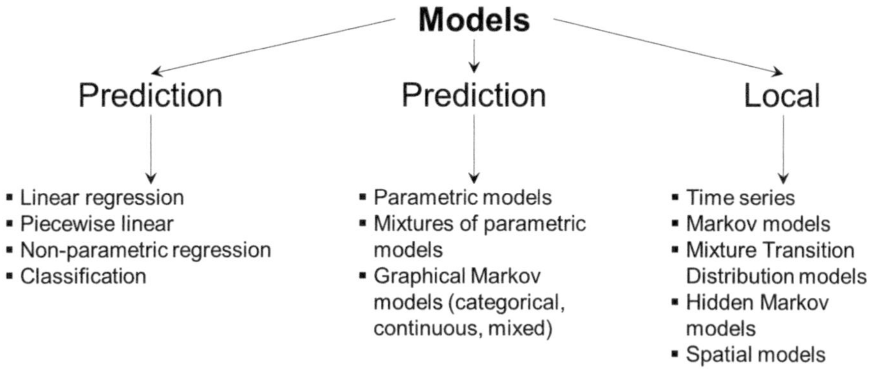

Fig. 7.2 Model classification scheme

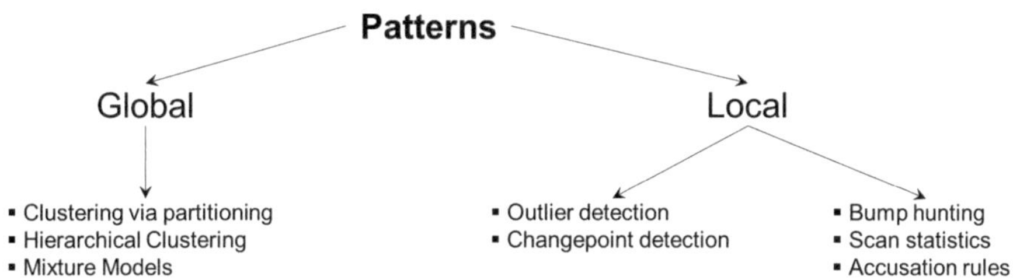

Fig. 7.3 Classifications in the descriptive data mining approach

(incl. Parametric models, mixtures of parametric models, graphical Markov models), and structured types (incl. Time series, Markov models, mixture transition distribution models, hidden Markov models). In this chapter, mining algorithms will be categorized into two clusters: (a) descriptive and (b) predictive. The former describes the dataset in a concise and brief manner and presents general data properties. The latter performs inference on the available dataset and predicts the outcome of the new datasets via the generation of one or more models.

Descriptive Data Mining

A descriptive data mining approach presents the main features, where data are randomly generated from a "good" descriptive model that has the same characteristics as the "real" data. In the descriptive data mining approach, patterns are evaluated either globally or locally, in accordance to the classification presented in Fig. 7.3. Global patterns include clustering methods via portioning, hierarchical clustering, and mixture modeling, and local patterns include outlier detection, changepoint detection pattern, “bump” hunting, scan statistics, and association rules.

Predictive Data Mining

Predictive data mining can be used to forecast explicit values, based on patterns determined from known results. Several techniques can be used for that purpose, ranging from typical approaches based on decision trees or k-nearest neighbors to more sophisticated ones that employ Artificial Neural Networks (ANNs), Support Vector Machines (SVMs), or Bayesian Networks (BNs). Hence, it is possible to perform inference on the available dataset, to perform a mere classification, and to study and unravel the feature attributes concealed in data (e.g., exposome). Various computational techniques, especially machine learning algorithms (Larranaga et al. 2003), are applied to select genes or proteins associated with the trait of interest and to classify different types of samples in gene expression of microarrays data (Allison et al. 2006) or mass spectrometry (MS)-based proteomics data (Aebersold and Mann 2003); to identify disease-associated genes, gene–gene interactions, and gene–environment interactions from Genome Wide Association (GWA) studies (Hirschhorn and Daly 2005); to recognize the regulatory elements in DNA or protein sequences (Zeng et al. 2009); to identify protein–protein interactions (Valencia and Pazos 2002); or to predict protein structure (Jones 2001). The aim of designing/using ensemble methods (Breiman 1996, 2001; Freund and Schapire 1996) is to achieve more accurate classification (on training data) as well as better generalization (on unseen data). However, this is often achieved at the expense of increased model complexity (i.e., decreased model interpretability) (Kuncheva 2004). A better generalization property of the ensemble approach is often explained using the classic bias–variance decomposition analysis (Webb and Zheng 2004). Specifically, previous studies pointed out methods like bagging (Fig. 7.4a) that improve generalization by decreasing variance (Breiman 1998) while methods similar to boosting (Fig. 7.4b) achieve this by decreasing bias (Schapire et al. 1998).

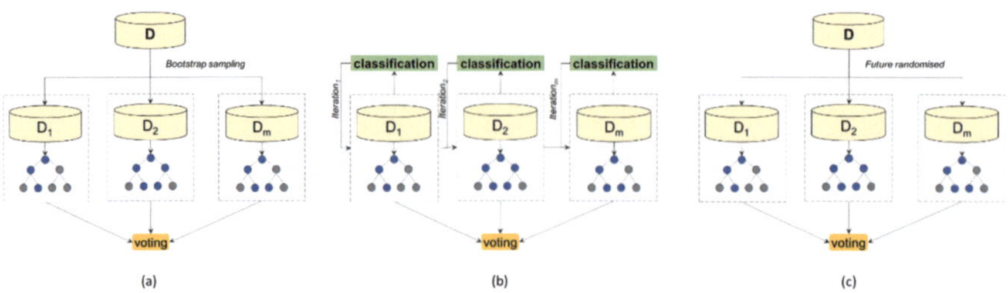

Fig. 7.4 Schematic illustration of the three popular ensemble methods. (**a**) Bagging. (**b**) Boosting. (**c**) Random forests

Data Mining Algorithms

Clustering

DNA microarray technology has made it possible to simultaneously monitor the expression levels of thousands of genes during important biological processes and across collections of related samples. Elucidating the patterns hidden in gene expression data offers a tremendous opportunity for enhanced understanding of functional genomics. However, the large number of genes and the complexity of biological networks greatly increase the challenges of comprehending and interpreting the resulting mass of data, which often consists of millions of measurements. In this regard, clustering techniques can be used to reveal the natural structures and identify the patterns in the underlying data. Specifically, cluster analysis seeks to partition a given dataset into groups based on specified features so that the data points within a group are more similar to each other than the points in different groups. Many conventional clustering algorithms are available to be adapted or directly applied to the gene expression data. Assuming that a two-dimensional matrix of gene expression is available, the row will represent the genes and the columns will represent different experiments. This representation corresponds to a gene expression profile, available for clustering. Figure 7.5 illustrates such an example, where rows represent the different exposure profiles and

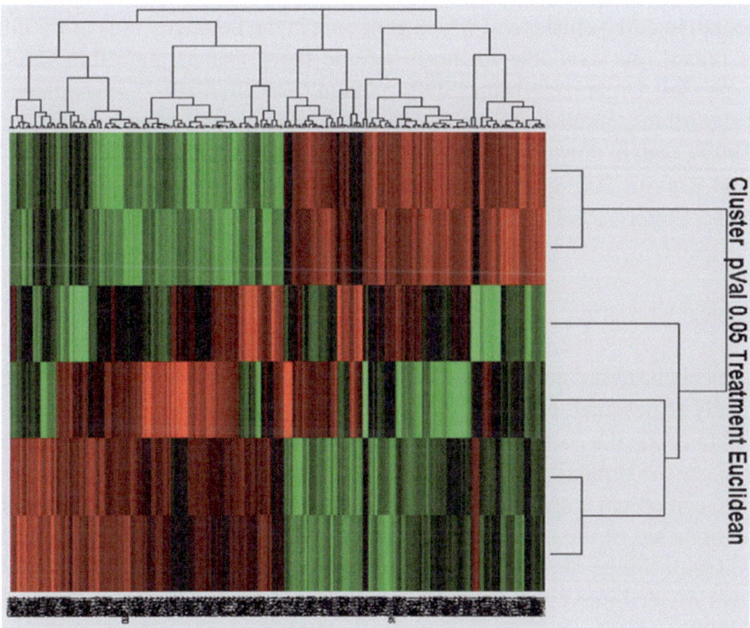

Fig. 7.5 Hierarchical clustering using Euclidean distance of a typical gene expression profile after multiple exposure treatments

columns the genes differentially expressed. The goal is to organize profiles into clusters so that the instances in the same cluster are highly similar to each other and the instances from different clusters have low similarity to each other.

Due to the special characteristics of gene expression data, and the particular requirements from the biological domain, gene-based clustering presents several challenges. Firstly, cluster analysis is typically in data mining and knowledge discovery. The purpose of clustering gene expression data is to reveal the natural data structures and gain some initial insights into data distribution. Therefore, a good clustering algorithm should depend as little as possible on prior knowledge, which is usually not available before cluster analysis. For example, a clustering algorithm which can accurately estimate the "true" number of clusters in the dataset would be more favored than the one requiring the predetermined number of clusters. Secondly, due to the complex procedures of microarray experiments, gene expression data often contain a huge amount of noise. Therefore, clustering algorithms for gene expression data should be capable of extracting useful information from a high level of background noise. Thirdly, gene expression data are often "highly connected" (Jiang et al. 2003a, b), and clusters may be highly intersected with each other or even embedded into one another (Jiang et al. 2003a, b). Therefore, algorithms for gene-based clustering should be able to effectively handle this situation. Finally, users of microarray data may not only be interested in the clusters of genes, but also be interested in the relationship between the clusters (e.g., which clusters are closer to each other, and which clusters are remote from each other) and the relationship between the genes within the same cluster (e.g., which gene can be considered as the representative of the cluster and which genes are at the boundary area of the cluster). In this regard, the available methods include the K-means algorithm (McQueen 1967), the Self-Organizing Map (SOP) method (Kohonen 1984), hierarchical clustering algorithms, graph-theoretical approaches (Shamir and Sharan 2000; Ben-Dor et al. 1999), pattern clustering methods (Agrawal and Srikant 1994; Han et al. 2003; Seno and Karypis 2001), and model-based clustering (Dasgupta and Raftery 1998; Fraley and Raftery 1998).

Hierarchical Clustering

Hierarchical clustering generates a hierarchical series of nested clusters which can be graphically represented by a tree, called a *dendrogram*, as seen in Fig. 7.6.

The branches of a *dendrogram* not only record the formation of the clusters but also indicate the similarity between the clusters. By cutting the *dendrogram* at some level, we can obtain a specified number of clusters. By reordering the objects such that the branches of the corresponding *dendrogram* do not cross, the dataset can be arranged with similar objects placed together. Hierarchical clustering algorithms can be further divided into agglomerative approaches and divisive approaches based on how the hierarchical *dendrogram* is formed. Agglomerative algorithms (bottom-up approach) initially regard each data object as an individual cluster and, at each step, merge the closest pair of clusters until all the groups are merged into one cluster.

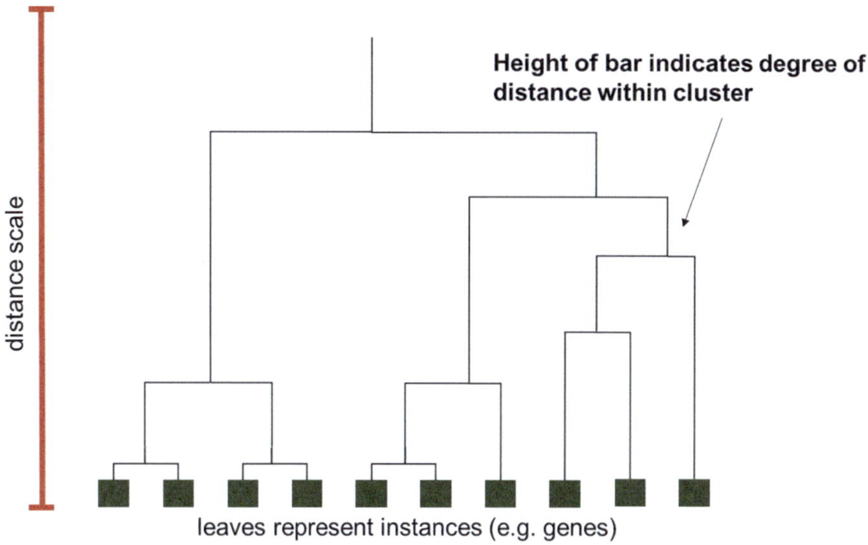

Fig. 7.6 An example of hierarchical clustering

Divisive algorithms (top-down approach) start with one cluster containing all the data objects and, at each step, split a cluster until only singleton clusters of individual objects remain. For agglomerative approaches, different measures of cluster proximity, such as single link, complete link, and minimum-variance (Dubes 1988; Kaufman and Rousseeuw 2008), derive various merging strategies. For divisive approaches, the essential problem is to decide how to split clusters at each step.

Pattern Discovery

Association rules can reveal biologically relevant associations between different genes or between environmental effects and gene expression. An association rule has the form LHS $\rightarrow$ RHS, where LHS and RHS are disjoint sets of items, the RHS set being likely to occur whenever the LHS set occurs. Items in gene expression data can include genes that are highly expressed or repressed, as well as relevant facts describing the cellular environment of the genes. It is hence a commonly used methodology for detecting local patterns in unsupervised learning systems and representing feature-value conditions among the data. Analysis of large genomic data has two important goals:

- To determine how the expression of any particular gene might affect the expression of other genes; the genes involved in this case could belong to the same gene regulatory network. By this we mean a set of genes being expressed together in a nonrandom pattern to code for specific proteins regulating biological function

- To determine what genes are expressed as a result of certain cellular conditions, for example, what genes are expressed in diseased cells that are not expressed in healthy cells

A Priori

The most basic join-based algorithm is the a priori method (Agrawal and Srikant 1994). The a priori approach uses a level-wise approach in which all frequent item sets of length k are generated before those of length $(k + 1)$. The main observation which is used for the a priori algorithm is that every subset of a frequent pattern is also frequent. Therefore, candidates for frequent patterns of length $(k + 1)$ can be generated from known frequent patterns of length k with the use of *joins*. A *join* is defined by pairs of frequent k-patterns that have at least $(k - 1)$ items in common. This method identifies the frequent items in the dataset: that is, those items with minimum *Support*. Here Support (S) is defined as the proportion of records in the dataset which contain the item set. An advantage of this approach is the fact that it is bottom-up; as such it can be extended to larger item sets that appear often in the dataset. Figure 7.7 shows the implementation of the a priori method in a database D.

FP-Growth

The most popular frequent item set mining algorithm is FP-Growth (Han et al. 2000). The main aim of this algorithm was to remove the bottlenecks of the a priori

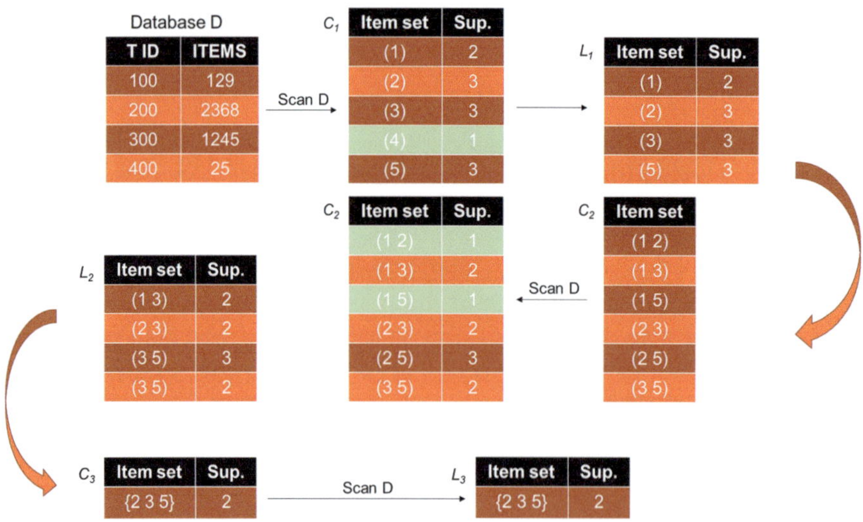

Fig. 7.7 Implementation of the a priori method on a database D

algorithm in generating and testing a candidate set. The problem of the a priori algorithm was dealt with, by introducing a novel compact data structure called frequent pattern tree, or FP-tree, then based on this structure an FP-tree-based pattern fragment growth method was developed. FP-growth uses a combination of the vertical and horizontal database layout to store the database in main memory. Instead of storing the cover for every item in the database, it stores the actual transactions from the database in a tree structure and every item has a linked list going through all transactions that contain that item. This new data structure is denoted by FP-tree (frequent-pattern tree) (Han et al. 2003). Essentially, all transactions are stored in a tree data structure. The advantages of this method are that it avoids the costly iterative scans and generations of a large number of candidate pattern sets and produces a more compact pattern set.

LPMiner

The LPMiner approach (Seno and Karypis 2001) is based on the FP-growth method, using false frequent patterns. The main idea behind is that the minimal frequency threshold decreases with the length of the pattern, then a smaller number of false frequent patterns is generated. Therefore, an additional constraint was introduced, the *length-decreasing support* constraint. Using this method, frequent items are found, the support of which decreases as a function of item length. This method combines the FP-tree data structure with the pruning process and it is known to be faster than the FP-growth method. However, it has been found to be problematic when dealing with short item sets.

Joint Pathway Analysis Toward Understanding Disease Mechanisms

The outcomes of transcriptomics, epigenomics, proteomics, and metabolomics analysis can be mapped onto regulatory pathways, which will be analyzed jointly using integrative systems biology software such as Agilent GeneSpring. The aim is to integrate multiple-omics responses into a mechanistic description of toxicity pathway interactions with biological plausibility in relation to internal exposure as described in Manrai et al. (2016). Advanced bioinformatics algorithms will be used to identify common, biologically associated nodes across several pathways to identify a limited number of Pathways of Toxicity (PoT). This approach is expected to identify the most critical regulatory pathway nodes regulating the onset of regulatory pathways beyond cellular homeostasis on different biological scales, thus identifying potential candidates for AOP, as well as to investigate the biological plausibility of the proposed AOPs. The latter will be obtained by developing biological pathway models for human health outcomes. In addition, joint pathway analysis is critical for overcoming a critical deficit of AOPs; AOPs are inherently pathway-specific and cannot predict toxic effects due to molecular perturbations beyond the key events that define them. However, often there are common nodes between several AOPs, even when modes of action (MoAs) of the respective chemical stressors are different. This is of importance when dealing with combined exposures, where several toxicological pathways should converge, integrated into larger AOP networks. The relevant toxicity pathways identified will result in the identification of the molecular initiating event that will in turn result in a quantitative understanding of the Key Event Relationships (KERs) related to the respective AOPs. Thus, in addition to the experimental data available from a cohort or an in vitro study, in the case of data gaps relevant to specific key events, these will be filled using computationally predicted AOP (cpAOP) nets, which can be probed to yield focused subnetworks, such as cpAOP subnets. The knowledge sources will be used for phenotypic anchoring using available in vivo data. AOPs are laid out in a linear fashion, progressing from the molecular initiating event to the adverse outcome. It cannot be assumed that identification of a molecular initiating event will always lead to the adverse outcome as a significant number of factors are required for a quantitative understanding of the link between key events and key event relationships. This problem will be addressed by investigating the dose-dependent activation of toxicity pathways rather than individual key events. Moreover, joint pathway analysis of multi-omics data obtained after different time lengths of exposure will allow us to distinguish adaptive from persistent responses linked to pathologies.

To use the AOP methodology for regulatory risk assessment, quantitative AOPs will need to be established and thresholds for molecular initiating events to trigger the downstream key events or adverse outcomes need to be defined. The concentration range (of the investigated mixtures) at which these occur could then be compared to the environmentally relevant exposure. KE thresholds would need to be defined with regard to not only the dose of the investigated mixtures, but also the duration of exposure necessary not only to trigger but also to sustain a particular key event. Integrating the data over the time of exposure, a biologically based dose–time response model will be developed related to the sustained key event. Finally, based on this information, a strategy for identification of new markers of effect for the relevant health endpoints can be developed. Derivation of predictive biomarkers that differentiate adaptive versus pathology-related molecular changes based on these data requires preprocessing the large amount of data produced, discovering specific data patterns and/or clusters, creating a data model based on a training dataset, and, finally, evaluating that model regarding its validity and prediction capacity on the basis of test data. Biomarker evidence will be used to verify the mechanistic hypotheses regarding the developed AOPs.

Application of Transcriptomics in the Frame of the Exposome

Introduction

Toxicogenomics, the integration of high-throughput gene expression technologies, bioinformatics, and toxicology, has benefitted from significant investment in the pharmaceutical industry for both predictive and mechanism-based toxicology in an effort to identify candidate drugs more quickly and economically (Lesko and Woodcock 2004; Yang et al. 2004). Regulatory agencies such as the US Food and Drug Administration (FDA) and the US Environmental Protection Agency (EPA) also recognize the potential of toxicogenomics and encourage the use and submission of complementary toxicogenomics data for potential inclusion in submitted applications and incorporation into regulatory decision-making (Hackett and Lesko 2003). In Europe, the new chemical safety regulation introduced in 2006 (REACH) in the preamble of the legal text encourages the scientific community to investigate how toxicogenomics may better inform risk assessment of chemical substances. However, the use of transcriptomics for informing risk assessment and, in particular, for a better understanding of the environmental risks associated to the multitude of components encountered in various environments and microenvironments still remains limited. A key advantage of transcriptomics is that it provides insights into the early effects of environmental exposure; this is of particular importance considering the advent of AOPs in recent years. In the AOP context, transcriptomics can effectively describe the molecular initiating events that trigger the activation of various AOPs. Toward this aim, in this study, the use of transcriptomics is demonstrated in the context of understanding the differences in early effects associated to two ubiquitous mixtures of putative health stressors, namely, indoor air and PAHs (mostly associated to outdoor air pollution). Indoor and outdoor air pollution has been linked through numerous epidemiological and toxicological studies to severe health effects such as respiratory and cardiovascular diseases, atherosclerosis, cancer, childhood asthma, and other allergic airway diseases (Brunekreef and Holgate 2002; Fiedler et al. 2005). The current scientific paradigm toward assessing the human health impact of air pollution to date is based on analysis of epidemiological data. Even though the body of epidemiological evidence for linking air pollution to ill health is steadily increasing, significant gaps still exist with regard to the actual biological mechanisms resulting in air pollution-related health effects (Fiedler et al. 2005). Available toxicological data on air pollutants, however, are often difficult to delineate due to the complexity of chemical mixtures present in the environment and the complexity of the biological responses associated with chemical-to-chemical differences in toxicokinetics and toxicodynamics. Although additivity models for a similar or independent mechanism of toxic action groupings exist, no comprehensive models have been developed to study interactions between chemicals present in complex heterogeneous mixtures such as those present in the air we breathe (Brunekreef and Holgate 2002). Thus, to better understand how the different sources of air pollution mixtures contribute to human health (Sarigiannis et al. 2015) and to set the frame for well-targeted cost-effective interventions, additional insights than the ones provided by conventional epidemiology or toxicology are required.

Material and Methods

Overall Study Design

In the study described by way of example herein, we applied a toxicogenomics approach to evaluate the modulation of gene expression in cultured human cells derived from the bronco-pulmonary system (A549) exposed to an indoor air chemical mixture representative of typical indoor air in Europe as defined in the INDEX review (Kotzias et al. 2005), or a mixture of polyaromatic hydrocarbons (PAHs) extracted from urban air samples from Milan, Italy. This approach takes advantage of the "whole genome microarrays" assay developed by Applied Biosystems. Specific biological processes and molecular functions were identified for each of the mixtures tested, suggesting that genomic technologies may be a useful tool to facilitate and expedite progress in understanding complex air mixture interactions. Figure 7.8 provides a schematic outline of the overall toxicogenomics methodology we followed.

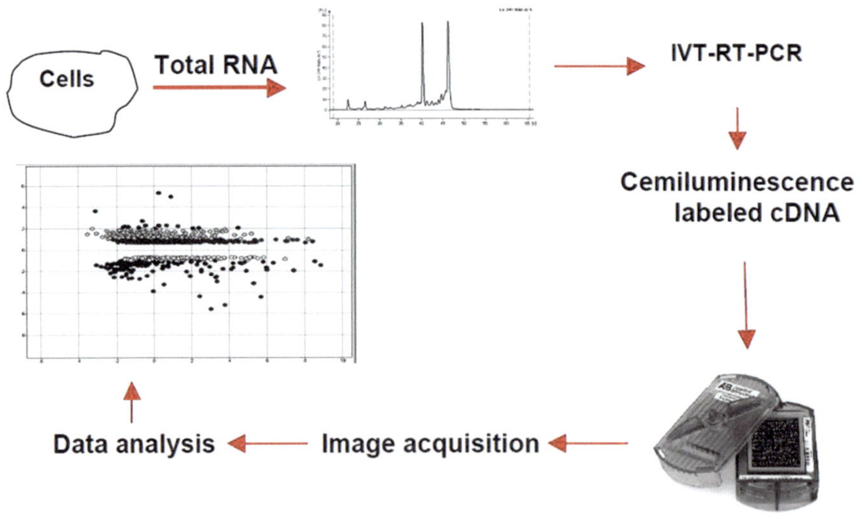

Fig. 7.8 Graphical description of the methodological steps followed in the toxicogenomics approach

Cell Cultures and Treatments

A549 cells were cultured in F12K medium with 10% fetal calf serum at 37 °C with 5% CO_2. The cells were stabilized overnight and then exposed to an indoor air chemical mixture (INDEX) or to a mixture of PAHs extracted from urban air. The chemical composition of the indoor air priority compounds identified in INDEX (Kotzias et al. 2005) and the actual doses of each individual chemical in the mixture used in the cell treatments are listed in Table 7.1. For each compound of the INDEX mixture, a 10X stock solution in DMSO was prepared. These solutions were then diluted to 1X work solution in the appropriate medium, and the reconstituted mixture was added to the cell cultures.

The composition of the urban PAH mixture from Milan is listed in Table 7.2.

A 1 g/ml stock solution of PAHs in DMSO was prepared. Immediately before use, a 0.5 mg/ml work solution was prepared, added to culture medium, and the latter

Table 7.1 List of volatile organic compounds identified in the INDEX review

Chemical families	Environmental concentration (μg/m^3)	Dose used in cell treatment (μg/l)
Aromatics		
Benzene	3–23	10
Naphthalene	2–46	20
Styrene	1–5	1
Toluene	25–130	50
m-, p-xylene	25–55	10
o-xylene	8–15	10
Aldehydes		
Acetaldehyde	8	6
Formaldehyde	21–31	25
Terpenes		
a-Pinene	7–18	12
Limonene	19–56	40

Table 7.2 Chemical composition of the PAH mixture in the ambient air of Milan, Italy

Chemical substance	Concentration (pg/m^3)
Fluorene	11
Phenanthrene	122
Anthracene	20
Fluoranthene	247
Pyrene	318
Benzo(a)anthracene	218
Chrysene	385
Benzo(k)fluoranthene	441
Benzo(a)pyrene	392
Indeno(1,2,3-c,d)pyrene	569
Dibenzo(a,h)anthracene	37
Benzo(g,h,i)perylene	823

applied to the cells. The final concentration of DMSO in cell solutions was 0.1%. Control samples in 0.1% DMSO were included in each incubation. After 24 h of treatment, the cells were collected, washed in 1X PBS, and immediately lysed in Lysis Buffer containing guanidine thiocyanate plus mercaptoethanol according to the RNA extraction protocol Qiagen RNeasy Mini Kits. The immediate lysis is an absolute prerequisite for reliable gene expression analysis. The lysate can be stored at −80 °C or proceeded immediately with RNA extraction.

RNA Preparation

Total RNA was extracted from cells using Qiagen RNeasy Mini Kits according to the manufacturer's protocol. The quality of the extracted RNA was verified by the Agilent 2100 Bioanalyzer (Agilent Technologies, USA). The RNA was quantified spectrophotometrically and then stored at −80 °C until use.

Microarray Hybridization

All microarray experiments were performed using the Applied Biosystems Expression Array System, which is based upon chemiluminescence detection. The Human Genome Survey Arrays contain approximately 32,000 60-mer oligonucleotides probes, together with about 1000 control probes.

Briefly, 1 μg of total RNA from each sample was reverse transcribed, amplified, and DIG-labeled (DIG-dUTP; Roche, Germany), using the Applied Biosystems Chemiluminescent NanoAmp RT-IVT Labeling Kit. Amount and quality of the DIG-labeled cRNA were controlled by Agilent 2100 Bioanalyzer.

Each microarray was first prehybridized at 55 °C for 1 h in hybridization buffer with blocking reagent, and then hybridized with 15 μg of DIG-labeled and fragmented cRNA at 55 °C for 16 h. An internal control target (ICT, 24-mer oligo labeled with LIZ fluorescent dye) was used.

After hybridization, the arrays were washed with hybridization wash buffer and chemiluminescence rinse buffer. Enhanced chemiluminescent signals were generated by first incubating arrays with anti-digoxigenin-alkaline phosphatase, enhanced with chemiluminescence enhancing solution, and finally adding chemiluminescence substrate.

Chemiluminescent signal detection, image acquisition, and image analysis of the microarrays were performed on the Applied Biosystems 1700 Chemiluminescent Microarray Analyzer. Images were auto-gridded and the chemiluminescent signals were quantified, corrected for background, and spot and spatially normalized.

Microarray Data Analysis

The Applied Biosystems Expression System software was used to extract Assay Signal, and Assay Signal to Noise ratio (S/N) values from the microarray images. Bad spots flagged by the software were removed from the analysis. Only microarrays showing average normalized signal intensity above 5000 and a median background below 600 were included in the study. Signal intensities were imported into AB1700Guide© (Integromics, S.L., Spain). Using this software, the Assay Signal of the remaining sets of genes was inter-quantile-normalized across arrays (Bioconductor, R Foundation for Statistical Computing, library ABarray). To select deferentially expressed genes, the genes were filtered by standard Expression Array System Signal to Noise threshold (S/N greater than 3 in at least one sample and quality flag <5000).

Genes with minimum p-value ($p < 0.05$, Student's t test) were identified as candidates for significantly differentially expressed genes. Fold change (treated vs. control assay signal ratio) data samples were calculated for each gene. The Spotfire DecisionSite for Microarray Analysis (Spotfire, Inc) was used for graphic representation of results.

Significant probe sets were evaluated for relevance to canonical pathway, molecular function, and biological function using the PANTHER (Protein Analysis THrough Evolutionary Relationships) Classification System (http://www.pantherdb.org) (Sarigiannis et al. 2009). The genes expressed were compared to the entire list of genes on the Applied Biosystems Human Genome Survey Array. Statistically significant over- and underrepresented annotation categories were determined by binomial statistics, using the observed number of expressed genes versus the numbers expected by chance within a certain annotation group. Categories with p-values $>10^{-2}$ were rejected.

Results and Discussion

Gene Expression Analysis in the Indoor Air Mixture Vs. the PAHs-Exposed A549 Cells

All RNA samples were of high quality based on an assessment of the Bioanalyzer output as provided by RNA Integrity Number (RIN). These values for all samples were 9.5 or greater on a scale of 1–10, where a value of 10 is considered completely intact.

The number of genes, which differed significantly in each treatment and grouped according to the fold-change value, is illustrated in Fig. 7.9. The number of genes commonly modulated by both treatments is also reported.

These genes were evaluated for relevance to canonical pathway, biological processes, and molecular function using PANTHER. The classification in significant biological processes of the modulated genes with fold change 2 is listed in Fig. 7.10.

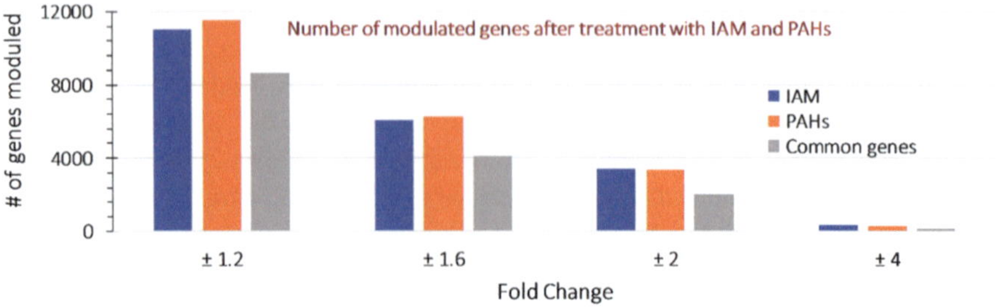

Fig. 7.9 Number of modulated genes after treatment with an Indoor Air Mixture (IAM) or PAH mixture (PAHs) in A549 cells. The genes are grouped according to the fold change value, and the number of genes commonly modulated by both treatments is also reported

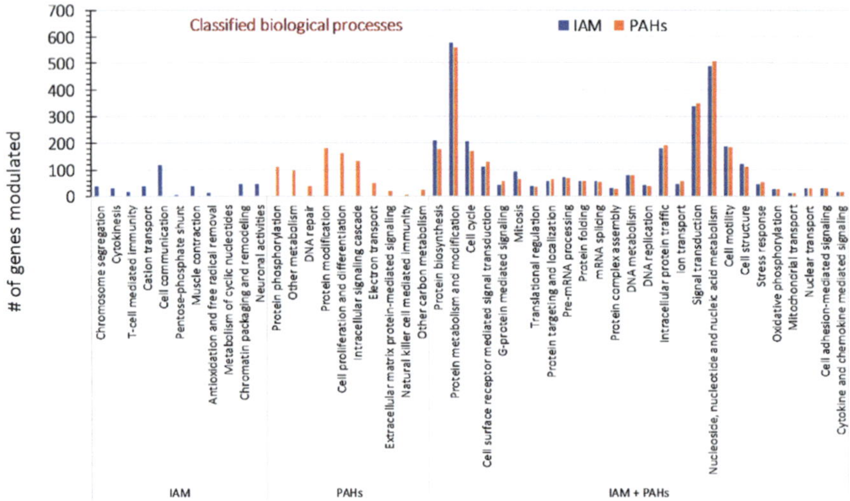

Fig. 7.10 List of Biological Processes classified by PANTHER analysis. The modulated genes identified through gene expression analysis with a twofold Change value were categorized by biological process. The numbers of genes involved in each biological process were considered and presented herein. Only those categories with p-values $<10^{-2}$ were considered and reported in bold characters

The categories with p-values $<10^{-2}$ were considered and reported in bold. Three separate groups of biological process categories are evident. The first group with 11 categories is specific for indoor air mixture treatments; the second group with 10 categories is specific for PAHs treatments; the last group with 26 categories comprises the significant biological processes in A549 cells treated with either of the two chemical air mixtures.

Out of the first group with cells treated with indoor air chemicals (IAM), the two most significant biological processes are chromosome segregation (38 genes,

p value = 6.8E-07) and cytokinesis (30 genes, p value = 6.0E-05); all these genes are involved in the process of cell division.

The most significant gene expression modifications associated with A549 cells exposed to the PAH mixture that is characteristic of urban ambient air were related to protein phosphorylation (110 genes, p = 7.5E-04) and protein modification (98 genes, p = 1.9E-04). Furthermore, genes associated with cell proliferation and differentiation and with cell–cell signaling involving extracellular matrix proteins are significantly modulated after these treatments.

Gene expression within 26 biological process categories was significantly modulated by both the indoor air mixture and PAH mixture treatments: protein biosynthesis, protein metabolism and modification, protein targeting and localization, protein complex assembly, protein folding, cell structure.

In order to identify the specific signatures of biological processes regulated by the IAM or PAHs mixtures, we focused our attention on those genes that were up- or down-regulated by more than fourfold after treatment. In spite of the strict cutoff we applied, the number of modulated genes remained high in both treatment groups: 325 significantly modulated genes after exposure to the mixture identified in the INDEX review and 289 modulated genes in cells treated with PAHs. Figure 7.11 shows the significant over- and underrepresented biological processes in A549 cells treated with either of the two chemical air mixtures. In this list of genes, the categories with p-values $<10^{-2}$ were considered and reported in bold characters.

From cells treated with a mixture of indoor air chemicals (IAM), the most significantly altered gene population (18 genes, p = 5.7E-05) belonged to a biological process category encompassing protein targeting and localization; this category includes processes through which proteins are distributed into the appropriate subcellular location including cytoskeletal organization. Protein folding and proteins that act as catalysts of the process of assembling protein complexes were associated with nine and six of the modulated genes (p = 8.9E-04 and 4.6E-04), respectively. Six of the genes (p = 3.4E-03) were associated with nuclear transport, and seven genes with muscle contraction (p = 8.2E-03). DNA metabolism and DNA

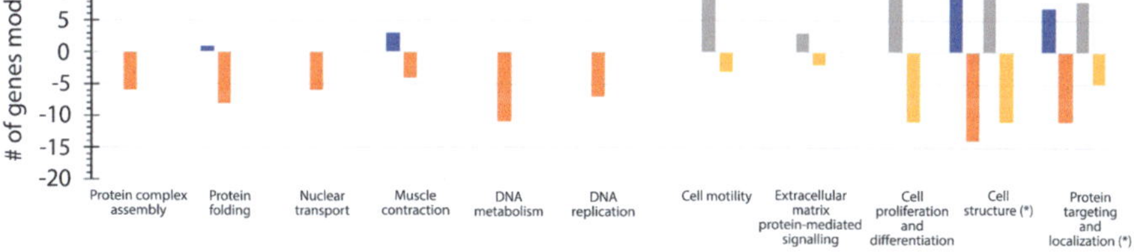

Fig. 7.11 List of Biological Processes classified by PANTHER analysis. The fourfold modulated genes identified through gene expression analysis were categorized by biological process. The numbers of up- and down-modulated genes are reported. Only those categories with p-values $<10^{-2}$ were considered and presented herein. Asterisk indicates a biological process effected by both treatments (indoor and outdoor)

replication processes involved 11 and 7 genes, respectively, with a p-value of about E-03. At least 23 genes ($p = 7.8E\text{-}03$) altered by IAM exposure, corresponded to a cell structure biological process category.

The most significant gene expression modifications associated with A549 cells exposed to a PAH mixture were related to cell motility (12 genes, $p = 4.6E\text{-}04$) and cell–cell signaling involving extracellular matrix proteins (five genes, $p = 6.4E\text{-}04$). Twenty of the modulated genes identified were classified in a cell proliferation and differentiation category ($p = 4.2E\text{-}03$), confirming the proliferating activity of PAHs in exposed cells (Svihalkova-Sindlerova et al. 2007). Gene expression within two of the biological process categories was significantly modulated by both the indoor air mixture and PAH mixture treatments: protein targeting and localization, and cell structure. At fold change of 4, there was a reorganization of the distribution of the same biological process categories such as protein complex assembly, protein folding, nuclear transport, DNA metabolism, DNA replication, and cell mortality. At fold change = 2, these categories were modulated by both the indoor air and PAHs mixtures. Using a higher fold change cutoff, the number of genes decreased and the significance of the biological process became more selective. In this way, we could identify the modulated genes that are specific for each one of the two different mixtures. A similar trend was observed considering the molecular functions of the significantly modulated genes with p-values $<10^{-2}$ (Fig. 7.12).

In the cells treated with indoor air mixtures, the 25 genes that code for cytoskeletal protein belong to molecular functions with a p-value of 3.5E-05 and the three genes that code for receptors were modulated with a p-value of 1.3E-05. The genes that code for chaperonin (11 genes, $p = 4.5E\text{-}03$) and for a family of chaperone (two genes, $p = 4.3E\text{-}03$) proteins acting as cytoplasmic proteins that bind to nascent or unfolded polypeptides and ensure their correct folding or transport are also modulated. Intermediate filament (seven genes, p-value = 3.4E-03) and microtubule binding motor protein (four genes, p-value = 6.6E-3) were the other molecular functions significantly modulated by indoor air mixture exposure. These results suggest that the most significantly altered molecular functions involved the genes

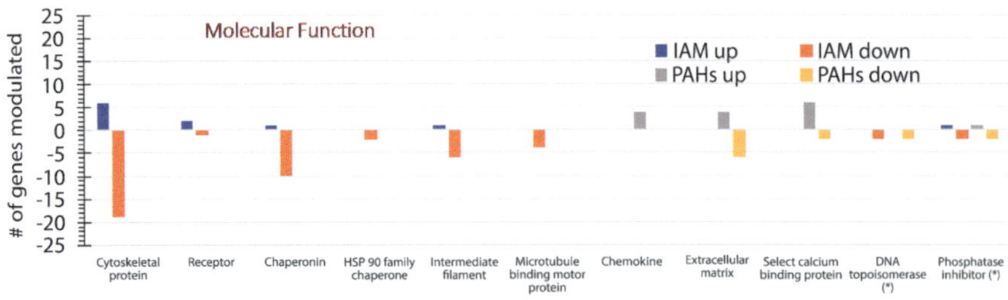

Fig. 7.12 List of Molecular Function resulted by PANTHER analysis. The numbers of up- and down-modulated genes with a fourfold change value involved in groups of molecular function are reported. Only those categories with p-values $<10^{-2}$ were considered and presented herein. Asterisk indicates a molecular function effected by both treatments (indoor and outdoor)

that form a flexible framework for the cell providing attachment points for organelles and assure cell-cell communication.

The most significant molecular function modifications associated with exposure to a PAH mixture were related to genes that code for chemokines (four genes, $p = 1.6E\text{-}03$), for extracellular matrix proteins (ten genes, $p = 5.0E\text{-}03$), and for select calcium-binding proteins (eight genes, $p = 9.2E\text{-}03$). Only two molecular functions (DNA topoisomerase and phosphatase inhibitor) were involved in both treatments at a p value of 1E-03.

The larger pattern of biochemical activities regulated by indoor air chemicals compared to PAHs seems to reflect the heterogeneous compositions of the IAM mixtures that include aromatic compounds, aldehydes, and terpenes. Interestingly, exposure to indoor air mixtures induces a more pronounced down-modulation of the genes than exposure to PAHs, confirming major differences in toxic mode of action of these two chemical mixtures. A more in-depth analysis of expression patterns of genes involved in key regulatory processes such as the p53 pathway (which effectively regulates cell life and apoptosis, it is thus intimately associated with the onset of carcinogenesis) shows significant differences after exposure to different chemical fractions of the overall indoor air mixture (see Fig. 7.13).

The comparative analysis of such complex gene regulatory networks in conjunction with the corresponding biological and physiological processes they regulate allows us to derive important conclusions regarding the molecular initiating and key events lying behind robust adverse outcome pathways. Even though a large number of genes with modulated expression levels exist after treatment with different chemical families and with the overall mixture, molecular fingerprints and synergistic effects among chemical mixture components at the level of modulated gene expression can be identified. What is even more interesting is the fact that such fingerprints are usually associated with specific biological processes (e.g., the induction of cell survival), that is, the potential for carcinogenicity is much higher after exposure to the overall indoor air mixtures (see the large number of red boxes linking to cell survival in Fig. 7.13) compared to any of the individual treatments with chemical fractions of the mixture. Vice versa, genes affected by exposure to terpenes (marked in blue in Fig. 7.13) are associated with regulation of apoptosis, inhibition of angiogenesis and metastasis, and with the induction of feedback loops in cell-to-cell signaling. Such analyses give the impetus for targeted multi-omics studies, thus enabling the effective unraveling of the intricate web that links the exposome with the advent of human disease.

Conclusions

The capabilities of multi-omics and the place of transcriptomics in the omics arena, as well as the proper statistical and computational methods, have been illustrated herein. The particular use of transcriptomics has been demonstrated in the example of a complete genomic response analysis of “early events” following exposure to

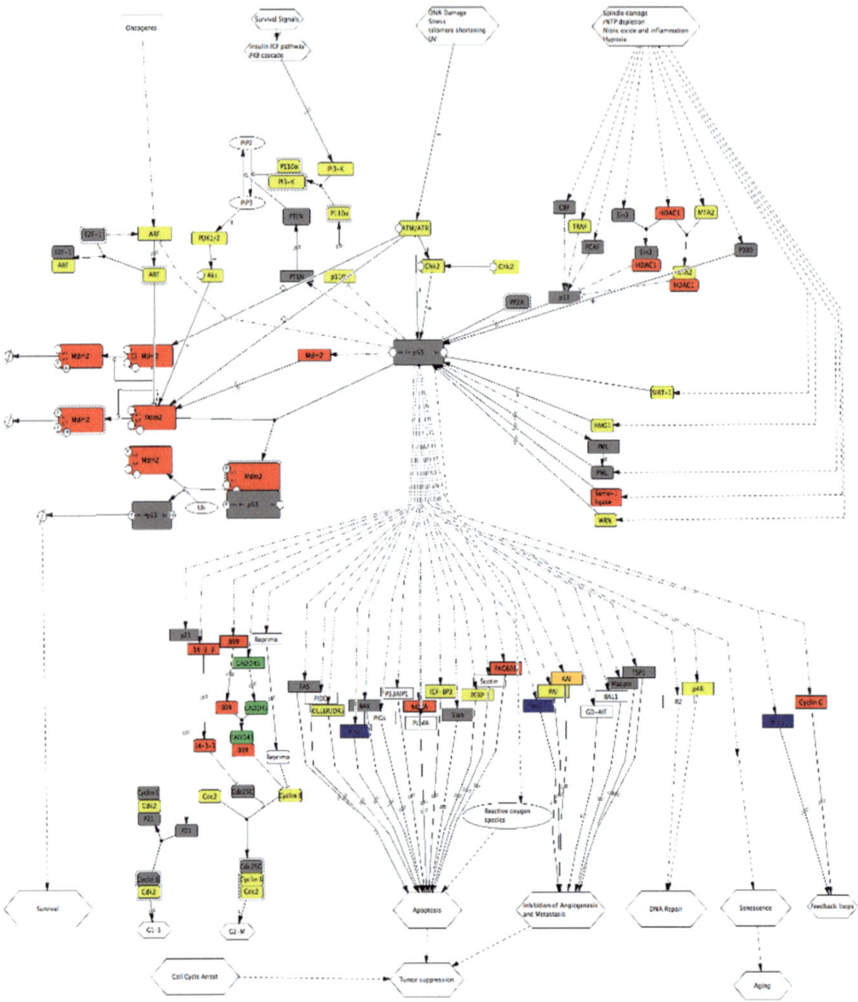

Fig. 7.13 Differential modulation of the expression of the gene set involved in the p53 regulatory pathway after exposure to the overall indoor air mixture and the main chemical families that compose it. In red are shown the genes with modulated expression only after exposure to the overall indoor air mix; in green the genes affected only by aromatic compounds; in orange the genes affected by aldehydes; and in blue the ones affected by terpenes only. In yellow are shown the genes that modulated their expression level after exposure to all treatments

complex chemical mixtures. The results reported here clearly show that transcriptomic methodologies would be relevant for understanding the wide range of biological processes and biochemical effects modulated by exposure to complex environmental mixtures such as indoor and outdoor air. The toxicogenomics approach followed in this study is in response to recommendations for the development of plausible descriptions of mechanisms and mode of action for chemical mixtures. In this context, a major objective is to identify "key events" involved in biological responses to better inform critical evaluations of risk from mixtures. The results of our study indicate that advanced molecular and systems biology-based approaches using toxicogenomics may facilitate the selection of appropriate sets of biomarkers to characterize responses induced by different mixtures of chemicals. The challenge that remains is to comprehensively integrate the disparate sources of chemical and biological data, including high-throughput technologies such as toxicogenomics, in order to elucidate the mechanisms and networks involved in toxicity and to develop quantitative models capable of accurately predicting thresholds and exposure-response functions.

Acknowledgments The author gratefully acknowledges the support of the European Commission through the grant No. 603946 (HEALS—Health and Environment-wide Associations via Large Population Studies) funded through the 7th Framework Program for Research and Technological Development of the EU.

References

Aebersold R, Mann M (2003) Mass spectrometry-based proteomics. Nature 422(6928):198–207

Agrawal R, Srikant R (1994) Fast algorithms for mining association rules in large databases. In: VLDB conference, pp 487–499

Agyeman AS, Chaerkady R, Shaw PG, Davidson NE, Visvanathan K, Pandey A, Kensler TW (2012) Transcriptomic and proteomic profiling of KEAP1 disrupted and sulforaphane-treated human breast epithelial cells reveals common expression profiles. Breast Cancer Res Treat 132 (1):175–187. https://doi.org/10.1007/s10549-011-1536-9

Allison DB, Cui X, Page GP, Sabripour M (2006) Microarray data analysis: from disarray to consolidation and consensus. Nat Rev Genet 7(1):55–65. https://doi.org/10.1038/nrg1749

Altman RB, Raychaudhuri S (2001) Whole-genome expression analysis: challenges beyond clustering. Curr Opin Struct Biol 11(3):340–347. https://doi.org/10.1016/s0959-440x(00)00212-8

Ankley GT, Bennett RS, Erickson RJ, Hoff DJ, Hornung MW, Johnson RD, Mount DR, Nichols JW, Russom CL, Schmieder PK, Serrrano JA, Tietge JE, Villeneuve DL (2010) Adverse outcome pathways: a conceptual framework to support ecotoxicology research and risk assessment. Environ Toxicol Chem 29(3):730–741

Audouze K, Juncker AS, Roque FJ, Krysiak-Baltyn K, Weinhold N, Taboureau O, Jensen TS, Brunak S (2010) Deciphering diseases and biological targets for environmental chemicals using toxicogenomics networks. PLoS Comput Biol 6(5):e1000788. https://doi.org/10.1371/journal.pcbi.1000788

Ben-Dor A, Shamir R, Yakhini Z (1999) Clustering gene expression patterns. J Comput Biol 63 (3/4):281–297

Boezio B, Audouze K, Ducrot P, Taboureau O (2017) Network-based approaches in pharmacology. Molecular Inform 36(10), Wiley-VCH Verlag GmbH & Co. KGaA, Weinheim

Borkowski K, Wrzesinski K, Rogowska-Wrzesinska A, Audouze K, Bakke J, Petersen RK, Haj FG, Madsen L, Kristiansen K (2014) Proteomic analysis of cAMP-mediated signaling during differentiation of 3 T3-L1 preadipocytes. Biochim Biophys Acta 1844(12):2096–2107. https://doi.org/10.1016/j.bbapap.2014.07.015

Breiman L (1996) Bagging predictors. Mach Learn 24(2):123–140

Breiman L (1998) Arcing classifiers (with discussion). Ann Stat 26(3):801–849

Breiman L (2001) Random forests. Mach Learn 45(1):5–32. https://doi.org/10.1023/a:1010933404324
Brunekreef B, Holgate ST (2002) Air pollution and health. Lancet 360(9341):1233–1242. https://doi.org/10.1016/s0140-6736(02)11274-8
Chang B, Halgamuge SK (2002) Protein motif extraction with neuro-fuzzy optimization. Bioinformatics 18(8):1084–1090
Ciriello G, Gatza ML, Beck AH, Wilkerson MD, Rhie SK, Pastore A, Zhang H, McLellan M, Yau C, Kandoth C, Bowlby R, Shen H, Hayat S, Fieldhouse R, Lester SC, Tse GM, Factor RE, Collins LC, Allison KH, Chen YY, Jensen K, Johnson NB, Oesterreich S, Mills GB, Cherniack AD, Robertson G, Benz C, Sander C, Laird PW, Hoadley KA, King TA, Perou CM (2015) Comprehensive molecular portraits of invasive lobular breast cancer. Cell 163(2):506–519. https://doi.org/10.1016/j.cell.2015.09.033
Dasgupta A, Raftery AE (1998) Detecting features in spatial point processes with clutter via model-based clustering. J Am Stat Assoc 93(441):294–302
Dong G, Zhang X, Wong L, Li J (1999) CAEP: classification by aggregating emerging patterns. In: Springer-Verlag (ed) Proceedings of the second international conference on discovery science, pp 30–42
Dubes R (1988) Algorithms for clustering data. Prentice Hall, Englewood Cliffs, NJ
Dumas ME, Domange C, Calderari S, Martinez AR, Ayala R, Wilder SP, Suarez-Zamorano N, Collins SC, Wallis RH, Gu Q, Wang Y, Hue C, Otto GW, Argoud K, Navratil V, Mitchell SC, Lindon JC, Holmes E, Cazier JB, Nicholson JK, Gauguier D (2016) Topological analysis of metabolic networks integrating co-segregating transcriptomes and metabolomes in type 2 diabetic rat congenic series. Genome Med 8(1):101. https://doi.org/10.1186/s13073-016-0352-6
Ebrahim A, Brunk E, Tan J, O'Brien EJ, Kim D, Szubin R, Lerman JA, Lechner A, Sastry A, Bordbar A, Feist AM, Palsson BO (2016) Multi-omic data integration enables discovery of hidden biological regularities. Nat Commun 7:13091. https://doi.org/10.1038/ncomms13091
Fayyad UM, Piatetsky-Shapiro G, Smyth P (1996) Knowledge discovery and data mining: towards a unifying framework. In: Proceedings of the second international conference on knowledge discovery and data mining, p 82
Fiedler N, Laumbach R, Kelly-McNeil K, Lioy P, Fan ZH, Zhang J, Ottenweller J, Ohman-Strickland P, Kipen H (2005) Health effects of a mixture of indoor air volatile organics, their ozone oxidation products, and stress. Environ Health Perspect 113(11):1542–1548
Fraley C, Raftery AE (1998) How many clusters? Which clustering method? Answers via model-based cluster analysis. Comput J 41(8):586–588
Freund Y, Schapire R (1996) Experiments with a new boosting algorithm. In: Proceedings of the thirteenth national conference on machine learning, pp 148–156
Garcia-Reyero N (2015) Are adverse outcome pathways here to stay? Environ Sci Technol 49 (1):3–9. https://doi.org/10.1021/es504976d
Gasch A, Eisen M (2002) Exploring the conditional corregulation of yeast gene expression through fuzzy k-means clustering. Genome Biol 3:1–22
Gevaert O, De Smet F, Timmerman D, Moreau Y, De Moor B (2006) Predicting the prognosis of breast cancer by integrating clinical and microarray data with Bayesian networks. Bioinformatics 22(14):e184–e190. https://doi.org/10.1093/bioinformatics/btl230
Hackett JL, Lesko LJ (2003) Microarray data--the US FDA, industry and academia. Nat Biotechnol 21(7):742–743. https://doi.org/10.1038/nbt0703-742
Han J, Pei H, Yin Y (2000) Mining frequent patterns without candidate generation. In: Conference on the management of data, ACM Press, Dalas
Han J, Pei J, Yin Y, Mao R (2003) Mining frequent patterns without candidate generation: a frequent-pattern tree approach. Data Min Knowl Discov 8:53–87
Hao Q, Yadav R, Basse AL, Petersen S, Sonne SB, Rasmussen S, Zhu Q, Lu Z, Wang J, Audouze K, Gupta R, Madsen L, Kristiansen K, Hansen JB (2015) Transcriptome profiling of brown adipose tissue during cold exposure reveals extensive regulation of glucose metabolism. Am J Phys Endocrinol Metab 308(5):E380–E392. https://doi.org/10.1152/ajpendo.00277.2014

Hirschhorn JN, Daly MJ (2005) Genome-wide association studies for common diseases and complex traits. Nat Rev Genet 6(2):95–108. https://doi.org/10.1038/nrg1521

Jiang D, Pei J, Zhang A (2003a) Interactive exploration of coherent patterns in time-series gene expression data. In: Proceedings of the ACM SIGKDD international conference on knowledge discovery and data mining, pp 565–570. https://doi.org/10.1145/956750.956820

Jiang D, Pei J, Zhang A (2003b) DHC: a density-based hierarchical clustering method for timeseries gene expression data. In: BIBE2003 (ed) 3rd IEEE international symposium on bioinformatics and bioengineering, Bethesda, Maryland, 10–12 Mar 2003

Jones DT (2001) Protein structure prediction in genomics. Brief Bioinform 2(2):111–125

Kaufman L, Rousseeuw PJ (2008) Finding groups in data: an introduction to cluster analysis. Wiley, New York

Kohonen T (1984) Self-organization and associative memory. Spring, Berlin

Kongsbak K, Vinggaard AM, Hadrup N, Audouze K (2014) A computational approach to mechanistic and predictive toxicology of pesticides. ALTEX 31(1):11–22. https://doi.org/10.14573/altex.1304241

Kotzias P, Koistinen K, Kephalopoulos S, Schlitt C, Carrer P, Maroni VI, Jantunen MJ, Cochet C, Kirchner S, Lindvall T, McLaughlin J, Molhave L, Fernandes E, Seifert B (2005) The INDEX project: critical appraisal of the setting and implementation of indoor exposure limits in the EU. EUR 21590 EN. doi:Cited By (since 1996) 1 Export Date 17 April 2012

Kuncheva L (2004) Combining pattern classifiers: methods and algorithms. Wiley, New York

Larranaga P, Calvo B, Santana R, Bielza C, Galdiano J (2003) Machine learning in bioinformatics. Brief Bioinform 7(1):86–112

Lesko LJ, Woodcock J (2004) Translation of pharmacogenomics and pharmacogenetics: a regulatory perspective. Nat Rev Drug Discov 3(9):763–769. https://doi.org/10.1038/nrd1499

Li T, Wernersson R, Hansen RB (2017) A scored human protein-protein interaction network to catalyze genomic interpretation. Nat Methods 14(1):61–64. https://doi.org/10.1038/nmeth.4083

Linkov I, Massey O, Keisler J, Rusyn I, Hartung T (2015) From "weight of evidence" to quantitative data integration using multicriteria decision analysis and Bayesian methods. ALTEX 32(1):3–8. https://doi.org/10.14573/altex.1412231

Manrai AK, Cui Y, Bushel PR, Hall M, Karakitsios S, Mattingly C, Ritchie M, Schmitt C, Sarigiannis DA, Thomas DC, Wishart D, Balshaw DM, Patel CJ (2016) Informatics and data analytics to support exposome-based discovery for public health. Annu Rev Public Health 38:279–294. https://doi.org/10.1146/annurev-publhealth-082516-012737

Martinez R, Collard M (2007) Extracted knowledge: interpretation in mining biological data, a survey. Int J Comput Sci Appl 1:1–21

McQueen JB (1967) Some methods for classification and analysis of multivariate observations. In: Press UoC (ed) Fifth Berkeley symposium on mathematical statistics and probability, University of California Press, Berkeley, pp 281–297

Perkins EJ, Antczak P, Burgoon L, Falciani F, Garcia-Reyero N, Gutsell S, Hodges G, Kienzler A, Knapen D, McBride M, Willett C (2015) Adverse outcome pathways for regulatory applications: examination of four case studies with different degrees of completeness and scientific confidence. Toxicol Sci 148(1):14–25. https://doi.org/10.1093/toxsci/kfv181

Pleil JD (2012) Categorizing biomarkers of the human exposome and developing metrics for assessing environmental sustainability. J Toxicol Environ Health B Crit Rev 15(4):264–280. https://doi.org/10.1080/10937404.2012.672148

Sarigiannis D, Gotti A, Cimino Reale G, Marafante E (2009) Reflections on new directions for risk assessment of environmental chemical mixtures. Int J Risk Assess Manag 13(3-4):216–241

Sarigiannis DA, Kermenidou M, Nikolaki S, Zikopoulos D, Karakitsios SP (2015) Mortality and morbidity attributed to aerosol and gaseous emissions from biomass use for space heating. Aerosol Air Qual Res 15(7):2496–2507

Saykin AJ, Shen L, Yao X, Kim S, Nho K, Risacher SL, Ramanan VK, Foroud TM, Faber KM, Sarwar N, Munsie LM, Hu X, Soares HD, Potkin SG, Thompson PM, Kauwe JS, Kaddurah-Daouk R, Green RC, Toga AW, Weiner MW (2015) Genetic studies of quantitative MCI and AD phenotypes in ADNI: progress, opportunities, and plans. Alzheimers Dement 11 (7):792–814. https://doi.org/10.1016/j.jalz.2015.05.009

Schapire R, Freund Y, Bartlett P, Lee WS (1998) Boosting the margin: a new explanation for the effectiveness of voting methods. Ann Stat 26(5):1651–1686

Seno M, Karypis G (2001) LPMiner: an algorithm for finding frequent itemsets using length-decreasing support constraint. In: 1st IEEE conference on data mining

Shamir R, Sharan R (2000) Click: a clustering algorithm for gene expression analysis. In: AAAI Press (ed) 8th international conference on intelligent systems for molecular biology (ISMB '00)

Shatkay H, Edwards S, Wilbur WJ, Boguski M (2000) Genes, themes, microarrays: using information retrieval for large-scale gene analysis. Proc Int Conf Intell Syst Mol Biol 8:340–347

Svihalkova-Sindlerova L, Machala M, Pencikova K, Marvanova S, Neca J, Topinka J, Sevastyanova O, Kozubik A, Vondracek J (2007) Dibenzanthracenes and benzochrysenes elicit both genotoxic and nongenotoxic events in rat liver 'stem-like' cells. Toxicology 232 (1–2):147–159. https://doi.org/10.1016/j.tox.2006.12.024

Szklarczyk D, Morris JH, Cook H, Kuhn M, Wyder S, Simonovic M, Santos A, Doncheva NT, Roth A, Bork P, Jensen LJ, von Mering C (2017) The STRING database in 2017: quality-controlled protein-protein association networks, made broadly accessible. Nucleic Acids Res 45 (D1):D362–d368. https://doi.org/10.1093/nar/gkw937

Taboureau O, Audouze K (2017) Human Environmental Disease Network: a computational model to assess toxicology of contaminants. ALTEX 34(2):289–300. https://doi.org/10.14573/altex.1607201

Taboureau O, Jacobsen UP, Kalhauge C, Edsgard D, Rigina O, Gupta R, Audouze K (2013) HExpoChem: a systems biology resource to explore human exposure to chemicals. Bioinformatics 29(9):1231–1232. https://doi.org/10.1093/bioinformatics/btt112

TCGA (2011) Integrated genomic analyses of ovarian carcinoma. Nature 474(7353):609–615. https://doi.org/10.1038/nature10166

TCGA (2014) Comprehensive molecular characterization of urothelial bladder carcinoma. Nature 507(7492):315–322. https://doi.org/10.1038/nature12965

Valencia A, Pazos F (2002) Computational methods for the prediction of protein interactions. Curr Opin Struct Biol 12(3):368–373. https://doi.org/10.1016/s0959-440x(02)00333-0

Villeneuve DL, Crump D, Garcia-Reyero N, Hecker M, Hutchinson TH, LaLone CA, Landesmann B, Lettieri T, Munn S, Nepelska M, Ottinger MA, Vergauwen L, Whelan M (2014) Adverse outcome pathway development II: best practices. Toxicol Sci 142(2):321–330. https://doi.org/10.1093/toxsci/kfu200

Vitkina TI, Yankova VI, Gvozdenko TA, Kuznetsov VL, Krasnikov DV, Nazarenko AV, Chaika VV, Smagin SV, Tsatsakis AM, Engin AB, Karakitsios SP, Sarigiannis DA, Golokhvast KS (2016) The impact of multi-walled carbon nanotubes with different amount of metallic impurities on immunometabolic parameters in healthy volunteers. Food Chem Toxicol 87:138–147. https://doi.org/10.1016/j.fct.2015.11.023

Webb G, Zheng Z (2004) Multistrategy ensemble learning: reducing error by combining ensemble learning techniques. IEEE Trans Knowl Data Eng 16(8):980–991

Weiner MW, Aisen PS, Jack CR Jr, Jagust WJ, Trojanowski JQ, Shaw L, Saykin AJ, Morris JC, Cairns N, Beckett LA, Toga A, Green R, Walter S, Soares H, Snyder P, Siemers E, Potter W, Cole PE, Schmidt M (2010) The Alzheimer's disease neuroimaging initiative: progress report and future plans. Alzheimers Dement 6(3):202–211.e207. https://doi.org/10.1016/j.jalz.2010.03.007

Yan J, Risacher SL, Shen L, Saykin AJ (2017) Network approaches to systems biology analysis of complex disease: integrative methods for multi-omics data. Brief Bioinform. https://doi.org/10.1093/bib/bbx066

Yang Y, Blomme EA, Waring JF (2004) Toxicogenomics in drug discovery: from preclinical studies to clinical trials. Chem Biol Interact 150(1):71–85. https://doi.org/10.1016/j.cbi.2004.09.013

Zeng J, Zhu S, Yan H (2009) Towards accurate human promoter recognition: a review of currently used sequence features and classification methods. Brief Bioinform 10(5):498–508. https://doi.org/10.1093/bib/bbp027

Part III
Measuring Sources of Exposure

Chapter 8
The Food Exposome

Augustin Scalbert, Inge Huybrechts, and Marc J. Gunter

Abstract Estimating the food exposome, that is, the totality of dietary exposures, represents a substantial challenge because of the wide nature of foods consumed and the variability of the amount and frequency of intake according to food preference, season, and other individual characteristics. Classical approaches to estimating dietary intake have used dietary assessment instruments such as questionnaires and food composition tables. However, these are subject to a number of biases and errors, including misreporting and recall bias. The use of dietary biomarkers provides a more objective approach to assess exposures to dietary compounds and their food sources. Approximately 150 dietary biomarkers have been measured so far in cohort or biomonitoring studies and many new candidate biomarkers are now regularly proposed following the rapid development of mass spectrometry techniques and metabolomics. Several 100 food-derived compounds can be simultaneously measured as part of the human exposome in very small human biospecimens collected in large populations. These approaches and their implementation in dietary-wide association studies are reviewed. This could significantly advance nutritional epidemiology and its utility in linking dietary exposures with disease outcomes.

Keywords Food · Foodome · Diet-disease associations · Dietary biomarkers

Introduction

Food is needed for our sustenance, nourishment, and enjoyment. Food is highly complex in its composition due to the high diversity of food products consumed by humans and to the considerable diversity of compounds contained within each food. These compounds include macro- and micronutrients essential for normal physiological functions, and a large diversity of other compounds that are often specific to a

A. Scalbert (✉) · I. Huybrechts · M. J. Gunter
Nutrition and Metabolism Section, International Agency for Research on Cancer (IARC), Lyon, France
e-mail: scalberta@iarc.fr; HuybrechtsI@iarc.fr; GunterM@iarc.fr

S. Dagnino, A. Macherone (eds.), *Unraveling the Exposome*,
https://doi.org/10.1007/978-3-319-89321-1_8

particular food species or food group. Several 1000 plant and animal species are consumed as foods by humans in various regions of the world, and approximately 30,000 different food compounds have been described in these various foods (University_of_Alberta 2016). Besides nutrients, some compounds found in plant and animal species commonly used as food can be toxic or may make food unpalatable (e.g., bitter or astringent compounds in many plants). Humans have learned how to avoid them through the selection of edible food species and, for a given food species, through the selection of varieties or cultivars containing low or no amounts of these compounds. Some food compounds, although devoid of acute toxicity, may still exert detrimental effects on health upon long-term exposures. Furthermore, foods may contain 500 food additives, 2000 food flavorings, and many food contaminants, either natural or man-made (e.g., mycotoxins, pesticides, veterinary drugs, etc.) (The_Joint_FAO/WHO_Expert_Committee_on_Food_Additives_(JECFA) 2017).

In addition, foods are often processed and cooked, resulting in the degradation of some food constituents and in the formation of others. Some typical examples of such chemical reactions during food processing or storage include enzymatic and nonenzymatic browning (Maillard reaction), lipid oxidation, starch hydrolysis, formation of trans-fatty acids, cross-linking and denaturation of proteins, gel formation, starch retrogradation, toughening and softening of meat texture, degradation of vitamins, development of food flavor and off-flavors, formation of carcinogenic compounds during cooking, migration of various chemicals from packaging materials to food and their interactions.

It is now well accepted that the food we consume does more than just providing energy and essential nutrients (i.e., nutrients that cannot be synthesized by the body and thus must be derived from the diet). Food constituents may exert some biological effects on the human organism that can be beneficial or detrimental and limit or increase the risk of various chronic diseases such as obesity, diabetes, cardiovascular diseases, cancer, osteoporosis, neurodegenerative diseases, and dental diseases. Current knowledge on these effects has formed the basis for dietary guidelines to better prevent chronic diseases. Variety in our diet is one of the core recommendations in almost all dietary guidelines and is built on the principles of (1) increasing our chances for adequately providing all essential nutrients to our body and (2) decreasing the risk of overexposure to toxic/detrimental food components. A global characterization of exposures to the many foods and dietary constituents is therefore needed to better assess the causes of chronic diseases, to define evidence-based dietary recommendations to better prevent diseases, and to monitor dietary exposures in populations.

Estimating the food exposome, that is, the totality of dietary exposures (Fig. 8.1), represents a substantial challenge when compared with other types of exposures described in this volume because of a number of characteristics that make dietary exposures unique. Firstly, the nature of foods consumed varies widely from one individual to another or from one country to another. Secondly, the amount and frequency of intake of a particular food or food compound can vary considerably between individuals according to food preference, season, and other individual characteristics. The chemical composition of a particular food is also highly variable

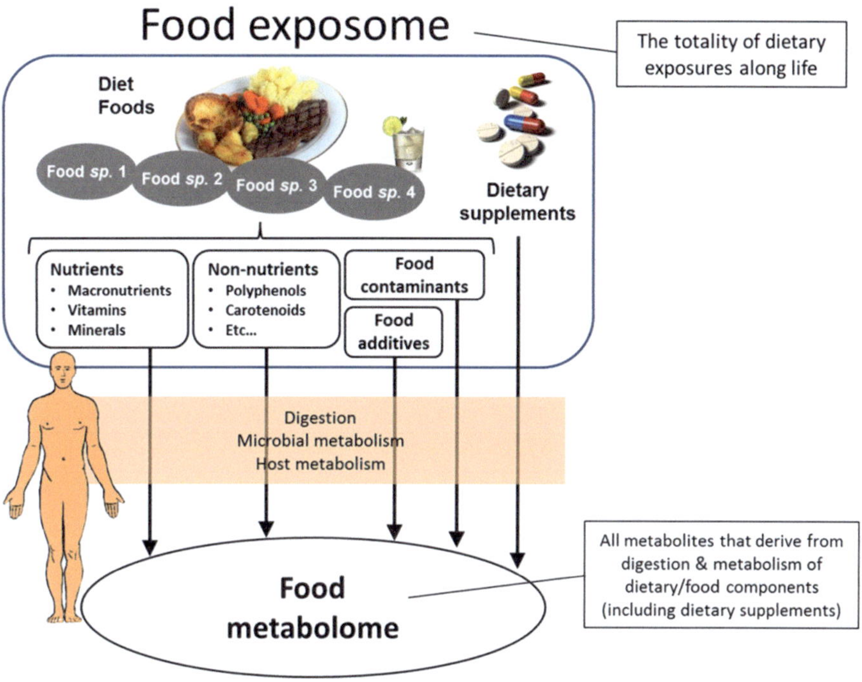

Fig. 8.1 The food exposome and the food metabolome. The food exposome represents an individual's lifelong exposure to all dietary components, consumed as part of the diet (foods or food constituents) or as dietary supplements. The food metabolome is the part of the human metabolome directly derived from the digestion of foods. It includes all dietary compounds and their metabolites

according to cultivars or species, mode of production, storage, processing, and cooking. Classical approaches for estimating dietary intake have used dietary assessment instruments such as questionnaires and food composition tables. However, these are subject to a number of biases and errors, including misreporting and recall bias (Kristal et al. 2005). Various food composition databases exist for the 40 most common nutrients but are not currently available for many other food compounds.

The use of dietary biomarkers measured in human biospecimens such as blood or urine provides a more objective approach to assess exposures to dietary compounds. Dietary biomarkers are based on the concept that their levels are highly correlated with food or nutrient intake over a given period of time. Modern mass spectrometers permit the measurement of hundreds of food-derived compounds in human biospecimens that may be used as indicators of food intake or of exposure to food compounds. Approximately 150 dietary biomarkers have been measured so far in cohort or biomonitoring studies (Neveu et al. 2017). Many new candidate biomarkers are now regularly proposed following the rapid development of mass spectrometry and metabolomics (Scalbert et al. 2014). Various approaches for the characterization of the food exposome in populations are reviewed with particular emphasis on biomarkers and their implementation in dietary-wide association studies (DWAS).

Measuring the Food Exposome Through Dietary Intake Assessment

The food exposome can be assessed through dietary questionnaires (asking the subject what he/she consumes) and through matching with food composition tables that provide detailed information on the nutritional composition of foods. Data from food composition tables are matched with food consumption data to assess the intake of nutrients and other food components (e.g., non-nutrients).

Dietary Intake Assessment with Questionnaires, Recalls, or Records

There are three widely used dietary assessment methods: food records (*syn* food diaries), 24-h dietary recalls (24HDR), and food frequency questionnaires (FFQ). Currently, there is no overall consensus on the best method to assess dietary intake for nutritional studies. Depending on the intended use, feasibility, and available funding, each method has its own strengths and weaknesses. These methods are used either individually or can be combined where resources allow. Because of its convenient use and applicability in large-scale populations, FFQs have been the most frequently used dietary assessment instrument in large-scale epidemiological studies.

Details on the description of food items depend on the dietary assessment method and the study objectives and designs. In addition, the number and/or types of food items included in the FFQ food list vary according to the purpose of the study and study population but the appropriateness of the food list is crucial as the full variability of an individual's diet, which includes many foods and mixed dishes, cannot be captured by a finite food list. Some FFQs include extra questions regarding usual food preparation methods (e.g., brewing method for coffee), identification of the type (e.g., caffeinated versus decaffeinated), and brand of certain types of foods. FFQs generally include between 50 and 150 (mostly generic) food items. The majority of the epidemiological studies investigating associations between dietary factors and disease outcomes used an FFQ to assess individual usual intake of food or of particular food components. The food frequency approach asks respondents to report their usual frequency of consumption of each food from a list of foods for a specific period of time, from several weeks to years. With such a long time frame,

FFQs avoid the need for repeated measurements and have the advantage that they do not affect the respondent's eating behavior.

Whether portion size information is required also depends on the study aims. Different types of FFQs can be distinguished depending on the portion size information inquired in the questionnaire (qualitative or quantitative information). FFQs are generally used for ranking subjects according to food or nutrient intake rather than for estimating absolute levels of intake, and are used widely in case–control and cohort studies to assess the association between dietary intake and disease risk (Willett 1998). The method is often used for estimating average intake of those foods, nutrients, and food components that have large day-to-day variability and for which there are relatively few significant food sources (e.g., alcohol, vitamin A, and vitamin C).

The 24HDR method is a structured interview intended to capture detailed information on all foods and beverages (and possibly dietary supplements) consumed by the respondent in the past 24 h. A key feature of the 24HDR is that, when appropriate, the respondent is asked for more detailed information than first reported (e.g., extra information regarding the type of food or brand name, or cooking method). 24HDRs are typically administered by a trained interviewer, but automated self-administered tools, such as the National Cancer Institute's Automated Self-Administered 24-Hour Dietary Assessment Tool (ASA24), also are available (Kirkpatrick et al. 2014). These 24HDR are less often used in large-scale epidemiological studies because of their higher cost (e.g., interviewer cost compared to self-completed FFQ) and due to their limitation that they only capture information about one single day which is not necessarily representative for a subject's usual dietary intake.

Food records are also a detailed dietary assessment method in which the respondent is asked to write down (or electronically enter) all the foods and drinks that he/she consumes on one or more selected days. Although this method allows very accurate reporting of portion sizes when foods and beverages are being weighed and measured, in many studies using food records, the portion sizes consumed are being estimated rather than measured because of the high respondent burden in measuring all the foods/beverages. Food records are more often used for assessing dietary intakes among children as their capability to recall food intake is often limited while a food diary can be completed by parents and/or caregivers. Like 24HDR, 1-day food records are not able to provide an estimate of the respondent's usual dietary intake, unless the recording is repeated over multiple days.

The high respondent burden in collecting multiple recalls or records limits the feasibility of using these tools as primary intake assessment methods in large-scale epidemiological surveys. However (repeated), 24HDR or dietary records are often used for calibration of FFQs in a subsample of large-scale epidemiological studies (Kaaks et al. 1995).

Dietary assessment methodologies are under continuous development in an attempt to improve the quality and validity of dietary exposure data further while taking profit of rapid evolutions in new innovative technologies such as mobile applications, scan and sensor-based technologies (Illner et al. 2012). A review of dietary assessment methodologies and technologies used in epidemiologic studies is beyond the scope of this report but can be found in several excellent reviews or books (Slimani et al. 2015; Thompson and Subar 2013; Schoeller and Westerterp-Plantenga 2017).

Databases on Food Constituents and Food Composition

A number of organizations worldwide maintain food composition tables to help in providing consumers, health professionals, and researchers with nutritional information about the food that we consume.

INFOODS (http://www.fao.org/infoods/infoods/tables-and-databases/en/) is the International Network of Food Data Systems that was established in 1984. It is a worldwide network of food composition experts aiming to improve the quality, availability, reliability, international harmonization, and use of food composition data. INFOODS provides food composition data from different regions, including Asia, Africa, Canada, Caribbean, United States, Europe, Latin America, Middle East, and Oceania. INFOODS further provides Global food composition databases for pulses, fish, and shellfish; a Global supplement database; a Food Density Database; and a Food Composition Database for Biodiversity.

EuroFIR AISBL is an international, member-based, nonprofit Association set up in 2009 to ensure sustained advocacy for food information in Europe, providing access to several national food composition tables. Its purpose is to develop, publish, and exploit food composition information, and promote international cooperation and harmonization of standards to improve data quality, storage, and access. In the meantime, EuroFIR AISBL draws food information from 26 compiler organizations in Europe, the USA, and Canada (FoodEXplorer) as well as validated information about bioactive compounds (eBASIS). It provides a list of food composition databases and up-to-date and scientifically validated food composition information (http://www.eurofir.org/food-information/food-composition-databases-2/).

Food safety authorities like the European Food Safety Authority (EFSA) and the United States Department of Agriculture (USDA) also collect and harmonize food composition data through their member states with a primary focus on food contaminants and additives that are regulated through upper safety limits to guarantee the safety of food products in their authorized region. Examples of food component databases being harmonized by EFSA are contaminants such as mycotoxins, pesticide residues, several metals (e.g., lead and cadmium), and food additive databases (e.g., food coloring agents). The USDA food composition databases (https://ndb.nal.usda.gov/ndb/) cover main nutrients (macro- and micronutrients), some secondary metabolites such as carotenoids, flavonoids, caffeine, and choline, as well as some food additives and contaminants (pesticide residues) (https://www.ams.usda.gov/datasets/pdp).

Most of the aforementioned databases include content data for many different food components (e.g., macro- and micronutrients). Information on the content of more specific food constituents can be found in some specialized databases. For example, Phenol-Explorer (http://phenol-explorer.eu/) is a comprehensive database on polyphenol contents in foods. The database contains more than 35,000 content values for 500 different polyphenols in over 400 foods. FooDB (The Food Database) (http://foodb.ca/) is the most comprehensive resource on food constituents, and includes data on their chemistry and biology. It provides information on macronutrients, micronutrients, and a large diversity of secondary metabolites from various chemical classes, comprising more than 26,000 compounds. Each chemical entry in FooDB contains more than 100 separate data fields covering detailed compositional, biochemical, and physiological information (obtained from the literature and other databases). However, food composition data is not comprehensive, particularly for secondary metabolites, and this data would not be sufficient to build food composition tables for these compounds.

Food Intake, Nutrient/Non-nutrient and Contaminant Intake

Food intake levels can be estimated for the foods included in the food list of a dietary questionnaire or for frequently consumed foods in 24HDR or food records. Given the open nature of most 24HDR and food records, these open methods often allow the assessment of a larger number of food items than dietary questionnaires. However, for assessing intakes of foods that are irregularly consumed (e.g., not on a daily basis) on an individual level, multiple assessments are needed to cover at least 1 day on which the particular food has been consumed.

As mentioned before, calculation of the intake of nutrients and non-nutrients also requires appropriate food composition databases to be matched with the food consumption data. When looking for the most optimal food composition data to be matched with the food consumption data, several factors should be considered such as the geographical region, eating culture, local regulations regarding fortification of foods, etc. Ideally, local/national food composition tables are used for the matching with dietary intakes from a particular region/country. However, when local/national food composition data is lacking then alternative databases from neighboring countries or populations with similar eating cultures can be used. When studying particular food components not necessarily available in national food composition tables (such as polyphenols, additives, or contaminants), then specific databases such as those described in the previous section can be used.

Limitations of Dietary Intake Assessments Via Questionnaires, Recalls, or Records

An important limitation when using dietary questionnaires such as FFQs for assessing dietary intakes is the limited level of detail available in many questionnaires. FFQs are often developed for achieving the aims of a specific study (e.g., comparing food intakes with food-based dietary guidelines) and are not always flexible for additional analyses that were not foreseen in the development of the questionnaires. The ability to quantify intake of specific dietary factors such as energy depends on the number of food items listed in the FFQ and whether or not portion sizes for foods/food groups are included (Block et al. 1986; Rimm et al. 1992). Furthermore, completion of an FFQ may be a cognitive challenge for some respondents and usual intakes of some foods may also be difficult to estimate precisely due to seasonality (Kristal et al. 2005).

Although the 24HDR and food record methods are more flexible regarding the food lists obtained from the subjects through its open nature (foods originally not foreseen in the food list can often be entered as a new item), their usefulness for assessing usual intakes of a food item on an individual level depends on the number of repeated assessments, particularly for foods less regularly consumed.

The lack of food composition data may also be a limiting factor. Some countries/regions (more particularly in low-income countries) may still be lacking accurate food composition data for the main food components in culture-specific foods and recipes. In addition, accurate food composition data for particular components such as certain food additives and contaminants may not be available or may be too variable in a given food (e.g., depending on plant varieties or animal breed, growing and farming conditions, food processing and cooking). Biomarkers may, in these cases, offer a suitable alternative for dietary exposure assessment.

The Food Metabolome, Definition, and Characterization

The food metabolome has been defined as the sum of all metabolites directly derived from the digestion of foods, their absorption in the gut, and biotransformation by the host tissues and microbiota (Fig. 8.1) (Scalbert et al. 2014). More than 25,000 different compounds have been described in various foods (University_of_Alberta 2016) and are commonly ingested with the diet and eventually absorbed through the gut. To add to this complexity, each food compound, either nutrient or non-nutrient, is often transformed into various metabolites during digestion in the gut and in the liver and other tissues. Many of these metabolites have been identified but many more have not yet been fully characterized. Knowledge on the identity of these metabolites and on the conditions of their formation is needed to understand the mechanisms linking dietary exposures to health effects or to validate their use as biomarkers of dietary exposure.

Currently, it remains difficult to identify all metabolites formed from the many ingested food compounds. Macronutrients are hydrolyzed during digestion into simple molecules such as sugars, amino acids, and fatty acids which are then rapidly metabolized in the host tissues. These metabolites most often cannot be distinguished from endogenous human metabolites. On the other hand, other compounds, such as secondary metabolites in plant foods, some amino acids or lipids in animal foods, food additives and contaminants, can be food-specific and clearly distinguishable from endogenous human metabolites. These compounds are recognized by the organism as xenobiotics and metabolized through various detoxification reactions (phase 1 and 2 enzymes) to facilitate their excretion in urine and in the bile, and thereby prevent their accumulation in the organism. Quite often, food xenobiotics are extensively metabolized and exclusively present in the form of various conjugated compounds (e.g., as glucuronides, sulfate esters, glycine conjugates, or mercapturates) and no parent compound can be detected in human biospecimens. Many of these compounds have never been isolated, are absent in most public metabolite databases such as the Human Metabolome Database (HMDB), and cannot be easily recognized in human biospecimens.

Much has been written on the metabolism of food xenobiotics in the host tissues or by the gut microbiota and hundreds of metabolites have been identified (Scheline 1978). Although major reactions involved are well known today, it is still difficult to have a global view on metabolites considering the dispersion of the information in thousands of publications. This information will be particularly important in future exposome-wide association studies (or nutrient-wide association studies) to identify food-derived compounds associated with disease outcomes (Rappaport et al. 2014; Tzoulaki et al. 2012). However, the current lack of a comprehensive database on the food metabolome makes the identification of food metabolites in blood or urinary profiles difficult.

Recently, some efforts have been made to curate information on specific groups of food metabolites in public databases. Some good examples are the Phenol-Explorer and the PhytoHub databases containing respectively information on the main 375 metabolites formed from dietary polyphenols or on about 300 metabolites formed from phytochemicals such as terpenoids and polyphenols (http://phytohub.eu/) (Rothwell et al. 2012). Food metabolites in these databases can be matched based on their accurate mass with signals detected in complex metabolic profiles from blood or urine measured by high-resolution mass spectrometry. The Phenol-Explorer database was used to annotate over 80 polyphenol metabolites in urine samples collected in 500 subjects from the European Prospective Investigation on Cancer and nutrition (EPIC) cohort (Edmands et al. 2015). However, many more food metabolites remain to be identified. For metabolites not presently included in any databases, mass fragmentation spectra can be systematically acquired and new software tools applied to the data to recognize specific molecular substructures characteristic of a chemical class or of a conjugated group like glucuronide, sulfate, or mercapturic acid (van der Hooft et al. 2016; Edmands et al. 2017; Yao et al. 2016).

Measuring the Food Exposome Through Biomarkers

Specific components of the food metabolome have been used for more than 30 years to assess nutrient status or to predict risk of associated diseases (Hunter 1998). These can be vitamins, minerals, and other food components. For essential nutrients, they can also be biochemical indicators of status that directly depend on the intake and bioavailability of the nutrient.

Dietary biomarkers can be divided into two main classes: recovery or concentration biomarkers, depending on the extent of recovery in the compartment where measured and on correlation between excretion or concentration levels and intake (Jenab et al. 2009). Recovery biomarkers (e.g., potassium and nitrogen in 24-h urine collections; Bingham 2002) show a high recovery in the biospecimen collected (urine) and a high correlation with intake. They provide an estimate of absolute intake levels and have been used to calibrate dietary questionnaires (Kaaks 1997). In contrast, concentration biomarkers (e.g., serum vitamins, blood lipids, urinary electrolytes) directly correlate with intakes of corresponding foods or nutrients, but their concentrations cannot be translated into absolute levels of intake as the strength of the correlation with intake measurements is lower (<0.6) than that expected for recovery biomarkers (>0.8) (Jenab et al. 2009). Since the errors associated with nutritional biomarkers are independent of the errors associated with dietary questionnaires, they can be considered as complementary methods for assessing dietary intakes.

Biomarkers, when compared to intake values as determined with questionnaires, may provide a more direct estimate of the internal dose. This may be key information when studying the biological effects of a nutrient or food compound as exposure may also be influenced by genetic factors, adiposity, and lifestyle factors other than diet, such as smoking and physical exercise—factors that are not taken into consideration when measuring intake through dietary questionnaires, recalls, or records. Biomarkers may also be used as surrogates of dietary intake when their concentrations are not substantially influenced by such confounding factors. These two different applications of dietary biomarkers are discussed in this section.

Biomarkers for Nutrients and Food Compounds

Examples of biomarkers of exposure to nutrients, non-nutrients, and contaminants are described below.

Biomarkers for Vitamins

Plasma/serum folate and vitamin B12 are often used as surrogate markers for folate and vitamin B12 intakes in cohort studies. The usefulness of these biomarkers as surrogate intake estimates is often investigated through associations between the measured plasma/serum levels and dietary intakes. Associations of natural food vitamin B12 and folate intakes with biomarker concentrations are rather weak (generally $r < 0.30$) (Mendonça et al. 2016; de Batlle et al. 2016; Verkleij-Hagoort et al. 2007; Bailey et al. 2015). Folate levels were clearly higher in countries with mandatory folic acid fortification, followed by those countries with voluntary fortification (Bailey et al. 2015).

Vitamin C is easily oxidized in foods and variability of its content in foods may result in significant errors when assessing its intake. Blood measures of vitamin C (ascorbate and dihydroascorbate) may thus be considered as a surrogate for dietary vitamin C intake. They were shown in a systematic analysis to correlate moderately with vitamin C intake, as assessed with FFQ ($r = 0.35$), 24HDR ($r = 0.46$), or weighted dietary records ($r = 0.39$) (Dehghan et al. 2007). However, correlations were also shown to vary importantly between studies and populations depending on lifestyle factors (e.g., smoking, physical exercise, chronic low-grade inflammation) which may reduce systemic vitamin C levels. In addition, the absorption efficiency is also known to vary according to the dose consumed. Key proteins for vitamin C uptake are saturated or down-regulated at high levels of intake (MacDonald et al. 2002), leading to reduced intestinal absorption and saturation of the dose–response curve (Levine et al. 1996). High intakes may thus not be well represented by plasma vitamin C measurements.

Serum 25-hydroxyvitamin D [25(OH)D] is a widely accepted marker of vitamin D status. However, [25(OH)D] is also largely produced by cutaneous synthesis catalyzed by sunlight and this limits its usefulness as a dietary biomarker of vitamin D intake (Saraff and Shaw 2016; Personne et al. 2013).

Carotenoids

Few carotenoids (mostly α-carotene, β-carotene, β-cryptoxanthin, canthaxanthin, lycopene, lutein, and zeaxanthin) can be found in significant amounts in blood (Crews et al. 2001a). Their measurement in blood as biomarkers of carotenoid intake is subject to many of the same issues detailed above for vitamin C. Their intestinal absorption is influenced by the lipid content of the diet, the nature of the food sources, competition with other carotenoids, degree of colonic fermentation, menstrual cycle, and hormonal factors (Jenab et al. 2009). Their intestinal absorption can range from less than 10% to over 50% depending on the source and type (Reboul et al. 2006), and their blood concentrations appear to be moderately correlated with diet (r ranges from ~0.2 to over 0.5) (Kaaks et al. 1997).

Fatty Acids

A high diversity of fatty acids has been measured in adipose tissue, plasma, serum, or red blood cells. Essential fatty acids such as docosahexaenoic acid (DHA) and eicosapentaenoic acid (EPA) originate from the diet. Serum or plasma levels of individual fatty acids can reflect intake over the last few days or meals and were shown to be sensitive indicators of changes in the polyunsaturated fat intakes of the diet (Arab 2003; Hedrick et al. 2012). Trans-fatty acids measured in blood were found to be a good indicator of intake of these fatty acids formed by industrial food processes (Chajes et al. 2011a). However, for other fatty acids, it is often not easy to discriminate the contribution of dietary fat from endogenously produced fats. Various factors (alcohol consumption and smoking, apolipoprotein levels, hormones) also influence deposition and mobilization of individual fats and these may vary within and between individuals (Arab 2003).

Polyphenols

Polyphenol biomarkers have been measured both in urine and plasma. High correlations were observed between habitual intake of some polyphenols like isoflavones in Asian populations and levels of the corresponding biomarkers ($r = 0.57$–0.72) (Zamora-Ros et al. 2016). Lower correlations were observed for other types of polyphenols ($r = 0.33$–0.52 for flavonols, 0.32–0.35 for flavanones, and 0.09–0.19 for lignans). Correlation values depend on pharmacokinetic properties (most polyphenols are excreted within 24 h following intake), frequency of intake of their main dietary sources, and accuracy of intake measurements. Correlations may also vary largely by geographical region and dietary habits. For isoflavones consumed with soyfoods, correlations were higher in Asian countries where soyfoods are frequently consumed than in Europe or North America where consumption of soyfoods is more limited.

Caffeine

Caffeine intake is not easily assessed via questionnaires or dietary recalls as caffeine concentrations show large variations in foods/beverages and are often lacking in food composition tables. Caffeine and their metabolites can therefore be used as indicators of exposure. Several metabolites of caffeine have been described and some of them showing stronger correlations with caffeine intake were also less influenced by genetic factors (Crews et al. 2001b).

Other nutrients and contaminants

Other nutrients not easily assessed via dietary intake measurements are iron, iodine, and sodium. All of these nutrients can be estimated with the use of biomarkers (Ovesen et al. 2002). These biomarkers have been proposed as indicators of exposure for health monitoring in Europe (Steingrimsdottir et al. 2002).

Exposures to food contaminants are also not easily measured via questionnaires because of a large variability of their concentrations in highly diverse foods. Biomarkers may provide better estimates of exposures for informing epidemiological investigations. Aflatoxin M1 (AFM1) and aflatoxin B1 (AFB1)-mercapturic acid are two metabolites of mycotoxins. Measured in urine, they are indicators of recent exposure to aflatoxins. The excretion of AFB1-N7-Gua in urine is an indicator of genetic damage due to AFB1 exposure. Fumonisin B1 and ochratoxin A are two other mycotoxins that have been measured in urine, serum, hair, breast milk, or cord blood, and that have been used as indicators of exposure (Zain 2011; de Nijs et al. 2016). Urinary deoxynivalenol (DON) was identified as being a reliable biomarker of exposure in several populations. The toxin was detectable in approx. 99% of the samples and the levels of DON in urine significantly correlated with the amount of contaminated cereals ingested one day prior to biospecimen collection (Turner et al. 2010).

Pesticide residues from diet are measured in urine samples. For example, a recent study in the Multi-Ethnic Study of Atherosclerosis cohort compared dietary pesticide residue levels in foods with urinary levels of dialkylphosphate (DAP) metabolites of organophosphate (OP) pesticides. Among conventional consumers, increasing tertiles of estimated dietary OP exposure was associated with higher DAP concentrations ($p < 0.05$). DAP concentrations were also significantly lower in groups reporting more frequent consumption of organic produce ($p < 0.02$). Thus, long-term dietary exposure to OPs estimated from dietary intake data seems to be consistent with DAP measurements (Curl et al. 2015).

Contaminants such as mycotoxins or organophosphates are usually found in low to very low levels in humans (Rappaport et al. 2014) and are measured in urine samples where they are present at higher concentrations when compared with blood samples. Many of these contaminants are also known to be quickly excreted after intake resulting in large intra-individual variations in urinary excretion (Bradman et al. 2013). Reliable estimates of individual exposures may thus require repeated measurements.

Biomarkers for Food Intake

Various components of the food metabolome have been used as biomarkers of individual food intake (Hedrick et al. 2012; Jenab et al. 2009; Scalbert et al. 2017). Such biomarkers are, in general, a compound relatively abundant in the food or food group considered or a metabolite of this compound formed in the body. Biomarkers of food intake have been mainly used to validate dietary questionnaires, in biomonitoring studies, in cohort studies as surrogates of food intake, or in intervention studies to assess compliance. Biomarkers are particularly useful when the corresponding exposure cannot be measured accurately with questionnaires (e.g., food not considered in the questionnaire, important measuring error due to misreporting or variability in the characteristics of a given food).

The accurate assessment of coffee consumption can be problematic since different brewing methods for coffee result in large variations in the dilution of coffee beverages which cannot easily be captured with questionnaires. Coffee biomarkers provide more direct estimates of exposure to coffee solutes. Different biomarkers have been proposed, and various polyphenols (chlorogenic acid, caffeic acid, and various metabolites), methylxanthines (1-methylxanthine, paraxanthine), trigonelline, *N*-2-furoylglycine, and cyclo(isoleucylprolyl) best discriminate high and low coffee consumers in free-living populations (Mennen et al. 2006; Edmands et al. 2015; Rothwell et al. 2014; Guertin et al. 2015). Their respective specificity and the most appropriate biospecimens for measurement still need to be more thoroughly evaluated (Rothwell et al. 2018).

Biomarkers for alcoholic beverages are also of great interest because of common underreporting when measuring intake with questionnaires. Biomarkers of alcohol drinking can include metabolites of alcohol such as ethyl glucuronide, ethyl sulfate, or phosphatidylethanol (Isaksson et al. 2011; Dahl et al. 2011) or other compounds specific for particular alcoholic beverages, such as hydroxytyrosol, resveratrol, gallic acid ethyl ester, or tartaric acid for wine (Edmands et al. 2015; Regueiro et al. 2014; Schroeder et al. 2009; Zamora-Ros et al. 2009) or isoxanthohumol, *N*-methyltyramine, pyroglutamyl proline, and 2-ethylmalate for beer (Quifer-Rada et al. 2014; Gurdeniz et al. 2016).

Plant species differ in their chemistry and often contain some secondary metabolites particularly abundant in one particular plant food and absent or present at low concentrations in all other foods. Many such compounds have been identified through the systematic analysis of the composition of plant foods, and through the analysis of correlations between candidate biomarkers and the consumption of the corresponding foods in cross-sectional studies. The following biomarkers are known for their specificity for intake of particular foods: alkylresorcinols for wholegrain foods (Landberg et al. 2009), proline betaine for citrus fruits, hesperetin for orange, naringenin for grapefruit (Pujos-Guillot et al. 2013; Heinzmann et al. 2010), phloretin for apple (Edmands et al. 2015), lycopene for tomato (Al-Delaimy et al. 2005), or *S*-methyl-L-cysteine sulfoxide for cruciferous vegetables (Edmands et al. 2011). Although traditionally single biomarkers have been used to characterize dietary exposures, a shift to more complex analytic approaches based on combinations of dietary constituents may improve the specificity of dietary exposure measurements. A systematic review on biomarkers of fruit and vegetable intake including over 90 intervention trials showed that a combination of biomarkers (α- and β-carotene, vitamin C, lutein, zeaxanthin, and β-cryptoxanthin) was a better indicator of compliance of intake in fruit and vegetable intervention trials than single biomarkers (Baldrick et al. 2011). With the possible exception of fruit-only intervention studies, where assessment of vitamin C status alone may be sufficient, it seems rarely possible to rely on assessment of a single biomarker as an indicator of change in fruit and vegetable intake.

Various compounds have also been proposed as biomarkers of intake for animal foods. Markers for meat intake can be amino acids (3-methylhistidine, β-alanine, hydroxyproline), dipeptides containing these amino acids (anserine, carnosine, ophionine), amino acid metabolites (creatinine, creatine), quaternary amines (carnitine), and acylcarnitines (Dragsted 2010; Ross et al. 2015). Their specificity for meat or particular meat products has often been overlooked. In particular, biomarkers like 3-methylhistidine and anserine (a dipeptide of 3-methylhistidine and β-alanine) appear specific for chicken, whereas carnosine (a dipeptide of β-alanine and histidine) and acylcarnitines (acetyl-, propionyl- or 2-methylbutyrylcarnitine) appear generic for all sorts of meats (Cheung et al. 2017). Phytanic acid (a metabolite of chlorophyll) and fatty acids with an odd number of carbons (pentadecylic acid, margaric acid) all formed in the rumen of ruminants have been proposed as biomarkers of milk and dairy food intake (Allen et al. 2008; Brevik et al. 2005; Hodson et al. 2008). Elaidic acid, a trans-fatty acid, measured in blood phospholipids, was found to be a good indicator of highly processed food intake (Chajes et al. 2011a).

Two biomarkers have been more particularly discussed with regard to fish intake estimation: n-3 polyunsaturated fatty acids [PUFAs; eicosapentaenoic acid (EPA) and docosahexaenoic acid (DHA)] and trimethylamine oxide (TMAO), an osmolyte abundant in various fish species. EPA and DHA appear more specific for fatty fish species, but these markers may be confounded by n-3 PUFA dietary supplements (Brantsaeter et al. 2010; Chung et al. 2008). TMAO is a common constituent of many lean and fatty fish species and appears as a more generic marker of fish intake (Cheung et al. 2017).

These few examples illustrate the diversity of components of the food metabolome that can be used as indicators of food intake. However, a number of these biomarkers have not been fully validated and more studies are needed to evaluate their specificity and sensitivity, dose–response relationship, and reproducibility over time (Scalbert et al. 2017).

Limitations of Biomarkers as Surrogates for Food Intake

Reproducibility over time is an important characteristic of biomarkers, in particular when applied to large cohort studies for which biospecimens were collected at baseline as a unique time point. Measurement of dietary biomarkers made at baseline should be representative of dietary exposures over the following years. Reproducibility depends on pharmacokinetic properties of the biomarker and on the frequency of dietary exposures (Scalbert et al. 2017). Some of the dietary biomarkers described above have relatively short half-lives of elimination, less than 12 h for many polyphenols, ethanol metabolites, and proline betaine, and are rapidly excreted after absorption. Other dietary biomarkers, particularly more lipophilic compounds such as carotenoids, fatty acids, or tocopherols, have longer half-lives (up to 60 days) and this explains their more common use in epidemiological studies. However, biomarkers with shorter half-lives are still useful to assess intake of foods or beverages frequently consumed, like coffee consumed on a daily basis.

Other possible limitations of biomarkers are saturation of dose–response curves at high or very high doses for some nutrients like ascorbic acid or folate (Levine et al.

1996; Gregory and Quinlivan 2002). Various interactions between nutrients in the gut have also been described, for example, between carotenoids and fat or between nonheme iron and ascorbic acid, and these interactions may modify dose–response relationships (Hunter 1998).

Further limitations of biomarkers as surrogate markers for dietary intakes may be linked to possible interactions with genetic and (patho-)physiological factors such as low-grade inflammation, body fluid excretion, and lifestyle factors such as alcohol intake or smoking (Hedrick et al. 2012; Jenab et al. 2009). Interactions between body mass index and intake of some lipophilic compounds such as carotenoids have also been reported (Vioque et al. 2007).

Lastly some important characteristics of the diet may also not be easily described with biomarkers. Meal patterns, a potentially important dietary factor in relation to several disease outcomes such as obesity and metabolic syndrome (Leech et al. 2015), cannot be measured through dietary biomarkers, underlining the importance of combining dietary intake assessment methods with dietary biomarkers.

The Food Exposome in Dietary-Wide Association Studies

There is a real benefit in measuring a large diversity of dietary exposures and in simultaneously evaluating their associations with disease outcomes or surrogate endpoints in what has been called an "exposome-wide association study" (EWAS) or "metabolome-wide association study" (MWAS) (Patel et al. 2010; Chadeau-Hyam et al. 2010). When assessed exposures are limited to nutrients, these studies have been called a "nutrient-wide association study" (NWAS) (Tzoulaki et al. 2012). In this section, we review association studies that simultaneously evaluate all dietary constituents, either foods or food constituents, and use the more generic term "dietary-wide association study" (DWAS) (Fig. 8.2) (Davis et al. 2014). An initial DWAS was conducted in 4680 subjects of the INTERMAP cross-sectional study to characterize associations between intake of 82 nutrients and urinary levels of 3 electrolytes with blood pressure. Several dietary factors (e.g., alcohol, sodium-to-potassium ratio, B vitamins) were found to be highly significantly associated with systolic or diastolic pressure (Tzoulaki et al. 2012). Similarly, in another DWAS conducted in the EPIC cohort with a replication step in the Netherlands Cohort Study, associations between 84 foods and nutrients and risk of endometrial cancer were studied and coffee was found to be inversely associated with development of this malignancy (Merritt et al. 2015). In these two early DWASs, a diversity of dietary exposures including nutrients, non-nutrients, and contaminants have been considered that were largely assessed through the use of dietary questionnaires.

Progress in analytical techniques and the increasing number of dietary biomarkers already identified make it possible to perform a DWAS in which many or most dietary exposures are measured with biomarkers. A first proof of principle study was carried out in the NHANES cohorts to study associations between up to 211 unique environmental factors measured in urine or in blood and fasting glucose (Patel et al.

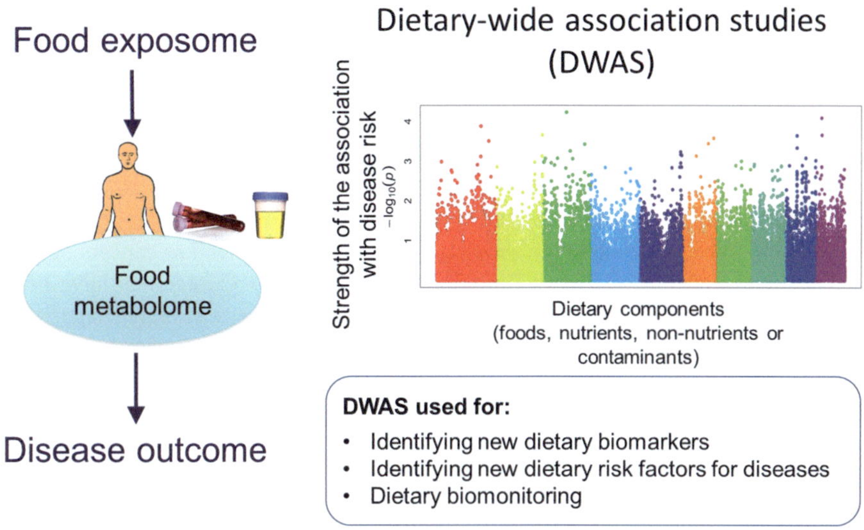

Fig. 8.2 Dietary-wide association studies (DWAS). The Manhattan plot shows components of the food exposome, foods, nutrients, non-nutrients, or contaminants measured with either questionnaires or biomarkers, associated with dietary exposures or disease outcomes to identify new dietary biomarkers or new dietary risk factors for diseases

2010). These factors were mainly contaminants and pollutants but also included over 30 compounds related to the diet, like vitamins, carotenoids, phytoestrogens, and markers of nutritional status (e.g., transferrin saturation). Associations with fasting glucose as a surrogate marker of type-2 diabetes were measured, and significant associations were found for two dietary compounds (β-carotene and α-tocopherol). These 211 compounds were measured using specific (bio)chemical assays. These analyses were made over several years at a relatively high cost and required significant volumes of blood or urine to carry out the various individual assays.

Modern analytical techniques based on mass spectrometry allow DWAS to be performed on larger numbers of dietary biomarkers analyzed in a single analytical run. These can be case–control or cohort studies aiming at identifying dietary factors associated with disease outcomes or cross-sectional studies and dietary intervention studies aiming at the identification of new dietary biomarkers. DWASs use two different analytical approaches, either targeted methods in which usually <100 dietary compounds known a priori are measured, or untargeted methods in which >1000 metabolites/compounds are measured with no a priori knowledge on their identity, and the data subsequently mined to extract the information on dietary compounds.

Targeted Analytical Approaches

Various targeted methods have been developed and used to measure compounds from specific classes of food constituents and study their associations with diseases or with dietary exposure. For example, 30–60 fatty acids, including several fatty acids directly derived from the diet like elaidic acid or n-3 polyunsaturated fatty acids, have been measured in blood by gas chromatography in cohort studies and were found to be associated with the risk of several diseases such as cancers or type 2 diabetes (Chajes et al. 2011b; Forouhi et al. 2016). Six carotenoids have been measured in blood in various cohorts and a number of associations have been reported with cancers (Al-Delaimy et al. 2004; Lu et al. 2001; Aune et al. 2012). Thirty-four polyphenols were measured in urine and found to be associated with the intake of various foods or with colorectal cancer outcomes (Achaintre et al. 2016; Zamora-Ros et al. 2016; Murphy et al. 2018). Five alkylresorcinols measured in blood showed associations with distal colon cancer or type 2 diabetes (Kyro et al. 2014; Biskup et al. 2016). Overall, these studies reveal new associations between specific compounds in their respective chemical classes and diseases. They provide new evidence on their contribution to disease risk or suggest a contribution of their main dietary sources to disease risk.

Untargeted Metabolomic Approaches

No attempt has been made so far to simultaneously quantify panels of dietary biomarkers from diverse classes of dietary compounds in the same analysis. However, although not the specific focus of untargeted metabolomic analyses, various components of the food metabolome are measured when using such agnostic approaches. Data acquired by NMR spectroscopy or mass spectrometry coupled to gas or liquid chromatography can be mined a posteriori to identify novel dietary biomarkers in dietary intervention and cross-sectional studies, or novel dietary factors associated with diseases in case–control or cohort studies. Metabolic fingerprints are compared through statistical approaches to identify compounds or metabolites characteristic of one or another group of subjects (e.g., consumers and nonconsumers of a particular food, or cases and controls). NMR spectroscopy is a robust technique that can be applied to a large series of samples (Soininen et al. 2015). However, no more than 200 metabolites are commonly reported because of the limited sensitivity of the technique. Only few food-derived compounds have been detected in NMR metabolic fingerprints from free-living subjects (e.g., proline betaine, acylcarnitine, and a few common organic acids) (Scalbert et al. 2017).

High-resolution mass spectrometry coupled to liquid or gas chromatography is a more sensitive analytical technique now widely used to detect hundreds to thousands of compounds in biospecimens such as blood or urine, including many compounds derived from the diet (Keski-Rahkonen et al. 2017). Peak intensities in the various samples are compared after aligning the different chromatograms through statistical approaches. These data based on peak intensities are "semiquantitative" and differ from quantitative data where results are expressed in concentration after calibration with authentic chemical standards (calibration is obviously impossible in untargeted analyses considering the large number of compounds measured). Peaks of interest are then identified through matching of their spectral characteristics (accurate mass, MS/MS spectra) with those of metabolites in metabolite databases.

Metabolomic studies aiming at discovering new dietary biomarkers were reviewed recently (Scalbert et al. 2017). Thirteen of these studies were cross-sectional studies on free-living subjects of particular relevance for the present discussion. Most used urine samples and high-resolution mass spectrometry for data acquisition. Dietary intake was measured with either 24-h dietary recalls (acute food intake) or with food frequency questionnaires (habitual intake). In an early study on 68 volunteers, associations between the urinary metabolome and habitual intake of 58 dietary factors were systematically explored, and several metabolites could be identified as potential biomarkers of intake for citrus, oily fish, coffee, and tomato (Lloyd et al. 2011). A similar study was conducted in 481 subjects of the EPIC cohort and various polyphenol metabolites in urine were found to be good predictors of intake for several polyphenol-rich foods and beverages including coffee, tea, wine, citrus fruit, apples, and pears (Edmands et al. 2015). Biomarkers of intake for chicken, beef, processed meat, and fish could also be identified in urine and plasma in the EPIC cohort using the same analytical approach (Cheung et al. 2017). Several metabolites of ellagitannins and fatty acids were identified in urine samples in the PREDIMED cohort, as biomarkers of walnut intake (Garcia-Aloy et al. 2014).

Specificity of these biomarkers identified in such cross-sectional studies can be further assessed through their analysis in potentially confounding food sources (Heinzmann et al. 2010; Cheung et al. 2017) or through the scrutiny of food composition databases when available (Edmands et al. 2015). In addition, dose–response and reproducibility over time is eventually evaluated (Sun et al. 2017; Scalbert et al. 2017). This information is currently being collated in the Exposome-Explorer database (Neveu et al. 2017), a new database that should be useful to select panels of dietary biomarkers most suitable for future DWASs.

In several other epidemiological studies using untargeted metabolomics, statistical analyses were limited to metabolites that could be identified a priori using an in-house metabolite database. Metabolomic data were acquired on blood or urine samples by Metabolon Inc., and up to 400 metabolites could be identified. Several of these metabolites were dietary compounds and their associations with intake of various foods were systematically explored (Pallister et al. 2016a, b; Zheng et al. 2014b; Playdon et al. 2016; Guertin et al. 2014). A number of expected correlations were observed (e.g., proline betaine with citrus fruits or trigonelline with coffee) (Guertin et al. 2014). Other compounds showing correlations with several foods or food groups may not be useful as dietary biomarkers unless used in combination with other compounds. Scyllo-Inositol level in blood was correlated with both citrus fruits and wine intake and this is consistent with the known occurrence of this compound in both citrus fruits and grapes (Sanz et al. 2004). Some significant and positive correlations could also be observed with some metabolites but the too limited data on food composition could not support their use as biomarker of exposure. Cyclo(leucylprolyl) in blood was positively correlated with alcohol intake (Guertin et al. 2014). It is a known compound of some fermented foods but its occurrence in alcoholic beverages is poorly documented (Axel et al. 2014). Finally, the distinction between exogenous or endogenous origin of some compounds found in blood or urine is often difficult and some observed associations may not be the direct result of a dietary exposure but would rather reflect biological effects caused by some dietary exposures. These associations may then suggest new mechanisms linking dietary exposures to diseases (Zheng et al. 2014a; Troche et al. 2016; Wurtz et al. 2016; Wittenbecher et al. 2015).

One limitation of these untargeted metabolomic studies focused on named metabolites is that these metabolites were not selected on purpose and may not be the best biomarkers of dietary exposures. Indeed various biomarkers known to be derived from a given food may show different ability to predict food intake depending on the variability of their content in the food of interest, and their possible occurrence in other foods that may partially confound the associations observed (Edmands et al. 2015). The best biomarkers may correspond to some unannotated signals that constitute the majority of the signals detected. However, metabolomic data can be further mined as more dietary biomarkers are discovered and validated. Data on metabolites specifically associated with particular foods or beverages can be extracted from untargeted metabolomic datasets and associations with diseases or surrogate endpoints characterized. Trigonelline and several other biomarkers directly related to coffee intake were shown to be inversely related to colorectal cancer risk (Guertin et al. 2015). TMAO, a biomarker for fish intake, was associated with aggressive prostate cancer risk in the ATBC cohort (Mondul et al. 2015) and with a higher risk of cardiovascular diseases in a cohort of patients (Wang et al. 2011).

Measurement of the food metabolome through metabolomics is not only useful to assess exposure to specific foods. The joint estimation of panels of dietary biomarkers in metabolomic experiments makes the approach particularly appropriate to characterize and monitor changes in dietary patterns. Urinary metabolite models were developed in an intervention study using NMR spectroscopy to characterize four diets differing in their concordance with healthy eating guidelines [Dietary Approaches to Stop Hypertension (DASH) score] (Garcia-Perez et al. 2017). These models were built from 28 metabolites including both food-derived metabolites (e.g., 3-methylhistidine, TMAO, hippuric acid, *N*-acetyl-*S*-methyl-L-cysteine sulfoxide, tartaric acid) and endogenous metabolites, and were validated in two different cohorts. In another study, a model based on 33 metabolites measured in plasma by GC-MS was developed to discriminate subjects on a healthy New Nordic Diet from those having an average Danish diet (Khakimov et al. 2016). These metabolites were either directly derived from the diets or were reflecting different effects on energy metabolism.

Conclusions

Considerable progress in analytical technologies made over the last decade now means that the simultaneous measurement of hundreds to thousands of highly diverse compounds that are part of the food exposome is possible. These analyses can be made in small volumes of biospecimens with a high throughput, an achievement that was still unimaginable 10 years ago. This could significantly advance nutritional epidemiology and its utility in linking dietary exposures with disease outcomes. More than 10 years ago, Kristal et al. (2005) published a paper entitled "Is it time to abandon the food frequency questionnaire?" in which they emphasized limitations of FFQs and made a number of propositions to improve measures of dietary intake. Dietary biomarkers received limited attention in the article. Over 10 years later, FFQs are still widely used and dietary biomarkers, although increasingly incorporated into epidemiological studies, are not seen as an alternative to FFQs but rather as complementary tools to assess dietary exposures.

Progress in the development of dietary biomarkers and DWAS face a number of challenges. The first is the need to further develop databases and software tools necessary for the annotation of the food metabolome in biomarker discovery experiments. A second challenge will be to identify those dietary biomarkers that are reproducible and stable over time. It will also be essential to invest more time and efforts to properly validate candidate dietary biomarkers and test their sensitivity and specificity in population studies. Lastly, new analytical methods will have to be developed to quantify large panels of chemically diverse dietary biomarkers. So far methods used in DWASs were mainly semiquantitative and applied to the analysis of no more than a few 100 samples. Quantitative and more reproducible methods that can be applied to larger sample sets will be needed to limit false discoveries when analyzing large panels of dietary biomarkers. Once all these conditions are met, we may expect DWASs to become widely used in nutritional epidemiology to identify new dietary risk factors for diseases, or to monitor dietary exposures in populations.

Conflict of Interest Authors have no conflict of interest.

Acknowledgments Support from the FoodBAll project funded by the BIO-NH call under the Joint Programming Initiative, "a Healthy Diet for a Healthy Life" (grant number 529051002), is acknowledged.

References

Achaintre D, Bulete A, Cren-Olive C, Li L, Rinaldi S, Scalbert A (2016) Differential isotope labeling of 38 dietary polyphenols and their quantification in urine by liquid chromatography electrospray ionization tandem mass spectrometry. Anal Chem 88(5):2637–2644

Al-Delaimy WK, van Kappel AL, Ferrari P, Slimani N, Steghens JP, Bingham S, Johansson I, Wallstrom P, Overvad K, Tjonneland A, Key TJ, Welch AA, Bueno-de-Mesquita HB, Peeters PH, Boeing H, Linseisen J, Clavel-Chapelon F, Guibout C, Navarro C, Quiros JR, Palli D,

Celentano E, Trichopoulou A, Benetou V, Kaaks R, Riboli E (2004) Plasma levels of six carotenoids in nine European countries: report from the European Prospective Investigation into Cancer and Nutrition (EPIC). Public Health Nutr 7(6):713–722

Al-Delaimy WK, Ferrari P, Slimani N, Pala V, Johansson I, Nilsson S, Mattisson I, Wirfalt E, Galasso R, Palli D, Vineis P, Tumino R, Dorronsoro M, Pera G, Ocke MC, Bueno-de-Mesquita HB, Overvad K, Chirlaque MAD, Trichopoulou A, Naska A, Tjonneland A, Olsen A, Lund E, Alsaker EHR, Barricarte A, Kesse E, Boutron-Ruault MC, Clavel-Chapelon F, Key TJ, Spencer E, Bingham S, Welch AA, Sanchez-Perez MJ, Nagel G, Linseisen J, Quiros JR, Peeters PHM, van Gils CH, Boeing H, van Kappel AL, Steghens JP, Riboli E (2005) Plasma carotenoids as biomarkers of intake of fruits and vegetables: individual-level correlations in the European Prospective Investigation into Cancer and Nutrition (EPIC). Eur J Clin Nutr 59 (12):1387–1396

Allen NE, Grace PB, Ginn A, Travis RC, Roddam AW, Appleby PN, Key T (2008) Phytanic acid: measurement of plasma concentrations by gas-liquid chromatography-mass spectrometry analysis and associations with diet and other plasma fatty acids. Br J Nutr 99(3):653–659

Arab L (2003) Biomarkers of fat and fatty acid intake. J Nutr 133(3):925S–932S

Aune D, Chan DSM, Vieira AR, Rosenblatt DAN, Vieira R, Greenwood DC, Norat T (2012) Dietary compared with blood concentrations of carotenoids and breast cancer risk: a systematic review and meta-analysis of prospective studies. Am J Clin Nutr 96(2):356–373. https://doi.org/10.3945/ajcn.112.034165

Axel C, Zannini E, Arendt EK, Waters DM, Czerny M (2014) Quantification of cyclic dipeptides from cultures of Lactobacillus brevis R2Delta by HRGC/MS using stable isotope dilution assay. Anal Bioanal Chem 406(9–10):2433–2444. https://doi.org/10.1007/s00216-014-7620-3

Bailey LB, Stover PJ, McNulty H, Fenech MF, Gregory JF 3rd, Mills JL, Pfeiffer CM, Fazili Z, Zhang M, Ueland PM, Molloy AM, Caudill MA, Shane B, Berry RJ, Bailey RL, Hausman DB, Raghavan R, Raiten DJ (2015) Biomarkers of nutrition for development-folate review. J Nutr 145(7):1636S–1680S. https://doi.org/10.3945/jn.114.206599

Baldrick FR, Woodside JV, Elborn JS, Young IS, McKinley MC (2011) Biomarkers of fruit and vegetable intake in human intervention studies: a systematic review. Crit Rev Food Sci Nutr 51 (9):795–815. https://doi.org/10.1080/10408398.2010.482217

Bingham SA (2002) Biomarkers in nutritional epidemiology. Public Health Nutr 5(6A):821–827

Biskup I, Kyro C, Marklund M, Olsen A, van Dam RM, Tjonneland A, Overvad K, Lindahl B, Johansson I, Landberg R (2016) Plasma alkylresorcinols, biomarkers of whole-grain wheat and rye intake, and risk of type 2 diabetes in Scandinavian men and women. Am J Clin Nutr 104 (1):88–96. https://doi.org/10.3945/ajcn.116.133496

Block G, Hartman AM, Dresser CM, Carroll MD, Gannon J, Gardner L (1986) A data-based approach to diet questionnaire design and testing. Am J Epidemiol 124(3):453–469

Bradman A, Kogut K, Eisen EA, Jewell NP, Quiros-Alcala L, Castorina R, Chevrier J, Holland NT, Barr DB, Kavanagh-Baird G, Eskenazi B (2013) Variability of organophosphorous pesticide metabolite levels in spot and 24-hr urine samples collected from young children during 1 week. Environ Health Perspect 121(1):118–124

Brantsaeter AL, Haugen M, Thomassen Y, Ellingsen DG, Ydersbond TA, Hagve TA, Alexander J, Meltzer HM (2010) Exploration of biomarkers for total fish intake in pregnant Norwegian women. Public Health Nutr 13(1):54–62. https://doi.org/10.1017/s1368980009005904

Brevik A, Veierød MB, Drevon CA, Andersen LF (2005) Evaluation of the odd fatty acids 15:0 and 17:0 in serum and adipose tissue as markers of intake of milk and dairy fat. Eur J Clin Nutr 59 (12):1417–1422

Chadeau-Hyam M, Ebbels TMD, Brown IJ, Chan Q, Stemler J, Huang CC, Daviglus ML, Ueshima H, Zhao LC, Holmes E, Nicholson JK, Elliott P, De Iorio M (2010) Metabolic profiling and the metabolome-wide association study: significance level for biomarker identification. J Proteome Res 9(9):4620–4627

Chajes V, Biessy C, Byrnes G, Deharveng G, Saadatian-Elahi M, Jenab M, Peeters PHM, Ocke M, Bueno-de-Mesquita HB, Johansson I, Hallmans G, Manjer J, Wirfalt E, Jakszyn P, Gonzalez

CA, Huerta J-M, Martinez C, Amiano P, Rodriguez Suarez L, Ardanaz E, Tjonneland A, Halkjaer J, Overvad K, Jakobsen MU, Berrino F, Pala V, Palli D, Tumino R, Vineis P, de Magistris MS, Spencer EA, Crowe FL, Bingham S, Khaw K-T, Linseisen J, Rohrmann S, Boeing H, Noeethlings U, Olsen KS, Skeie G, Lund E, Trichopoulou A, Zilis D, Oustoglou E, Clavel-Chapelon F, Riboli E, Slimani N (2011a) Ecological-level associations between highly processed food intakes and plasma phospholipid elaidic acid concentrations: results from a cross-sectional study within the European prospective investigation into cancer and nutrition (EPIC). Nutr Cancer 63(8):1235–1250. https://doi.org/10.1080/01635581.2011.617530

Chajes V, Jenab M, Romieu I, Ferrari P, Dahm CC, Overvad K, Egeberg R, Tjonneland A, Clavel-Chapelon F, Boutron-Ruault M-C, Engel P, Teucher B, Kaaks R, Floegel A, Boeing H, Trichopoulou A, Dilis V, Karapetyan T, Mattiello A, Tumino R, Grioni S, Palli D, Vineis P, Bueno-de-Mesquita HB, Numans ME, Peeters PHM, Lund E, Navarro C, Ramon Quiros J, Sanchez-Cantalejo E, Barricarte Gurrea A, Dorronsoro M, Regner S, Sonestedt E, Wirfaelt E, Khaw K-T, Wareham N, Allen NE, Crowe FL, Rinaldi S, Slimani N, Carneiro F, Riboli E, Gonzalez CA (2011b) Plasma phospholipid fatty acid concentrations and risk of gastric adenocarcinomas in the European Prospective Investigation into Cancer and Nutrition (EPIC-EURGAST). Am J Clin Nutr 94(5):1304–1313. https://doi.org/10.3945/ajcn.110.005892

Cheung W, Keski-Rahkonen P, Assi N, Ferrari P, Freisling H, Rinaldi S, Slimani N, Zamora-Ros R, Rundle M, Frost G, Gibbons H, Carr E, Brennan L, Cross AJ, Pala V, Panico S, Sacerdote C, Palli D, Tumino R, Kühn T, Kaaks R, Boeing H, Floegel A, Mancini F, Boutron-Ruault M-C, Baglietto L, Trichopoulou A, Naska A, Orfanos P, Scalbert A (2017) A metabolomic study of biomarkers of meat and fish intake. Am J Clin Nutr 105(3):600–608. https://doi.org/10.3945/ajcn.116.146639

Chung H, Nettleton JA, Lemaitre RN, Barr RG, Tsai MY, Tracy RP, Siscovick DS (2008) Frequency and type of seafood consumed influence plasma (n-3) fatty acid concentrations. J Nutr 138(12):2422–2427

Crews H, Alink G, Andersen R, Braesco V, Holst B, Maiani G, Ovesen L, Scotter M, Solfrizzo M, van den Berg R, Verhagen H, Williamson G (2001a) A critical assessment of some biomarker approaches linked with dietary intake. Br J Nutr 86(Suppl 1):S5–S35

Crews HM, Olivier L, Wilson LA (2001b) Urinary biomarkers for assessing dietary exposure to caffeine. Food Addit Contam 18(12):1075–1087. https://doi.org/10.1080/02652030110056630

Curl CL, Beresford SA, Fenske RA, Fitzpatrick AL, Lu C, Nettleton JA, Kaufman JD (2015) Estimating pesticide exposure from dietary intake and organic food choices: the Multi-Ethnic Study of Atherosclerosis (MESA). Environ Health Perspect 123(5):475–483. https://doi.org/10.1289/ehp.1408197

Dahl H, Carlsson AV, Hillgren K, Helander A (2011) Urinary ethyl glucuronide and ethyl sulfate testing for detection of recent drinking in an outpatient treatment program for alcohol and drug dependence. Alcohol Alcohol 46(3):278–282. https://doi.org/10.1093/alcalc/agr009

Davis MA, Gilbert-Diamond D, Karagas MR, Li Z, Moore JH, Williams SM, Frost HR (2014) A dietary-wide association study (DWAS) of environmental metal exposure in US children and adults. PLoS One 9(9):e104768. https://doi.org/10.1371/journal.pone.0104768

de Batlle J, Matejcic M, Chajes V, Moreno-Macias H, Amadou A, Slimani N, Cox DG, Clavel-Chapelon F, Fagherazzi G, Romieu I (2016) Determinants of folate and vitamin B12 plasma levels in the French E3N-EPIC cohort. Eur J Nutr 57(2):751–760. https://doi.org/10.1007/s00394-016-1365-z

de Nijs M, Mengelers MJB, Boon PE, Heyndrickx E, Hoogenboom LAP, Lopez P, Mol HGJ (2016) Strategies for estimating human exposure to mycotoxins via food. World Mycotoxin J 9 (5):831–845. https://doi.org/10.3920/WMJ2016.2045

Dehghan M, Akhtar-Danesh N, McMillan CR, Thabane L (2007) Is plasma vitamin C an appropriate biomarker of vitamin C intake? A systematic review and meta-analysis. Nutr J 6:41. https://doi.org/10.1186/1475-2891-6-41

Dragsted LO (2010) Biomarkers of meat intake and the application of nutrigenomics. Meat Sci 84 (2):301–307

Edmands WM, Ferrari P, Rothwell JA, Rinaldi S, Slimani N, Barupal DK, Biessy C, Jenab M, Clavel-Chapelon F, Fagherazzi G, Boutron-Ruault MC, Katzke VA, Kuhn T, Boeing H, Trichopoulou A, Lagiou P, Trichopoulos D, Palli D, Grioni S, Tumino R, Vineis P, Mattiello A, Romieu I, Scalbert A (2015) Polyphenol metabolome in human urine and its association with intake of polyphenol-rich foods across European countries. Am J Clin Nutr 102 (4):905–913. https://doi.org/10.3945/ajcn.114.101881

Edmands WMB, Beckonert OP, Stella C, Campbell A, Lake BG, Lindon JC, Holmes E, Gooderham NJ (2011) Identification of human urinary biomarkers of cruciferous vegetable consumption by metabonomic profiling. J Proteome Res 10(10):4513–4521. https://doi.org/10.1021/pr200326k

Edmands WMB, Petrick L, Barupal DK, Scalbert A, Rappaport SM (2017) compMS2Miner: an automatable metabolite identification, visualization and data-sharing R package for high-resolution LC-MS datasets. Anal Chem 89(7):3919–3928

Forouhi NG, Imamura F, Sharp SJ, Koulman A, Schulze MB, Zheng J, Ye Z, Sluijs I, Guevara M, Huerta JM, Kroger J, Wang LY, Summerhill K, Griffin JL, Feskens EJ, Affret A, Amiano P, Boeing H, Dow C, Fagherazzi G, Franks PW, Gonzalez C, Kaaks R, Key TJ, Khaw KT, Kuhn T, Mortensen LM, Nilsson PM, Overvad K, Pala V, Palli D, Panico S, Quiros JR, Rodriguez-Barranco M, Rolandsson O, Sacerdote C, Scalbert A, Slimani N, Spijkerman AM, Tjonneland A, Tormo MJ, Tumino R, van der AD, van der Schouw YT, Langenberg C, Riboli E, Wareham NJ (2016) Association of plasma phospholipid n-3 and n-6 polyunsaturated fatty acids with Type 2 diabetes: the EPIC-InterAct case-cohort study. PLoS Med 13(7): e1002094. https://doi.org/10.1371/journal.pmed.1002094

Garcia-Aloy M, Llorach R, Urpi-Sarda M, Tulipani S, Estruch R, Martinez-Gonzalez MA, Corella D, Fito M, Ros E, Salas-Salvado J, Andres-Lacueva C (2014) Novel multimetabolite prediction of walnut consumption by a urinary biomarker model in a free-living population: the PREDIMED study. J Proteome Res 13(7):3476–3483

Garcia-Perez I, Posma JM, Gibson R, Chambers ES, Hansen TH, Vestergaard H, Hansen T, Beckmann M, Pedersen O, Elliott P, Stamler J, Nicholson JK, Draper J, Mathers JC, Holmes E, Frost G (2017) Objective assessment of dietary patterns by use of metabolic phenotyping: a randomised, controlled, crossover trial. Lancet Diabetes Endocrinol 5 (3):184–195. https://doi.org/10.1016/s2213-8587(16)30419-3

Gregory JF, Quinlivan EP (2002) In vivo kinetics of folate metabolism. Annu Rev Nutr 22:199–220. https://doi.org/10.1046/annurev.nutr.22.120701.083554

Guertin KA, Moore SC, Sampson JN, Huang W-Y, Xiao Q, Stolzenberg-Solomon RZ, Sinha R, Cross AJ (2014) Metabolomics in nutritional epidemiology: identifying metabolites associated with diet and quantifying their potential to uncover diet-disease relations in populations. Am J Clin Nutr 100:208–217. https://doi.org/10.3945/ajcn.113.078758

Guertin KA, Loftfield E, Boca SM, Sampson JN, Moore SC, Xiao Q, Huang W-Y, Xiong X, Freedman ND, Cross AJ, Sinha R (2015) Serum biomarkers of habitual coffee consumption may provide insight into the mechanism underlying the association between coffee consumption and colorectal cancer. Am J Clin Nutr 101:1000–1011. https://doi.org/10.3945/ajcn.114.096099

Gurdeniz G, Jensen MG, Meier S, Bech L, Lund E, Dragsted LO (2016) Detecting beer intake by unique metabolite patterns. J Proteome Res 15(12):4544–4556. https://doi.org/10.1021/acs.jproteome.6b00635

Hedrick VE, Dietrich AM, Estabrooks PA, Savla J, Serrano E, Davy BM (2012) Dietary biomarkers: advances, limitations and future directions. Nutr J 11:109. https://doi.org/10.1186/1475-2891-11-109

Heinzmann SS, Brown IJ, Chan Q, Bictash M, Dumas M-E, Kochhar S, Stamler J, Holmes E, Elliott P, Nicholson JK (2010) Metabolic profiling strategy for discovery of nutritional biomarkers: proline betaine as a marker of citrus consumption. Am J Clin Nutr 92(2):436–443

Hodson L, Skeaff CM, Fielding BA (2008) Fatty acid composition of adipose tissue and blood in humans and its use as a biomarker of dietary intake. Prog Lipid Res 47(5):348–380

Hunter D (1998) Biochemical indicators of dietary intake. In: Willett WC (ed) Nutritional epidemiology, 2nd edn. Oxford University Press, Oxford, pp 174–243

Illner A-K, Freisling H, Boeing H, Huybrechts I, Crispim S, Slimani N (2012) Review and evaluation of innovative technologies for measuring diet in nutritional epidemiology. Int J Epidemiol 41(4):1187–1203. https://doi.org/10.1093/ije/dys105

Isaksson A, Walther L, Hansson T, Andersson A, Alling C (2011) Phosphatidylethanol in blood (B-PEth): a marker for alcohol use and abuse. Drug Test Anal 3(4):195–200. https://doi.org/10.1002/dta.278

Jenab M, Slimani N, Bictash M, Ferrari P, Bingham SA (2009) Biomarkers in nutritional epidemiology: applications, needs and new horizons. Hum Genet 125(5–6):507–525

Kaaks R, Riboli E, Vanstaveren W (1995) Calibration of dietary intake measurements in prospective cohort studies. Am J Epidemiol 142(5):548–556

Kaaks R, Riboli E, Sinha R (1997) Biochemical markers of dietary intake. IARC Sci Publ 142:103–126

Kaaks RJ (1997) Biochemical markers as additional measurements in studies of the accuracy of dietary questionnaire measurements: conceptual issues. Am J Clin Nutr 65(4 Suppl):1232S–1239S

Keski-Rahkonen P, Rothwell JA, Scalbert A (2017) Metabolomic techniques to discover food biomarkers. In: Schoeller DA, Westerterp-Plantenga M (eds) Advances for in the assessment of dietary intake. CRC Press, Boca Raton

Khakimov B, Poulsen SK, Savorani F, Acar E, Gurdeniz G, Larsen TM, Astrup A, Dragsted LO, Engelsen SB (2016) New Nordic diet versus average Danish diet: a randomized controlled trial revealed healthy long-term effects of the new Nordic Diet by GC-MS blood plasma metabolomics. J Proteome Res 15(6):1939–1954

Kirkpatrick SI, Subar AF, Douglass D, Zimmerman TP, Thompson FE, Kahle LL, George SM, Dodd KW, Potischman N (2014) Performance of the Automated Self-Administered 24-hour Recall relative to a measure of true intakes and to an interviewer-administered 24-h recall. Am J Clin Nutr 100(1):233–240

Kristal AR, Peters U, Potter JD (2005) Is it time to abandon the food frequency questionnaire? Cancer Epidemiol Biomark Prev 14(12):2826–2828. https://doi.org/10.1158/1055-9965.epi-12-ed1

Kyro C, Olsen A, Landberg R, Skeie G, Loft S, Aman P, Leenders M, Dik VK, Siersema PD, Pischon T, Christensen J, Overvad K, Boutron-Ruault M-C, Fagherazzi G, Cottet V, Kuehn T, Chang-Claude J, Boeing H, Trichopoulou A, Bamia C, Trichopoulos D, Palli D, Krogh V, Tumino R, Vineis P, Panico S, Peeters PH, Weiderpass E, Bakken T, Asli LA, Argueelles M, Jakszyn P, Sanchez M-J, Amiano P, Huerta JM, Barricarte A, Ljuslinder I, Palmqvist R, Khaw K-T, Wareham N, Key TJ, Travis RC, Ferrari P, Freisling H, Jenab M, Gunter MJ, Murphy N, Riboli E, Tjonneland A, Bueno-de-Mesquita HB (2014) Plasma alkylresorcinols, biomarkers of whole-grain wheat and rye intake, and incidence of colorectal cancer. J Natl Cancer Inst 106(1): djt352. https://doi.org/10.1093/jnci/djt352

Landberg R, Aman P, Friberg LE, Vessby B, Adlercreutz H, Kamal-Eldin A (2009) Dose response of whole-grain biomarkers: alkylresorcinols in human plasma and their metabolites in urine in relation to intake. Am J Clin Nutr 89(1):290–296

Leech RM, Worsley A, Timperio A, McNaughton SA (2015) Understanding meal patterns: definitions, methodology and impact on nutrient intake and diet quality. Nutr Res Rev 28 (1):1–21. https://doi.org/10.1017/S0954422414000262

Levine M, Conry-Cantilena C, Wang Y, Welch RW, Washko PW, Dhariwal KR, Park JB, Lazarev A, Graumlich JF, King J, Cantilena LR (1996) Vitamin C pharmacokinetics in healthy volunteers: evidence for a recommended dietary allowance. Proc Natl Acad Sci U S A 93 (8):3704–3709

Lloyd AJ, Beckmann M, Fave G, Mathers JC, Draper J (2011) Proline betaine and its biotransformation products in fasting urine samples are potential biomarkers of habitual citrus fruit consumption. Br J Nutr 106(6):812–824. https://doi.org/10.1017/s0007114511001164

Lu QY, Hung JC, Heber D, Go VLW, Reuter VE, Cordon-Cardo C, Scher HI, Marshall JR, Zhang ZF (2001) Inverse associations between plasma lycopene and other carotenoids and prostate cancer. Cancer Epidemiol Biomark Prev 10(7):749–756

MacDonald L, Thumser AE, Sharp P (2002) Decreased expression of the vitamin C transporter SVCT1 by ascorbic acid in a human intestinal epithelial cell line. Br J Nutr 87(2):97–100

Mendonça N, Mathers JC, Adamson AJ, Martin-Ruiz C, Seal CJ, Jagger C, Hill TR (2016) Intakes of folate and vitamin B12 and biomarkers of status in the very old: the Newcastle 85+ Study. Nutrients 8(10):pii: E604. https://doi.org/10.3390/nu8100604

Mennen L, Sapinho D, Ito H, Galan P, Hercberg S, Scalbert A (2006) Urinary flavonoids and phenolic acids as biomarkers of intake for polyphenol-rich foods. Br J Nutr 96:191–198

Merritt MA, Tzoulaki I, Tworoger SS, De Vivo I, Hankinson SE, Fernandes J, Tsilidis KK, Weiderpass E, Tjonneland A, Petersen KE, Dahm CC, Overvad K, Dossus L, Boutron-Ruault MC, Fagherazzi G, Fortner RT, Kaaks R, Aleksandrova K, Boeing H, Trichopoulou A, Bamia C, Trichopoulos D, Palli D, Grioni S, Tumino R, Sacerdote C, Mattiello A, Bueno-de-Mesquita HB, Onland-Moret NC, Peeters PH, Gram IT, Skeie G, Quiros JR, Duell EJ, Sanchez MJ, Salmeron D, Barricarte A, Chamosa S, Ericson U, Sonestedt E, Nilsson LM, Idahl A, Khaw KT, Wareham N, Travis RC, Rinaldi S, Romieu I, Patel CJ, Riboli E, Gunter MJ (2015) Investigation of dietary factors and endometrial cancer risk using a nutrient-wide association study approach in the EPIC and Nurses' Health Study (NHS) and NHSII. Cancer Epidemiol Biomark Prev 24(2):466–471

Mondul AM, Moore SC, Weinstein SJ, Karoly ED, Sampson JN, Albanes D (2015) Metabolomic analysis of prostate cancer risk in a prospective cohort: the alpha-tocolpherol, beta-carotene cancer prevention (ATBC) study. Int J Cancer 137(9):2124–2132. https://doi.org/10.1002/ijc.29576

Murphy N, Achaintre D, Zamora-Ros R, Jenab M, Boutron-Ruault MC, Carbonnel F, Savoye I, Kaaks R, Kühn T, Boeing H, Aleksandrova K, Tjønneland A, Kyrø C, Overvad K, Quirós JR, Sánchez MJ, Altzibar JM, María Huerta J, Barricarte A, Khaw KT, Bradbury KE, Perez-Cornago A, Trichopoulou A, Karakatsani A, Peppa E, Palli D, Grioni S, Tumino R, Sacerdote C, Panico S, Bueno-de-Mesquita B, Peeters PH, Rutegård M, Johansson I, Freisling H, Noh H, Cross AJ, Vineis P, Tsilidis K, Gunter MJ, Scalbert A (2018) A prospective evaluation of plasma polyphenol levels and colon cancer risk. Int J Cancer: in press. https://https://doi:10.1002/ijc.31563

Neveu V, Moussy A, Rouaix H, Wedekind R, Pon A, Knox C, Wishart DS, Scalbert A (2017) Exposome-explorer: a manually-curated database on biomarkers of exposure to dietary and environmental factors. Nucleic Acids Res 45(D1):D979–D984. https://doi.org/10.1093/nar/gkw980

Ovesen L, Boeing H, EFCOSUM Group (2002) The use of biomarkers in multicentric studies with particular consideration of iodine, sodium, iron, folate and vitamin D. Eur J Clin Nutr 56(Suppl 2):S12–S17. https://doi.org/10.1038/sj.ejcn.1601424

Pallister T, Haller T, Thorand B, Altmaier E, Cassidy A, Martin T, Jennings A, Mohney RP, Gieger C, MacGregor A, Kastenmüller G, Metspalu A, Spector TD, Menni C (2016a) Metabolites of milk intake: a metabolomic approach in UK twins with findings replicated in two European cohorts. Eur J Nutr 56(7):2379–2391. https://doi.org/10.1007/s00394-016-1278-x

Pallister T, Jennings A, Mohney RP, Yarand D, Mangino M, Cassidy A, MacGregor A, Spector TD, Menni C (2016b) Characterizing blood metabolomics profiles associated with self-reported food intakes in female twins. PLoS One 11(6):e0158568

Patel CJ, Bhattacharya J, Butte AJ (2010) An environment-wide association study (EWAS) on type 2 diabetes mellitus. PLoS One 5(5):e10746

Personne V, Partouche H, Souberbielle JC (2013) Vitamin D insufficiency and deficiency: epidemiology, measurement, prevention and treatment. Presse Med 42(10):1334–1342. https://doi.org/10.1016/j.lpm.2013.06.013

Playdon MC, Sampson JN, Cross AJ, Sinha R, Guertin KA, Moy KA, Rothman N, Irwin ML, Mayne ST, Stolzenberg-Solomon R, Moore SC (2016) Comparing metabolite profiles of habitual diet in serum and urine. Am J Clin Nutr 104(3):776–789. https://doi.org/10.3945/ajcn.116.135301

Pujos-Guillot E, Hubert J, Martin J-F, Lyan B, Quintana M, Claude S, Chabanas B, Rothwell JA, Bennetau-Pelissero C, Scalbert A, Comte B, Hercberg S, Morand C, Galan P, Manach C (2013) Mass spectrometry-based metabolomics for the discovery of biomarkers of fruit and vegetable intake: citrus fruit as a case study. J Proteome Res 12(4):1645–1659. https://doi.org/10.1021/pr300997c

Quifer-Rada P, Martinez-Huelamo M, Chiva-Blanch G, Jauregui O, Estruch R, Lamuela-Raventos RM (2014) Urinary isoxanthohumol is a specific and accurate biomarker of beer consumption. J Nutr 144(4):484–488. https://doi.org/10.3945/jn.113.185199

Rappaport SM, Barupal DK, Wishart D, Vineis P, Scalbert A (2014) The blood exposome and its role in discovering causes of disease. Environ Health Perspect 122(8):769–774. https://doi.org/10.1289/ehp.1308015

Reboul E, Richelle M, Perrot E, Desmoulins-Malezet C, Pirisi V, Borel P (2006) Bioaccessibility of carotenoids and vitamin E from their main dietary sources. J Agric Food Chem 54 (23):8749–8755. https://doi.org/10.1021/jf061818s

Regueiro J, Vallverdu-Queralt A, Simal-Gandara J, Estruch R, Lamuela-Raventos RM (2014) Urinary tartaric acid as a potential biomarker for the dietary assessment of moderate wine consumption: a randomised controlled trial. Br J Nutr 111(9):1680–1685. https://doi.org/10.1017/s0007114513004108

Rimm EB, Giovannucci EL, Stampfer MJ, Colditz GA, Litin LB, Willett WC (1992) Reproducibility and validity of an expanded self-administered semiquantitative food frequency questionnaire among male health professionals. Am J Epidemiol 135(10):1114–1126

Ross AB, Svelander C, Undeland I, Pinto R, Sandberg AS (2015) Herring and beef meals lead to differences in plasma 2-aminoadipic acid, beta-alanine, 4-hydroxyproline, cetoleic acid, and docosahexaenoic acid concentrations in overweight men. J Nutr 145(11):2456–2463

Rothwell J, Madrid-Gambin F, Garcia-Aloy M, Gao Q, Andres-Lacueva C, Logue C, Gallagher AM, Mack C, Kulling S, Dragsted L, Scalbert A (2018) Biomarkers of intake for coffee, tea and sweetened beverages. Genes Nutr, In Revision

Rothwell JA, Fillâtre Y, Martin J-F, Lyan B, Pujos-Guillot E, Fezeu L, Hercberg S, Comte B, Galan P, Touvier M, Manach C (2014) New biomarkers of coffee consumption identified by the non-targeted metabolomic profiling of cohort study subjects. PLoS One 9(4):e93474. https://doi.org/10.1371/journal.pone.0093474

Rothwell JR, Urpi-Sarda M, Boto-Ordonez M, Knox C, Llorach R, Eisner R, Cruz J, Neveu V, Wishart D, Manach C, Andres-Lacueva C, Scalbert A (2012) Phenol-Explorer 2.0: a major update of the Phenol-Explorer database integrating data on polyphenol metabolism and pharmacokinetics in humans and experimental animals. Database 2012:bas031

Sanz ML, Villamiel M, Martínez-Castro I (2004) Inositols and carbohydrates in different fresh fruit juices. Food Chem 87(3):325–328. https://doi.org/10.1016/j.foodchem.2003.12.001

Saraff V, Shaw N (2016) Sunshine and vitamin D. Arch Dis Child 101(2):190–192. https://doi.org/10.1136/archdischild-2014-307214

Scalbert A, Brennan L, Manach C, Andres-Lacueva C, Dragsted LO, Draper J, Rappaport SM, van der Hooft JJJ, Wishart DS (2014) The food metabolome—a window over dietary exposure. Am J Clin Nutr 99:1286–1308

Scalbert A, Rothwell JA, Keski-Rahkonen P, Neveu V (2017) The food metabolome and dietary biomarkers. In: Schoeller DA, Westerterp-Plantenga M (eds) Advances in the assessment of dietary intake. CRC Press, Boca Raton

Scheline R (1978) Mammalian metabolism of plant xenobiotics. Academic Press, New York

Schoeller DA, Westerterp-Plantenga M (eds) (2017) Advances in the assessment of dietary intake. CRC Press, Boca Raton

Schroeder H, de la Torre R, Estruch R, Corella D, Angel Martinez-Gonzalez M, Salas-Salvado J, Ros E, Aros F, Flores G, Civit E, Farre M, Fiol M, Vila J, Fernandez-Crehuet J, Ruiz-

Gutierrez V, Lapetra J, Saez G, Covas M-I, Investigators PS (2009) Alcohol consumption is associated with high concentrations of urinary hydroxytyrosol. Am J Clin Nutr 90 (5):1329–1335
Slimani N, Freisling H, Illner AK, Huybrechts I (2015) Methods to determine dietary intake. In: Lovegrove JA, Hodson L, Sharma S, Lanham-New SA (eds) Nutrition research methodologies. Wiley, Hoboken
Soininen P, Kangas AJ, Würtz P, Suna T, Ala-Korpela M (2015) Quantitative serum nuclear magnetic resonance metabolomics in cardiovascular epidemiology and genetics. Circ Cardiovasc Genet 8(1):192–206. https://doi.org/10.1161/circgenetics.114.000216
Steingrimsdottir L, Ovesen L, Moreiras O, Jacob S (2002) Selection of relevant dietary indicators for health. Eur J Clin Nutr 56(Suppl 2):S8–S11. https://doi.org/10.1038/sj.ejcn.1601423
Sun Q, Bertrand KA, Franke AA, Rosner B, Curhan GC, Willett WC (2017) Reproducibility of urinary biomarkers in multiple 24-h urine samples. Am J Clin Nutr 105(1):159–168. https://doi.org/10.3945/ajcn.116.139758
The_Joint_FAO/WHO_Expert_Committee_on_Food_Additives_(JECFA) (2017) Chemical risks and JECFA. Food and Agriculature Organisation (FAO), http://www.fao.org/food/food-safety-quality/scientific-advice/jecfa/en/. Accessed 15 Feb 2017
Thompson FE, Subar AF (2013) Dietary assessment methodology. In: Coulston AM, Boushey CJ, Ferruzzi MG (eds) Nutrition in the prevention and treatment of disease, 3rd edn. Elsevier, San Diego, CA, pp 5–46
Troche JR, Mayne ST, Freedman ND, Shebl FM, Guertin KA, Cross AJ, Abnet CC (2016) Alcohol consumption-related metabolites in relation to colorectal cancer and adenoma: two case-control studies using serum biomarkers. PLoS One 11:e0150962. https://doi.org/10.1371/journal.pone.0150962
Turner PC, Hopton RP, Lecluse Y, White KL, Fisher J, Lebailly P (2010) Determinants of urinary deoxynivalenol and de-epoxy deoxynivalenol in male farmers from Normandy, France. J Agric Food Chem 58(8):5206–5212. https://doi.org/10.1021/jf100892v
Tzoulaki I, Patel CJ, Okamura T, Chan Q, Brown IJ, Miura K, Ueshima H, Zhao L, Van Horn L, Daviglus ML, Stamler J, Butte AJ, Ioannidis JPA, Elliott P (2012) A nutrient-wide association study on blood pressure. Circulation 126(21):2456–2464. https://doi.org/10.1161/circulationaha.112.114058
University_of_Alberta (2016) FooDB—the food component database. http://www.foodb.ca/. Accessed 11 Dec 2016
van der Hooft JJ, Wandy J, Barrett MP, Burgess KE, Rogers S (2016) Topic modeling for untargeted substructure exploration in metabolomics. Proc Natl Acad Sci U S A 113 (48):13738–13743. https://doi.org/10.1073/pnas.1608041113
Verkleij-Hagoort AC, de Vries JHM, Stegers MPG, Lindemans J, Ursem NTC, Steegers-Theunissen RPM (2007) Validation of the assessment of folate and vitamin B-12 intake in women of reproductive age: the method of triads. Eur J Clin Nutr 61(5):610–615
Vioque J, Weinbrenner T, Asensio L, Castello A, Young IS, Fletcher A (2007) Plasma concentrations of carotenoids and vitamin C are better correlated with dietary intake in normal weight than overweight and obese elderly subjects. Br J Nutr 97(5):977–986
Wang Z, Klipfell E, Bennett BJ, Koeth R, Levison BS, DuGar B, Feldstein AE, Britt EB, Xiaoming F, Yoon-Mi C, Yuping W, Schauer P, Smith JD, Hooman A, Wilson Tang WH, DiDonato JA, Lusis AJ, Hazen SL (2011) Gut flora metabolism of phosphatidylcholine promotes cardiovascular disease. Nature 472(7341):57–63
Willett WC (ed) (1998) Nutritional epidemiology, 2nd edn. Oxford University Press, Oxford
Wittenbecher C, Muehlenbruch K, Kroeger J, Jacobs S, Kuxhaus O, Gel AF, Fritsche A, Pischon T, Prehn C, Adamski J, Joost H-G, Boeing H, Schulze MB (2015) Amino acids, lipid metabolites, and ferritin as potential mediators linking red meat consumption to type 2 diabetes. Am J Clin Nutr 101(6):1241–1250. https://doi.org/10.3945/ajcn.114.099150
Wurtz P, Cook S, Wang Q, Tiainen M, Tynkkynen T, Kangas AJ, Soininen P, Laitinen J, Viikari J, Kahonen M, Lehtimaki T, Perola M, Blankenberg S, Zeller T, Mannisto S, Salomaa V, Jarvelin

MR, Raitakari OT, Ala-Korpela M, Leon DA (2016) Metabolic profiling of alcohol consumption in 9778 young adults. Int J Epidemiol 45(5):1493–1506. https://doi.org/10.1093/ije/dyw175

Yao YY, Wang PG, Shao G, Del Toro LVA, Codero J, Giese RW (2016) Nontargeted analysis of the urine nonpolar sulfateome: a pathway to the nonpolar xenobiotic exposome. Rapid Commun Mass Spectrom 30(21):2341–2350. https://doi.org/10.1002/rcm.7726

Zain ME (2011) Impact of mycotoxins on humans and animals. J Saudi Chem Soc 15(2):129–144. https://doi.org/10.1016/j.jscs.2010.06.006

Zamora-Ros R, Urpi-Sarda M, Lamuela-Raventos RM, Estruch R, Martinez-Gonzalez MA, Bullo M, Aros F, Cherubini A, Andres-Lacueva C (2009) Resveratrol metabolites in urine as a biomarker of wine intake in free-living subjects: the PREDIMED Study. Free Radic Biol Med 46(12):1562–1566

Zamora-Ros R, Achaintre D, Rothwell JA, Rinaldi S, Assi N, Ferrari P, Leitzmann M, Boutron-Ruault M-C, Fagherazzi G, Auffret A, Kühn T, Katzke V, Boeing H, Trichopoulou A, Naska A, Vasilopoulou E, Palli D, Grioni S, Mattiello A, Tumino R, Ricceri F, Slimani N, Romieu I, Scalbert A (2016) Urinary excretions of 34 dietary polyphenols and their associations with lifestyle factors in the EPIC cohort study. Sci Rep 6:26905. https://doi.org/10.1038/srep26905

Zheng Y, Yu B, Alexander D, Steffen LM, Boerwinkle E (2014a) Human metabolome associates with dietary intake habits among African Americans in the atherosclerosis risk in communities study. Am J Epidemiol 179(12):1424–1433. https://doi.org/10.1093/aje/kwu073

Zheng Y, Yu B, Alexander D, Steffen LM, Nettleton JA, Boerwinkle E (2014b) Metabolomic patterns and alcohol consumption in African Americans in the Atherosclerosis Risk in Communities Study. Am J Clin Nutr 99(6):1470–1478. https://doi.org/10.3945/ajcn.113.074070

Chapter 9
The Dust Exposome

I. R. Schultz, S. Cade, and L. J. Kuo

Abstract Contact with indoor dust has been shown to be an important source of contaminant exposure in humans. The composition of dust and its associated contaminants varies by a number of factors including location, nearby objects, and activities. Indoor locations that contain furniture, appliances, and other electronic goods produce dust enriched with chemical additives used in the manufacturing process. Classes of chemicals commonly found in dust include plasticizers, perfluorocarbons, flame retardants, lubricants, and metals used in electronics, batteries, power supplies, and imaging devices. Other types of hazardous chemicals associated with dust are contaminants that are formed after burning or combustion. These include polycyclic aromatic hydrocarbons (PAHs), chlorinated and brominated PAHs, halogenated dioxins, and dibenzofurans. These contaminants are established components of dust and are frequently the focus of environmental monitoring efforts. However, nontargeted chemical analysis of some workplace dust suggests the presence of other organic and inorganic compounds, which remain to be characterized. In addition, many of these contaminants are environmentally persistent and bioaccumulate in seafood and other types of food, complicating exposure assessments and the importance of dust as an exposure source. However, recent research has begun to identify differences in the contaminant profiles between dust and food that can allow better source appointment for some classes such as PBDEs.

Keywords Occupational inorganic exposures · Household exposures · Dust · Organic exposures

I. R. Schultz (✉) · S. Cade · L. J. Kuo
Lynker Technologies under contract to Northwest Fisheries Science Center, National Marine Fisheries Service, National Oceanic and Atmospheric Administration, Seattle, WA, USA

Pacific Northwest National Laboratory, Sequim, WA, USA
e-mail: irvins3@uw.edu

S. Dagnino, A. Macherone (eds.), *Unraveling the Exposome*,
https://doi.org/10.1007/978-3-319-89321-1_9

Introduction

Indoor dust or the particulate matter found in homes and workplaces has been the focus of chemical monitoring studies for several decades. The sources and composition of dust are varied but in general consist of inorganic and organic matter derived from dead skin, soil (brought indoors from foot traffic), insects, plants, microorganisms, and fungal spores along with particulates shed from housing materials (surface coatings on ceilings and walls) and material released from furnishings and appliances. Because of the diverse sources of dust, its composition continually changes depending on room location, nearby objects, and activities (e.g., kitchen/cooking or occupation related). Other well-known factors that influence the composition of dust are the type of air circulation, heating and cooling systems employed, and seasonality (Butte and Heinzow 2002). For example, the amount of indoor dust tends to increase in the summer or warmer months of the year (Edwards et al. 1998). In addition to the total quantity of dust present on indoor surfaces, the size range of dust particles is broad and has been the focus of numerous studies. A particularly thorough characterization of indoor dust particles was reported by Lewis et al. (1999) who examined dust recovered from vacuum cleaners used in 25 middle-class homes in North Carolina. Approximately 57% of the total mass of dust was comprised of particulates >2 mm in diameter with the next two most abundant size fractions being 53–106 μm and <25 μm in diameter (23 and 21% of total, respectively; Lewis et al. 1999). Thus, both relatively large and small sized particles can be well represented in household dust. This broad size distribution of dust particles also influences the routes of human exposure with inhalation of suspended particles, dermal absorption (through direct contact with settled dust), and oral ingestion (e.g., hand to mouth activity) all demonstrated to be important routes of exposure (Butte and Heinzow 2002).

Contaminant exposure resulting from contact with dust is an important contributor to the exposome for several reasons. First, indoor dust acts as a repository for a diverse range of anthropogenic and naturally occurring contaminants. Nearly all classes of known or established contaminants have now been identified in dust. Secondly, contact with indoor dust has been shown to be an important source of contaminant exposure in humans, particularly infants and young children (Moya and Phillips 2014). Recent studies have suggested the average daily dust ingestion rates for infants are nearly 20 times higher than for adults due to greater hand-to-mouth contact and contact with flooring due to crawling (Wilson et al. 2013). These differences occur without consideration of body size, thus contaminant exposure from dust ingestion when normalized for body weight is even higher for infants and toddlers than for adults. From an exposome perspective, indoor dust likely represents one of the first sources of contaminant exposure for humans. Finally, many of the contaminants found in dust, such as various types of flame retardants, have been associated with a variety of adverse health outcomes such as endocrine disruption (Hwang et al. 2008).

Dust Collection

Indoor dust collection methods can use a passive or active approach. An example of a passive approach is to place a non-electrostatic plate in a room for a predefined period of time and then recover dust that has naturally settled on the plate (Edwards et al. 1998). Active sampling approaches have become more common in recent years, and typically involve a vacuum device with an in-line trap to collect dust. Many types of vacuums and dust traps have been used. Other approaches may use moistened hand wipes to remove dust and possibly contaminants that are adsorbed to floor or counter surfaces. There is no generally accepted “best” approach to dust sampling and each approach has its benefits. However, it is important to recognize that dust sampling methods have a direct influence on the recovery and size range of dust particles collected (for example, passive methods tend to collect smaller particles while vacuum methods recover more dust from carpets; Edwards et al. 1998; Reynolds et al. 1997) and can influence the measurement of contaminants in dust. For the most complete characterization of contaminants in indoor dust, multiple sampling methods should probably be employed.

In recent years there has been greater emphasis on active approaches to dust sample collection, which typically involve either self-reporting by study participants of dust samples taken from household vacuum cleaners or more controlled dust collection. Although self-reporting is cost effective and may be the only practical approach for large cross-sectional studies, it suffers from variations in efficiency of dust collection. More controlled methods involve trained investigators using portable vacuum cleaners that are fitted with cellulose extraction thimbles to trap dust. This offers advantages in that dust collection can be more easily standardized and contamination from vacuum components or other sources is minimized (Allen et al. 2008). The use of standardized protocols and single use traps helps facilitate comparison of results and prevent cross-contamination dust between sites (Watkins et al. 2012). A simple but commonly used protocol is to vacuum a 2 m $\times$ 2 m area for approximately 5 min capturing dust from the entire surface area of the accessible floor space. After sample collection, thimbles or dust traps are removed and wrapped in aluminum foil (previously baked at 450 °C), sealed in polypropylene bags and stored at −20°C or lower until processing. Blank dust samples can be easily generated by vacuuming sodium sulfate powder from a sheet of aluminum foil (previously baked overnight). Upon analysis, dust samples are usually sieved through metal sieves to isolate specific size fractions. The sieved dust sample can then be divided into aliquots as needed and then extracted and processed for chemical analysis according to established analytical protocols.

Types of Contaminants and Levels in Dust

Many of the contaminants identified in indoor dust reflect the classes of chemicals intentionally added to furniture and appliances found in homes or workplaces. This would include plasticizers, flame retardants, lubricants, perfluorocarbons, UV stabilizers, dielectric fluids and metals used in circuit boards, batteries, power supplies, and imaging devices. Other classes of hazardous chemicals associated with dust include polycyclic aromatic hydrocarbons (PAHs), chlorinated and brominated PAHs, halogenated dioxins, and dibenzofurans, which are typically formed as combustion by-products (Robinson 2009). These established components of dust continue to be the focus of many environmental monitoring studies; however, other classes of contaminants have recently been identified such as brominated azo-dyes (Peng et al. 2016). This suggests dust likely contains other types of contaminants of potential health concern that have not yet been characterized.

Household electronic appliances such as computers and televisions contain relatively high levels of flame retardants, which are considered necessary to meet fire safety standards. Synthetic PBDEs have been the most widely used flame retardants and are among the most frequently identified chemicals in dust. The use of all PBDEs is now banned although they have been replaced with newer types of brominated flame retardants (BFRs). Non-brominated flame retardants have also been introduced as alternatives. The content of PBDEs and other BFR/non-BFRs in consumer products ranges from 3% to 33% (Arias 2001; Alaee et al. 2003). There was a gradual reduction in the commercial use of PBDEs beginning in 2004 due in part to bans on specific types of PBDE mixtures (LaGuardia et al. 2006; Betts 2008). This led to the use of PBDE products largely composed of PBDE-209, which was itself banned in 2013. However, other types of structurally related BFRs were introduced as replacements (Covaci et al. 2011) and recent estimates of annual worldwide use of BFRs and non-BFRs has been 2,000,000 tons (www.flameretardants-online.com). Thus, PBDEs, other BFRs, and non-BFRs continue to represent important classes of organic contaminants identified in dust.

In addition to flame retardants and other organic contaminants, several studies have reported elemental analysis of household dust. The most abundant elements are calcium, sodium (Na), magnesium (Mg), iron (Fe), and aluminum (Al; Seifert et al. 2000; Rasmussen et al. 2001) although levels can vary many fold between sites and particle sizes. Other more toxicologically important metals including lead (Pb), chromium (Cr), and mercury (Hg) are found at much lower levels and in the case of cadmium (Cd) or Beryllium (Be), are normally found at extremely low levels (Beamer et al. 2012). Some rare metals associated with electronic goods, such as antimony, indium, and vanadium, have also been identified with dust particles, which may be the primary source of exposure for these elements (Julander et al. 2014).

A summary of contaminant levels found in dust is shown in Table 9.1. These values represent the mean or average levels reported in different studies. Often, dust concentrations for a specific contaminant or contaminant class can vary more than

Table 9.1 Overview and approximate abundance of selected common and persistent compounds identified in household dust

Chemical class	µg/g-dust
Phthalate esters (di-2-ethylhexyl-phthlate most abundant)	10–3200 (Mercier et al. 2011)
PBDE-209	10–340 (Mercier et al. 2011; Schultz et al. unpublished observations)
Non-phthalate plastic additives (acetyl tributyl citrate, di-isononyl-1,2-cyclohexane-dicarboxylate, trioctyl trimellitate, Tris (2-ethylhexyl)trimellitate	<1–504 (Fromme et al. 2016)
Non-brominated replacement flame retardants	<1–30 (Mercier et al. 2011; Tao et al. 2016)
Brominated Azo-dyes	0.7~30 (Peng et al. 2016)
Inorganics Na, Mg, Al, Fe	5–>1000 (Rasmussen et al. 2001; Seifert et al. 2000)
Inorganics Pb, Cu, total Hg	22–139,124 1–11 (Rasmussen et al. 2001; Seifert et al. 2000; Sun et al. 2013)
Polyaromatic hydrocarbons	0.5–5.0 (Mercier et al. 2011)
PBDEs (all congeners except PBDE 209)	0.5–2.0 (Mercier et al. 2011; Schultz et al. unpublished observations; Tao et al. 2016)
Brominated replacement flame retardants	1–6 (Mercier et al. 2011; Tao et al. 2016)
Hexabromocyclododecane (HBCD)	0.6–1.3 (Mercier et al. 2011)
UV stabilizers/corrosion inhibitors	0.6 (Wang et al. 2013)
Phenolics (OH-PCBs, OH-PBDEs, triclosan)	0.7–2 (Mercier et al. 2011)
PCBs	~1 or less (Mercier et al. 2011)
Polychlorinated dibenzodioxins/dibenzofurans (majority as octachlorodibenzodioxin)	0.001–6 (Hinwood et al. 2014; Deziel et al. 2012)
N-ethyl perfluorooctane sulfonamidoethanol (EtFOSE)	0.1–1 (Goosey and Harrad 2011)
Perfluorinated compounds	0.5–2 (Goosey and Harrad 2011)
Brominated dibenzo-pdioxins/dibenzofurans (mainly PBDFs)	0.001–0.01 (Suzuki et al. 2010)
Inorganics Cd, Be	<0.5 (Beamer et al. 2012)

10–100 fold among sites. Thus, a range of values is provided to give the reader a sense of the variation in levels. For brevity, major classes of contaminants are listed that combine individual chemicals within a class (phthalates) or congeners (PCBs and PBDEs). An exception is made for PBDE-209, which is the predominant PBDE identified in dust. A more complete list of chemicals identified in indoor dust has been reported by Mercier et al. (2011), who provided dust concentrations for approximately 455 different contaminants. Table 9.1 and similar lists of this type reflect the concentrations of known contaminants determined using a targeted analytical approach, which permits a high degree of precision due to the use of technical standards and/or closely related structural analogues. However, there have been recent attempts at nontargeted analysis of indoor dust using novel analytical workflows. For example, Rager et al. (2016) used liquid chromatography time-of-flight mass spectrometry (LC–TOF/MS) with electrospray ionization (ESI) to characterize chemicals extracted from 56 dust samples obtained through self-reporting methods. Analysis of the ion chromatograms suggested the presence of at least 978 distinct chemicals that could be associated with previously identified or predicted chemical structures. However, identifiable chemicals represented a small minority with the majority (e.g., >90%) of chemicals not linked to any known chemical structures (Rager et al. 2016). Another recent study used a LC-MS/MS-Orbitrap system with atmospheric-pressure photoionization to perform nontargeted analysis of brominated compounds in 23 indoor dust samples obtained using a portable vacuum and standardized protocol (Peng et al. 2016). The results indicated at least 1008 distinct brominated contaminants were present, with approximate molecular weights ranging from 229 to 1000 (Peng et al. 2016). A subset of these contaminants was assigned chemical structures, which revealed that 78 of the most abundant compounds were brominated azo dyes (Peng et al. 2016). These results were particularly significant as the dyes appeared to originate from pigmented clothing and possessed mutagenic activity as assessed in an Ames II test (Peng et al. 2016). These and other experimental findings make it clear that indoor dust contains a complex mixture of thousands of organic and inorganic contaminants, most of which remain to be characterized.

Conclusion and Contribution to the Exposome

The exposome concept was introduced to represent the complete life history of exposures to endogenous and exogenous agents and serves as the environmental compliment to the genome (Wild 2005). Many of the contaminants in indoor dust are also found in air, food, and water making it likely that the exposome represents a combination of exposures from multiple matrices. However, dust is an important contributor to the exposome for several reasons. (1) Due to the proportionally greater intake of dust by infants and toddlers, dust is the predominant contaminant-laden environmental matrix humans are first exposed to. Thus, much of the exposome during early childhood is likely to be associated with dust exposure. (2) The dust exposome is gradually being recognized to contain an extremely diverse and rich source of contaminants, perhaps greater than for any other environmental matrix humans are regularly exposed to. (3) Dust can be the primary source of exposure for some contaminants. A good example is PBDE-209, which is generally observed to be absent from food but is consistently found at high levels in dust (Table 9.1). As the exposome concept evolves, future studies will benefit from the continued chemical characterization of indoor dust, which is helping to reveal the multitude of contaminants humans are exposed to.

References

Alaee M, Arias P, Sjödin A, Bergman Å (2003) An overview of commercially used brominated flame retardants, their applications, their use patterns in different countries/regions and possible modes of release. Environ Int 29:683–689

Allen JG, McClean MD, Stapleton HM, Webster T (2008) Critical factors in assessing exposure to PBDEs via house dust. Environ Int 34:1085–1091

Arias P (2001) Brominated flame retardants—an overview. In: the second international workshop on brominated flame retardants, AB Firmatryck, Stockholm, pp 17–19

Beamer PI, Elish CA, Roe DJ, Loh MM, Layton DW (2012) Differences in metal concentration by particle size in house dust and soil. J Environ Monit 14(3):839–844

Betts KS (2008) New thinking on flame retardants. Environ Health Perspect 116:A210–A213

Butte W, Heinzow B (2002) Pollutants in house dust as indicators of indoor contamination. Rev Environ Contam Toxicol 175:1–46

Covaci A, Harrad S, Abdallah MAE, Ali N, Law RJ, Herzke D, de Wit CA (2011) Novel brominated flame retardants: a review of their analysis, environmental fate and behaviour. Environ Int 37(2):532–556

Deziel NC, Nuckols JR, Colt JS, De Roos AJ, Pronk A, Gourley C, Severson RK, Cozen W, Cerhan JR, Hartge P, Ward MH (2012) Determinants of polychlorinated dibenzo-p-dioxins and polychlorinated dibenzofurans in house dust samples from four areas of the United States. Sci Total Environ 433:516–522

Edwards RD, Yurkow EJ, Uoy P (1998) Seasonal deposition of house dusts onto household surfaces. Sci Total Environ 224(1–3):69–80

Fromme H, Schuetze A, Lahrz T, Kraft M, Fernbacher L, Siewering S, Burkardt R, Dietrich S, Koch HM, Voelkel W (2016) Non-phthalate plasticizers in German daycare centers and human biomonitoring of DINCH metabolites in children attending the centers (LUPE 3). Int J Hyg Environ Health 219(1):33–39

Goosey E, Harrad S (2011) Perfluoroalkyl compounds in dust from Asian, Australian, European, and North American homes and UK cars, classrooms, and offices. Environ Int 37:86–92

Hinwood AL, Callan AC, Heyworth J, Rogic D, de Araujo J, Crough R, Mamahit G, Piro N, Yates A, Stevenson G, Odland JO (2014) Polychlorinated biphenyl (PCB) and dioxin concentrations in residential dust of pregnant women. Environ Sci Processes Impacts 16 (12):2758–2763

Hwang H, Park EK, Young TM (2008) Occurrence of endocrine-disrupting chemicals in indoor dust. Sci Total Environ 404(1):26–35

Julander A, Lundgren L, Skare L, Grandér M, Palma B, Vahter M, Lidéna C (2014) Formal recycling of e-waste leads to increased exposure to toxic metals: an occupational exposure study from Sweden. Environ Int 73:243–251

LaGuardia MJ, Hale RC, Harvey E (2006) Detailed polybrominated diphenyl ether (PBDE) congener composition of the widely used penta-, octa-, and deca-PBDE technical flame-retardant mixtures. Environ Sci Technol 40(20):6247–6254

Lewis RG, Fortune CR, Willis RD, Camann DE, Antley JT (1999) Distribution of pesticides and polycyclic aromatic hydrocarbons in house dust as a function of particle size. Environ Health Perspect 107(9):721–726

Mercier F, Glorennec P, Thomas O, Le Bot B (2011) Organic contamination of settled house dust, a review for exposure assessment purposes. Environ Sci Technol 45:6716–6727

Moya J, Phillips L (2014) A review of soil and dust ingestion studies for children. J Expo Sci Environ Epidemiol 24(6):545–554

Peng H, Saunders DMV, Sun J, Jones PD, Wong CKC, Liu HL, Giesy JP (2016) Mutagenic azo dyes, rather than flame retardants, are the predominant brominated compounds in house dust. Environ Sci Technol 50(23):12669–12677

Rager JE, Strynar MJ, Liang S, McMahen RL, Richard AM, Grulke CM, Wambaugh JF, Isaacs KK, Judson R, Williams AJ, Sobus JR (2016) Linking high resolution mass spectrometry data with

exposure and toxicity forecasts to advance high-throughput environmental monitoring. Environ Int 88:269–280
Rasmussen PE, Subramanian KS, Jessiman BJ (2001) A multi-element profile of housedust in relation to exterior dust and soils in the city of Ottawa, Canada. Sci Total Environ 267 (1–3):125–140
Reynolds SJ, Etre L, Thorne PS (1997) Laboratory comparison of vacuum, OSHA, and HUD sampling methods for lead in household dust. Am Ind Hyg Assoc J 58(6):439–446
Robinson BH (2009) E-waste: an assessment of global production and environmental impacts. Sci Total Environ 408:183–191
Seifert B, Becker K, Helm D (2000) The German Environmental Survey 1990/1992 (GerES II): reference concentrations of selected environmental pollutants in blood, urine, hair, house dust, drinking water and indoor air. J Expo Anal Environ Epidemiol 10(6):552–565
Sun GY, Li ZG, Bi XY, Chen YP, Lu SF, Yuan X (2013) Distribution, sources and health risk assessment of mercury in kindergarten dust. Atmos Environ 73:169–176
Suzuki GO, Someya M, Takahashi S (2010) Dioxin-like activity in Japanese indoor dusts evaluated by means of in vitro bioassay and instrumental analysis: brominated dibenzofurans are an important contributor. Environ Sci Technol 44(21):8330–8336
Tao F, Abdallah MA, Harrad S (2016) Emerging and legacy flame retardants in UK indoor air and dust: evidence for replacement of PBDEs by emerging flame retardants? Environ Sci Technol 50 (23):13052–13061
Wang L, Asimakopoulos AG, Hyo-Bang M, Nakata H, Kannan K (2013) Benzotriazole, benzothiazole, and benzophenone compounds in indoor dust from the United States and East Asian Countries. Environ Sci Technol 47:4752–4759
Watkins DJ, McClean MD, Fraser AJ, Weinberg J, Stapleton HM, Sjodin A (2012) Impact of dust from multiple microenvironments and diet on PentaBDE body burden. Environ Sci Technol 46:1192–1200
Wild CP (2005) Complementing the genome with an "exposome": the outstanding challenge of environmental exposure measurement in molecular epidemiology. Cancer Epidemiol Biomark Prev 14:1847–1850
Wilson R, Jones-Otazo H, Petrovic S (2013) Revisiting dust and soil ingestion rates based on hand-to-mouth transfer. Hum Ecol Risk Assess 19(1):158–188

Chapter 10
From the Outside In: Integrating External Exposures into the Exposome Concept

Yuxia Cui and David Balshaw

Abstract The exposome comprises exposures arising from both endogenous processes and the external environment, which include not only chemical compounds but also nutrients, drugs, infectious agents, the microbiome, physical stress, and psychosocial stress. But where do those exposures come from, and what can we do about them? This chapter will focus on two distinct but interrelated questions: (1) Why understanding the environment outside of the body is critical for unraveling the exposome; and (2) How to measure the exposome from the outside the body. The external environment is critical to unraveling the exposome for it provides a depth of context that analysis of biological samples cannot, it also informs data interpretation by providing linkage between external sources of exposure and the internal dose, and in some cases, external assessment is the only way to objectively evaluate an exposure such as social and built environment as well as behavioral factors. The grand challenge lies in how to provide a comprehensive and rigorous characterization of the external environment on the exposome scale. This chapter will discuss

- Scientific advances, emerging technologies, and novel opportunities to unravel the exposome from the external environment
- The importance of multi-scale data integration in assessing the temporal and spatial dynamics of the exposome
- Strategies and the value of integrating external and internal environments in studies of the exposome
- The opportunity of applying the exposome to inform citizen science and disease prevention

Keywords Integrating external exposures with exposomics · Multi-scale data integration

Y. Cui · D. Balshaw (✉)
Exposure, Response, and Technology Branch, National Institute of Environmental Health Sciences, Durham, NC, USA
e-mail: cuiy2@niehs.nih.gov; balshaw@niehs.nih.gov

S. Dagnino, A. Macherone (eds.), *Unraveling the Exposome*,
https://doi.org/10.1007/978-3-319-89321-1_10

Introduction

In 2005, Chris Wild proposed the exposome concept to draw attention to the importance of developing methodologies for more accurate and complete exposure assessment in epidemiological studies (Wild 2005). In 2012, he further defined that the exposome (all nongenetic exposures) can be categorized into three domains based on the nature of the exposure: internal exposures such as metabolism, endogenous hormones, gut microbiome, and oxidative stress; specific external exposures such as radiation, environmental pollutants, diet and lifestyle factors; and general external exposures such as social and built environments (Wild 2012). While external exposures impact human health through interactions with internal biological processes and many can be estimated by analysis of biological samples, measuring the exposure from outside the body remains an essential approach to understanding complex environmental exposures and sometimes the only way to objectively evaluate an exposure.

Why Do We Need to Measure the exposome from Outside the Body?

Increasing evidence is showing that the causes of complex human diseases are more complicated and likely due to the combination of exposures, both external and internal, within the context of dose and time of exposure(s) in given individual genetic environment. Recent development in mass spectrometry and omics technologies offers opportunities to detect and quantify environmental chemicals, metabolites of environmental chemicals, or molecular signatures of environmental exposures in biological samples (Athersuch 2016; Dennis et al. 2016; Rappaport 2012). As discussed in this book, these methods have become the major focus of efforts to implement the exposome due largely to their inherent ability to comprehensively measure multiple exposure-related signals simultaneously. However, measuring exposures in biological samples only provides snapshots of postexposure events inside the body. It does not tell when the exposures occur, how long they last, or where they come from. It is impractical to know the sequence of the exposures, which may be critically important for understanding the effects of complex mixtures. External exposure assessment, on the other hand, can provide a depth of context that analysis of biological samples cannot, such as time and location of the exposure, and the dynamic characteristics of the exposure. Temporally and spatially resolved environmental exposure data not only enables more accurate association between exposures and adverse health events, when combined with assessment of the internal exposure and biological response, it allows for better characterization of the exposure biology and thus improved understanding of how environmental exposures lead to diseases.

In addition, external exposure assessment provides the possibility to precisely pinpoint the environmental sources of disease-causing exposures, which is vital for effective public health intervention and prevention strategies. Studies using environmental and personal sensors to track school-aged children's exposure to particulate matter (PM) revealed that diesel exhaust from school buses is a significant source of PM exposure for children who are riding a bus and in adjacent school buildings (Hochstetler et al. 2011). These led to a series of local recommendations as well as national programs to reduce school bus diesel exhaust and children's exposure to air pollution, including EPA's Clean School Bus Program (EPA 2017).

In some other cases, external measurement may be the only way to capture the exposure when a direct internal correlate of an exposure is not available, such as exposure to noise, heat, greenness, and other social and built environments (Peters et al. 2012). Increasing evidence suggests that these nonchemical environmental exposures play a significant role in human health and disease. How to assess and quantify these exposures in epidemiological studies to allow for more comprehensive understanding of a person's environmental exposure remains a significant challenge.

What Exposure Can We Measure from the External?

Most environmental exposures are originated from outside the body and many can be measured in a person's environment, including environmental chemicals, dietary exposure, consumer products, and social and built environment. There is a reality that the ability to measure *multiple* exposures in the external environment is currently limited and this is a significant challenge for integrating external exposure assessment into the exposome concept. Here, we briefly discuss a few categories of environmental factors, the characterization of which relies heavily on external measurement approaches, and why it is important to characterize them in environmental health research.

Air pollution has been associated with various adverse human health effects in numerous population studies (Beelen et al. 2014; Hoek et al. 2013; Shah et al. 2013). According to the World Health Organization (WHO), 92% of the world's population is living in places where air quality levels exceed WHO limits, making it the largest single environmental health risk (WHO 2016). Traditionally, personal exposure to air pollution is estimated according to data collected by stationed regulatory monitors and computational models based upon the regulatory monitor data. The last decade has seen a dramatic evolution of the capabilities of external exposure assessment spurred largely by the publication of the results of the Relationship of Indoor, Outdoor, and Personal Air (RIOPA) study and the funding of the NIH Genes, Environment, and Health Initiative to develop improved and cost-effective tools for personal air pollution monitoring (Weisel et al. 2005; Turpin et al. 2007; National Institutes of Health 2007). Today, with the emergence of personal and low-cost air monitors, more personal exposure data with high spatiotemporal resolution is available for more accurate assessment of exposure to air pollution and incorporation in environmental health studies.

Lifestyle and behavior factors, such as dietary intake, alcohol and drug use, and physical activity, are major contributors to complex human disease (Lopresti et al. 2013; Phillips et al. 2013; Weston et al. 2014). Approaches to assess these risk factors from the external include questionnaires, face-to-face interviews, and, more recently, devices to record physical activity and mobile technology-enabled methodologies for real-time self-report of personal behavior. The real-time data capture function of these newly developed technologies is highly valuable in validating and completing self-reported lifestyle and behavioral assessment. In addition, real-time data capture also provides contextual information that measurement of the internal biomarker cannot provide such as when, where, and how the behavior is triggered, which may offer opportunities for intervention (Grenard et al. 2013; Thomas et al. 2011).

There is growing evidence suggesting that neighborhood conditions, including green space, walkability, neighborhood safety, and accessibility to resources, can impact a person's health outcome (Chiu et al. 2016; Cohen-Cline et al. 2015; Dadvand et al. 2015). Although some effects can be explained by participants' behavioral change promoted by neighborhood conditions such as green space promoting physical activity, there are implications that the built and social environment may have broader health effects. Therefore, there is growing interest in characterizing neighborhood conditions in order to understand the relationship between the built and social environment and human health, and identifying specific characteristics that affect a person's well-being. A set of emerging tools and methodologies for characterizing the neighborhood environment have been reviewed by Schootman et al. including Google Street View, webcams, crowdsourcing, remote sensing, social media, and more (Schootman et al. 2016).

Above is just a short list of exemplary environmental exposures that are best to be characterized by external measurements. This chapter will provide an overview of the scientific advances, emerging technologies, and novel opportunities that can be used to unravel the exposome from the external, and discuss how to utilize multiscale data integration to assess the temporal and spatial dynamics of the exposome as well as novel opportunities to inform citizen science and disease prevention offered by the exposome.

Scientific Advances, Emerging Technologies, and Their Applications in Characterizing the External Environment

Traditionally, environmental epidemiology studies rely heavily on questionnaires and surveys for the lack of technologies capable of quantifying external exposures at the personal level. Sometimes, questionnaires are administered repeatedly to collect longitudinal exposure information. However, data scarcity and recall bias in self-reported data often make it difficult to get an accurate and comprehensive view of an individual's external environment. Recent advances in environmental sensing technologies and novel methodologies enable improved estimate of personal exposure, many of which also allow for real-time exposure assessment of an expanded range of external environmental stressors. When combined with measurements of related physiological responses and health outcomes, this provides valuable insight into the dynamic nature of the exposure–disease relationship and disease causality.

Remote Sensing

Remote sensing generally refers to the use of satellite- or aircraft-based sensor technologies to collect information about the surface of the earth. The new generation of pollution-monitoring satellites began about two decades ago. In 1999, the National Aeronautics and Space Administration (NASA) launched the remote sensing satellite Terra that equipped with three sensor instruments: the Moderate-Resolution Imaging Spectroradiometer (MODIS), the Multi-angle Imaging SpectroRadiometer (MISR), and the Measurements of Pollution in the Troposphere (MOPITT) instrument, the first two of which measure particular matter (PM) and the third measures carbon monoxide (CO) (Streets et al. 2013). In the following years, more satellites were launched with additional sensing instruments such as the Ozone Monitoring Instrument (OMI) and the Tropospheric Emission Spectrometer (TES) on Aura. Together, these sensing instruments provide improved understanding of earth's atmospheric chemistry and valuable insights into the spatial distribution and temporal trends of many air pollutants, including nitrogen dioxide (NO_2), sulfur dioxide (SO_2), CO, PM, and more (Boys et al. 2014; Krotkov et al. 2016). For example, using the column aerosol optical depth (AOD) retrieved from two satellite instruments, MISR and SeaWiFS, Boys et al. (2014) were able to generate a 15-year global time series (1998–2012) of ground-level PM2.5 concentrations and examine the annual trends. Satellite-derived remote sensing data has also been used to estimate individual-level environmental exposure in epidemiological studies. Chan et al. (2015) reported the use of satellite-based NO_2 data and satellite-based land use data in the Sister Study for estimation of long-term residential air pollution and its impacts on cardiovascular morbidity. VoPham et al. (2015) described the use of Landsat remote sensing images for estimation of agricultural pesticide exposure in California. The Landsat method incorporates four datasets to predict individual-level pesticides exposure in a geographic information system (GIS), including the Landsat satellite images, ground truth data, pesticide application data, and geocoded residential information.

The major advantage of remote sensing is the profound insight into the spatial and temporal trends across large geographical areas and time spans, which can be incorporated into large-scale epidemiological studies to reconstruct historical exposure and study the correlation between long-term exposure and chronic disease outcomes. The limitations, however, are the low resolutions in temporal spatial coverage and the inability to capture fine-scale variations and indoor concentrations, which are important for personal exposures. Limited by the retrieval return time of low earth orbit, where the satellites are located, as well as retrieval noise, useful data may only be obtained in a low temporal resolution and less likely to capture short-term peak exposures. In addition, satellite-retrieved exposure data is subject to a number of uncertainties, such as cloud contamination, temperature, and surface properties. Unmanned aerial vehicles (UAVs), or drones, can fill the niche at some level by overcoming some of the limitations in satellite data but cannot replace satellite sensing. A hybrid approach that incorporates other surface properties such as land use, road network, and ground ambient monitoring data has the potential to better predict personal exposure level (Beckerman et al. 2013). With recent advancement of satellite sensing technologies, higher spatial resolution data is now available and AOD-derived PM2.5 estimates at 1 km $\times$ 1 km resolution have been reported, showing great promise for assessment of personal-level exposure in epidemiological studies, especially in regions where ground mentoring is sparse (Just et al. 2015; Sorek-Hamer et al. 2016).

Personal Exposure Assessment

Environmental exposures vary across time and space and a person's exposure is highly correlated with the individual's time activity and behavioral patterns. Personal environmental measurement captures exposure levels in the immediate proximity of a person and enables more accurate exposure estimate. There is a wide range of sensing technologies and novel methodologies for assessment of an individual's personal environment.

Personal Sensors

Recent development in wearable sensors offers the opportunity to measure various environmental factors at the personal level including air pollution (PM, ozone, and toxic gases), UV, noise, temperature, and physical activity. Nieuwenhuijsen et al. (2014) searched the market and put together a personal sensor set, which included a dozen individual personal sensors. Collectively, this sensor set can continuously measure a person's exposure to UV, noise, black carbon, temperature, green space, and physical activity as well as location and physiological parameters such as blood pressure and heart rate over a 24-h period. Results showed not only considerable variability in exposure levels for all parameters measured during the time of the experiment, but also clear differences in variability and exposure levels when comparing daytime exposure with nighttime exposure, or comparing indoor exposure with outdoor exposure. This further proves that an individual's environmental exposure is highly variable across time and space and correlated with a person's behavioral and activity pattern. In addition, a person's body burden of air pollutants is highly correlated with physical activity and ventilation volume. Rodes et al. (2012) demonstrated the inclusion of accelerometers in personal samplers would enable a temporally synchronized estimate of ventilation rate and potential dose when making exposure measurements to air pollution.

In recent years, with increased awareness and adoption of the exposome concept, there is a growing need for personal sensors that are capable of measuring multiple environmental pollutants and a few passive sampling technologies showed capability of capturing multiple environmental exposures simultaneously. Particularly notable is a passive sampling technology using silicone wristband sampler that was developed by the research group at Oregon State University. Using the technology, a variety of environmental chemicals including polycyclic aromatic hydrocarbons (PAHs), consumer products, personal care products, pesticides, and phthalates can be measured through laboratory analysis after wearing the wristband for a period of time (Donald et al. 2016; O'Connell et al. 2014). The silicone band is also a good surrogate for determining the internal dose of exposure to certain organic compounds because of its similar absorption profiles as humans (O'Connell et al. 2015).

Despite the rapid development in personal sensing technologies, there are many hurdles to overcome for personal sensors to be applied in population studies. Battery life is often a challenge for many wearable sensors, which requires periodical charging by the user, and this may even be a major limiting factor for developing a more wearable device in some cases (miniaturized and weight-reduced). Data quality and reproducibility is also a concern, which requires devices to be validated both in the laboratory and in the field to ensure reliability before deployment in large population studies. The types of technologies and devices to be used for personal monitoring should be prioritized according to the purpose of the data collection and the data quality requirements. Practical considerations may also be factored in to improve compliance. In recent years, as many small sensors are commercially available at an affordable price, personal exposure monitoring is increasingly being used in citizen science and community research, the implication of which will be discussed later in the chapter.

Smartphone-Enabled Personal Sensing

A smartphone is a mobile phone with computing power and features such as a touch screen, Internet access, global positioning system (GPS), camera and more. In the United States, 77% of adults own a smartphone and they are nearly ubiquitous among younger adults (Pew Reserch Center 2017). As the use of smartphones and other mobile communication devices, such as tablets, smartwatches, and smart bands, becomes part of our daily routine, it has transformed nearly every aspect of our daily practice. Particularly notable is the recent advancement of mobile health (mHealth), the use of mobile phones and other wireless technologies in health care. Incorporation of wearable biometric sensors, lab-on-a-chip, and imaging technologies in mobile platforms enables acute disease diagnosis and chronic condition management to take place outside typical health care facilities (Steinhubl et al. 2015). A systematic literature review reveals a wide application mHealth in the research and management of a number of health conditions including diabetes, asthma, depression, hearing loss, and more (Martinez-Perez et al. 2013). In the last decade, mHealth has grown exponentially and remains to be a hot spot in health care innovation.

Smartphones and personal mobile communication devices are also being utilized as a platform to facilitate measurement of personal exposure to air pollution. Chen et al. developed a wearable sensor that uses a smartphone as the interface for the assessment of personal exposure to volatile organic compounds (VOCs) both in indoor and outdoor environments, including aromatic, alkyl, and chlorinated hydrocarbons (Chen et al. 2012). Through an application specifically developed for communication with the sensor, the smartphone was able to receive the exposure data, process the data, and display the results. In another study, a cell phone-based monitoring system was used to measure black carbon exposure, where a miniaturized aerosol filter sampler collects black carbon and the image of the filter is then captured, and transmitted by the cell phone for determination of black carbon level in real time (Ramanathan et al. 2011). Nanotechnology-based, compact, and low-cost sensor chips are also available for incorporation into a smartphone for sensitive detection of toxic gases in the environment (Hannon et al. 2016).

Dietary intake and physical activity are considered important contributing environmental factors to many human chronic diseases. Use of a mobile phone to assess dietary intake and physical activity has been widely implemented for more efficient measurement at the individual level (Sharp and Allman-Farinelli 2014). Daugherty et al. (2012) described a method using the mobile telephone food record (mpFR) to facilitate dietary assessment and tested the proficiency and accuracy of using the mpFR in adult and adolescent populations. The mpFR facilitates dietary assessment by capturing an image of food and beverages, along with a scale marker (for size reference) before and after consumption. Energy and nutrition intake can then be estimated by linking the information from image analysis and volume estimation to a nutrient database. Recently, a more advanced smartphone application has been developed to capture energy intake and total energy expenditure in addition to mobile phone-enabled image analysis (Svensson and Larsson 2015). Smartphones have also been widely used in physical activity studies because the built-in motion sensors in the phone can help recognize activity patterns given that people carry the phone around in their everyday lives. Today, a smartphone can be equipped with a number of sensors including an accelerometer, a gyroscope, a magnetometer, and more. Using the accelerometer alone or, in combination with other motion sensors, a smartphone can classify a variety of physical activities, including walking, running, climbing stairs, descending stairs, driving, cycling, and being inactive (Shoaib et al. 2014; Anjum and Ilyas 2013).

Smartphones and other mobile communication-enabled sensing technologies are increasingly being employed in environmental health research. Mobile applications are being developed to take advantage of the built-in sensors. In a recent study of student behavior change, smartphone-enabled sensing methods were used to assess more complicated activity and social behavior patterns, where the smartphone accelerometer sensor was used to track daily activity, and the microphone sensor was used to evaluate sociability, both of which provide inference regarding consequential student outcomes (Harari et al. 2017). In another study, the smartphone accelerometer was used to detect and compute heartbeat rate by positioning an iPhone6 on the thorax (Landreani et al. 2016).

Self-Reported Environmental Exposures

Although subject to recall bias and misclassification, self-reported data via environmental questionnaires and surveys provide retrospective information on personal environment characteristics (home, work, neighborhood, and social stress) that might be difficult to capture by other personal sensing technologies. In particular, large population studies rely heavily on questionnaires and surveys because they are inexpensive and can cover a broad set of environmental variables to provide the researcher some basic classification of the population being studied. The PhenX Toolkit led by RTI International and the National Human Genome Research Institute is a catalog of recommended, standard measures of phenotypes and environmental exposures for use in biomedical research (Hamilton et al. 2011). A large portion of the measurements regarding the external environment is questionnaire-based, either self-administered or administered by interviewers. The standardized and free available questionnaires on selected measures provide a start point for researchers to develop more sophisticated and subject-specific questionnaires and a possibility for data harmonization and cross-study comparisons.

Traditionally, self-reported data are collected through paper, phone, or face-to-face interview, which could be time-consuming and challenging in post-survey data processing. With the increased access to Internet and mobile communication technologies, web-based questionnaires and applications are more frequently deployed in collecting self-reported exposure information. Web-based or computer-aided methods can provide an interactive and user-friendly interface, where questionnaires can be tailored according to the user's response to allow personal exposure data to be collected in a more efficient way, and the response can be rapidly incorporated into datasets. Computerized adaptive tests (CATs) have been adapted by the Patient-Reported Outcomes Measurement Information System (PROMIS) to collect self-reported health information. Item Response Theory (IRT) is utilized to ensure test items administered are appropriate for the persons' trait level, thus better precision than traditional measures (Cella et al. 2007). In the last decade, PROMIS has been widely used to assess the physical, mental, and social health in adults and children with differencing conditions. Developing and incorporating environmental variables in PROMIS, especially variables of the built and natural environment, may provide valuable insight into how the environment influences these conditions and offer opportunities for health intervention (Heinemann et al. 2016; Ananthakrishnan et al. 2013).

Ecological momentary assessment (EMA) was originally developed to reduce recall bias in self-report-based clinical psychology assessment. EMA collects real-word data in real time in the context of a subject's microenvironment. The repeated sampling and data collection using an electronical diary provides a more complete picture of a subject's experience and behavioral variations over time (Shiffman et al. 2008). Dunton et al. (2012) described the use of EMA in characterizing children's physical and social environment and how that affects children's physical activity patterns. Jones et al. (2016) explored the feasibility of using EMA to study how behavioral patterns affect obesity in African American woman.

Novel Opportunities for Environmental Sensing

The advancement of technologies and social media also provides novel opportunities for unraveling the exposome from outside the body, and a few selected are described below regarding their utility in characterizing an individual's external environment.

Crowdsourcing and Sensor Networks

Crowdsourcing, the collection of information from a volunteered and usually undefined large group of people, is showing great promise in leveraging environmental data generated by the public. The concept of crowdsourcing is not new but it has been taken to a new level by the widespread use of Internet and smartphones, the readily available low-cost environmental sensors, and the built-in global positioning system (GPS) on smartphones. With voluntary GPS information from participating users, crowdsourced environmental data can be used to create interactive exposure maps to understand in real time both a person's microenvironment and the environment of a specific geographical area. For example, the AirCasting project led by HabitatMap consists of a wearable sensor, a smartphone equipped with an AirCasting Android app, and a website for exposure data mapping and sharing (AirCasting 2017). Real-world exposure data, such as fine particulate matter, CO, and CO_2, is sent to the website via the smartphone to produce exposure heat maps. With thousands of users reporting real-time exposure data collected from their personal environment, the exposure heat maps can provide high temporal and spatial resolution data regarding current exposure in specific geographical areas. Exposure mapping using crowdsourced environmental sensing data has been implemented to inform personal exposure and decision-making in several cases, including the SafeCast program for radiation measurement and communication after Fukushima, and the NoiseTube app that was used for mapping noise pollution in the city of Zagreb (Brown et al. 2016; Posloncec-Petric et al. 2016).

Crowdsourcing is clearly a valuable tool for engaging the public and a powerful way to increase real-word measurements of the environment. This is particularly important in regions where regulatory monitoring data is scarce or unavailable.

However, data quality is a serious concern over the subsequent use of crowdsourced environmental data. Validation and calibration against other type of measurements, such as ground monitoring, as well as appropriate quality control procedures are needed to ensure data accuracy.

Free Web Mapping Services

Using free web mapping services such as Google Earth, Google Street View, and Bing Maps, one can browse and search many streets and cities worldwide. The satellite and aerial images, bird's-eye views, and street side views of these online maps provide rich information regarding the characteristics of the built environment on a finer spatial scale. Free web mapping services have been explored as potential tools for characterizing the built environment and neighborhood conditions, such as land use, transport infrastructure, walkability, health facilities, and green space (Brookfield and Tilley 2016; Charreire et al. 2014; Kurka et al. 2016; Li et al. 2015). Comparisons between online map-based viral assessment and direct observation reveal that the agreement is generally higher for assessment of items that could be measured objectively (e.g., presence of infrastructure and equipment) and lower for subjectively assessed items (e.g., aesthetics, safety, street and housing conditions). The major advantage of using free online maps in assessing the built environment is that, comparing to field audits, it is free of charge, easy to access and use. The limitation, however, is the temporal and special coverage of the maps. Sometimes, web mapping services may not be available for the time period or location of research interest (Charreire et al. 2014).

Social Media

Today, an average person will spend nearly 2 h on social media every day (SocialMediaToday 2017). People share a variety of information on social media including pictures and comments related to personal daily activates. As the use of social media becomes part of the daily routine for many people, it potentially can be used as a data source for studying people's environmental and social exposure and their health impacts. Chen and Yang (2014) described the use of Twitter messages, or tweets, to study food environment exposure and dietary behavior. In their study, crowdsourced, food-themed tweets were geocoded and used as an indicator of healthful or unhealthful food choice. By mapping these geocoded tweets with the distribution of different types of food retailers (grocery stores and fast-food restaurants) in a geographical information system, they found the prevalence of grocery stores in a person's neighborhood may influence food choice significantly. In another study, geocoded food tweets were coordinately mapped with geospatial distribution of economically disadvantaged and low-access census track (food desert) to test the hypothesis that unhealthy foods are more common in disadvantaged areas (Widener and Li 2014). Geotagged data from Twitter and other social media outlets has also been used for studying many other aspects of the environment and human health including physical activity, neighborhood conditions, the outbreak of infectious disease (Nguyen et al. 2016; Ye et al. 2016).

The advantage of social media data is the real-time, penetrative, and dynamic nature of the data. However, social media data is volunteered, self-reported, and unverified. In addition, not all social media data is publicly available and only a small percent of the available data is geotagged. Therefore, data accuracy and representativeness need to be factored in when making inference from this type of analysis.

Multi-scale Data Integration in Assessing the Temporal and Spatial Dynamics of the External Exposure

Integration of data across temporal and spatial scales is the key to a comprehensive understanding of the exposome. Geographic information system (GIS) enables the construction of a more complete picture of the exposure by providing a framework for integration of geospatial information from disparate data sources. These include, but not limited to, remote sensing data on air pollution and earth surface, ground air quality data, geological survey data on water and soil and other census data, online maps, and environmental data generated by community and citizen science sensing networks. In epidemiological studies, GIS has been used to combine satellite-based remote sensing data with earth surface characteristics and other types of local observations for computational exposure modeling and estimation of personal exposure to air pollution, pesticides, and greenness (Chan et al. 2015; VoPham et al. 2015; Almanza et al. 2012). GIS is also utilized to assemble a range of neighborhood environmental characteristics, including built environment, food environment, activity space, and walkability, and to link neighborhood conditions to individual's health behaviors and outcomes (Chen and Yang 2014; Thornton et al. 2011). Furthermore, with increased access to global positioning system (GPS) tracking and other personal sensing devices, personal time, space, and activity patterns obtained by these devices can be incorporated into the GIS framework. Incorporation of such personal tracking information in exposure assessment enables better characterization of the temporal and spatial dynamics of the exposome and provides valuable insights into how an individual's activity pattern affects one's exposure (Chaix et al. 2013; Perchoux et al. 2013; Su et al. 2015).

However, personal tracking data is not always available, especially in retrospective studies. In environmental epidemiology, residential location is frequently used as a surrogate in combination with GIS tools for assigning relevant exposures and reconstructing exposure history. For example, Gallagher et al. (2010) developed a historical groundwater model in the context of GIS and incorporated participants' residential histories and drinking water sources to estimate personal exposure to contaminated water and the association with breast cancer. To make meaningful estimates of historical exposure, epidemiologists must account for participants' residential mobility, which is a great challenge due to the cost and recall error of self-reported questionnaire data. As the importance of the residential history in long-term health outcome studies has been increasingly recognized, researchers explored the possibility of generating residential histories using public records (Jacquez et al. 2011; Wheeler and Wang 2015). Comparisons with questionnaire-based residential information indicate that residential histories generated by the LexisNexis public record match fairly well with survey data and can potentially replace the relatively expensive survey collection in some studies.

Data integration and data analysis is a significant challenge in general in exposome research because humans are exposed to numerous environmental stressors through their lives that can potentially impact health and disease (Manrai et al. 2016). With various GIS tools, computational exposure models, and personal tracking technologies as well as increased access to residential histories, it is promising to address the temporal and spatial dynamics of the exposome through integration of diverse data streams across a wide range of environmental attributes and expanded time span.

Integrating External and Internal Environments in Studies of the Exposome

For the convenience of discussion, the exposome is divided into internal exposure and external exposure, the latter of which is the focus of this chapter. In reality, external environment impacts human health via a network of connected events through the exposure-to-outcome continuum, and linking the external exposure to what happens in the internal environment is required for clear understanding of the causal pathway. In addition, integrating external and internal environments informs exposure modeling and prediction as well as risk assessment and policy making for improved public health.

By targeting an environmental stressor both externally and internally, one can correlate an external exposure with its internal dose (or metabolites) and associated biological responses. Many epidemiology studies use surveys and questionnaires or direct measurement of home environment (e.g., phthalates and flame retardants in house dust) to get a broad idea about external exposures and then correlate that information with measurements in biological samples for causal inference and pinpointing the source (Cequier et al. 2015; Le Cann et al. 2011). The employment of personal sensors in studies of environmental health makes it possible for more accurate correlation of external exposure with internal environment (real-time or similar). For example, in a study of urban adolescent cohort, researchers investigated the protective effect of physical activity on black carbon (BC) exposure-related airway inflammation. Using a personal BC sampler and a wrist mounted accelerometer, they were able to record BC exposure, physical activity in real-time on a daily basis, link the external exposure with measurements of a biological marker of airway inflammation, and reveal that high BC exposure offsets the protective relationship between physical activity and airway inflammation, a relationship that otherwise would be difficult to discover without accurate assessment of both the external exposures and internal biomarker (Lovinsky-Desir et al. 2016). The Human Early-Life Exposome (HELIX) project is a large exposome initiative consisting of six participating cohorts and 32,000 mother–child pairs (Vrijheid et al. 2014). To gain insight into the connections between external environment and internal exposure, a nested design was adopted, where a subcohort of 1200 representative mother–child pairs were selected for full characterization of the exposome including exposure biomarkers, omics profiles, and social and behavioral characterization, and smaller panels within the sub-cohort were targeted for additional personal sensing and dietaries to facilitate the integration of external exposure with the internal environment. It would be interesting to see in the future how insight gained from the sub-cohort can be extrapolated to the parent cohorts.

Laboratory models have played a key role in understanding the pathogenesis of a variety of human diseases and are powerful tools to fill knowledge gaps in the causal pathway of external exposure to health outcomes. Laboratory discoveries of biomarkers of exposure and response, once validated, can become useful tools for internal measurement of exposure or early detection of adverse health effects. Toxicokinetic/dynamic models of a chemical exposure observed in animal models may inform the dose–response relationships of the exposure in humans. Last but not least, laboratory models offer incredible flexibilities in mechanistic understanding of diseases processes. Recent advancement of gene targeting technologies and increased availability of genetically diverse mouse models provide even more opportunities for investigating gene environment interactions and how that may contribute to human disease (Churchill et al. 2012; Tu et al. 2015; Welsh et al. 2012).

Humans are exposed to tens of thousands potentially toxic chemicals. Yet resources and technologies can be limited for exposure measurement from either the external or internal, or both. Large biomonitoring databases and high-throughput toxicity assay as well as computational models provide another resource for predicting the relationship between external exposure, internal environment, and health outcomes. The ExpoCast program led by EPA incorporates far-field production volume and consumer use information to predict human exposure potential to chemicals (Wambaugh et al. 2014). It provides a computational framework for evaluating whether a chemical is likely to be a target of further data collection. The ToxCast and Tox21 program, on the other hand, utilizes a variety of technologies, such as computational chemistry, high-throughput screening, and various toxicogenomic technologies, to predict potential for toxicity and provide insight on toxicity pathways, which can be leveraged in prediction of in vivo biological processes (Tice et al. 2013). A few additional data sources such as the Toxin-Toxin-Target Database (T3DB), the Exposome-Explorer, and the National Health and Nutrition Examination Survey (NHAES) biomonitoring data are also available, providing reference information regarding human internal exposure to a variety of chemicals and across large populations (Wishart et al. 2015; Neveu et al. 2017; Centers for Disease Control and Prevention 2015).

Applying the Exposome to Inform Citizen Science and Disease Prevention

Public participation and community engagement are critical components of environmental health research because of their important roles in health intervention and disease prevention. The exposome research benefits citizen science and community research in two ways. First, the holistic view of the environment in exposome research provides opportunities for discovery of environmental factors that affect people's health that are previously unknown to the public. Second, the emerging new tools and technologies developed for unraveling the exposome, such as personal sensors, smartphone-enabled sensing, and crowdsourcing, enable scientists to collaborate with volunteers (i.e., citizen scientists) and community members for better understanding of human–environment interaction, which eventually contributes to more effective intervention and disease prevention. A few examples mentioned in this chapter, such as the AirCasting project for air pollution mapping, and the SafeCast program for radiation measurement and communication after Fukushima, are great examples of data-intensive citizen science enabled by real-time personal sensing and crowdsourcing to help identify exposure hot spots and alert the community on exposures that can potentially affect their health adversely. Data-intensive citizen science shows great promise in environmental surveillance monitoring and health intervention and prevention. However, it is also facing challenges including data quality control, data interpretation, and risk communication as well as privacy issues regarding personal-level environmental and health data.

The exposome concept also changes the public's perspective of the environment and has been used by citizen scientists and community members to promote healthier city environments in urban design. It becomes more and more accepted that environmental factors impacting people's health and disease include not only air pollution, heat, and noise but also lifestyle factors and the physical environment that can affect people's behavioral and activity such as transportation in the city and greenness. Information collected by citizen scientists on these environmental factors helps bring together different parties and decision-makers in urban and transport planning to create healthier urban environment for improved public health (Nieuwenhuijsen 2016).

Conclusions

One major demand of incorporating the exposome concept into environmental health research is advancing our technological capabilities and infrastructure. Likewise, although many tools and technologies exist for external exposure assessment, how to apply these tools on the exposome scale for a comprehensive and rigorous characterization of the external environment remains a grand challenge. There is a great need for refining, validating, and scaling up existing tools before they can be reliably deployed. In particular, versatile devices with the capability of measuring multiple analytes are still lacking. As a result, an array of complementary technologies and approaches are required for comprehensive assessment of the external environment, which poses two significant challenges, the financial cost and burden to both investigators and participants, and data integration and analysis across the different data types and temporal-spatial scales. Therefore, developing novel and cost-effective tools and methods for detection and quantitation of multiple external environmental stressors, and integration of multiple data streams are both crucial for incorporation of the external environment into the exposome research. Last but not least, the exposome is a continuum that spans from the external source to the internal environment and through a person's life span. External exposure assessment must be integrated with the internal environment in the context of exposure history for a comprehensive understanding of the causal relationship between the environment and human health and disease.

References

AirCasting (2017) The AirCasting Platform. http://www.aircasting.org/. Accessed 28 Feb 2017

Almanza E, Jerrett M, Dunton G, Seto E, Pentz MA (2012) A study of community design, greenness, and physical activity in children using satellite, GPS and accelerometer data. Health Place 18(1):46–54. https://doi.org/10.1016/j.healthplace.2011.09.003

Ananthakrishnan AN, Long MD, Martin CF, Sandler RS, Kappelman MD (2013) Sleep disturbance and risk of active disease in patients with crohn's disease and ulcerative colitis. Clin Gastroenterol Hepatol 11(8):965–971. https://doi.org/10.1016/j.cgh.2013.01.021

Anjum A, Ilyas MU (2013) Activity recognition using smartphone sensors. In: IEEE consumer communications and networking conference, IEEE, New York

Athersuch T (2016) Metabolome analyses in exposome studies: profiling methods for a vast chemical space. Arch Biochem Biophys 589:177–186. https://doi.org/10.1016/j.abb.2015.10.007

Beckerman BS, Jerrett M, Serre M, Martin RV, Lee SJ, van Donkelaar A, Ross Z, Su J, Burnett RT (2013) A hybrid approach to estimating national scale spatiotemporal variability of PM2.5 in the contiguous United States. Environ Sci Technol 47(13):7233–7241. https://doi.org/10.1021/es400039u

Beelen R, Raaschou-Nielsen O, Stafoggia M, Andersen ZJ, Weinmayr G, Hoffmann B, Wolf K, Samoli E, Fischer P, Nieuwenhuijsen M, Vineis P, Xun WW, Katsouyanni K, Dimakopoulou K, Oudin A, Forsberg B, Modig L, Havulinna AS, Lanki T, Turunen A, Oftedal B, Nystad W, Nafstad P, De Faire U, Pedersen NL, Ostenson CG, Fratiglioni L, Penell J, Korek M, Pershagen G, Eriksen KT, Overvad K, Ellermann T, Eeftens M, Peeters PH, Meliefste K,

Wang M, Bueno-de-Mesquita B, Sugiri D, Kramer U, Heinrich J, de Hoogh K, Key T, Peters A, Hampel R, Concin H, Nagel G, Ineichen A, Schaffner E, Probst-Hensch N, Kunzli N, Schindler C, Schikowski T, Adam M, Phuleria H, Vilier A, Clavel-Chapelon F, Declercq C, Grioni S, Krogh V, Tsai MY, Ricceri F, Sacerdote C, Galassi C, Migliore E, Ranzi A, Cesaroni G, Badaloni C, Forastiere F, Tamayo I, Amiano P, Dorronsoro M, Katsoulis M, Trichopoulou A, Brunekreef B, Hoek G (2014) Effects of long-term exposure to air pollution on natural-cause mortality: an analysis of 22 European cohorts within the multicentre ESCAPE project. Lancet 383(9919):785–795. https://doi.org/10.1016/s0140-6736(13)62158-3

Boys BL, Martin RV, van Donkelaar A, MacDonell RJ, Hsu NC, Cooper MJ, Yantosca RM, Lu Z, Streets DG, Zhang Q, Wang SW (2014) Fifteen-year global time series of satellite-derived fine particulate matter. Environ Sci Technol 48(19):11109–11118. https://doi.org/10.1021/es502113p

Brookfield K, Tilley S (2016) Using virtual street audits to understand the walkability of older adults' route choices by gender and age. Int J Environ Res Public Health 13(11):12. https://doi.org/10.3390/ijerph13111061

Brown A, Franken P, Bonner S, Dolezal N, Moross J (2016) Safecast: successful citizen-science for radiation measurement and communication after Fukushima. J Radiol Prot 36(2):S82–S101. https://doi.org/10.1088/0952-4746/36/2/s82

Cella D, Yount S, Rothrock N, Gershon R, Cook K, Reeve B, Ader D, Fries JF, Bruce B, Rose M, Grp PC (2007) The patient-reported outcomes measurement information system (PROMIS). Progress of an NIH roadmap cooperative group during its first two years. Med Care 45(5):S3–S11. https://doi.org/10.1097/01.mlr.0000258615.42478.55

Centers for Disease Control and Prevention (2015) The national health and nutrition examination survey. https://www.cdc.gov/nchs/nhanes/nhanes_questionnaires.htm. Accessed 28 Feb 2017

Cequier E, Sakhi AK, Marce RM, Becher G, Thomsen C (2015) Human exposure pathways to organophosphate triesters—a biomonitoring study of mother-child pairs. Environ Int 75:159–165. https://doi.org/10.1016/j.envint.2014.11.009

Chaix B, Meline J, Duncan S, Merrien C, Karusisi N, Perchoux C, Lewin A, Labadi K, Kestens Y (2013) GPS tracking in neighborhood and health studies: a step forward for environmental exposure assessment, a step backward for causal inference? Health Place 21:46–51. https://doi.org/10.1016/j.healthplace.2013.01.003

Chan SH, Van Hee VC, Bergen S, Szpiro AA, DeRoo LA, London SJ, Marshall JD, Kaufman JD, Sandler DP (2015) Long-term air pollution exposure and blood pressure in the sister study. Environ Health Persp 123(10):951–958. https://doi.org/10.1289/ehp.1408125

Charreire H, Mackenbach JD, Ouasti M, Lakerveld J, Compernolle S, Ben-Rebah M, McKee M, Brug J, Rutter H, Oppert JM (2014) Using remote sensing to define environmental characteristics related to physical activity and dietary behaviours: a systematic review (the SPOTLIGHT project). Health Place 25:1–9. https://doi.org/10.1010/j.healthplace.2013.09.017

Chen C, Campbell KD, Negi I, Iglesias RA, Owens P, Tao NJ, Tsow F, Forzani ES (2012) A new sensor for the assessment of personal exposure to volatile organic compounds. Atmos Environ 54:679–687. https://doi.org/10.1016/j.atmosenv.2012.01.048

Chen X, Yang XN (2014) Does food environment influence food choices? A geographical analysis through "tweets". Appl Geogr 51:82–89. https://doi.org/10.1016/j.apgeog.2014.04.003

Chiu M, Rezai MR, Maclagan LC, Austin PC, Shah BR, Redelmeier DA, Tu JV (2016) Moving to a highly walkable neighborhood and incidence of hypertension: a propensity-score matched cohort study. Environ Health Perspect 124(6):754–760. https://doi.org/10.1289/ehp.1510425

Churchill GA, Gatti DM, Munger SC, Svenson KL (2012) The diversity outbred mouse population. Mamm Genome 23(9–10):713–718. https://doi.org/10.1007/s00335-012-9414-2

Cohen-Cline H, Turkheimer E, Duncan GE (2015) Access to green space, physical activity and mental health: a twin study. J Epidemiol Community Health 69(6):523–529. https://doi.org/10.1136/jech-2014-204667

Dadvand P, Nieuwenhuijsen MJ, Esnaola M, Forns J, Basagana X, Alvarez-Pedrerol M, Rivas I, Lopez-Vicente M, Pascual MD, Su J, Jerrett M, Querol X, Sunyer J (2015) Green spaces and

cognitive development in primary schoolchildren. Proc Natl Acad Sci U S A 112 (26):7937–7942. https://doi.org/10.1073/pnas.1503402112
Daugherty BL, Schap TE, Ettienne-Gittens R, Zhu FQM, Bosch M, Delp EJ, Ebert DS, Kerr DA, Boushey CJ (2012) Novel technologies for assessing dietary intake: evaluating the usability of a mobile telephone food record among adults and adolescents. J Med Internet Res 14(2):12. https://doi.org/10.2196/jmir.1967
Dennis KK, Marder E, Balshaw DM, Cui Y, Lynes MA, Patti GJ, Rappaport SM, Shaughnessy DT, Vrijheid M, Barr DB (2016) Biomonitoring in the era of the exposome. Environ Health Perspect 125(4):502–510. https://doi.org/10.1289/EHP474
Donald CE, Scott RP, Blaustein KL, Halbleib ML, Sarr M, Jepson PC, Anderson KA (2016) Silicone wristbands detect individuals' pesticide exposures in West Africa. R Soc Open Sci 3 (8):160433. https://doi.org/10.1098/rsos.160433
Dunton GF, Kawabata K, Intille S, Wolch J, Pentz MA (2012) Assessing the social and physical contexts of children's leisure-time physical activity: an ecological momentary assessment study. Am J Health Promot 26(3):135–142. https://doi.org/10.4278/ajhp.100211-QUAN-43
EPA (2017) Clean school bus. https://www.epa.gov/cleandiesel/clean-school-bus. Accessed 27 Feb 2017
Gallagher LG, Webster TF, Aschengrau A, Vieira VM (2010) Using residential history and groundwater modeling to examine drinking water exposure and breast cancer. Environ Health Persp 118(6):749–755. https://doi.org/10.1289/ehp.0901547
Grenard JL, Stacy AW, Shiffman S, Baraldi AN, MacKinnon DP, Lockhart G, Kishu-Sakarya Y, Boyle S, Beleva Y, Koprowski C, Ames SL, Reynolds KD (2013) Sweetened drink and snacking cues in adolescents. A study using ecological momentary assessment. Appetite 67:61–73. https://doi.org/10.1016/j.appet.2013.03.016
Hamilton CM, Strader LC, Pratt JG, Maiese D, Hendershot T, Kwok RK, Hammond JA, Huggins W, Jackman D, Pan HQ, Nettles DS, Beaty TH, Farrer LA, Kraft P, Marazita ML, Ordovas JM, Pato CN, Spitz MR, Wagener D, Williams M, Junkins HA, Harlan WR, Ramos EM, Haines J (2011) The PhenX Toolkit: get the most from your measures. Am J Epidemiol 174 (3):253–260. https://doi.org/10.1093/aje/kwr193
Hannon A, Lu YJ, Li J, Meyyappan M (2016) A sensor array for the detection and discrimination of methane and other environmental pollutant gases. Sensors 16(8):11. https://doi.org/10.3390/s16081163
Harari GM, Gosling SD, Wang R, Chen FL, Chen ZY, Campbell AT (2017) Patterns of behavior change in students over an academic term: a preliminary study of activity and sociability behaviors using smartphone sensing methods. Comput Hum Behav 67:129–138. https://doi.org/10.1016/j.chb.2016.10.027
Heinemann AW, Lai JS, Wong A, Dashner J, Magasi S, Hahn EA, Carlozzi NE, Tulsky DS, Jerousek S, Semik P, Miskovic A, Gray DB (2016) Using the ICF's environmental factors framework to develop an item bank measuring built and natural environmental features affecting persons with disabilities. Qual Life Res 25(11):2775–2786. https://doi.org/10.1007/s11136-016-1314-6
Hochstetler HA, Yermakov M, Reponen T, Ryan PH, Grinshpun SA (2011) Aerosol particles generated by diesel-powered school buses at urban schools as a source of children's exposure. Atmos Environ 45(7):1444–1453. https://doi.org/10.1016/j.atmosenv.2010.12.018
Hoek G, Krishnan RM, Beelen R, Peters A, Ostro B, Brunekreef B, Kaufman JD (2013) Long-term air pollution exposure and cardio- respiratory mortality: a review. Environ Health 12(1):43
Jacquez GM, Slotnick MJ, Meliker JR, AvRuskin G, Copeland G, Nriagu J (2011) Accuracy of commercially available residential histories for epidemiologic studies. Am J Epidemiol 173 (2):236–243. https://doi.org/10.1093/aje/kwq350
Jones KK, Zenk SN, McDonald A, Corte C (2016) Experiences of African-American women with smartphone-based ecological momentary assessment. Public Health Nurs 33(4):371–380. https://doi.org/10.1111/phn.12239

Just AC, Wright RO, Schwartz J, Coull BA, Baccarelli AA, Tellez-Rojo MM, Moody E, Wang YJ, Lyapustin A, Kloog I (2015) Using high-resolution satellite aerosol optical depth to estimate daily PM2.5 geographical distribution in Mexico City. Environ Sci Technol 49(14):8576–8584. https://doi.org/10.1021/acs.est.5b00859

Krotkov NA, McLinden CA, Li C, Lamsal LN, Celarier EA, Marchenko SV, Swartz WH, Bucsela EJ, Joiner J, Duncan BN, Boersma KF, Veefkind JP, Levelt PF, Fioletov VE, Dickerson RR, He H, Lu ZF, Streets DG (2016) Aura OMI observations of regional SO2 and NO2 pollution changes from 2005 to 2015. Atmos Chem Phys 16(7):4605–4629. https://doi.org/10.5194/acp-16-4605-2016

Kurka JM, Adams MA, Geremia C, Zhu WF, Cain KL, Conway TL, Sallis JF (2016) Comparison of field and online observations for measuring land uses using the microscale audit of pedestrian streetscapes (MAPS). J Transp Health 3(3):278–286. https://doi.org/10.1016/j.jth.2016.05.001

Landreani F, Martin-Yebra A, Casellato C, Frigo C, Pavan E, Migeotte PF, Caiani EG (2016) Beat-to-beat heart rate detection by smartphone accelerometers. Eur Heart J 37:859–860

Le Cann P, Bonvallot N, Glorennec P, Deguen S, Goeury C, Le Bot B (2011) Indoor environment and children's health: recent developments in chemical, biological, physical and social aspects. Int J Hyg Environ Health 215(1):1–18. https://doi.org/10.1016/j.ijheh.2011.07.008

Li XJ, Zhang CR, Li WD, Ricard R, Meng QY, Zhang WX (2015) Assessing street-level urban greenery using Google Street View and a modified green view index. Urban For Urban Green 14 (3):675–685. https://doi.org/10.1016/j.ufug.2015.06.006

Lopresti AL, Hood SD, Drummond PD (2013) A review of lifestyle factors that contribute to important pathways associated with major depression: diet, sleep and exercise. J Affect Disord 148(1):12–27. https://doi.org/10.1016/j.jad.2013.01.014

Lovinsky-Desir S, Jung KH, Rundle AG, Hoepner LA, Bautista JB, Perera FP, Chillrud SN, Perzanowski MS, Miller RL (2016) Physical activity, black carbon exposure and airway inflammation in an urban adolescent cohort. Environ Res 151:756–762. https://doi.org/10.1016/j.envres.2016.09.005

Manrai AK, Cui Y, Bushel PR, Hall M, Karakitsios S, Mattingly CJ, Ritchie M, Schmitt C, Sarigiannis DA, Thomas DC, Wishart D, Balshaw DM, Patel CJ (2016) Informatics and data analytics to support exposome-based discovery for public health. Annu Rev Public Health. https://doi.org/10.1146/annurev-publhealth-082516-012737

Martinez-Perez B, de la Torre-Diez I, Lopez-Coronado M (2013) Mobile health applications for the most prevalent conditions by the world health organization: review and analysis. J Med Internet Res 15(6):19. https://doi.org/10.2196/jmir.2600

National Institutes of Health (2007) Genes, environment and health initiative invests in genetic studies, environmental monitoring technologies. https://www.nih.gov/news-events/news-releases/genes-environment-health-initiative-invests-genetic-studies-environmental-monitoring-technologies. Accessed 14 Mar 2017

Neveu V, Moussy A, Rouaix H, Wedekind R, Pon A, Knox C, Wishart DS, Scalbert A (2017) Exposome-explorer: a manually-curated database on biomarkers of exposure to dietary and environmental factors. Nucleic Acids Res 45(D1):D979–D984. https://doi.org/10.1093/nar/gkw980

Nguyen QC, Kath S, Meng HW, Li DP, Smith KR, VanDerslice JA, Wen M, Li FF (2016) Leveraging geotagged Twitter data to examine neighborhood happiness, diet, and physical activity. Appl Geogr 73:77–88. https://doi.org/10.1016/j.apgeog.2016.06.003

Nieuwenhuijsen MJ (2016) Urban and transport planning, environmental exposures and health-new concepts, methods and tools to improve health in cities. Environ Health-Glob 15:11. https://doi.org/10.1186/s12940-016-0108-1

Nieuwenhuijsen MJ, Donaire-Gonzalez D, Foraster M, Martinez D, Cisneros A (2014) Using personal sensors to assess the exposome and acute health effects. Int J Environ Res Public Health 11(8):7805–7819. https://doi.org/10.3390/ijerph110807805

O'Connell SG, Kind LD, Anderson KA (2014) Silicone wristbands as personal passive samplers. Environ Sci Technol 48(6):3327–3335. https://doi.org/10.1021/es405022f

O'Connell SG, Kerkvliet NI, Carozza S, Rohlman D, Pennington J, Anderson KA (2015) In vivo contaminant partitioning to silicone implants: implications for use in biomonitoring and body burden. Environ Int 85:182–188. https://doi.org/10.1016/j.envint.2015.09.016

Perchoux C, Chaix B, Cummins S, Kestens Y (2013) Conceptualization and measurement of environmental exposure in epidemiology: accounting for activity space related to daily mobility. Health Place 21:86–93. https://doi.org/10.1016/j.healthplace.2013.01.005

Peters A, Hoek G, Katsouyanni K (2012) Understanding the link between environmental exposures and health: does the exposome promise too much? J Epidemiol Community Health 66 (2):103–105. https://doi.org/10.1136/jech-2011-200643

Pew Reserch Center (2017) Record shares of Americans now own smartphones, have home broadband. http://www.pewresearch.org/fact-tank/2017/01/12/evolution-of-technology/. Accessed 28 Feb 2017

Phillips CM, Dillon C, Harrington JM, McCarthy VJC, Kearney PM, Fitzgerald AP, Perry IJ (2013) Defining metabolically healthy obesity: role of dietary and lifestyle factors. PLoS One 8(10):13. https://doi.org/10.1371/journal.pone.0076188

Posloncec-Petric V, Vukovic V, Franges S, Bacic Z (2016) Voluntary noise mapping for smart city. In: Zlatanova S, Laurini R, Baucic M, Rumor M, Ellul C, Coors V (eds) First international conference on smart data and smart cities, 30th Udms, vol 4–4. International Archives of the Photogrammetry Remote Sensing and Spatial Information Sciences, vol W1. Copernicus Gesellschaft Mbh, Gottingen, pp 131–137. https://doi.org/10.5194/isprs-annals-IV-4-W1-131-2016

Ramanathan N, Lukac M, Ahmed T, Kar A, Praveen PS, Honles T, Leong I, Rehman IH, Schauer JJ, Ramanathan V (2011) A cellphone based system for large-scale monitoring of black carbon. Atmos Environ 45(26):4481–4487. https://doi.org/10.1016/j.atmosenv.2011.05.030

Rappaport SM (2012) Biomarkers intersect with the exposome. Biomarkers 17(6):483–489. https://doi.org/10.3109/1354750x.2012.691553

Rodes CE, Chillrud SN, Haskell WL, Intille SS, Albinali F, Rosenberger ME (2012) Predicting adult pulmonary ventilation volume and wearing compliance by on-board accelerometry during personal level exposure assessments. Atmos Environ 57:126–137. https://doi.org/10.1016/j.atmosenv.2012.03.057

Schootman M, Nelson EJ, Werner K, Shacham E, Elliott M, Ratnapradipa K, Lian M, McVay A (2016) Emerging technologies to measure neighborhood conditions in public health: implications for interventions and next steps. Int J Health Geogr 15(1):20. https://doi.org/10.1186/s12942-016-0050-z

Shah ASV, Langrish JP, Nair H, McAllister DA, Hunter AL, Donaldson K, Newby DE, Mills NL (2013) Global association of air pollution and heart failure: a systematic review and meta-analysis. Lancet 382(9897):1039–1048. https://doi.org/10.1016/S0140-6736(13)60898-3

Sharp DB, Allman-Farinelli M (2014) Feasibility and validity of mobile phones to assess dietary intake. Nutrition 30(11–12):1257–1266. https://doi.org/10.1016/j.nut.2014.02.020

Shiffman S, Stone AA, Hufford MR (2008) Ecological momentary assessment. Annu Rev Clin Psychol 4:1–32. https://doi.org/10.1146/annurev.clinpsy.3.022806.091415

Shoaib M, Bosch S, Incel OD, Scholten H, Havinga PJM (2014) Fusion of smartphone motion sensors for physical activity recognition. Sensors 14(6):10146–10176. https://doi.org/10.3390/s140610146

SocialMediaToday (2017) How much time do people spend on social media. http://www.socialmediatoday.com/marketing/how-much-time-do-people-spend-social-media-infographic. Accessed 28 Feb 2017

Sorek-Hamer M, Just AC, Kloog I (2016) Satellite remote sensing in epidemiological studies. Curr Opin Pediatr 28(2):228–234. https://doi.org/10.1097/mop.0000000000000326

Steinhubl SR, Muse ED, Topol EJ (2015) The emerging field of mobile health. Sci Transl Med 7 (283):6. https://doi.org/10.1126/scitranslmed.aaa3487

Streets DG, Canty T, Carmichael GR, de Foy B, Dickerson RR, Duncan BN, Edwards DP, Haynes JA, Henze DK, Houyoux MR, Jacobi DJ, Krotkov NA, Lamsal LN, Liu Y, Lu ZF, Martini RV,

Pfister GG, Pinder RW, Salawitch RJ, Wechti KJ (2013) Emissions estimation from satellite retrievals: a review of current capability. Atmos Environ 77:1011–1042. https://doi.org/10.1016/j.atmosenv.2013.05.051
Su JG, Jerrett M, Meng YY, Pickett M, Ritz B (2015) Integrating smart-phone based momentary location tracking with fixed site air quality monitoring for personal exposure assessment. Sci Total Environ 506:518–526. https://doi.org/10.1016/j.scitotenv.2014.11.022
Svensson A, Larsson C (2015) A mobile phone app for dietary intake assessment in adolescents: an evaluation study. JMIR mHealth uHealth 3(4):15–35. https://doi.org/10.2196/mhealth.4804
Thomas JG, Doshi S, Crosby RD, Lowe MR (2011) Ecological momentary assessment of obesogenic eating behavior: combining person-specific and environmental predictors. Obesity 19(8):1574–1579. https://doi.org/10.1038/oby.2010.335
Thornton LE, Pearce JR, Kavanagh AM (2011) Using Geographic Information Systems (GIS) to assess the role of the built environment in influencing obesity: a glossary. Int J Behav Nutr Phys Act 8:9. https://doi.org/10.1186/1479-5868-8-71
Tice RR, Austin CP, Kavlock RJ, Bucher JR (2013) Improving the human hazard characterization of chemicals: a Tox21 update. Environ Health Persp 121(7):756–765. https://doi.org/10.1289/ehp.1205784
Tu ZC, Yang WL, Yan S, Guo XY, Li XJ (2015) CRISPR/Cas9: a powerful genetic engineering tool for establishing large animal models of neurodegenerative diseases. Mol Neurodegener 10:8. https://doi.org/10.1186/s13024-015-0031-x
Turpin BJ, Weisel CP, Morandi M, Colome S, Stock T, Eisenreich S, Buckley B (2007) Relationships of Indoor, Outdoor, and Personal Air (RIOPA): part II. Analyses of concentrations of particulate matter species. Res Rep Health Eff Inst 130(Pt 2):1–77. discussion 79–92
VoPham T, Wilson JP, Ruddell D, Rashed T, Brooks MM, Yuan JM, Talbott EO, Chang CCH, Weissfeld JL (2015) Linking pesticides and human health: a geographic information system (GIS) and Landsat remote sensing method to estimate agricultural pesticide exposure. Appl Geogr 62:171–181. https://doi.org/10.1016/j.apgeog.2015.04.009
Vrijheid M, Slama R, Robinson O, Chatzi L, Coen M, van den Hazel P, Thomsen C, Wright J, Athersuch TJ, Avellana N, Basagana X, Brochot C, Bucchini L, Bustamante M, Carracedo A, Casas M, Estivill X, Fairley L, van Gent D, Gonzalez JR, Granum B, Grazuleviciene R, Gutzkow KB, Julvez J, Keun HC, Kogevinas M, McEachan RRC, Meltzer HM, Sabido E, Schwarze PE, Siroux V, Sunyer J, Want EJ, Zeman F, Nieuwenhuijsen MJ (2014) The human early-life exposome (HELIX): project rationale and design. Environ Health Persp 122 (6):535–544. https://doi.org/10.1289/ehp.1307204
Wambaugh JF, Wang A, Dionisio KL, Frame A, Egeghy P, Judson R, Setzer RW (2014) High throughput heuristics for prioritizing human exposure to environmental chemicals. Environ Sci Technol 48(21):12760–12767. https://doi.org/10.1021/es503583j
Weisel CP, Zhang J, Turpin BJ, Morandi MT, Colome S, Stock TH, Spektor DM, Korn L, Winer AM, Kwon J, Meng QY, Zhang L, Harrington R, Liu W, Reff A, Lee JH, Alimokhtari S, Mohan K, Shendell D, Jones J, Farrar L, Maberti S, Fan T (2005) Relationships of Indoor, Outdoor, and Personal Air (RIOPA). Part I. Collection methods and descriptive analyses. Res Rep Health Eff Inst 130(Pt 1):1–107. discussion 109–127
Welsh CE, Miller DR, Manly KF, Wang J, McMillan L, Morahan G, Mott R, Iraqi FA, Threadgill DW, de Villena FPM (2012) Status and access to the collaborative cross population. Mamm Genome 23(9–10):706–712. https://doi.org/10.1007/s00335-012-9410-6
Weston KS, Wisloff U, Coombes JS (2014) High-intensity interval training in patients with lifestyle-induced cardiometabolic disease: a systematic review and meta-analysis. Br J Sports Med 48(16):1227–U1252. https://doi.org/10.1136/bjsports-2013-092576
Wheeler DC, Wang AB (2015) Assessment of residential history generation using a public-record database. Int J Environ Res Public Health 12(9):11670–11682. https://doi.org/10.3390/ijerph120911670

WHO (2016) Ambient air pollution: a global assessment of exposure and burden of disease. http://apps.who.int/iris/bitstream/10665/250141/1/9789241511353-eng.pdf?ua=1. Accessed 27 Feb 2017

Widener MJ, Li WW (2014) Using geolocated Twitter data to monitor the prevalence of healthy and unhealthy food references across the US. Appl Geogr 54:189–197. https://doi.org/10.1016/j.apgeog.2014.07.017

Wild CP (2005) Complementing the genome with an "exposome": the outstanding challenge of environmental exposure measurement in molecular epidemiology. Cancer Epidemiol Biomark Prev 14(8):1847–1850. https://doi.org/10.1158/1055-9965.epi-05-0456

Wild CP (2012) The exposome: from concept to utility. Int J Epidemiol 41(1):24–32. https://doi.org/10.1093/ije/dyr236

Wishart D, Arndt D, Pon A, Sajed T, Guo AC, Djoumbou Y, Knox C, Wilson M, Liang Y, Grant J, Liu Y, Goldansaz SA, Rappaport SM (2015) T3DB: the toxic exposome database. Nucleic Acids Res 43(Database issue):D928–D934. https://doi.org/10.1093/nar/gku1004

Ye XY, Li SW, Yang XN, Qin CL (2016) Use of social media for the detection and analysis of infectious diseases in China. ISPRS Int Geo-Inf 5(9):17. https://doi.org/10.3390/ijgi5090156

Part IV
Data Analysis for the Exposome

Chapter 11
Statistical Models to Explore the Exposome: From OMICs Profiling to 'Mechanome' Characterization

Marc Chadeau-Hyam and Roel Vermeulen

Abstract Over the past decade, high-resolution molecular profiles using OMICS technologies have accumulated and have given rise to an unprecedented source of information to explore the effective biological effects of external stressors and to detect drivers of subsequent disease risk. Although the volume, dimensionality, and complexity of OMICs data are constantly increasing, several methods enabling their analysis are now available. The exploration of these data relies on statistical approaches including univariate models coupled with multiple testing correction, dimensionality reduction techniques, and variable selection approaches. While these methods are established, their application in an exposome context is raising specific methodological challenges. In addition, the isolated exploration of an OMIC profile offers the possibility to capture stressor-induced biological/biochemical alterations, potentially impacting individual risk profiles, but this may only yield a fractional picture of the complex molecular events involved, therefore limiting our understanding of the effective mechanisms mediating the effect of the exposome. Despite efficient developments over systems biological approaches, such integrations remain at best data-specific, usually disease-specific, and more systematically restricted to the exploration of (few) predefined hypotheses. The challenging task of exploring the 'mechanome' as defined by the ensemble of stressor-induced molecular mechanisms occurring throughout the life course and determining the individual's risk of developing adverse conditions can be decomposed in three interdependent streams

M. Chadeau-Hyam (✉)
MRC/PHE Centre for Environment and Health, Department of Epidemiology and Biostatistics, School of Public Health, Imperial College London, London, UK

Institute for Risk Assessment Sciences (IRAS), Utrecht University, Utrecht, The Netherlands
e-mail: m.chadeau@imperial.ac.uk

R. Vermeulen
MRC/PHE Centre for Environment and Health, Department of Epidemiology and Biostatistics, School of Public Health, Imperial College London, London, UK

Institute for Risk Assessment Sciences (IRAS), Utrecht University, Utrecht, The Netherlands

Department of Molecular Epidemiology, Julius Center, University Medical Center Utrecht, Utrecht, The Netherlands

S. Dagnino, A. Macherone (eds.), *Unraveling the Exposome*,
https://doi.org/10.1007/978-3-319-89321-1_11

focusing on (1) OMICs profiling, (2) OMICs data integration, and (3) the exploration of molecular mechanisms involved in the exposure effect mediation towards (chronic) disease development.

Keywords Statistical models · Omics · Mechanome · Bioinformatics

Exposome and Statistical Challenges

The original definition of the exposome (Wild 2005) is broad and comprises all external stressors, as defined by any nongenetic factor affecting, possibly at different life stages, the individual risk of chronic diseases. Following-up on the exposome concept, Rappaport and Smith (2010) complemented this definition by assuming that any effective component of the exposome should be detectable in the internal environment, hence defining the internal exposome. The internal exposome comprises biological responses to the external environment and interactions with exogenous compounds as well as exogenous compounds themselves entering the internal environment.

With the rapid advances in molecular biology, cost-effective technologies have emerged and enable the generation of high-resolution and high-quality individual molecular profiles in large-scale studies. Resulting OMICs profiles can be defined as high-throughput biochemical measurements of the abundance and/or structural features of molecules involved in the main biological processes such as metabolism and its regulation. As depicted in Fig. 11.1, OMICs data are supported by a large range of molecules including DNA, RNA, proteins, as well as (potentially inorganic) small molecules. By nature, OMICs data are heterogeneous in terms of their

1. *Dimension*: From tens to millions of features being measured
2. *Nature*: OMICs data are either binary, categorical, counts, or continuous.

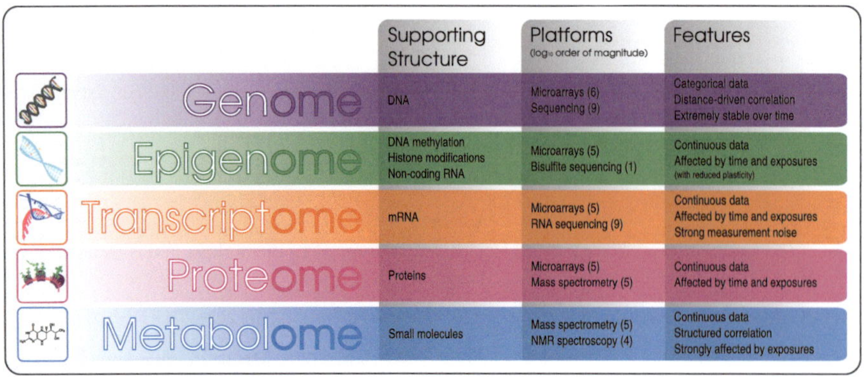

Fig. 11.1 Overview of the main types of OMICs data available (from Chadeau-Hyam et al. 2013)

3. *Complexity:* Both the intensity and the complexity of the correlation patterns between features assayed within single OMICs can be simply distance-driven, or involve distant and multivariate correlations.
4. *Stability/volatility*: OMICs data do not respond with the same intensity and dynamics to external stresses.

These sources of heterogeneity define specific statistical challenges to enable the full exploration of OMICs data. However, it also confers OMICs data a large level of biological complementarity, which, coupled with their high dimensionality, provides an agnostic view of the cellular activity and its regulation at different molecular levels, and therefore has the potential to identify effective and functional alterations induced by external stressors.

The characterization of the internal exposome then first relies, through OMICs profiling, on the identification of biological/physiological signals that are associated to the development of adverse health outcomes and, second, on the identification of potential (sets of) exposures driving these alterations. Hence, to fully characterize the internal exposome it needs to be complemented by the external exposome, which includes the full set of external exposures and experiences potentially triggering the biology and health-relevant factors.

The analysis of OMICs profiles in relation to exposome factors imposes specific methodological constraints first relating to the fact that (sets of) exposures are expected to have subtle and complex effects, hence challenging the statistical power of the screening models. To improve statistical power, complex study designs featuring multiple observations per participants such as experimental crossover designs are warranted (Vineis et al. 2016). In addition, exposure mixtures are supposed to play a more important role than each of the exposures taken separately (Carlin et al. 2013; Dominici et al. 2010; Rider et al. 2013), which calls for these models to be able to handle multivariate predictors, to account for the complex biological responses to sets of exposures, multivariate outcomes, and possibly include interactions between exposures.

As a result, a full exposome dataset is complex and, for a given health outcome and a given age range, it comprises three compartments: one health outcome, one set of exposures, and one set of OMICs profiles (Fig. 11.2).

The full exploration of an exposome dataset involves three main sets of analyses. First, Exposures–health outcome analyses aiming at identifying (sets of) exposures that are involved in the development of an adverse condition. Similarly, OMICs–health outcome profiling is seeking to identify molecular alterations that are related to the development of the condition of interest. Resulting associations comprise putative early disease manifestations, effective physiological changes affecting the risk of the outcome, or exposure-related marks that are (directly or within a pathway) affecting the health outcome of interest. Finally, OMICs–exposure profiling aims at characterizing the internal exposome as defined by the physiological response to sets of external stressors.

To fully exploit the entire dataset, several levels of integration are warranted. First, when multiple OMICs profiles are available in the same individuals, ‘cross-

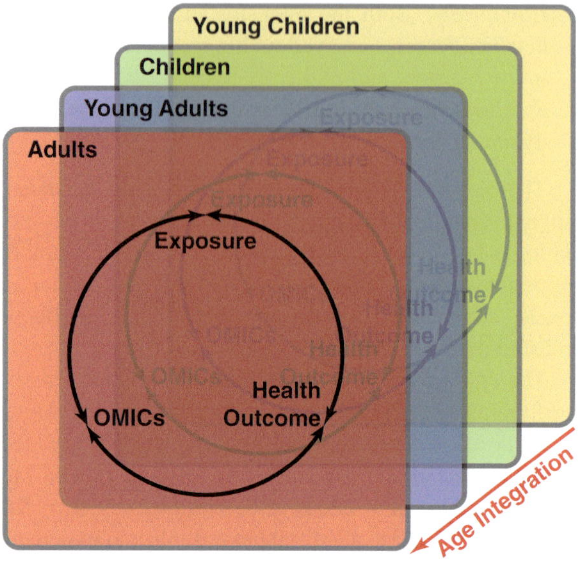

Fig. 11.2 Schematic representation of an exposome dataset for a given health outcome

OMICs' analyses investigate how disease or exposure-related signals found at one molecular level correlate to those found at another level. As such, these multi-OMIC analyses have the potential to disentangle redundant and complementary molecular signals, and have the potential to inform the molecular cascades triggered by exposure(s) and/or involved in the development of an adverse condition. A second level of integration involves the investigation of the common information between markers of exposures and markers of the health outcome. This idea has been formalized as the "meet-in-the-middle" concept (Assi et al. 2015; Vineis and Perera 2007; Chadeau-Hyam et al. 2011) and may help identifying intermediate biomarkers that are on the molecular pathway linking exposure to health outcome.

The complexity of the exposome characterization is further challenged by its dynamic nature. First, exposure levels naturally vary in time and at different time resolution scales: for example, chronically, within a day or following environmental changes, or across historical periods. Furthermore, the effect of exposures may respond to different dynamics (e.g., acute or chronic effect), and there are possible age-related effect modifications and susceptibility functions that may regulate the effect of exposures during the life course, defining critical life stages where exposures have a greater effect.

To account for the dynamics of the exposome, study designs must be adapted. This includes the implementation of experimental designs in which the participants are exposed to different levels of exposure, to enable the exploration of acute effects of exposures (Font-Ribera et al. 2010; McCreanor et al. 2007). Personal exposure monitoring campaigns capture effect of exposures at the scale of days and weeks, and long-term effect of exposure relies on exposure modeling applied to cohort studies.

Age-related susceptibility functions and effect modification could be formally modeled, but in the absence of either a single study covering all ages or studies with multiple measurements in the same individuals at different life stages, it could be restricted to investigating OMICs-Exposure-Health outcomes relationship in each age class separately (see Fig. 11.2) and subsequently seeking for signals that are common or specific to certain age classes.

Altogether, the analysis of a full exposome dataset can be decomposed in three main analytical streams which will be detailed in the reminder of this chapter: OMICs and Exposure profiling techniques, methods to improve results interpretation notably through the integration of prioritized OMICs signals, and methodological developments to ensure higher dimensional OMICs integration. In a concluding section of this chapter, we will define the notion of the "mechanome" and will present possible analytical framework enabling its exploration.

Main Approaches for OMIC-Exposure Profiling

Over the past decade, technological developments in molecular biology have given rise to a large amount of complex OMICs datasets enabling in-depth investigation of both physiological responses to external exposures (e.g., biomarkers of exposure and early effect), and of internal signatures of health outcomes (e.g., biomarkers of disease risk and onset). Concurrently to the emergence of this high-resolution data, strong methodological efforts have been carried out to enable their exploration. Resulting methods are now established and have been reviewed (Balding 2006; Chadeau-Hyam et al. 2013; Agier et al. 2016).

As mentioned above, one key feature to be accounted for while analyzing OMICs and exposome data is the correlation structure existing across the variables. For OMICs data, both the strength and the complexity of correlation structure are heterogeneous across different types of OMICs data. For instance, correlation between genetic markers is mainly distance-driven resulting in nearby genetic variants being strongly correlated. For other OMICs data such as NMR-based metabolomics data, the correlation patterns respond to more complex patterns including (1) a local component resulting in two nearby features relating to the same compound, (2) a nonlocal component resulting from the fact that a single compound could be reflected at different spectral regions, and (3) a functional component reflecting that an observed biological phenomenon is unlikely to be driven by a single compound. Similarly, external exposome data comprise large ranges of data sets including for instance biomarkers, exposure measurements, and behavioral factors. As a result, the correlation patterns within and across families of exposures are complex (Fig. 11.3).

Despite their specificity, most OMICs data, and in a lesser extent, external exposome data share a high-dimensional nature corresponding to the so-called "small n, large p" situation, in which the number of measurements p is large and can even exceed the number of observations n (Fig. 11.4).

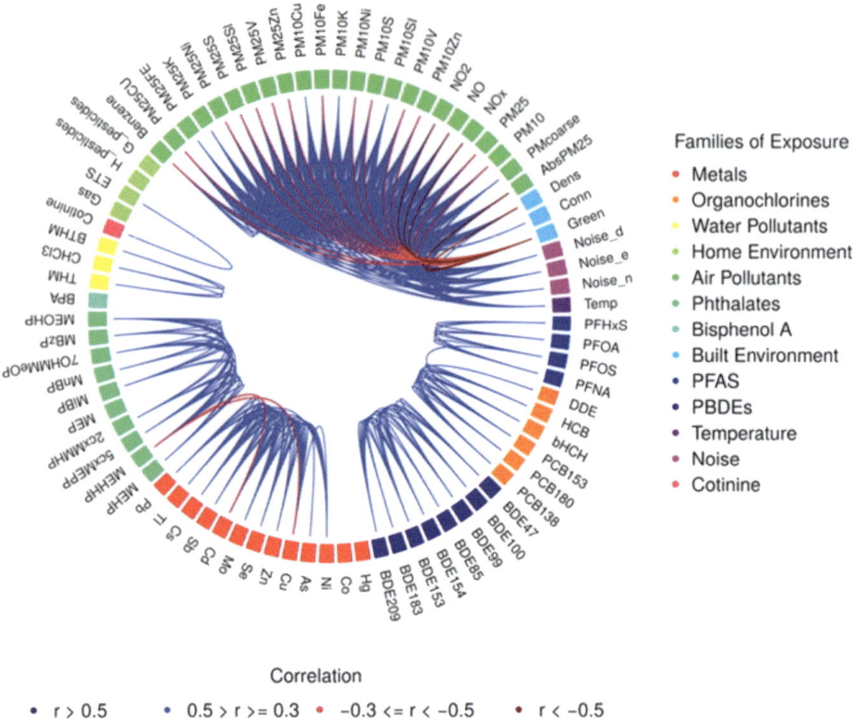

Fig. 11.3 Circos plot representation of the correlations within external exposome data (from Robinson et al. 2015)

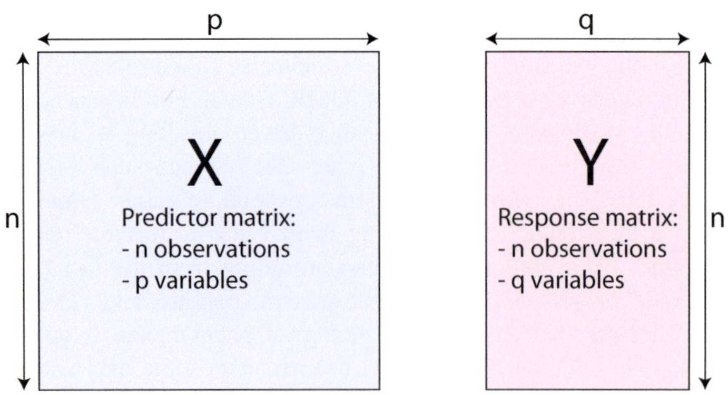

Fig. 11.4 Illustration of the "small n, large p" situation (from Chadeau-Hyam et al. 2013)

In this setup, statistical inferences based on classical approaches are at best biased, and generally numerically intractable. Three main approaches have been proposed to accommodate this situation:

1. *Univariate approaches* assessing separately the association between each variable in the predictor matrix (exposures or OMICs data) and the outcome of interest. These are coupled with multiple testing correction strategies.
2. *Dimensionality reduction techniques* building upon the correlation within the data to construct a summary of lower dimension at the cost of a minimal loss of information.
3. *Variable selection techniques* assuming that not all predictors are relevant to the outcome of interest and seek for a sparse subset of predictors mostly related to the outcome.

Univariate Models and Multiple Testing Correction

A first approach to accommodate the “large p, small n” situation consists in considering each variable in the predictor matrix separately. The association between each predictor and outcomes of interest is assessed and tested using the same statistical model. For a set of p predictors regressed against a single outcome, a total of p tests are then being performed, each with the same per-test significance level α' measuring the risk of rejecting the null hypothesis of no association while it is true (type I error). Conducting p tests results in an inflated number of expected false positive findings, which, assuming independence of the tests, equals $\alpha' \times p$. In order to control the inflated number of associations that are falsely declared as statistically significant across all performed tests, techniques to control the Family-Wise Error Rate (FWER) and the False Discovery Rate (FDR) have been proposed.

The outcome of p tests can be summarized by the following table, where V is the number of false positive findings: the number of times the null hypothesis has been rejected while it was true, S is the number of true positive findings, and R is the number of positive calls across the p tests performed.

	H_0 True	H_0 False	Total
H_0 rejected	V	S	R
H_0 not rejected	U	T	W

The FWER is defined as

$$\text{FWER} = p(V \geq 1).$$

It represents the probability of drawing at least one false positive conclusion across the p tests. Methods to control FWER can involve the calculation of the per-test significance level α' to be applied to each of the p tests to ensure that the actual FWER $\leq \alpha$, which is equivalent to ensure that 1-FWER, the probability not to draw any false positive, is greater than 1- α. In order to control the FWER, one intuitive approach is to adjust (i.e., reduce) the per-test significance level α' to be applied to each test, as proposed by the Bonferroni or Šidák corrections where

$\alpha' = \frac{\alpha}{p}$ and $\alpha' = 1 - (1 - \alpha)^{\frac{1}{m}}$, respectively. Both correction strategies are known to be stringently protecting against the type I error. However, due to the correlations existing across the predictors (OMICs data or exposome features), across the actual p tests performed, the same information is (at least partially) tested several times. Less conservative corrections account for this redundancy and some rely on calculation of the effective number of tests (ENT) as defined by the virtual number of independent tests performed across the actual p tests performed. One direct way to approach the ENT estimation is to perform eigen analysis of the variance–covariance matrix and subsequently apply either a Bonferroni or a Šidák correction using the estimated ENT (Patterson et al. 2006). While this approach is computationally efficient, it is numerically limited as the estimated ENT will be upper-bounded by the number of observations (Schafer and Strimmer 2005). As an alternative method, which is scalable to smaller sample sizes, estimation via resampling procedures such as permutations have been proposed (Castagne et al. 2017; Chadeau-Hyam et al. 2010; Hoggart et al. 2008; Westfall and Young 1993). In practice, for a given set of predictors and a given statistical model, the responses(s) are randomly shuffled. The resulting permuted data set mimics the null hypothesis of no association. Regressing the p predictors against the permuted (set of) outcome, the minimal p-value (noted q) represents the maximal per-test significance to be applied not to draw any false positive conclusion. Repeating that permutation procedure, one can estimate the distribution of q, from which the per-test significance to ensure a control of the FWER at a desired level can be derived. This type of approach has been successfully applied in genetics (Zou et al. 2004; Dudbridge and Gusnanto 2008; Hoggart et al. 2008) and metabolomics (Chadeau-Hyam et al. 2010) but can become computationally intensive.

The False Discovery Rate is the proportion of false positive findings among the significant association:

$$\mathrm{FDR} = E\left(\frac{V}{R}\right),$$

and its control defines the expected proportion of positive findings that are allowed to be false. Methods to control the FDR are iterative procedures looking at the ordered set of p p-values and comparing them to a cutoff value which depends on the rank of the considered p-value. For instance, Benjamini and Hochberg (1995) procedure compares p-values sorted in descending order to the increasingly stringent cutoff value. It returns a list of associations declared statistically significant ensuring that the FDR is upper-bounded by the desired value.

It can be shown that controlling the FWER also provides control of the FDR, and, intuitively, the FDR control is less stringent than that of the FWER. If one runs 100 experiments controlling the FWER at a 0.05 level, on average, less than 5 experiments will result in one or more false positive findings, while the FDR control over these experiments will allow that all experiments can include on average 5 false positives.

The use of univariate approaches using (generalized) linear models to analyze high dimensional data has been extremely successful mainly because (1) they are computationally efficient; (2) they are extremely flexible and can accommodate a large range of parametric and nonparametric relationships; (3) they can accommodate all types of predictors and outcomes; (4) they are readily available in most statistical packages.

However, by design, these approaches only assess the marginal effect of each predictor on the outcome of interest, and do not consider a potential joined effect of the predictors.

Multivariate Models: Dimensionality Reduction and Variable Selection

Owing to the complexity of the effect of exposures and the possible pleiotropy of their downstream consequences, there is a need to model jointly the OMICs or exposome data in relation to exposures and/or health outcomes.

Dimensionality Reduction Techniques

Dimensionality reduction techniques use the correlation among the predictors to define summary of the original variables into a few(er) synthetic variables (the principal components, PCs) that capture the latent structure of the data. This is achieved by searching for linear combinations of variables that optimize some measure of diversity among observations. The *i*th principle component PC_i is then defined as

$$\mathrm{PC}_i = \alpha_{i1}X_1 + \alpha_{i2}X_2 + \ldots + \alpha_{ip}X_p,$$

where $X_1, \ldots, X_p$ denote the p original variables in the predictor matrix and $\alpha_{i1}, \ldots, \alpha_{ip}$, the vector of linear coefficients or weights defining the contribution of each of the original variables to the *i*th component.

Hence, the use of dimensionality reduction techniques can be reduced to the search of the p weights vectors, which relies on eigen analysis. While considering p components, a simple rotation of the original dataset is performed (i.e., no loss of information). The principle of dimensionality reduction techniques is to identify the fewest components that minimally distort the original dataset. Across the dimensionality reduction techniques, different measures of the information are considered: the variance for Principal Component Analysis (PCA) (Hotelling 1933a, b; Pearson 1901), or the χ^2 distance for Correspondence Analysis (CA) (Greenacre 1984). From the eigen decomposition, it is possible to rank the components with respect to the proportion of information they explain. For instance, for PCA, scree plots (see

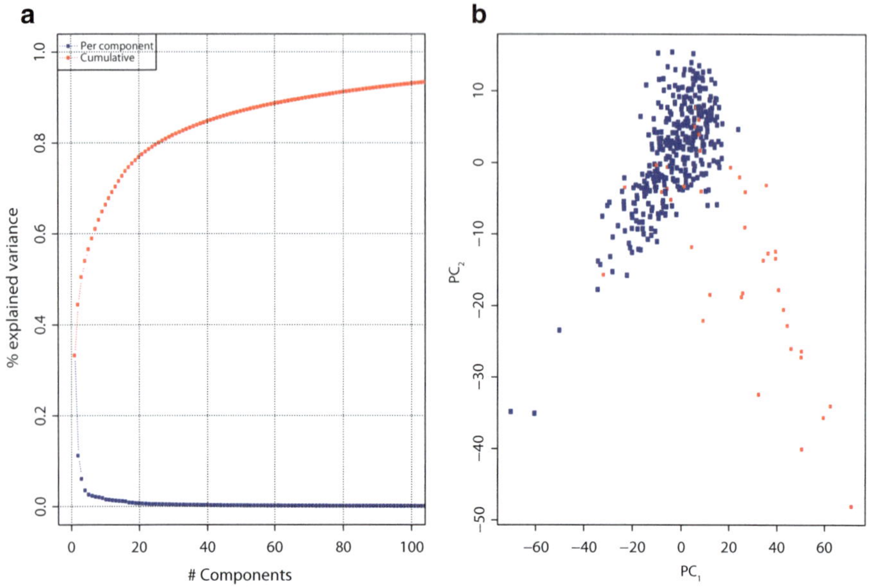

Fig. 11.5 Scree plot of a PCA for 745 transcripts found associated with risk of chronic lymphocytic leukemia (**a**). In the score plot (**b**), individuals are colored according to their prospective disease status: cases in red and controls in blue (from Chadeau-Hyam et al. 2014b)

Fig. 11.5) represents the information restitution, as measured by the proportion of the variance of the original dataset explained by each component.

From Fig. 11.5a, it is then possible to identify the number of PCs that are required to explain more than a certain proportion of the original variance: 25 and 72 components are necessary to explain 80% and 90% of the variance, respectively. The individual contribution of each PC shows that from the 4th component onward, the proportion of the variance explained is marginal. Projection of the data on the two first components (score plot in Fig. 11.5b) suggests that the 45% explained variance by these two sole components yield good discrimination between cases and controls.

Those latent variables can subsequently be used in a (possibly multivariable) regression model to assess how the main drivers of the variation in the original set of variables are related to an outcome of interest. While these may result in numerically tractable inferences, the characterization of potential associations remains conditional on the interpretability of the components. Loadings plots (Fig. 11.6) precisely represent the contribution of each of the original variables to the PCs and can help in understanding the latent structure captured by each component.

In the example presented in Fig. 11.6, while the first two component discriminated cases from controls, none of the first two components were driven by a specific set of transcripts, which hampers the biological understanding of the underlying associations. One way to improve interpretability of the components is to ensure sparsity in the loadings coefficients which can be done through penalization (see below).

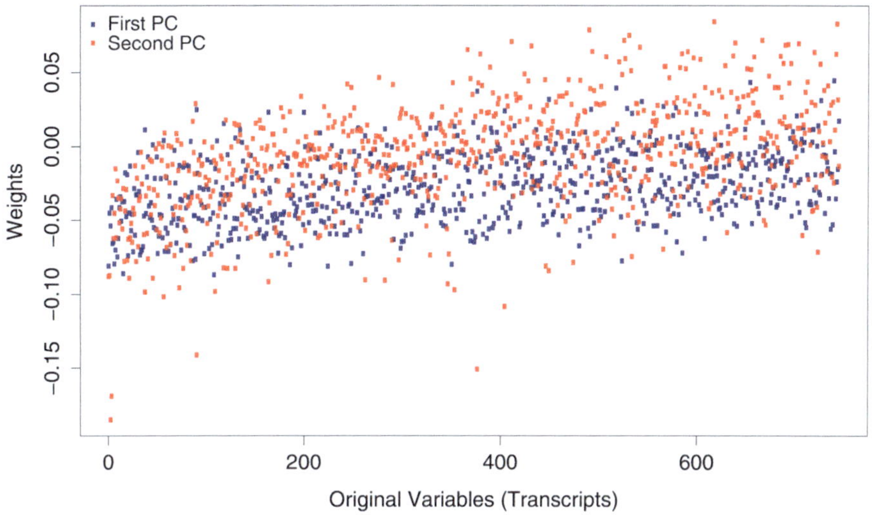

Fig. 11.6 Loadings plots for the first two componenents of the PCA of the 745 CLL-associated transcripts (from Chadeau-Hyam et al. 2014b)

Principal Components Analysis (PCA) (Hotelling 1933a, b; Pearson 1901) has proved useful in genetics (Jombart et al. 2009) and has become a standard in genome-wide association studies (GWAS), where it is used to correct for population stratification (Price et al. 2006; Reich et al. 2008). PCA can accommodate both continuous and discrete data, and is not affected by the potential correlation between predictors or by a larger number of variables than observations.

However, while PCA has proven efficient in summarizing large datasets in (far) fewer dimensions, representing the latent structures in the data driving most of the variability in the original dataset, nothing guarantees that this variation is relevant to an outcome of interest. As a supervised alternative to PCA, Partial Least-Square (PLS) approaches have been proposed (Wold et al. 1984). PLS components are defined such that they maximize the covariance between the predictors and response variables. As such, PLS components not only capture as much variance of the original variable as possible, but also focus on the variance that is relevant to the outcome of interest. PLS-based methods are extremely popular in chemometrics and have been successfully applied in metabolomics (Holmes et al. 2008; Fonville et al. 2010; Yap et al. 2010). Application to other OMICs data were successful in epigenomics (Belshaw et al. 2010), transcriptomics (Musumarra et al. 2011; Fasoli et al. 2012), and proteomics (Wang et al. 2011).

Irrespective of the type of analysis, the use of latent variables in a regression context requires the specification of the number of latent variables to be considered, which usually relies on cross-validation procedures aiming at the identification of the number of components that optimizes both interpretability and prediction error.

Penalization

As mentioned above, one intuitive way to improve results interpretability in multivariate models is to ensure that the number of variables found relevant to the outcome is as sparse as possible. Inducing sparsity hence relies on favoring a minimal set of nonredundant variables jointly associated to the outcome and penalizes irrelevant and/or irrelevant ones.

Penalization techniques have been introduced in regression models to induce sparsity in the vector of regression coefficients. The principle of penalized regression is to estimate all regression coefficients under the constraint that a function of these regression coefficients, the penalty, is bounded by a fixed value.

Of the different penalized regression approaches, Ridge Regression uses as penalty the L^2 norm (Hoerl and Kennard 1970):

$$L^2 = \sum_{i=1}^{p} \beta_i^2,$$

where β_i^2 is the ith regression coefficient linking the outcome and the ith variable in the predictor matrix. It can be demonstrated that the least influential predictors will see their effect size estimate shrunk toward 0, while the estimates for the most important variables will remain unchanged.

LASSO models use the L^1 norm as defined by the sum of the absoluted value of the p regression coefficients (Tibshirani 1996):

$$L^1 = \sum_{i=1}^{p} |\beta_i|$$

Intuitively, upper-bounding the L^1 penalty is more demanding than constraining the L^2 norm (in which regression coefficients are squared). It can be demonstrated that LASSO penalization, for geometric reasons, is able to perform variable selection and shrinks irrelevant regression coefficients exactly to 0. Conversely, while ridge regression provides stable coefficient estimates for ill-posed problems (i.e., when $n < p$), it does not guarantee sparsity. Moreover, one major limitation of LASSO methods is that the number of non-penalized variables is upper-bounded by the number of observations: LASSO model cannot select more than n variables (i.e., provide more than n variables with nonzero regression coefficients). To provide a more general approach, combining advantages of both Ridge and LASSO regression, the Elastic Net has been developed (Zou and Hastie 2005) and uses a penalty which is a weighted sum of the L^1 and L^2 norms according to a calibration parameter λ.

In practice, penalized regression requires a preliminary calibration of the penalty parameter(s), which directly affects the number of selected variables, the estimates of the regression coefficients, and thus the statistical performance of the models.

Calibration procedures usually involve the minimization of the mean square error of prediction by cross-validation.

The calibrated model will return a list of shrunk regression coefficients, and for LASSO and Elastic Net, the variables with non-null shrunk coefficients are those found to be jointly associated to the outcome. Generalized version of both LASSO and Elastic Net accommodates linear, logistic (binary responses) and multinomial (categorical responses), Poisson (count data as response), and Cox (survival models) regression models (Friedman et al. 2010; Simon et al. 2011).

The penalization paradigm applies outside the scope of linear regression models and has been used to provide shrunk loadings coefficients for dimensionality reduction techniques, hence defining sparse versions of both PCA and PLS. These models ensure a sparse definition of the latent variables, thus improving their interpretability. Sparse PCA (sPCA) and sparse PLS (sPLS) have been used to analyze OMICs data (Zou et al. 2006; Shen and Huang 2008; Witten et al. 2009; Boulesteix and Strimmer 2007; Le Cao et al. 2008, 2009; Chun and Keles 2009, 2010).

Bayesian Variable Selection Approaches

Full Bayesian variable selection (BVS) procedures have been proposed to cope with the “large p, small n” framework (West 2003), and these rely on the estimation of the posterior distribution of the following latent binary vector:

$$\gamma = \left(\gamma_{1,}\gamma_2, \ldots, \gamma_p\right) \in \{0, 1\}^p,$$

where γ_i is a binary variable indicating if the ith variable is in the model.

In that setup, a model is defined by the subset of variables it includes, and hence by a given vector gamma (e.g., the null model corresponds to a vector γ comprising p 0’s).

The objective of BVS is to identify the best sets of variables that jointly associate to the outcome of interest, and the main challenge underlying BVS inference resides in the dimensionality of the space in which to search for the best models, which grows exponentially with p (2^p). While there is an extensive theoretical literature on the parameterization and numerical estimation of BVS models, few implementations offer a ready-to-use interface for high-dimensional profiling. These rely on linear or generalized linear models. Shotgun Stochastic Search (SSS) was one of the first packages to be made available that enabled Bayesian large-scale genome screening through BVS (Hans et al. 2007). The search in the model space is iteratively performed such that, if at a given iteration q features are selected, all the models (1) of size (q-1); (2) of size q (i.e., replacing any of the variables in the current model by any of the p-k remaining ones); (3) of size ($q + 1$) will be evaluated and compared. In order to favor the evaluation of meaningful models, piMASS (Guan and Stephens 2011) optimizes the search strategy by specifying a proposal distribution (i.e., defining, for a given

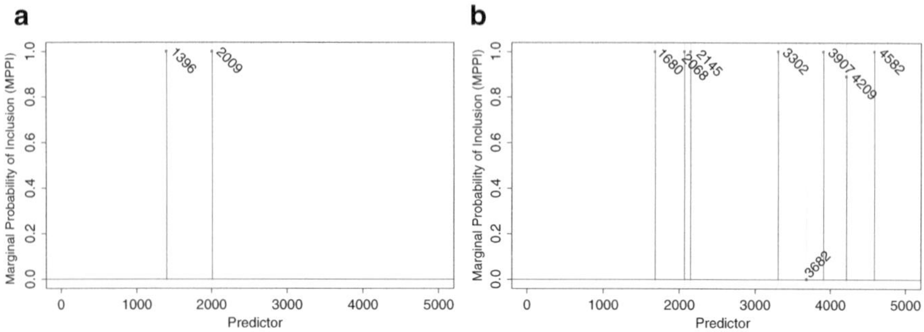

Fig. 11.7 Marginal Posterior Probability of Inclusion (MPPI) of the variables in one of the best models, from the R2GUESS model. Results are presented for 2 simulations using 5,000 SNP of which 2 (**a**) and 8 (**b**) are assumed to have an effect on the simulated outcome (from Liquet et al. 2016a)

predictor, the probability to be proposed for inclusion in a model) that accounts for the correlation between the predictor and outcome.

Both SSS and piMASS are able to accommodate binary, categorical, and continuous outcomes, and SSS additionally implements survival models.

GUESS and its R implementation R2GUESS is an alternative approach which accommodates multivariate outcomes through the use of a multivariate linear model (Liquet et al. 2016a; Bottolo et al. 2011). Its search algorithm is based on multiple-chain genetic algorithms, and its latest version makes use of graphics processing unit (GPU) linear algebra libraries, thus enabling its application to genome-wide scale analyses (hundreds of thousands of predictors simultaneously) (Bottolo et al. 2013).

In practice, BVS models return a list of models that have higher posterior probability. Integrating the posterior probabilities of the best models, it is also possible to infer the marginal probability of inclusion (MPPI) of a given variable in the best models. These MPPIs (see Fig. 11.7) can be viewed as a measure of the strength of association of the (set of) predictors with the outcome of interest.

In Fig. 11.7a, it is apparent that both SNPs used for the simulation were detected by the model and were always included in the best models. For the simulation using eight SNPs (Fig. 11.7b), only seven SNPs were recovered (with MPPI > 0.95), and one SNP was missed, due, notably, to its correlation with the other included SNPs.

OMIC–Exposure Profiling in Practice: Use and Extensions

Quantitative Assessment of the Profiling Techniques

In order to quantitatively assess statistical performances of the main profiling approaches, simulation studies were conducted (Agier et al. 2016; Liquet et al.

2016a). Of these, one exposome-focused study used real correlation matrices among 237 prioritized exposome features from the INMA study (Guxens et al. 2012). The wide range of simulation scenarios comprised (1) several levels of explained variance, (2) different number of "causal" exposures $k = 1, 2, 3, 5, 10$, or 25, and (3) different effect sizes.

Across all scenarios investigated, each of the simulated dataset was analysed using a preselected set of 6 methods borrowed from the three main families defined in the previous section:

1. Univariate models:
   (a) "GWAS"-type approaches, EWAS with multiple testing correction via FDR control
   (b) EWAS followed by a multivariate regression step to limit confounding: EWAS-MLR
2. Dimensionality Reduction Techniques:
   (a) Sparse Partial Least-Square s-PLS
3. Variable Selection Approaches:
   (a) Elastic Net E-NET
   (b) Bayesian Variable Selection R2GUESS
   (c) Deletion/Substitution/Addition algorithm DSA, a penalized stepwise model selection approach (Haight et al. 2010).

In that simulation context, the statistical performances of the different models can be quantified and compared based on both their ability to identify the "true" predictors, used to simulate the outcome (sensitivity), and their ability not to include irrelevant exposures (specificity). In Fig. 11.8, the latter is measured using the False Discovery Proportion (FDP).

Simulation showed that across the large number of scenarios investigated, multivariate methods systematically outperformed univariate approaches to explore external exposome–outcome relationships in the presence of complex correlations (Fig. 11.8). Although these approaches did not achieve low false discovery performances, they yielded a better balance between sensitivity and FDP. Based on refined performance metrics (i.e., accounting for the correlation among predictors), DSA and R2GUESS were identified as providing somewhat better performances. From this work, it was also concluded that in real case analyses, methodological choices should also be guided by computational complexity and feasibility, as well as flexibility considerations such as the ability to accommodate confounders.

These results are in-line with additional simulation studies carried out for OMICs data (i.e., larger number of variables) in relation to simpler outcomes (Liquet et al. 2016a). These simulations showed the superiority of R2GUESS over multivariate alternatives but showed better performances overall of methods under investigation, in the absence of complex correlation across predictors (Fig. 11.9).

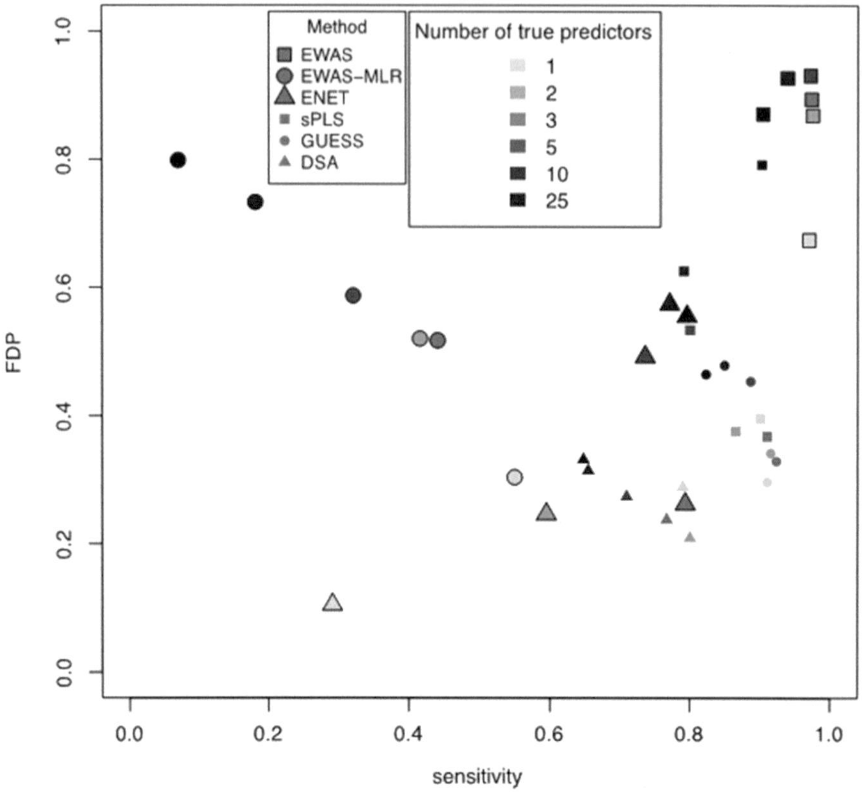

Fig. 11.8 Sensitivity and False Discovery Proportion (FDP) yielded by six different approaches to perform exposome–outcome association study. Results are based on simulated data using realistic exposome data and considering a number of "true" predictors ranging from 1 to 25 (from Agier et al. 2016)

Modeling and Correcting for Nuisance Variation

As mentioned above, both the acquisition of OMICs profiles and exposure measurements are prone to measurement error, resulting in part of the observed variability being related to technical factors during data generation rather than to the variables of interest. As a consequence, such technical variability has the potential to dilute the effects of interest and is referred to as nuisance variation. Several preventive approaches can be implemented to limit such dilution effects. First, during data acquisition through careful randomization of the samples, and notably ensuring that analytical entities (e.g., case–control pairs) are processed within the same analytical batch, and that entities are randomly distributed across batches. Second, through the preprocessing of the data, using specific normalization techniques, generally relying on quality control samples.

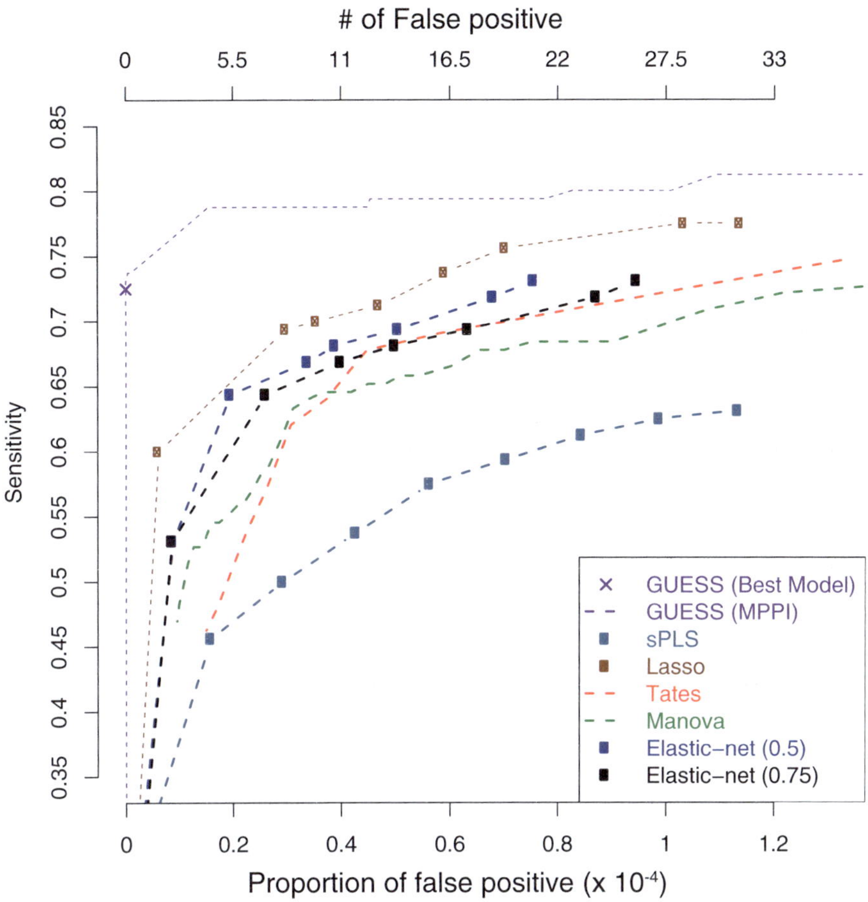

Fig. 11.9 ROC curves comparing the statistical performances of established multivariate OMICs profiling approaches. Results were obtained using 20 independent simulated datasets including 273,675 SNPs in 3122 individuals and setting 8 "true" predictors (from Liquet et al. 2016a)

Despite these preventive measures, technically induced noise can persist in the data, and may therefore be accounted for during the statistical analyses of the data.

Linear mixed models can be implemented to account for nuisance variation (McHale et al. 2011; Chadeau-Hyam et al. 2014b). These include random effects in the statistical model to account for a structure in the variance of observations. This structure and its strength are assumed to depend on nuisance parameters (e.g., technical covariates) and enter the linear model in the form of additional terms that depend on the nuisance parameters. These additional terms can be estimated by likelihood or restricted likelihood maximization (Lindstrom and Bates 1990). One possible formulation of mixed models for a given variable (Y) (e.g., one OMICs measurement) is, for one individual i:

$$Y^i = (\alpha + u_{A^i}) + (\beta_1 + \beta_{A^i})X^i + \beta_2 \mathrm{FE}^i + \varepsilon^i \ (1),$$

where α is the intercept of the model, ε^i is the residual error, and X^i is the outcome of interest (e.g., case–control status). The resulting effect size estimate $\beta_1$can be interpreted as the change in the response variable (Y^i) per unit change in the variable of interest X^i. FE^i is a vector of fixed effects (typically confounders) observed for individual i and corresponding regression coefficients are compiled in the vector β_2. Nuisance variation is modeled through a random intercept u_{A^i} and a random slope β_{A^i}, where the grouping factor A^i compiles the technical factors describing how data from sample i were generated. In that setup, the random intercept captures a systematic shift in the measurement of Y^i, which is related to experimental conditions, while the random slopes would account for a systematic experimentally induced attenuation (or amplification) of the relationship linking the measurement and the variable of interest. Nuisance variation is generally modeled through a random intercept (i.e., neglecting the random slope), and the random intercept u_{A^i} represents the shift associated to A^i, the value of the random effect variable(s) A observed for individual i. For example, in a study of microarray-based gene expression data (Chadeau-Hyam et al. 2014b), the dates of the three main steps of sample processing were used as random effect variables: RNA isolation, hybridization, and dye labeling.

Random intercept estimates over all assayed transcripts can be summarized by their variance, and more specifically the number of times each of the isolation, hybridization, and labeling steps was estimated to generate marginal noise (i.e., null variance). In this example, the proportion of null variance estimates was 19, 13, and 2% for isolation, labeling, and hybridization, respectively, suggesting that hybridization generated more noise than the two other processing steps. Random effect estimates can be further investigated by analyzing, for each of the three random effect covariates, the ranking of each date with respect to the estimated random intercept, as depicted in Fig. 11.10.

Estimates suggest that some dates were yielding higher variances (i.e., more noise), for instance, it seems that samples whose RNA was isolated on the May 12, 2010, were associated to higher noise. The impact of the nuisance variation on subsequent inferences can be assessed by comparing the distributions of p-values from the linear mixed models to those from a linear model (i.e., setting the random intercept to 0). In Fig. 11.11, linear models exhibit a typically null p-value distribution while the inclusion of a random intercept provides a stronger support for the alternative hypothesis, with a sharper peak at smaller p-values.

The inclusion of random effects to model and correct for nuisance variation is not restricted to univariate models and can be used for penalized regression, and BVS. While theoretically, these models could natively be used to account for nuisance variation, they may become computationally demanding and may yield convergence and calibration issues. Moreover, there is no integrative solution to explicitly model technical confounding while using dimensionality reduction techniques. In both cases, one possibility is to adopt a two-step strategy first fitting a linear mixed model as defined in (1). Once fitted, the model will return estimates of the random effects, and subtracting from the observed value (Y^i) the random effect estimates u_{A^i}

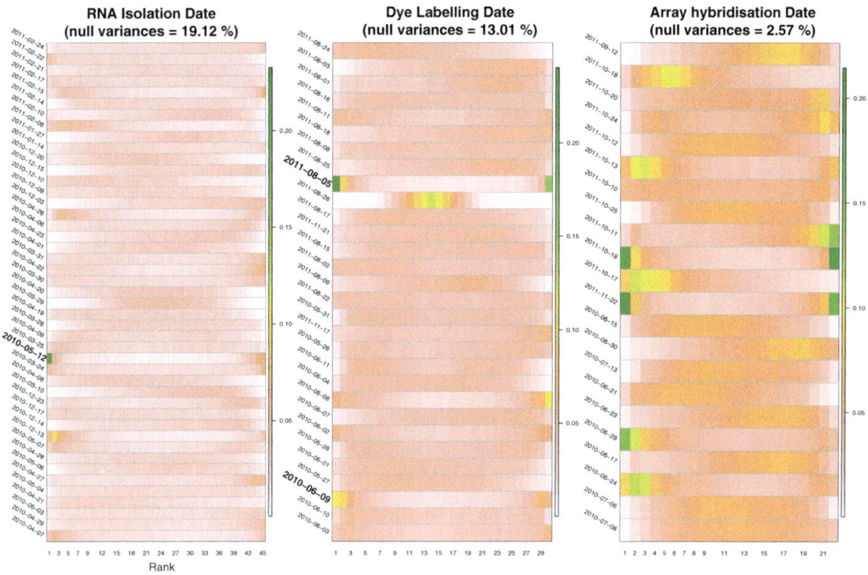

Fig. 11.10 Summary of the random intercept estimates from a linear mixed model investigating the relationship between (N = 29,662) transcripts and future onset of lymphoma. Random effects variables are the dates of the three main processing steps of each biosample underwent. Estimates are summarized by the proportion (across the 29,662 estimates) of null variance estimated, and modality of each grouping factor (i.e., dates) is characterized by the rank of their estimated variance (from Chadeau-Hyam et al. 2014b)

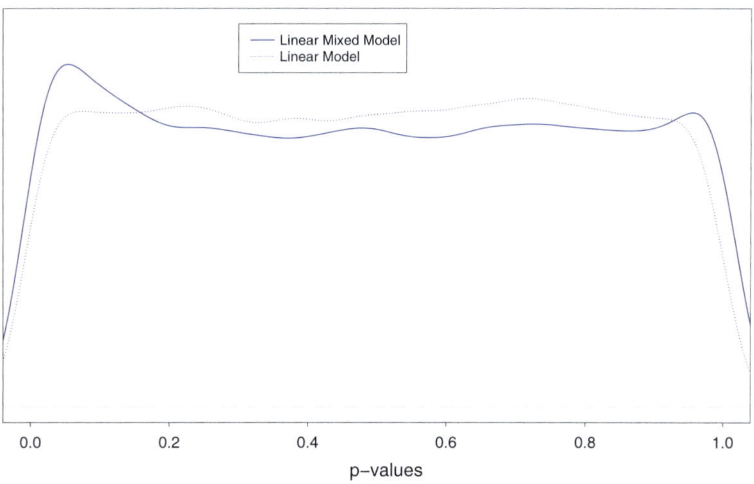

Fig. 11.11 *P*-value distribution from a linear mixed model (plain line) and a linear model (dotted line) assessing the relationship between gene expression level and future onset of lymphoma. Results are based on the 29,662 *p*-values from both models (from Chadeau-Hyam et al. 2014b)

provides measurement in which the effect of the potential technical confounder has been removed. Resulting "denoised" data can subsequently enter any statistical model and will provide results corrected for nuisance variation (Castagne et al. 2016; Chadeau-Hyam et al. 2014b; Guida et al. 2015).

Accommodating Complex Study Designs

In order to improve statistical power to detect the possibly complex (and hence multivariate) biological responses to external stressors, more complex study designs are warranted. These include intervention studies where participants are placed in several controlled environments, presenting exposure contrasts. Developments over these designs can also account for differential response time and may feature in each environment, multiple bio-sampling at different time points.

Irrespective of the detailed design, all such study features repeated measures (of exposure and/or OMICs data) for each participant. One natural way of modeling repeated measurement data is to adopt, in a univariate context, a linear mixed model approach, setting the participant ID as a random effect variable. This model will decompose the within and across individual variation to identify OMICs changes related to changes in exposure. In that context, k observations per participants are considered. The linear mixed model setting the individual ID as random effect assumes a simple variance–covariance structure across the k observations which only depends on the individual (see below for $k = 6$).

$$\begin{pmatrix} \sigma^2 & \delta & \delta & \delta & \delta & \delta \\ \delta & \sigma^2 & \delta & \delta & \delta & \delta \\ \delta & \delta & \sigma^2 & \delta & \delta & \delta \\ \delta & \delta & \delta & \sigma^2 & \delta & \delta \\ \delta & \delta & \delta & \delta & \sigma^2 & \delta \\ \delta & \delta & \delta & \delta & \delta & \sigma^2 \end{pmatrix}$$

The advantage of the linear mixed model approach is that it can be scaled to any other type of models using linear mixed versions of univariate models or variable selection approaches, or adopting a multilevel extension for dimensionality reduction techniques (Liquet et al. 2012).

Additional flexibility could be gained by adopting a multivariate normal (MVN) model where the variance–covariance matrix depends not only on individuals, but also on each of the k experimental conditions:

$$\begin{pmatrix} \sigma_{11}^2 & \delta_{12} & \delta_{13} & \delta_{14} & \delta_{15} & \delta_{16} \\ \delta_{21} & \sigma_{22}^2 & \delta_{23} & \delta_{24} & \delta_{25} & \delta_{26} \\ \delta_{31} & \delta_{32} & \sigma_{33}^2 & \delta_{34} & \delta_{35} & \delta_{36} \\ \delta_{41} & \delta_{42} & \delta_{43} & \sigma_{44}^2 & \delta_{45} & \delta_{46} \\ \delta_{51} & \delta_{52} & \delta_{53} & \delta_{54} & \sigma_{55}^2 & \delta_{56} \\ \delta_{61} & \delta_{62} & \delta_{63} & \delta_{64} & \delta_{65} & \sigma_{66}^2 \end{pmatrix}$$

Both approaches have been successfully used to identify OMICs biomarkers related to acute changes in environmental exposures. Despite limited sample sizes, the MVN approach is showed to be efficient in identifying metabolomic, transcriptomic, and inflammatory changes due to acute experimentally induced changes in exposure to water disinfection by-products (van Veldhoven et al. 2017; Vlaanderen et al. 2017; Espin-Perez et al. 2018).

Exposome studies investigate complex effects of exposures that may act jointly to generate a complex biological response. Such exploration calls for the use of methods handling multivariate exposures and multivariate responses. PLS approaches are able to handle that situation, and multilevel extensions of these approaches further accommodate multiple observations per participant. In essence, multilevel PLS first decomposes the observed variability into within- and between-individual variability (Liquet et al. 2012). The former captures differences between individuals, including confounding factors, and the within-individual variability is measuring the changes in both exposures and responses between the different measurements, hence measuring the effect of the experiment. The latter is subsequently included into a standard PLS model to identify linear combinations of exposures that are best able to explain effects on responses. A recent proof of principle analysis focused on the inflammatory response (measured through blood levels of 13 inflammatory-related proteins) to acute exposure to disinfection by-products (DBP) while swimming in a chlorinated pool (Jain et al. 2018).

PLS identifies the latent variable in X capturing most of the variability in exposures that is relevant to the inflammatory profile. The PLS components of exposures can be ordered as per their relevance to the full set of proteins, and symmetrically, the components of proteins are defined and ordered according to their covariance with exposures. The relevance of each component can be quantified by the proportion of variation it explains. The proportion of variance in X (or Y) explained by a given component of X (or Y) measures how accurately that single component summarizes the entire information contained in the original X (or Y) matrix. The percentage of variance of Y explained by the components of X measures the relevance of the information summary provided by the PLS components to the outcome matrix. This does not only depend on the quality of the component summary but also on the correlation between X and Y. In addition, Variable Importance in Projection (VIP) scores quantify the contribution of each original predictor (here exposure) to the overall explanatory performances of a given PLS component. Empirically, a VIP score < 1 indicates low-to-moderate contribution of a variable (Fig. 11.12a). Sparsity can be induced through penalization in the definition of the PLS components (Chun and Keles 2010). When penalization is applied to PLS

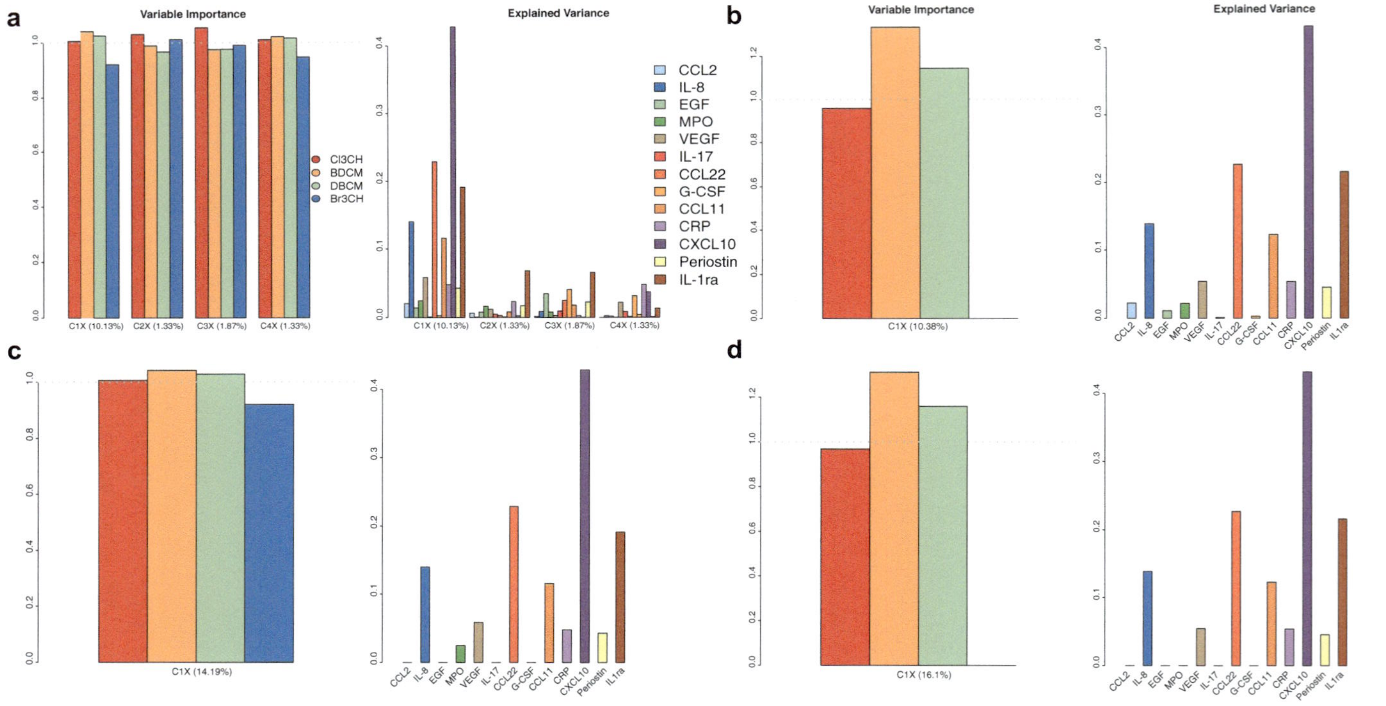

Fig. 11.12 Results from a multilevel PLS analysis of the exposure to 4 DBP while swimming in a chlorinated pool in relation to the blood level of 13 inflammatory-related proteins. Variable Importance in Projection (VIP) plots and proportion of variance explained by protein. Results are presented for PLS model (**a**), for sparse PLS performing variable selection on exposures (**b**), on proteins (**c**), and both on exposures and proteins (**d**) (from Jain et al. Submitted)

components of X (Fig. 11.12b), the resulting sparse PLS (sPLS) model shrinks the loadings coefficients towards zero for the least informative variables (exposures) and hence helps identifying the exposures mostly affecting inflammatory profiles. Symmetrically, variable selection can be performed on the responses (proteins, Fig. 11.12c), in order to identify the proteins whose expression is mostly affected by exposures. In a final step, variable selection can be performed on both exposures and proteins (Fig. 11.12d).

These results suggest that BrCH3 contributed less than other exposures to the inflammatory response to the swimming experiment and that among the 13 assayed proteins, 8 were more affected by the experiment. The application of multilevel-PLS models was successful, despite strong correlation and co-occurrence of the exposures, in identifying the most relevant exposures, and the proteins mostly affected by exposures.

From Improved Interpretability to Mechanome Characterization

Full exploitation of the rich sets of results yielded by OMICs and exposome profiling techniques relies on their biological interpretation (and potentially, validation). While for some molecular entities, interpretation can be eased by knowing the functional role of the molecule (e.g., proteins, or in a lesser extent, transcripts), interpretability can become challenging when the function of the putative biomarker is unknown. Interpretability becomes even more challenging when OMICs profiles are related to complex exposures as they encompass a multivariate combination of environmental and biochemical factors.

To address this issue, a natural approach is to adopt the "meet-in-the-middle" paradigm (Vineis and Perera 2007) and explore as exhaustively as possible which factors may affect the level of exposure-related biomarkers. This two-step strategy can help in identifying molecular alterations that are associated to external stressors and health outcomes. While a natural way to identify "meet-in-the-middle" associations is through univariate models, several examples and recent developments used multivariate approaches (Assi et al. 2015; Chadeau-Hyam et al. 2011).

Biological interpretability of molecular alterations/features identified in high-throughput profiling is highly dependent on the functional characterization of the assayed molecules. Typically, interpretation of the results of a targeted proteomic assay is easier as the biological functionality is documented. While gene expression profiles govern RNA translation, the overall gene expression regulation may be multivariate (i.e., involve several transcripts) and pleiotropic (i.e., through a complex cascade involving other genes and transcripts). Hence, direct interpretation of the results from a transcriptome-wide association study (i.e., based only on the genomic location of the identified transcripts) should be complemented by an investigation of biological pathways potentially affected by differential expression. Ontology-based tools interrogating existing databases are rich sources of information to infer biological pathways corresponding to the candidate biomarkers identified. Specifically, gene enrichment analyses assess if, and to what extent, the list of candidate transcripts is enriched for specific pathways, basically checking if the distribution of identified transcripts is significantly different from what would be expected if candidate transcripts were chosen at random (The Gene Ontology Consortium 2017; Ashburner et al. 2000; Huang et al. 2008).

For metabolomic data, feature annotation and signal interpretation can also rely on database interrogation and pathway identification. Recent developments include an efficient and reliable tool (Li et al. 2013), which has proven efficient in identifying molecules and corresponding pathways from full-resolution mass-spectrometry profiles.

For other OMIC profiles, such as DNA methylation data, results interpretation can be more challenging, as there is, to date, no established database linking the CpG site-specific levels to their downstream consequences or to sets of general biological pathways. In the absence of such information, biological interpretation of CpG sites found differentially methylated can be done by linking them to other OMIC data measured in the same individuals, and whose functional role is better characterized.

OMICs Integration: An Intuitive Approach

One intuitive approach to OMICs integration relies on the following two-step strategy corresponding to a targeted OMIC integration (Fig. 11.13):

1. Regressing the first matrix (X_1, e.g., methylation data) against the outcome (Y) in order to identify $X_{1\gamma}$, a subset of variable of size $p_{1\gamma}$ associated to Y. That list of $p_{1\gamma}$ candidate biomarkers can be derived from p_1 univariate models or a single multivariate model.
2. Regressing the $p_{1\gamma}$ outcome-associated OMICs (e.g., smoking-related CpG sites) against the full-resolution second OMIC profile (X_2, e.g., transcripts). Adopting a univariate approach that would correspond to $p_{1\gamma} \times p_2$ tests.

This strategy has been applied in several studies, and in particular in the study of smoking-induced DNA methylation alterations (Guida et al. 2015), where blood-derived DNA methylation profiles from Illumina Infinium HumanMethylation450 BeadChip from ($N = 745$) participants of EPIC and NOWAC cohorts were used to identify 751 differentially methylated CpG sites in relation to smoking history. In a subset of that study population ($N = 271$), genome-wide gene expression profiles obtained from the Illumina HumanWG-6 array (assaying $N = 8952$ genes) were also available.

Using a univariate linear approach, each of the $751 \times 8952 = 6.72 \times 10^6$ pairwise associations were tested and 5,636 CpG-transcript pairs (corresponding to

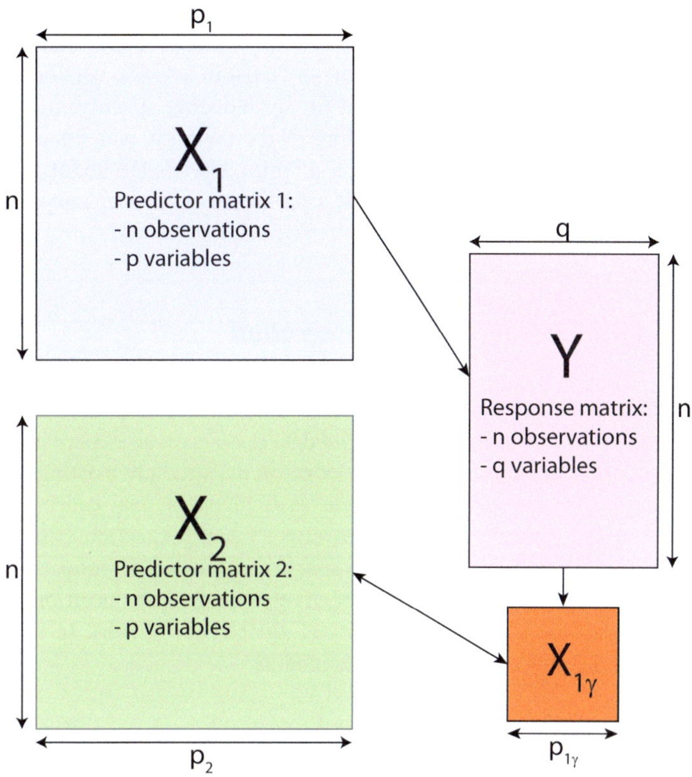

Fig. 11.13 OMICs integration, a two-step strategy

265 unique CpG sites, and 426 genes) were found significantly associated. Most CpG-transcript pairs were inversely associated suggesting that hypermethylation is associated to a gene down-regulation, and in case a CpG site was associated to several transcripts, the sign on these associations is generally consistent across transcripts.

The list of transcripts found associated to the smoking-related differentially methylated sites can be used as input for gene enrichment analyses, which, in that example, identified relevant biological pathways involved in the effect mediation (through methylation changes) of the exposure to tobacco smoke.

Another striking result from this analysis is that of the 5,636 statistically significant CpG transcript pairs, only 5 involved a CpG and a transcript located on the same gene. This suggests that regulatory cascades affected by exposure-induced methylation changes are complex and involve distant (trans) associations.

This supports the fact that there is no justification for reducing OMICs integration to local interactions or correlations, and there is a clear need to extend screening approaches to explore long-distance relationships. In that context, dimensionality reduction techniques, and Bayesian variable selection approaches handling multivariate X and Y, may be considered, and already exist in a sparse version. However, while interpretability is classically sought for by inducing sparsity, it may not be sufficient to ensure a detailed understanding of the complex patterns exhibited by OMICs data integration, and the inclusion of prior knowledge about functionally relevant structures within and across OMICs profiles may be necessary.

Further Approaches to OMICs Integration

PLS algorithms have proven efficient in the task of selecting correlated sets of signals across two blocks of high-throughput data (Le Cao et al. 2008; Parkhomenko et al. 2009). In order to exploit prior knowledge on the structure existing in the data, potential grouping of the covariates within each block of data can be envisaged (Zhou et al. 2010). Recently, based on a novel penalty function controlling the number of groups to be selected and the sparsity within each group, a group and sparse group PLS (gPLS and sgPLS, respectively) method has been proposed that improves both sparsity and interpretability (Liquet et al. 2016b). In practice, for sparse PLS models, the components are defined as

$$C^X = \underbrace{\alpha_1 X_1}_{\neq 0} + \underbrace{\alpha_2 X_2}_{=0} + \ldots + \underbrace{\alpha_k X_k}_{\neq 0} + \underbrace{\alpha_{k+1} X_{k+1}}_{\neq 0} + \ldots + \underbrace{\alpha_k X_k}_{=0}$$

where loadings for the unimportant variables are shrunk to 0 (e.g., X_2 and X_k in the example). For group PLS, the model is given *a priori* a group structure and will select the entire group (groups i and j in the example below):

$$C^X = \overbrace{\underbrace{\alpha_1 X_1}_{=0} + \underbrace{\alpha_2 X_2}_{=0} + \underbrace{\alpha_3 X_3}_{=0}}^{\text{group } 1} + \ldots + \overbrace{\underbrace{\alpha_k X_k}_{\neq 0} + \underbrace{\alpha_{k+1} X_{k+1}}_{\neq 0}}^{\text{group } i}$$
$$+ \ldots + \overbrace{\underbrace{\alpha_{p-2} X_{p-2}}_{\neq 0} + \underbrace{\alpha_{p-1} X_{p-1}}_{\neq 0} + \underbrace{\alpha_p X_p}_{\neq 0}}^{\text{group } j} .$$

For sparse group PLS models (sgPLS), the component will be defined by selecting or not a given group, and within the selected groups it will only select the most relevant variables (X_k in group i and X_p in group j in the example below):

$$C^X = \overbrace{\underbrace{\alpha_1 X_1}_{=0} + \underbrace{\alpha_2 X_2}_{=0} + \underbrace{\alpha_3 X_3}_{=0}}^{\text{group 1}} + \ldots + \overbrace{\underbrace{\alpha_k X_k}_{\neq 0} + \underbrace{\alpha_{k+1} X_{k+1}}_{=0}}^{\text{group } i}$$
$$+ \ldots + \overbrace{\underbrace{\alpha_{p-2} X_{p-2}}_{=0} + \underbrace{\alpha_{p-1} X_{p-1}}_{=0} + \underbrace{\alpha_p X_p}_{\neq 0}}^{\text{group } j} .$$

These features are key to improve the interpretability of results from OMICs profiling and can facilitate the integration of data arising from different platforms. In order to realistically and functionally define these groups, developments can be considered to use external information (e.g., empirical information from external studies), or to automatically define discriminant groupings of the features within each of the OMICs data. That could involve a preliminary topological investigation of the network within each type of OMICs data. The combined use of variable grouping and penalization can also be applied to penalized regression (e.g., through the group LASSO) (Simon et al. 2013).

OMICs profiling and their integrative alternatives produce a prioritized list of (multi-)OMIC markers that jointly reflect the molecular effects of exposures. Insight into their mode of action could be gained by exploring their inter-connections, the regulatory cascades they are involved in.

The inference of network topologies can identify nodes, as defined by (combination of) exposure-related biomarkers, which will be interconnected if they are related (typically with high pairwise correlations). Supervised alternatives, as defined by differential networks (Salamanca Beatriz et al. 2014; Valcarcel et al. 2011, 2014), will account for differences in subpopulations by linking two nodes if their relation is not the same in the two populations (e.g., cases and controls).

As a proof-of-principle example, we applied these models to the smoking-related markers (265 differentially methylated sites and 425 associated transcripts) of smoking exposures presented above (Chadeau-Hyam, personnal communication). The application of the differential networks requires to carefully choose the metrics used to measure pairwise correlation (e.g., Spearman correlation, partial correlation, shrinkage correlation), as well as the way to select influential edges (e.g., significance assessment via permutations, or stability analyses). Once these choices are set, differential networks provide a visualization of the features that are differentially related in two subpopulations of interest. After strong shrinkage of the network topology, the analysis of smoking-related CpG sites and associated transcripts showed sets of independent modules combining CpG sites and transcripts (Fig. 11.14), hence facilitating their functional interpretation.

This type of methodology can directly be generalized to experimental studies. For instance, differential network methodology can be applied to the metabolites and transcripts found associated to DBP exposure while swimming in a chlorinated pool (van Veldhoven et al. 2017). The preliminary screening for these two sets of OMICs

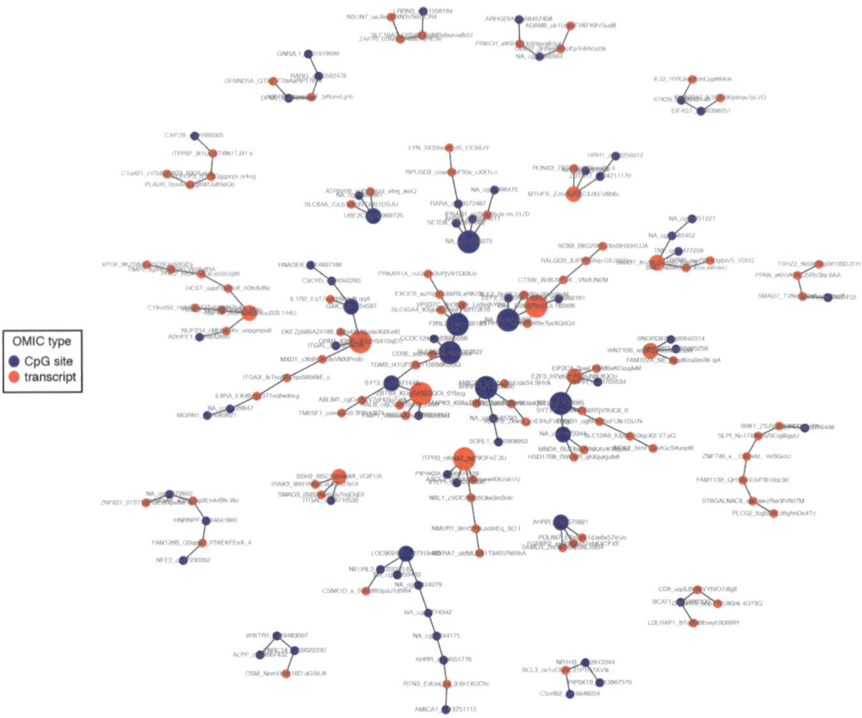

Fig. 11.14 Differential network including 265 smoking-related differentially methylated CpG sites and 426 correlated transcripts

profiles typically gives rise to two correlation heatmaps before and after swimming. In both heatmaps, correlation levels within each class of biomarkers (metabolic features on the one hand, and transcripts on the other hand) are higher than across classes and seem to be strengthened after the swimming experiment (Fig. 11.15).

The application of differential network comparing the correlation before and after the swim clearly shows no evidence of correlated/interacting responses at both molecular levels to the experimentally induced changes in exposure (Fig. 11.16).

Perspectives: Toward Mechanome Characterization

The set of methods described in this chapter represents a non-exhaustive list of approaches that have successfully been used in exposome studies. One primary methodological challenge raised by exposome characterization relates to high dimensionality of the data and the complexity of the effects of interest, which are usually multivariate and pleotropic. Models handling the dimensionality of such data are now established and have been successfully applied. Models to integrate the

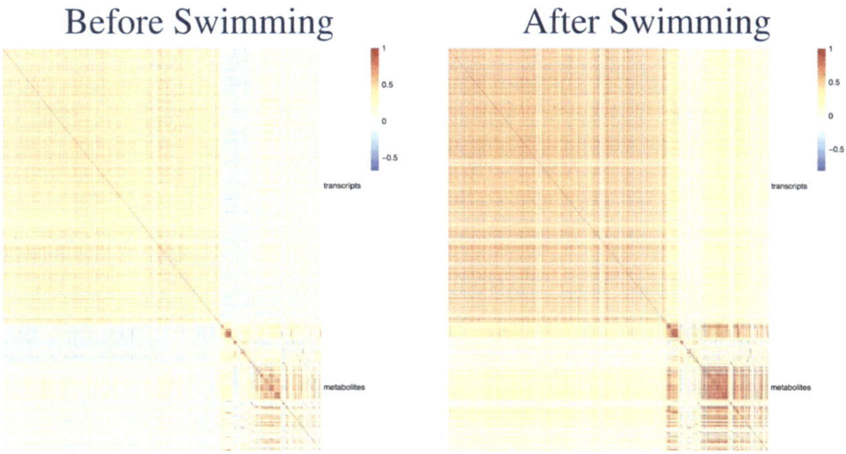

Fig. 11.15 Correlation heatmaps of the ($N = 293$) metabolic features and ($N = 721$) transcripts found associated to DBP exposure during an experimental swim in a chlorinated pool

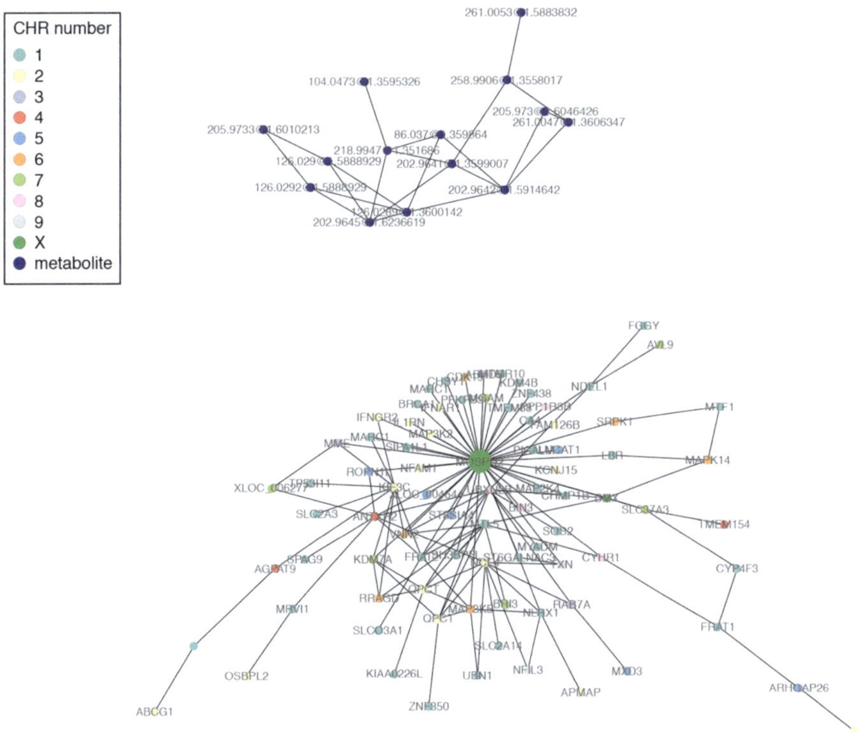

Fig. 11.16 Differential network of the metabolic features and transcripts found associated to DBP exposure during an experimental swim in a chlorinated pool

Fig. 11.17 Schematic representation of the sequential approach to explore the mechanome. The approach can start from exposures (E) and/or disease endpoint (D)

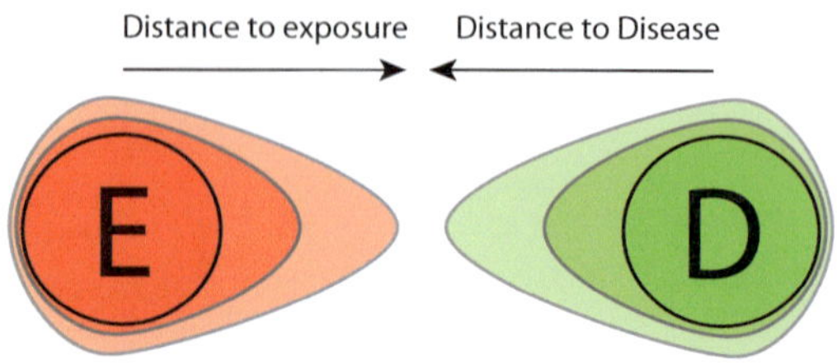

diversity of exposome data are developing and have been applied in a targeted setting (i.e., applied to preselected sets of exposures and biomarkers). While, from a computational standpoint, these approaches could be scaled up to full-resolution data, interpretability of the results will remain rate limiting in the absence of additional information describing the functional role of molecular markers of exposures.

The mechanome can be defined as the ensemble of exposure-triggered regulatory cascades affecting the individual's risk of developing an adverse condition. While its full exploration is not feasible, the combined use of profiling techniques and network inference as presented in this chapter have the potential to provide a prioritized list of biomarkers involved in the mechanome of a given condition.

Assuming that the strength of correlation/association among the OMICs signals is reflecting their functional proximity in the molecular pathway, sequential approaches can be used to identify a prioritized, sparse and nonredundant subset of OMICs signals potentially involved in the molecular mechanisms of interest (Fig. 11.17).

Starting with either the exposure or the health outcome, using the same pool of profiling techniques, one can identify a core set of OMICs markers of exposure and outcome, respectively. By using conditional modeling approaches, "first order" sets of biomarkers can be defined as those associated to at least one core biomarker but not directly to either exposure or disease endpoint. Repeating this sequential procedure will generate a list of ordered sets of biomarkers with respect to their "distance" to exposure or disease, which can potentially inform the structure about causal structures and relationships.

These structures can be explored by means of network topological investigations first within classes of markers to inform about the multivariate physiological (1) response to exposure, and (2) changes leading to disease onset, at different stages of the molecular pathway and at different molecular levels, through the identification hubs playing a pivotal role in functional translation of the effect of exposure or leading to increased risk of the outcome.

Topological investigations can be extended across classes of biomarkers by considering each order of biomarkers as a distinct subnetwork and seeking for the most likely (or the shortest) path across classes of markers. The identified path (s) linking central nodes across classes from core exposure, to core disease biomarkers, can provide a visualization of the molecular pathways involved in the exposure-induced development of the outcome. In that setting, OMICs integration will be achieved by either (1) running the full approach on each set of OMICs data separately and combining each OMIC specific network using, for instance, a multilayer network (Kivelä et al. 2013), or (2) pooling OMICs data (or sets of prioritized thereof) in a single network.

The resulting list of prioritized and ordered sets of biomarkers can subsequently be fed into probabilistic graphical models, where the inclusion of directed edges within the graph can help in addressing causality. In this setting, the mechanistic exploration of the data can be viewed as a search for the most relevant causal graph (s). To ensure computational feasibility, numerical algorithms to efficiently explore such a vast model space (e.g., stochastic search algorithms, and importance sampling) can be implemented.

Longitudinal data provide a gold-standard setting to investigate mechanisms as they enable the explicit modeling of the processes leading to the observation and allow formal causal assessment. Longitudinal models include multistate models, which are defined by a set of ordered states reflecting the evolution of the health status, and can be fully characterized by the set of transition probabilities between each compartment. Model estimation aims at quantifying the transitions ensuring the best reconstruction of the pathological trajectories in each subject, hence adding to the classification problem (discriminating healthy and diseased subjects) a dynamic component (estimating the time of onset). While these models were initially developed to accommodate data from longitudinal studies, they can fruitfully be generalized to cross-sectional data and history of exposure to external stressors (e.g., smoking history), as exemplified by a recent proof-of-principle publication on smoking-induced lung cancer (Chadeau-Hyam et al. 2014a). Including biomarkers in the definition of the transition probabilities may help identify the step(s) of the pathological pathway they may exert their effect on, and may therefore help in understanding their functional role.

This methodological framework will also be able to accommodate OMICs trajectories in case of repeated measurement of OMICs profiles. As such, this approach will define a quantitative complement to trajectory classification procedures (e.g., manifold and dynamic time warping algorithms) to identify OMICs evolution patterns that are characteristic of the exposure to external stressors and/or of future disease risk, by leveraging off the information brought about by the autocorrelation structure embedded in OMICs trajectories.

Overall the characterization of the mechanome relies on the elucidation of causal structures existing among prioritized sets of exposure-related biomarkers. Because formal causal assessment can only be achieved in a longitudinal setup, a deeper understanding of the mechanisms involved in the exposome, as formalized in the mechanome concept, will undoubtedly rely on the generation and exploitation of exposome data measured in the same individuals at different life stages. The inclusion of this temporal component in exposome data will define a new set of statistical challenges that should represent one of the key methodological priorities of exposome research in the coming years.

References

Agier L, Portengen L, Chadeau-Hyam M, Basagana X, Giorgis-Allemand L, Siroux V, Robinson O, Vlaanderen J, Gonzalez JR, Nieuwenhuijsen MJ, Vineis P, Vrijheid M, Slama R, Vermeulen R (2016) A systematic comparison of linear regression-based statistical methods to assess exposome-health associations. Environ Health Perspect 124(12):1848–1856. https://doi.org/10.1289/EHP172

Ashburner M, Ball CA, Blake JA, Botstein D, Butler H, Cherry JM, Davis AP, Dolinski K, Dwight SS, Eppig JT, Harris MA, Hill DP, Issel-Tarver L, Kasarskis A, Lewis S, Matese JC, Richardson JE, Ringwald M, Rubin GM, Sherlock G (2000) Gene ontology: tool for the unification of biology. Nat Genet 25:25. https://doi.org/10.1038/75556

Assi N, Fages A, Vineis P, Chadeau-Hyam M, Stepien M, Duarte-Salles T, Byrnes G, Boumaza H, Knüppel S, Kühn T, Palli D, Bamia C, Boshuizen H, Bonet C, Overvad K, Johansson M, Travis R, Gunter M, Lund E, Dossus L, Elena-Herrmann B, Riboli E, Jenab M, Viallon V, Ferrari P (2015) A statistical framework to model the meeting-in-the-middle principle using metabolomic data: application to hepatocellular carcinoma in the EPIC study. Mutagenesis 30 (6):743–753

Balding DJ (2006) A tutorial on statistical methods for population association studies. Nat Rev Genet 7(10):781–791. https://doi.org/10.1038/nrg1916

Belshaw NJ, Pal N, Tapp HS, Dainty JR, Lewis MPN, Williams MR, Lund EK, Johnson IT (2010) Patterns of DNA methylation in individual colonic crypts reveal aging and cancer-related field defects in the morphologically normal mucosa. Carcinogenesis 31(6):1158–1163. https://doi.org/10.1093/carcin/bgq077

Benjamini Y, Hochberg Y (1995) Controlling the false discovery rate: a practical and powerful approach to multiple testing. J R Stat Soc Series B Stat Methodol 57:289–300

Bottolo L, Chadeau-Hyam M, Hastie DI, Langley SR, Petretto E, Tiret L, Tregouet D, Richardson S (2011) ESS++: a C++ objected-oriented algorithm for Bayesian stochastic search model exploration. Bioinformatics 27(4):587–588. https://doi.org/10.1093/bioinformatics/btq684

Bottolo L, Chadeau-Hyam M, Hastie DI, Zeller T, Liquet B, Newcombe P, Yengo L, Wild PS, Schillert A, Ziegler A, Nielsen SF, Butterworth AS, Ho WK, Castagne R, Munzel T, Tregouet D, Falchi M, Cambien F, Nordestgaard BG, Fumeron F, Tybjaerg-Hansen A, Froguel P, Danesh J, Petretto E, Blankenberg S, Tiret L, Richardson S (2013) GUESS-ing polygenic associations with multiple phenotypes using a GPU-based evolutionary stochastic search algorithm. PLoS Genet 9(8):e1003657. https://doi.org/10.1371/journal.pgen.1003657

Boulesteix AL, Strimmer K (2007) Partial least squares: a versatile tool for the analysis of high-dimensional genomic data. Brief Bioinform 8(1):32–44. https://doi.org/10.1093/bib/bbl016

Carlin DJ, Rider CV, Woychik R, Birnbaum LS (2013) Unraveling the health effects of environmental mixtures: an NIEHS priority. Environ Health Perspect 121(1):A6–A8

Castagne R, Kelly-Irving M, Campanella G, Guida F, Krogh V, Palli D, Panico S, Sacerdote C, Tumino R, Kleinjans J, de Kok T, Kyrtopoulos SA, Lang T, Stringhini S, Vermeulen R, Vineis P, Delpierre C, Chadeau-Hyam M (2016) Biological marks of early-life socioeconomic experience is detected in the adult inflammatory transcriptome. Sci Rep 6:38705. https://doi.org/10.1038/srep38705

Castagne R, Boulange CL, Karaman I, Campanella G, Santos Ferreira DL, Kaluarachchi MR, Lehne B, Moayyeri A, Lewis MR, Spagou K, Dona AC, Evangelos V, Tracy R, Greenland P, Lindon JC, Herrington D, Ebbels TMD, Elliott P, Tzoulaki I, Chadeau-Hyam M (2017) Improving visualization and interpretation of metabolome-wide association studies: an application in a population-based cohort using untargeted 1h nmr metabolic profiling. J Proteome Res 16(10):3623–3633. https://doi.org/10.1021/acs.jproteome.7b00344

Chadeau-Hyam M, Ebbels TMD, Brown IJ, Chan Q, Stemler J, Huang CC, Daviglus ML, Ueshima H, Zhao L, Holmes E, Nicholson JK, Elliott P, De Iorio M (2010) Metabolic profiling and the metabolome-wide association study: significance level for biomarker identification. J Proteome Res 9(9):4620–4627. https://doi.org/10.1021/pr1003449

Chadeau-Hyam M, Athersuch TJ, Keun HC, De Iorio M, Ebbels TMD, Jenab M, Sacerdote C, Bruce SJ, Holmes E, Vineis P (2011) Meeting-in-the-middle using metabolic profiling - a strategy for the identification of intermediate biomarkers in cohort studies. Biomarkers 16 (1):83–88. https://doi.org/10.3109/1354750x.2010.533285

Chadeau-Hyam M, Campanella G, Jombart T, Bottolo L, Portengen L, Vineis P, Liquet B, Vermeulen RC (2013) Deciphering the complex: methodological overview of statistical models to derive OMICS-based biomarkers. Environ Mol Mutagen 54(7):542–557. https://doi.org/10.1002/em.21797

Chadeau-Hyam M, Tubert-Bitter P, Guihenneuc-Jouyaux C, Campanella G, Richardson S, Vermeulen R, De Iorio M, Galea S, Vineis P (2014a) Dynamics of the risk of smoking-induced lung cancer: a compartmental hidden Markov model for longitudinal analysis. Epidemiology 25(1):28–34. https://doi.org/10.1097/EDE.0000000000000032

Chadeau-Hyam M, Vermeulen RC, Hebels DG, Castagne R, Campanella G, Portengen L, Kelly RS, Bergdahl IA, Melin B, Hallmans G, Palli D, Krogh V, Tumino R, Sacerdote C, Panico S, de Kok TM, Smith MT, Kleinjans JC, Vineis P, Kyrtopoulos SA, EnviroGenoMarkers project consortium (2014b) Prediagnostic transcriptomic markers of chronic lymphocytic leukemia reveal perturbations 10 years before diagnosis. Ann Oncol 25(5):1065–1072. https://doi.org/10.1093/annonc/mdu056

Chun H, Keles S (2009) Expression quantitative trait loci mapping with multivariate sparse partial least squares regression. Genetics 182(1):79–90. https://doi.org/10.1534/genetics.109.100362

Chun H, Keles S (2010) Sparse partial least squares regression for simultaneous dimension reduction and variable selection. J R Stat Soc Series B Stat Methodol 72:3–25

Dominici F, Peng RD, Barr CD, Bell ML (2010) Protecting human health from air pollution: shifting from a single-pollutant to a multipollutant approach. Epidemiology 21(2):187–194. https://doi.org/10.1097/EDE.0b013e3181cc86e8

Dudbridge F, Gusnanto A (2008) Estimation of significance thresholds for genomewide association scans. Genet Epidemiol 32(3):227–234. https://doi.org/10.1002/gepi.20297

Espin-Perez A, Font-Ribera L, van Veldhoven K, Krauskopf J, Portengen L, Chadeau-Hyam M, Vermeulen R, Grimalt JO, Villanueva CM, Vineis P, Kogevinas M, Kleinjans JC, de Kok TM (2018) Blood transcriptional and microRNA responses to short-term exposure to disinfection by-products in a swimming pool. Environ Int 110:42–50. https://doi.org/10.1016/j.envint.2017.10.003

Fasoli M, Dal Santo S, Zenoni S, Tornielli GB, Farina L, Zamboni A, Porceddu A, Venturini L, Bicego M, Murino V, Ferrarini A, Delledonne M, Pezzotti M (2012) The grapevine expression atlas reveals a deep transcriptome shift driving the entire plant into a maturation program. Plant Cell 24(9):3489–3505. https://doi.org/10.1105/tpc.112.100230

Font-Ribera L, Kogevinas M, Zock JP, Gomez FP, Barreiro E, Nieuwenhuijsen MJ, Fernandez P, Lourencetti C, Perez-Olabarria M, Bustamante M, Marcos R, Grimalt JO, Villanueva CM (2010) Short-term changes in respiratory biomarkers after swimming in a chlorinated pool. Environ Health Perspect 118(11):1538–1544. https://doi.org/10.1289/ehp.1001961

Fonville JM, Richards SE, Barton RH, Boulange CL, Ebbels TMD, Nicholson JK, Holmes E, Dumas ME (2010) The evolution of partial least squares models and related chemometric approaches in metabonomics and metabolic phenotyping. J Chemom 24(11–12):636–649. https://doi.org/10.1002/cem.1359

Friedman J, Hastie T, Tibshirani R (2010) Regularization paths for generalized linear models via coordinate descent. J Stat Softw 33(1):1–22

Greenacre M (1984) Theory and applications of correspondence analysis. Academic Press, London

Guan YT, Stephens M (2011) Bayesian variable selection regression for genome-wide association studies and other large-scale problems. Ann Appl Stat 5(3):1780–1815. https://doi.org/10.1214/11-aoas455

Guida F, Sandanger TM, Castagne R, Campanella G, Polidoro S, Palli D, Krogh V, Tumino R, Sacerdote C, Panico S, Severi G, Kyrtopoulos SA, Georgiadis P, Vermeulen RCH, Lund E, Vineis P, Chadeau-Hyam M (2015) Dynamics of smoking-induced genome-wide methylation changes with time since smoking cessation. Hum Mol Genet 24(8):2349–2359. https://doi.org/10.1093/hmg/ddu751

Guxens M, Ballester F, Espada M, Fernandez MF, Grimalt JO, Ibarluzea J, Olea N, Rebagliato M, Tardon A, Torrent M, Vioque J, Vrijheid M, Sunyer J, Project I (2012) Cohort profile: the INMA--INfancia y Medio Ambiente--(environment and childhood) project. Int J Epidemiol 41 (4):930–940. https://doi.org/10.1093/ije/dyr054

Haight TJ, Wang Y, van der Laan MJ, Tager IB (2010) A cross-validation deletion-substitution-addition model selection algorithm: application to marginal structural models. Comput Stat Data Anal 54(12):3080–3094. https://doi.org/10.1016/j.csda.2010.02.002

Hans C, Dobra A, West M (2007) Shotgun stochastic search for "large p" regression. J Am Stat Assoc 102(478):507–516. https://doi.org/10.1198/016214507000000121

Hoerl AE, Kennard RW (1970) Ridge regression—biased estimation for nonorthogonal problems. Technometrics 12(1):661–676. https://doi.org/10.2307/1267351

Hoggart CJ, Clark TG, De Lorio M, Whittaker JC, Balding DJ (2008) Genome-wide significance for dense SNP and resequencing data. Genet Epidemiol 32(2):179–185

Holmes E, Loo RL, Stamler J, Bictash M, Yap IK, Chan Q, Ebbels T, De Iorio M, Brown IJ, Veselkov KA, Daviglus ML, Kesteloot H, Ueshima H, Zhao L, Nicholson JK, Elliott P (2008) Human metabolic phenotype diversity and its association with diet and blood pressure. Nature 453(7193):396–400

Hotelling H (1933a) Analysis of complex statistical variables into principal components. J Educ Psychol 24(6):417–441

Hotelling H (1933b) Analysis ofc omplex statistical variables into principal components. J Educ Psychol 24(7):498–520

Huang DW, Sherman BT, Lempicki RA (2008) Systematic and integrative analysis of large gene lists using DAVID bioinformatics resources. Nat Protoc 4:44. https://doi.org/10.1038/nprot.2008.211

Jain P, Vineis P, Liquet B, Vlaanderen J, Bodinier B, van Veldhoven K, Kogevinas M, Athersuch TJ, Font-Ribera L, Villanueva CM, Vermeulen R, Chadeau-Hyam M (2018) A multivariate approach to investigate the combined biological effects of multiple exposures. J Epidemiol Community Health 72(7):564–571. https://doi.org/10.1136/jech-2017-210061

Jombart T, Pontier D, Dufour AB (2009) Genetic markers in the playground of multivariate analysis. Heredity 102(4):330–341. https://doi.org/10.1038/hdy.2008.130

Kivelä M, Arenas A, Barthelemy M, Gleeson J, Moreno Y, Porter M (2013) Multilayer networks. J Complex Netw 2(3):203–271

Le Cao KA, Rossouw D, Robert-Granie C, Besse P (2008) A sparse PLS for variable selection when integrating omics data. Stat Appl Genet Mol Biol 7(1):35

Le Cao KA, Martin PGP, Robert-Granie C, Besse P (2009) Sparse canonical methods for biological data integration: application to a cross-platform study. BMC Bioinformatics 10:34. https://doi.org/10.1186/1471-2105-10-34

Li S, Park Y, Duraisingham S, Strobel FH, Khan N, Soltow QA, Jones DP, Pulendran B (2013) Predicting network activity from high throughput metabolomics. PLoS Comput Biol 9(7): e1003123. https://doi.org/10.1371/journal.pcbi.1003123

Lindstrom MJ, Bates DM (1990) Nonlinear mixed effects models for repeated measures data. Biometrics 46(3):673–687. https://doi.org/10.2307/2532087

Liquet B, Le Cao K-A, Hocini H, Thiebaut R (2012) A novel approach for biomarker selection and the integration of repeated measures experiments from two assays. BMC Bioinformatics 13 (1):325

Liquet B, Bottolo L, Campanella G, Richardson S, Chadeau-Hyam M (2016a) R2GUESS: a graphics processing unit-based R package for Bayesian variable selection regression of multivariate responses. J Stat Softw 69(2). https://doi.org/10.18637/jss.v069.i02

Liquet B, Lafaye de Micheaux P, Hejblum B, Thiebaut R (2016b) Group and sparse group partial least square approaches applied in genomics context. Bioinformatics 32(1):35–42
McCreanor J, Cullinan P, Nieuwenhuijsen MJ, Stewart-Evans J, Malliarou E, Jarup L, Harrington R, Svartengren M, Han IK, Ohman-Strickland P, Chung KF, Zhang J (2007) Respiratory effects of exposure to diesel traffic in persons with asthma. N Engl J Med 357 (23):2348–2358. https://doi.org/10.1056/NEJMoa071535
McHale CM, Zhang LP, Lan Q, Vermeulen R, Li GL, Hubbard AE, Porter KE, Thomas R, Portier CJ, Shen M, Rappaport SM, Yin SN, Smith MT, Rothman N (2011) Global gene expression profiling of a population exposed to a range of benzene levels. Environ Health Perspect 119 (5):628–634. https://doi.org/10.1289/ehp.1002546
Musumarra G, Condorelli DF, Fortuna CG (2011) OPLS-DA as a suitable method for selecting a set of gene transcripts discriminating RAS- and PTPN11-mutated cells in acute lymphoblastic leukaemia. Comb Chem High Throughput Screen 14(1):36–46
Parkhomenko E, Tritchler D, Beyene J (2009) Sparse canonical correlation analysis with application to genomic data integration. Stat Appl Genet Mol Biol 8:1. https://doi.org/10.2202/1544-6115.1406
Patterson N, Price AL, Reich D (2006) Population structure and eigenanalysis. PLoS Genet 2(12): e190. https://doi.org/10.1371/journal.pgen.0020190
Pearson K (1901) On lines and planes of closest fit to systems of points in space. Philos Mag 2 (6):559–572
Price AL, Patterson NJ, Plenge RM, Weinblatt ME, Shadick NA, Reich D (2006) Principal components analysis corrects for stratification in genome-wide association studies. Nat Genet 38(8):904–909. https://doi.org/10.1038/ng1847
Rappaport SM, Smith MT (2010) Environment and disease risks. Science 330(6003):460–461. https://doi.org/10.1126/science.1192603
Reich D, Price AL, Patterson N (2008) Principal component analysis of genetic data. Nat Genet 40 (5):491–492. https://doi.org/10.1038/ng0508-491
Rider CV, Carlin DJ, Devito MJ, Thompson CL, Walker NJ (2013) Mixtures research at NIEHS: an evolving program. Toxicology 313(2–3):94–102. https://doi.org/10.1016/j.tox.2012.10.017
Robinson O, Basagana X, Agier L, de Castro M, Hernandez-Ferrer C, Gonzalez JR, Grimalt JO, Nieuwenhuijsen M, Sunyer J, Slama R, Vrijheid M (2015) The pregnancy exposome: multiple environmental exposures in the INMA-Sabadell birth cohort. Environ Sci Technol 49 (17):10632–10641. https://doi.org/10.1021/acs.est.5b01782
Salamanca Beatriz V, Ebbels Timothy MD, Iorio Maria D (2014) Variance and covariance heterogeneity analysis for detection of metabolites associated with cadmium exposure. Stat Appl Genet Mol Biol 13:191–201. https://doi.org/10.1515/sagmb-2013-0041
Schafer J, Strimmer K (2005) A shrinkage approach to large-scale covariance matrix estimation and implications for functional genomics. Stat Appl Genet Mol Biol 4:32. https://doi.org/10.2202/1544-6115.1175
Shen HP, Huang JHZ (2008) Sparse principal component analysis via regularized low rank matrix approximation. J Multivar Anal 99(6):1015–1034. https://doi.org/10.1016/j.jmva.2007.06.007
Simon N, Friedman J, Hastie T, Tibshirani R (2011) Regularization paths for cox's proportional hazards model via coordinate descent. J Stat Softw 39(5):1–13
Simon N, Friedman J, Hastie T, Tibshirani R (2013) A sparse-group lasso. J Comput Graph Stat 22 (2):231–245. https://doi.org/10.1080/10618600.2012.681250
The Gene Ontology Consortium (2017) Expansion of the gene ontology knowledgebase and resources. Nucleic Acids Res 45(D1):D331–D338. https://doi.org/10.1093/nar/gkw1108
Tibshirani R (1996) Regression shrinkage and selection via the lasso. J R Stat Soc Series B Stat Methodol 58(1):267–288. https://doi.org/10.2307/2346178
Valcarcel B, Wurtz P, al Basatena NKS, Tukiainen T, Kangas AJ, Soininen P, Jarvelin MR, Ala-Korpela M, Ebbels TM, de Iorio M (2011) A differential network approach to exploring differences between biological states: an application to prediabetes. PLoS One 6(9):e24702. https://doi.org/10.1371/journal.pone.0024702

Valcarcel B, Ebbels TMD, Kangas AJ, Soininen P, Elliot P, Ala-Korpela M, Jarvelin MR, de Iorio M (2014) Genome metabolome integrated network analysis to uncover connections between genetic variants and complex traits: an application to obesity. J R Soc Interface 11 (94):20130908. https://doi.org/10.1098/rsif.2013.0908

van Veldhoven K, Keski-Rahkonen P, Barupal DK, Villanueva CM, Font-Ribera L, Scalbert A, Bodinier B, Grimalt JO, Zwiener C, Vlaanderen J, Portengen L, Vermeulen R, Vineis P, Chadeau-Hyam M, Kogevinas M (2017) Effects of exposure to water disinfection by-products in a swimming pool: a metabolome-wide association study. Environ Int 111:60–70. https://doi.org/10.1016/j.envint.2017.11.017

Vineis P, Perera F (2007) Molecular epidemiology and biomarkers in etiologic cancer research: the new in light of the old. Cancer Epidemiol Biomark Prev 16(10):1954–1965

Vineis P, Chadeau-Hyam M, Gmuender H, Gulliver J, Herceg Z, Kleinjans J, Kogevinas M, Kyrtopoulos S, Nieuwenhuijsen M, Phillips DH, Probst-Hensch N, Scalbert A, Vermeulen R, Wild CP (2016) The exposome in practice: design of the EXPOsOMICS project. Int J Hyg Environ Health 220(2 Pt A):142–151. https://doi.org/10.1016/j.ijheh.2016.08.001

Vlaanderen J, van Veldhoven K, Font-Ribera L, Villanueva CM, Chadeau-Hyam M, Portengen L, Grimalt JO, Zwiener C, Heederik D, Zhang X, Vineis P, Kogevinas M, Vermeulen R (2017) Acute changes in serum immune markers due to swimming in a chlorinated pool. Environ Int 105:1–11. https://doi.org/10.1016/j.envint.2017.04.009

Wang H, Gottfries J, Barrenäs F, Benson M (2011) Identification of novel biomarkers in seasonal allergic rhinitis by combining proteomic, multivariate and pathway analysis. PLoS One 6(8): e23563. https://doi.org/10.1371/journal.pone.0023563

West M (2003) Bayesian factor regression models in the “large p, small n” paradigm. Bayesian statistics 7. Clarendon Press, Oxford

Westfall P, Young S (1993) Resampling-based multiple testing: examples and methods for *p*-value adjustment (Wiley Series in Probability and Statistics). Wiley-Interscience

Wild CP (2005) Complementing the genome with an ‘exposome’: the outstanding challenge of environmental exposure measurement in molecular epidemiology. Cancer Epidemiol Biomark Prev 14(8):1847–1850. https://doi.org/10.1158/1055-9965.EPI-05-0456

Witten DM, Tibshirani R, Hastie T (2009) A penalized matrix decomposition, with applications to sparse principal components and canonical correlation analysis. Biostatistics 10(3):515–534. https://doi.org/10.1093/biostatistics/kxp008

Wold S, Ruhe A, Wold H, Dunn WJ (1984) The collinearity problem in linear-regression - the partial least-squares (PLS) approach to generalized inverses. SIAM J Sci Stat Comput 5 (3):735–743. https://doi.org/10.1137/0905052

Yap IKS, Brown IJ, Chan Q, Wijeyesekera A, Garcia-Perez I, Bictash M, Loo RL, Chadeau-Hyam M, Ebbels T, Iorio MD, Maibaum E, Zhao L, Kesteloot H, Daviglus ML, Stamler J, Nicholson JK, Elliott P, Holmes E (2010) Metabolome-wide association study identifies multiple biomarkers that discriminate north and south chinese populations at differing risks of cardiovascular disease: INTERMAP study. J Proteome Res 9(12):6647–6654. https://doi.org/10.1021/pr100798r

Zhou H, Sehl ME, Sinsheimer JS, Lange K (2010) Association screening of common and rare genetic variants by penalized regression. Bioinformatics 26(19):2375–2382. https://doi.org/10.1093/bioinformatics/btq448

Zou F, Fine JP, Hu J, Lin DY (2004) An efficient resampling method for assessing genome-wide statistical significance in mapping quantitative trait loci. Genetics 168(4):2307–2316. https://doi.org/10.1534/genetics.104.031427

Zou H, Hastie T (2005) Regularization and variable selection via the elastic net. J R Stat Soc Series B Stat Methodol 67(2):301–320. https://doi.org/10.1111/j.1467-9868.2005.00503.x

Zou H, Hastie T, Tibshirani R (2006) Sparse principal component analysis. J Comput Graph Stat 15 (2):265–286. https://doi.org/10.1198/106186006x113430

Chapter 12
Exposome-Wide Association Studies: A Data-Driven Approach for Searching for Exposures Associated with Phenotype

Chirag J. Patel

Abstract The promise of a unified way to measure the human exposome is the discovery of novel environmental factors associated with and potentially causative of disease. The human exposome has been tentatively defined as the totality of environmental exposures such as dietary nutrients, pharmaceutical drugs, infectious agents, and pollutants encountered from birth to death. Much as human genetics has benefited from high-throughput profiling in the form of genome-wide association studies (GWAS), a data-driven paradigm for the exposome is needed to systematically and reproducibly discover the environmental determinants of disease. In this chapter, we describe methods for associating the exposome with phenotypic state such as disease. Specifically, this chapter will describe hands-on analytic examples and data to search the exposome for correlates with phenotype, called the "environment/exposome-wide association study" (EWAS). First, we will describe the philosophy behind such a study, including transparency and mitigation of the chances for selection biases. Second, we will describe how to mitigate chances of type 1 error and investigate the possibility of true signals in a sea of possible false positives. We will describe open-source tools for visualization and display of correlated data to enable investigators to efficiently ascertain patterns in phenotypic associations. We end by describing a few success stories of the approach.

Keywords EWAS · Exposome-phenotype associations · Open source tools

Introduction and Overview

In a new era of high-throughput environmental exposure assessment of the exposome, there is an urgent need for new analytics approaches to drive discovery of new exposures associated with disease and phenotype (Patel and Ioannidis 2014a). This is particularly critical as many burdensome diseases are complex and

C. J. Patel (✉)
Department of Biomedical Informatics, Harvard Medical School, Boston, MA, USA
e-mail: Chirag_Patel@hms.harvard.edu

S. Dagnino, A. Macherone (eds.), *Unraveling the Exposome*,
https://doi.org/10.1007/978-3-319-89321-1_12

a combination of both hereditary factors and environmental factors. For example, the heritability—or the phenotypic variation in the population attributable to inherited factors—is on average 50%, leaving the other 50% of phenotypic variation to specific influences of the environment (Polderman et al. 2015). However, we lack analytic tools and data to discover new exposures to explain missing phenotypic variation in the population. In contrast, genomic investigations have accelerated the pace for discovery of inherited factors in disease at rapid pace and the same advances should be applied to discover exposures in disease (Manrai et al. 2017).

In this chapter, we present approaches called "exposome-wide association studies" or equivalently "environment-wide association studies" (EWASs) as one way to drive discovery of new exposures in disease and the missing phenotypic variation in the population (e.g., Patel et al. 2010, 2012, 2013; McGinnis et al. 2016). As will be described, EWASs offer a few advantages, including explicitly mitigating false positive findings and assessing the entire database of potential environmental correlates systematically to avoid a fragmented literature of associations (Ioannidis 2016; Ioannidis et al. 2011; Patel et al. 2015a). While far from causal (and observational), the associations that emerge are those that can be prioritized to investigate in biological experiments. For example, Patel et al. (2016b), after performing an EWAS-like approach to find exposure factors putatively correlated with telomere length, investigated how the exposure factors potentially influence changes in gene expression using publicly available data from the *Gene Expression Omnibus*. Exposures must influence changes in biological function if causal (Gibson 2008) and gene expression investigations are but one approach to take to decipher causal routes to disease.

However, the big data of the exposome poses analytical challenges. EWASs are observational studies and prone to biases such as confounding and reverse causality (e.g., disease coming before the exposure) (Ioannidis et al. 2009). While there are analytic and epidemiological study designs that attempt to mitigate these biases, they have yet to be harnessed in high-throughput exposure investigations. Yet another issue includes multiple significant associations with small effect sizes. For example, Patel et al. (2016a) recently performed a "prescription-wide association study" in time-to-cancer in a large and comprehensive Swedish database and found of the >500 drugs prescribed, almost a quarter of them were associated with cancer with small effect. Furthermore, these effects changed as a function of analytic method chosen (e.g., case-crossover vs. Cox proportional hazard), leaving an investigator with many potential false positives to sift through. This may be the norm in exposome-wide analytics, where correlations may be small but seemingly correlated with everything else and it will be a challenge to ascertain what causes what (Patel and Ioannidis 2014b; Patel and Manrai 2015).

Despite these challenges, the exposome paradigm may bear fruit in enhancing discoveries to identify factors that have yet been elusive in disease risk. Moving forward, new technologies (as covered in previous chapters) to measure environmental exposures in large populations in a cost-effective manner are needed, along with harnessing sound epidemiological designs to extract potential signals from noise, with data deposited in the public domain for investigators of all ilk to analyze results (Patel et al. 2016c). In the following chapter, we give an overview of the off-the-shelf methods one can utilize, datasets, and tools for accelerating the process of discovery while in the face of a multiplicity of exposures to potentially prioritize. We encourage readers to try hands on examples and code to associate exposures with phenotype on their own here: http://www.chiragjpgroup.org/exposome-analytics-course/.

From Genetic to Environmental Architecture of Disease and Phenotype

Phenotypes are the essence of examination in biomedical investigations and are defined as the manifestation of a trait, such as eye color, hair color, height, weight, disease, and even how genes are expressed. Further, phenotypes are, to a varying degree, influenced by inherited (genetic) and environmental factors, the primary subject of this book. Therefore, the canonical equation biologists might refer to describe this relationship includes $P = G + E$, where P is phenotype (e.g., height, weight, eye color, diabetes), G are inherited genetic factors transmitted through the genome, and E are environmental factors, such as diet, drugs, and infection.

For example, think about your family: You might recall how your own children might share eye color with your parents, or how diabetes might be pervasive among your relatives. Eye color and diabetes in these examples are influenced by both the factors that are passed down through generations and encoded in our genomes, or passed down through generations just by living together (known as shared environment). However, as will be described later, much of the influence in complex traits—or phenotypes that have *multiple* genes and exposures interacting to influence a phenotype, such as expression of a disease—is often due to *specific* environment, or environmental exposures individuals face from birth to death.

The phenomena of environmental exposure is complex (Patel and Ioannidis 2014a, b). As discussed throughout this book, it is a priority to identify *multiple* factors associated with disease in human populations. However, most associational and epidemiological studies consider only a handful exposures at a time and we currently lack data-driven methods to associate, and discover, multiple environmental exposures with phenotype and disease. This is of concern as complex diseases are multifactorial—caused by multiple genetic and environmental factors operating additively or in concert (Schwartz and Collins 2007). For example, the Genome-Wide Association Study (GWAS) is a low-cost, commoditized, and popular framework used by researchers to evaluate genetic factors in disease or phenotype along the entirety of the genome (Visscher et al. 2012; Hindorff et al. 2009). Due to wide accessibility of genome-scale GWAS assays, which can ascertain millions of genetic variants simultaneously, human geneticists have now moved from studying a handful of genetic variants at a time to a more data-driven, comprehensive, systematic, and agnostic reporting of robust genetic associations and their replication in independent populations. As a result, >500 GWAS have been published, often over 20 for specific diseases such as T2D (Welter et al. 2014) in sample sizes now reaching in the hundreds of thousands (e.g., Wood et al. 2014). GWAS has strengthened the epidemiological process and methodology of screening and validating genetic variants, debunking the candidate gene approach where variants were assessed in phenotype based on prior hypothesis. It is said most candidate gene studies are not reproducible (Ioannidis et al. 2001, 2009). Here, we claim analogous methods for environmental exposure can enable data-driven discovery of multiple environmental exposures that reflect a mixture of individual factors associated with disease.

Specifically, in this chapter, we propose an analogous framework to GWAS, called "Environment-wide association study" or equivalently exposome-wide association studies (EWASs), to search for and analytically validate environmental factors associated with continuous phenotypes or discrete ones such as disease. This type of question is different from a hypothesis-driven approach in which candidates or handful of environmental factors are chosen a priori and tested individually in their association to a phenotype and analogous to questions facilitated by GWAS. For a broad overview of the role of data-driven analytics and informatics in high-throughput exposure biology, the reader is encouraged to read Manrai et al. (2017).

But first, what accelerated genomic discovery through GWAS? The primary driver for GWAS was accessibility of low-cost, high-throughput, and unified genome-wide assays, giving human geneticists and genetic epidemiologists (Wild 2012; Wild et al. 2013) a more comprehensive way to search for genetic factors in disease. Currently, this is elusive in environmental epidemiological studies today (Patel and Ioannidis 2014a). To meet this challenge, as is the primary subject of this book, scientists have called for efforts to elucidate and measure the *exposome*, the environmental exposure analog to the genome, whereby a higher-throughput battery of environmental exposures are ascertained in humans from infection, pollutants, and nutrients simultaneously throughout the lifecourse (Dennis et al. 2017; Rappaport et al. 2014; Rappaport 2012; Louis et al. 2012; Miller and Jones 2014).

As such, we have few epidemiological cohorts that ascertain the *exposome* to facilitate the identification of mixtures associated with disease. However, one cross-sectional study, the National Health and Nutrition Examination Survey (e.g., Centers for Disease Control and Prevention (CDC 2013a, b) and http://www.cdc.gov/nchs/nhanes.htm) is one epidemiological survey to enable exposome and mixtures research. In fact, readers can download the data to their own computers and begin assessing the correlation between the exposome and phenotype (Patel et al. 2016c) and http://www.chiragjpgroup.org/exposome-analytics-course/.

The NHANES is a cross-sectional survey representative of the United States. It is comprised of both health questionnaire and laboratory and clinical data using a multistage probability sampling design, and the Centers for Disease Control and Prevention—the administrators of the survey—collected information through in-person interviews, physical measurement at mobile examination centers, and human samples. To date, the NHANES is a gold standard for ascertainment of quantitative exposome measurements in human tissue, consisting of over 300 individual biomarkers of environmental exposure. For a subset of the population, longer-term follow-up information such as cause of death and utilization of Medicare is also available to the public.

In fact, we have used NHANES to conduct EWASs in multiple phenotypes, including to identify in a data-driven manner, multiple exposures associated with type 2 diabetes (Patel et al. 2010), blood pressure (Tzoulaki et al. 2012; McGinnis et al. 2016), serum lipid levels (Patel et al. 2012) (Fig. 12.1), all-cause mortality (Patel et al. 2013) (Fig. 12.2), mothers with preterm birth (Patel et al. 2014), and even correlates of income (Patel et al. 2015b). These analyses provide illustrations of the potential for identifying multiple factors in populations and we encourage readers to read them for broader details. We begin our description of EWAS by briefly introducing the genomic complement and analog, GWAS. Second, we describe the EWAS framework and we describe the current EWAS methodology. Finally, we discuss our results and posit ways to extend the EWAS methodology.

Methods Background

Genome-Wide Association to Disease

With the sequencing of the genome and projects that characterized common genetic variation such as the HapMap, investigators are now able to interrogate how genome-wide genetic differences are associated with disease and disease-related phenotypes on an epidemiological scale (Hardy and Singleton 2009). These revolutionary studies, known as "genome-wide association studies" (GWAS), have enabled investigators to ask what common genetic loci are associated to a phenotype in an agnostic, systematic, and comprehensive way with explicit control of multiple comparisons.

Specifically, during the HapMap project, common single-nucleotide (SNP) variants and have been catalogued on basis of their population frequency ($\geq$10% population frequency), and major and minor allele versions (Manolio et al. 2008). The location of each SNP along the genome is referred to as a "locus" and the presence of variation at a particular locus denotes a "polymorphism" or a "polymorphic" locus. "Common" polymorphisms are those that occur at approximately greater than 5–10% in the population. Thus, by definition, a "common" SNP must reside at a polymorphic locus. There are greater than one million common SNPs in the genome (International HapMap, Consortium 2005). While SNPs are the most common type of polymorphism in the genome accounting for 90% of genetic variation, many other types of genetic variation exist, such as copy number variants (e.g., insertions and deletions of segments of the genome).

GWAS relate traits to variation at each—or a large subset of—common polymorphic locus in the genome and are enabled by genomic technologies, known as "SNP microarrays," which can assay greater than one million loci simultaneously for an individual. These microarrays are now mere commodity items, like computers, making accessible genome-wide measurements on a large number of individuals.

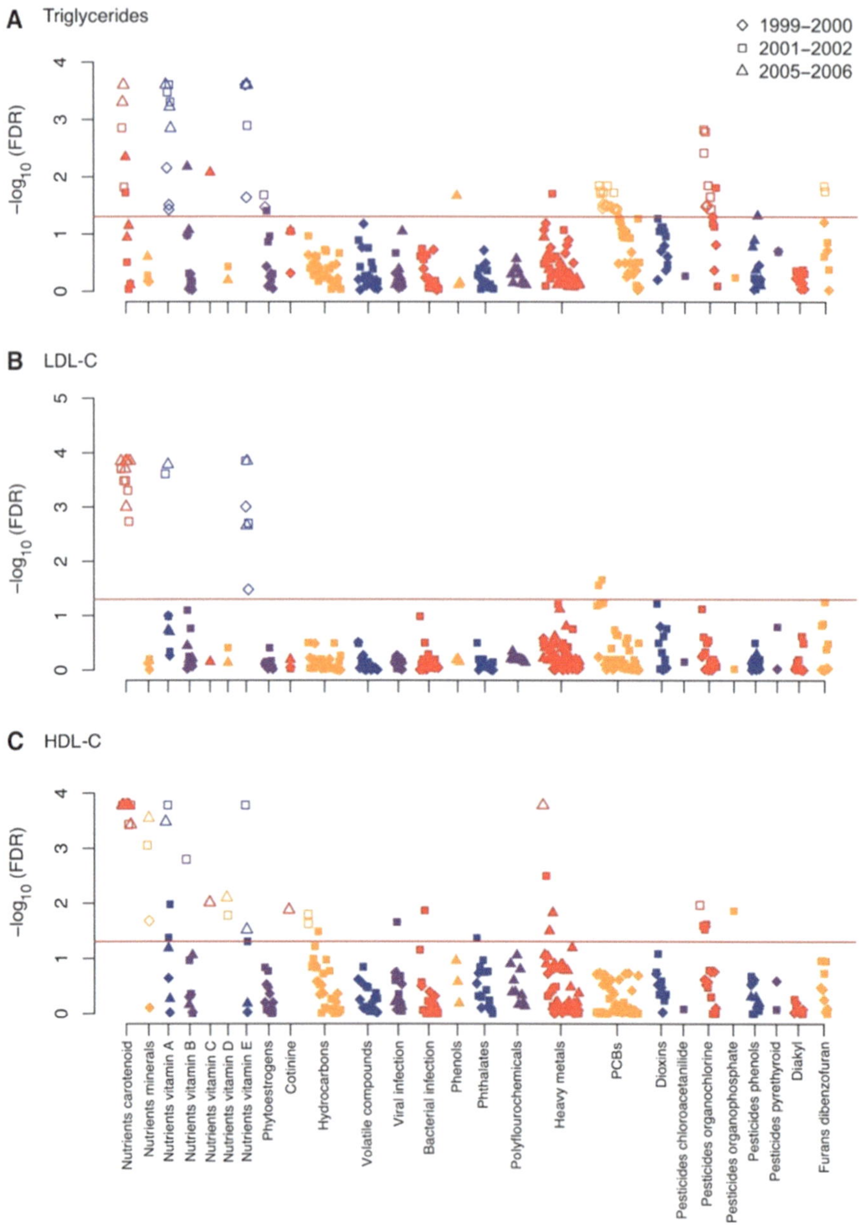

Fig. 12.1 Manhattan plots of exposures associated with lipid levels (reproduced with permission from Patel et al. 2012) in the NHANES participants. Significance of association [−log10(FDR) (Benjamini and Hochberg 1995)] for each of 188 factors by survey in association to (**a**) triglycerides, (**b**) LDL-C, (**c**) HDL-C. Y-axis indicates −log10(FDR) of the adjusted linear regression coefficient for each of the environmental factors. Colors represent different environmental categories. Red line corresponds to an FDR of 0.05. Findings replicated in an independent survey are seen in the open markers

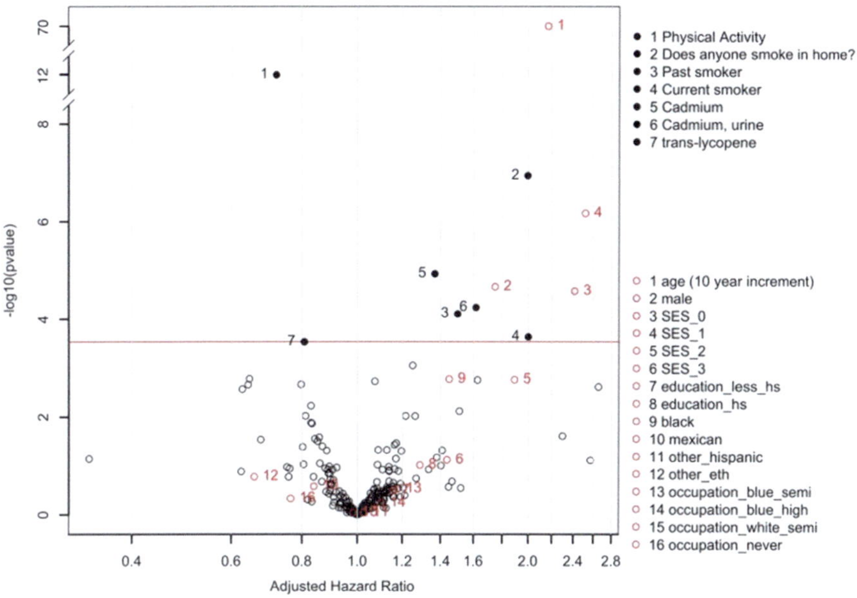

Fig. 12.2 Volcano plot of exposures and behaviors associated with death (reproduced from Patel et al. 2013). Volcano plot of 249 exposome-related and behavioural factor associations with all-cause mortality. Red horizontal line denotes FDR-adjusted level of statistical significance (FDR = 5%, *P*-value = 0.0003). Red points show the standard demographic and socio-economic (external exposome) factors considered as adjustments. For SES: SES_0: 1st quantile of SES, SES_1: 2nd quantile of SES, SES_2: 3rd quantile of SES, SES_3: 4th quantile of SES; SES HR are relative to highest quintile of SES. For education: education_hs: high school education, education_less_hs: less than high school education, education HR relative to greater than high school education. For occupation: occupation_blue_semi: semi blue collar, occupation_blue_high: high blue-collar, occupation_white_semi: semi white-collar, occupation_never: never worked. Filled black markers denote validated factors. –log10(*P*-value) for physical activity and age are annotated in parentheses, since they are extreme. Y-axis is discontinuous to accommodate higher –log10(*P*-values) for physical activity and age. Exposome-rleated factors including lycopene (an indicator of nutrition) was associated with longer survival; however. Exposome factors such as serum and urine cadmium were associated with hazards of greater than 1.5, indicative of a 50% increased risk for death versus individuals with 1SD lower levels of the logged exposure value

Further, these technology platforms are known to have very low measurement error (Ioannidis 2016; Ioannidis et al. 2009).

GWAS are constructed by recruiting thousands of individuals with ("cases") and without a trait or disease ("controls"). Genotype frequency at each locus across the genome is then compared between cases and controls using common statistical tests such as chi-squared test (Pearson and Manolio 2008), assuming the independence between each locus. Continuous traits, such as levels of a biomarker, may also be related to genetic variation by modeling the continuous phenotype in a linear regression model (Frayling et al. 2007). Multiple comparisons are accounted for through conservative Bonferroni adjustment, and significant loci are validated in independent populations.

Preceding GWAS were "candidate gene studies," a hypothesis-driven study to correlate a handful of genetic variants with a trait of interest using a "smaller" sample size. As a consequence of lack of power (small sample sizes) and prohibitive genotyping cost, the agnostic, comprehensive, and systematic analytical and validation procedure of GWAS eluded traditional genetic association studies (NCI-NHGRI Working Group on Replication in Association Studies et al. 2007; Manolio et al. 2009; Goldstein 2009). To facilitate discussion regarding "Environment-wide association," we describe these "agnostic," "systematic," and "comprehensive" characteristics of GWAS.

Exposome-Wide Association to Disease

In the following, we propose a study design analogous to GWAS, called "or equivalently, "exposome-wide association study" (EWAS), to search for and analytically validate environmental factors associated with complex diseases and phenotypes. EWAS assumes similar "data structure" to that of GWAS. Recall that in GWAS, multiple genetic factors are assayed along with phenotypic information on each individual. In other words, the genetic factors are the independent variables, and the phenotype is the dependent variable.

Specifically, the quantity or presence of environmental factors is directly measured on each individual, such as the amount of a chemical in bodily tissue, or a proxy measure, such as self-report historical exposure. Second, the environment is a dynamic entity, unlike the data structure of GWAS. Thus, the dimension of time may also be added to the structure of EWAS data structure, framing it in a longitudinal context.

GWAS variables are "binned" by their chromosomal location, facilitating the description of their correlation structure—known as Linkage Disequilibrium (LD)—when visualizing associations. Specifically, LD is the correlation of the two loci in the genome. Further, LD is a function of the relative location of the two loci, that is, the closer together two loci are on a chromosome, in general, the higher their LD. Suppose we are considering one locus: In this scenario, we inherit alleles from our parents, one from the mother and one from the father. The genotype at one locus is a random event and is dependent on the frequency of alleles present at that one locus in the mother and the father. Now suppose two loci (two sets of genotypes) are in "LD." This means that their pattern of inheritance are correlated, that is, the occurrence of a particular allele "A" at a locus *A* and "B" at a locus *B* are *nonrandom*, or dependent with respect to one another. In other words, the presence of one allele can predict the presence of another. LD among different populations has been characterized by the HapMap project and is ongoing with the 1000 Genomes project (International HapMap, Consortium 2005). In GWAS, LD structure is important. First, since we are but only assaying a prevalent subset of polymorphic loci, LD allows us to narrow down what variants might be causal; for example, given an association signal for a variant, the *causal* variant might be one in strong LD with it. LD also gives us an internal gauge of validity; for example, given a strong association signal of a variant at loci *X*, one would expect measured common variants that are also in LD with *X* to also harbor some signal.

At present, LD in EWAS is qualitative, not quantitative as in GWAS. In our applications, we binned factors according to categories that described the compound "class," had shared environmental health "relevance," or described some other arbitrary shared characteristic as a group of factors. We anticipate, as investigators characterize the exposome, that these categories will encompass assays for stress, microbial flora, drugs, noise, and ecological measures. A research effort will be to fully characterize the LD of the exposome including their correlation/covariance structure and population-wide prevalence as has been done with the HapMap.

EWAS achieves the agnostic, systematic, and comprehensive qualities that characterize GWAS. First, instead of testing a few environmental associations at a time, EWAS evaluates multiple environmental factors agnostically. EWAS is comprehensive in that each factor measured is associated with phenotype. Next, associations are systematically adjusted for multiplicity of comparisons. Further, EWAS calls for validation of significant associations in an independent population.

The EWAS framework calls for systematic and comprehensive sensitivity analyses of highly significant or validated factors. Specifically, all possible measured confounders are included in final models and their effect on the estimate of the environmental factor is assessed. Finally, given the dense web of correlation for nongenetic measures, such as between environmental factors and clinical measures, the correlation structure between validated environmental factors and risk factors is systematically computed and visualized to understand the degree of their interdependence called "exposome globes" (Patel and Ioannidis 2014b; Patel and Manrai 2015). By visualizing relationships in this way, we can infer mixtures of nonindependent exposures associated with phenotype, similar to "relevance network" or clustering analyses (Butte and Kohane 2000).

Conceptual Challenges in Data-Driven Studies of Environmental Exposures in Disease

Challenges and biases related to observational studies influence all association studies, be it from hypothesis-driven candidate factor study, GWAS, or EWAS. In contrast to "gold-standard" randomized trial study data, both environmental epidemiological studies rely on observational study data, such as longitudinal cohort, case–control, or cross-sectional data. Both types of epidemiological studies are subject to confounding biases that hinder causal inference and are avoided, to some degree, in randomized studies; however, the gold-standard scenario of a clinical trial is not suited to agnostic study of the exposome as it is impossible to randomize such a matrix of factors.

"Confounding" is used to describe a scenario in which a variable is correlated with both the factor of interest (the independent variable) and phenotype (dependent variable) (Greenland and Morgenstern 2011); in our analyses, the factor acts as a "proxy" to the confounding variable, resulting in a false association between the dependent and independent variables. A partial solution to this type of bias is including the confounder as a covariate in the statistical model, or "controlling" for the confounder. This, of course, is only possible when the confounder is known and measured.

Famous examples include associations derived from observational studies later contradicted by randomized control trials (RCTs): (1) b-carotene, thought to have mute risk for smoking-induced cancer (Peto et al. 1981), only to be refuted by a RCT later (Omenn et al. 1996); (2) the same with vitamin E and decreased risk of coronary heart disease (CHD) (Hooper et al. 2001); and (3) even for vitamin C and CHD, relative risks between observational studies and RCTs had even switched direction (Davey and Ebrahim 2005)!

Another source of "bias" includes "reverse causality": Reverse causality leads to the failure to infer proper "forward" direction between the independent variable and the dependent variable, the phenotype. Specifically, it occurs when the independent variable comes directly or indirectly as a result of the dependent variable. For example of this, it includes a sample-wide behavioral shift due to the dependent variable, such as increased intake of a vitamin due to an adverse phenotype. If we were to associate the environmental factor, the vitamin, with the phenotype as the dependent variable, the interpretation of the model as is suggests that change in vitamin exposure leads to change in phenotype when in fact the opposite is true. These biases are especially present in case–control or cross-sectional studies in which individuals are measured at one point in time. A way to take into account the dynamic nature of nongenetic variables and biases such as reverse causality includes conducting a longitudinal study in which we may observe jointly changes in phenotype and exposure pattern as a function of time.

The nature of the environmental factor itself also biases results. First, the assessment of the quantity of environmental factors in blood and serum is subject to measurement error (Ioannidis et al. 2009) and self-report variables are subject to recall bias. Further, physiological characteristics of factors themselves influence estimates, including the variability of the kinetics of chemical factors, such as how long they are retained in accessible body tissue. For example, chemical compounds that are easily measured include those that are lipophilic, persistent in fatty tissue. As adiposity is related to both the measurement of the factor and often the phenotype of interest (e.g., metabolic syndrome), a positive correlation might indicate confounding. On the other hand, many types of factors are excreted quickly, also affecting their measurement and association to the phenotype of interest; however, "steady-state" or constant exposure might allay a kinetic effect of environmental chemicals (Bartell et al. 2004).

Importantly, current epidemiological investigations may be fraught with issues in multiplicity and these issues may be exacerbated by the dense correlational web of environmental exposure. Multiplicity signifies the space of hypothesis tests that can occur: for example, given a database of 300 exposures equates to 300 hypothesis tests for a disease phenotype. Testing multiple variables for associations with other exposures and outcomes makes possible the prospect of making discoveries, but the cost of multiplicity can lead to type I error, or false positives. Multiplicity needs to be taken into consideration when identifying multiple exposures associated with disease, and we recommend that adjustment for multiple hypotheses take place to "pay the cost" for searching for numerous exposures. One way to adjust includes the simple, but constraining, Bonferroni correction and yet another includes the false discovery rate (FDR) approach. We will describe these topics at length in the following.

EWAS Method

The EWAS methodology and analysis framework is analogous to that utilized in GWAS. First, we conduct an initial scan for environmental factors associated with a phenotype of interest through general linear modeling, such as logistic or linear regression. Since environmental association occurs in the observational (vs. randomized scenario), these models include variables that adjust for known confounders, such as clinical risk factors. Second, we account for multiple hypotheses by estimating the false discovery rate (FDR). Third, factors that we deem significantly associated with the phenotype beyond the region of false discovery are "validated" in independent cohorts. Factors that are validated are considered true discoveries that should be followed in future studies.

The EWAS framework also calls for systematic sensitivity analyses, whereby validated factors are modeled under different assumptions or with additional covariates. Further, the pairwise correlation between each validated factor is computed and examined to determine their dependence, which can be interpreted as potential evidence for route of exposure or confounding through display of the correlation of identified factors (and presence of a mixture associated with a phenotype). Each step is described further below.

Stage 1: Linear Modeling

Each environmental factor is associated with a phenotype of interest using general linear models, for example, each associated with disease status (i.e., case or control) using logistic regression. Normally distributed continuous phenotypes are correlated with environmental factors with linear regression. Common risk, demographic, and clinical factors are added as adjusting variables, such as age, sex, ethnicity, socioeconomic status, as phenotypic states and environmental factors are confounded by these variables. Thus, for an environmental factor X_i in our list of measured factors $X_i \ldots X_p$, we model the disease state (Y) as a linear function of environmental factors and adjustment variables (represented by Z):

$$Y = \alpha + b_I X_i + \zeta Z$$

X_i corresponds to the environmental factor and b_i corresponds to the effect size of that factor, adjusted by other variables.

The strength of association is computed by the two-sided p-value for b_i, which tests the "null hypothesis" that b_i is equal to zero. When modeling the phenotype as the logit (logistic regression), the exponentiation of b_i serves as the odds ratio, or the change in the odds in disease versus un-diseased status for a unit change of the factor. In the linear regression setting, b_I can be interpreted as the change in phenotype per unit change of the factor. P-values are computed through common tests of significance, such as Wald tests.

Continuous factors are z-transformed (centered about the mean and divided by their standard deviation) in order to compare the effect sizes. Many factors measured in tissue have a right skew and thus are log-transformed prior to z-transformation. Binary factors (such as the presence or absence of a factor) are standardized such that effect size reflects a unit change between exposed and unexposed status; that is, the referent is consistently the "negative" result of a binary test. Ordinal factors are left untransformed.

Stage 2: Controlling for Multiple Hypotheses by Estimating the False Discovery Rate

Given a set of "discoveries", or a list of potentially significant factors, how can we deem those that are false discoveries? In the GWAS setting, Bonferroni correction is utilized to adjust for multiple comparisons. The Bonferroni adjustment is straightforward: It simply divides the significance threshold a for the total number of tests conducted. This adjustment guarantees the "family-wide error rate"—the probability of having one or more false positives in a set of results is equivalent to a setting in which only one hypothesis was tested at level a. However, the threshold is conservative and therefore we lose power for detection.

To account for multiple comparisons, in EWAS we compute an empirical estimate of the False Discovery Rate (FDR) derived through permutations of the phenotype multiple times, effectively creating a "null distribution" of test statistics. In contrast to the Bonferroni correction, the FDR provides a quantitative estimate of the number of false positives in a set of "discoveries." The FDR is less conservative and therefore more powerful than the Bonferroni correction (Noble 2009). Further, since our estimate of the FDR utilizes the data itself, it inherently considers the covariance structure of the data, an important quality given the dense correlation of nongenetic factors (Noble 2009).

The FDR is the estimated proportion of false discoveries made versus the number of real discoveries made for a given significance level a, to control for multiple hypothesis testing. To estimate the number of false discoveries, we create a "null distribution" of regression test statistics shuffling the phenotype a large number of times (100–1000) and refit the regression models. The FDR is the ratio of the proportion of results that were called significant at a given level a in the null distribution and the proportion of results called significant from our real tests. We use a significance level that corresponds to FDR of 5–10% to select associations.

The pseudo-code to compute the FDR follows:

```
1. Do: Stage 1, Linear Modeling.
2. nullPvalues <- NewList()
3. For i in [1...numberPermutations]:
4 randomPheno <- permutePhenotypeWithoutReplacement(phenotype)
5. For xi in [X1...Xp]:
6. Modi <-GeneralLinearModel(randomPheno,xi,Xses,Xeth,Xsex,
Xage)
7. ListAppend(nullPvalues, getPvalue(Modeli, xi))
8. fdrRaw <- []
9. for pvalue in Pvalues:
10. numerator <- sum(nullPvalues < pvalue)/numberPermutations
11. denominator <- sum(Pvalues < pvalue)
12. listAppend(fdrRaw, numerator/denominator)
13. fdrs <- []
14. for I in [1...p]:
15. fdr <- min(rawFdr[i...p])
16. ListAppend(fdrs, fdr)
```

Algorithm 1 Computing the FDR (*q*-value) for each *p*-value during Stage 1 of EWAS.

To begin Algorithm 1, we need to have established the *p*-values for association for each factor. Then, for a number of permutations, we refit the regression model for the random phenotype for each environmental factor and collect all of these "null" *p*-values (line 3–7). For each *p*-value computed, we compute the raw FDR, or the ratio if raw number of results that are exceed that *p*-value threshold in the permuted data and the number of results that exceed that *p*-value in stage 1 (line 11, 12, 13). As FDR should be a monotonically increasing function of the *p*-value, we ensure that the FDR for a *p*-value is the minimum of the FDRs for all *p*-values equal to or greater than that *p*-value (line 15). The resulting array of FDR values corresponds to the FDR for each *p*-value computed in Stage 1.

The original method for estimating the FDR can be used (Benjamini and Hochberg 1995), eliminating the need for Algorithm 1. However, as discussed earlier, estimating the FDR through permutations of the dependent variable is preferred in the scenario in which the variables are correlated. In addition, much has been documented about what variables to permute or bootstrap. For example, it has been suggested that model residuals, the difference between the predicted and true values, should be permuted (or bootstrapped) as opposed to the original outcome variables (replacing line 4 in the algorithm). In our experience, we had similar estimates of the FDR under different documented methods of permuting. The reader is advised to refer to Manly, Efron, and Westfall and Young for more in this area (Manly 2007; Westfall and Young 1993; Efron 2010).

Stage 3: Validation

Findings deemed significant corresponding to some nominal FDR level are validated or replicated in one or more additional independent cohorts with a nominal p-value (e.g., $p = 0.05$ in the independent replication cohort). Importantly, the sign of the effect size in the validation cohort must be equivalent to that in the initial screen.

Stage 4: Sensitivity Analyses

Confounding and reverse causality influence the strength of association, bias the effect size estimate, and, in general, affect causal inference of environmental factors to phenotypes. Thus, we propose a method to begin to measure these biases approximately. However, we cannot claim to find these biases nor eliminate confounding; nevertheless, we describe methods to assess bias given that they were measured.

In the first, we systematically comb through all measured variables that were not considered in our list of environmental factors—but could influence the association—and sequentially add them to the linear model as an additional covariate. Then the p-value of association and effect size corresponding to the environmental factor calculated from the extended model is compared to original model computed in Stage 2. The difference between the extended and original factor coefficients quantifies the approximate bias due to the new variable.

Types of variables that might bias associations depend on the phenotype and environmental factors under study, but often include knowledge of clinical status (e.g., diagnosis of a disease), recent food, supplement, or drug intake, and physical activity. For example, knowledge regarding one's disease state might induce behavioral change, resulting in increased exposure to foods high in vitamins and certain nutrients; an association between these vitamin factors and disease might then be attributed to reverse causality. Or, use of a drug might induce phenotypic change, biasing estimated effects toward the null. This method is dependent on a multitude of measured potential confounders. Large epidemiological datasets arising from the public domain or of large consortia often measure many of these other clinical and behavioral nongenetic variables which can be utilized to test the "sensitivity" of the final validated effects of environmental factors associated with a phenotype.

Stage 5: Correlation Globes

The correlation/covariance structure between nongenetic measures is known to be "dense," and this structure also influences our ability to infer the independent effect of factors on phenotype as discovered in EWAS. Furthermore, our initial screen methodology assumes independence between factors and we therefore have little idea about their correlation. One can view exposome globes for individuals taking part in the NHANES here: http://www.chiragjpgroup.org/exposome_correlation/html/

Concretely, given a list of discovered factors, their joint association to the phenotype of interest might be due to their correlation, such as similar routes of exposure. We assess the degree of dependency between validated factors by computing their raw correlation coefficient (Spearman's ρ) and visualizing this with a correlation "globe" (Patel and Manrai 2015) (Fig. 12.3). Briefly, correlations between exposures allow analysts to describe how exposures can lead to other, for example, many nutrients are consumed together. A nonoptimal diet (however it may be defined) may lead to a deficiency in a whole group of vitamins and nutrients. As another example, individuals who are exposed to air pollution may have high levels of hydrocarbons, volatile compounds, and heavy metal levels.

Many methods have been proposed to describe the correlation between multiple variables, and have been used successfully in the genomics field (e.g., Langfelder and Horvath 2008; Eisen et al. 1998). We utilize the exposome globe to visualize clusters of exposures correlated with exposures identified in EWAS ("EWAS-identified exposure", Fig. 12.3b). We hypothesize it is possible to attain a broader and more interpretable view of EWAS-identified exposures with an exposome globe.

Briefly, to compute and visualize an exposome globe, a nonparametric correlation coefficient between each pair of environmental factors is estimated. These coefficients are biserial coefficients between pair of binary factors and Spearman correlations for continuous factors. There are many ways to compute correlations between variables. We chose a nonparametric metric as to not make any distributional assumptions regarding the environmental factors. Next, a permutation-based approach—similar to that described in Algorithm 1 —is used to estimate the two-sided p-value of significance for each pair of correlations. Specifically, each environmental factor is randomly permuted (sampled without replacement) and the correlations were recomputed to create a set of correlations that reflected the null distribution of no correlation. The p-value for an individual correlation is then estimated by counting the number of permuted correlations (corresponding to the null distribution) exceeding the correlation in question. Next, the FDR is estimated using the Benjamini–Hochberg step-down approach (Benjamini and Hochberg 1995).

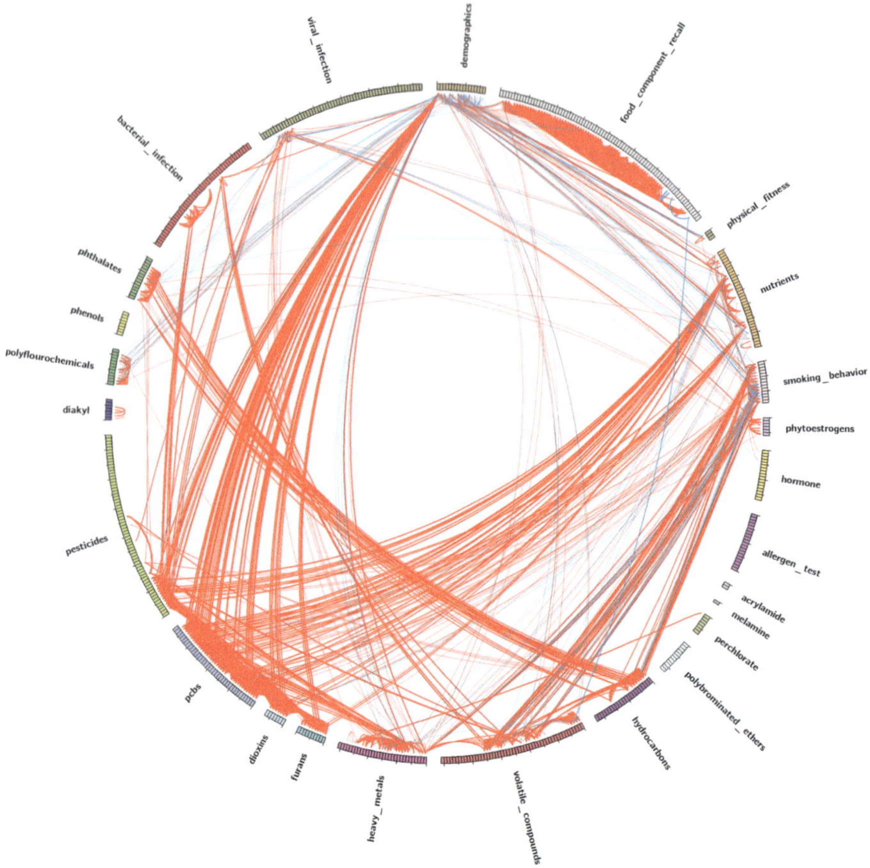

Fig. 12.3 Correlation globe of the exposome (reproduced from Patel and Manrai 2015). Each tickmark on the globe is an exposure measured in NHANES participants and an edge depicts a correlation between two exposures (see section on "Stage 5: Correlation Globes"). (**a**) All pariwise and replicated exposure correlations in NHANES. Association *p*-values from EWAS are shown as a separate track ("EWAS track") above each exposure (red points denote EWAS validated associations with positive effect size [indicating risk] blue points indicate an EWAS validated negative effect size [indicating protective]). Validated EWAS associations for T2D and all-cause mortality are offset in labeled in red or blue text. Only "first-degree" correlations (correlations for validated EWAS findings) are displayed in the globes and displayed in black text. *Acryl.* acrylamide, *Mel* melamine, *VoC* volatile organic compounds, *PCBs* polychlorinated biphenyls, *PFCs* polyfluorinated compounds. (**b**) Filtering edges for those exposures found in an EWAS on all-cause mortality. Association −log10(*p*-values) from EWAS (Fig. 12.2) are shown as a separate track above each exposure (red points denote EWAS-replicated associations with positive association size and blue points indicate an EWAS-replicated negative association size). Replicated EWAS associations for all-cause mortality (time to death) are offset in labeled in red or blue text. Only "first-degree" correlations (correlations for validated EWAS findings) are displayed in the globes and displayed in black text (as oposed to Fig. 12.3a, where all associations across NHANES are shown)

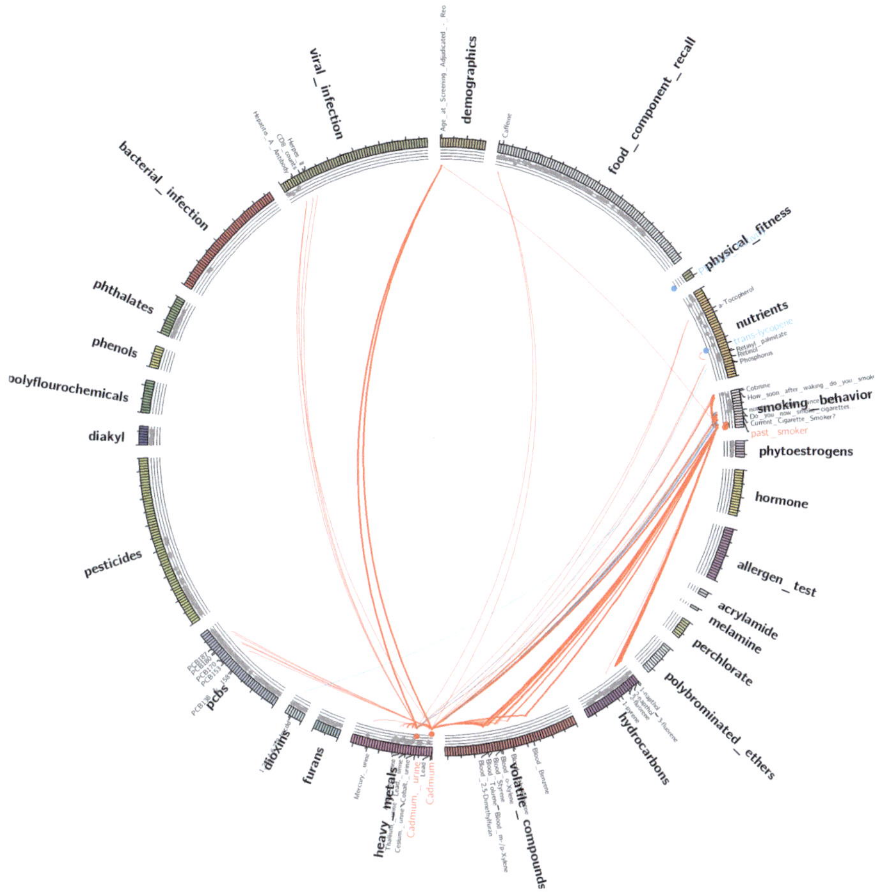

Fig. 12.3 (continued)

This exercise leaves us with an array of pairwise exposures whose interdependency is captured by a correlation coefficient. Each pairwise correlation is visualized with the Circos visualization tool kit version 0.67 (Krzywinski et al. 2009). Each individual environmental factor is grouped and arranged in a circle. Lines between factors on the inside of the circle depict replicated correlations between factors, and the thicknesses of the lines depict the absolute values of the correlations. Red and blue lines represent positive and negative correlations, respectively. By visualizing relationships in this way, we can infer nonindependent exposures associated with phenotype (Fig. 12.3).

Code to estimate correlation globes in the National Health and Nutrition Examination Survey (NHANES) can be found here: https://github.com/chiragjp/exposome_correlation. Further, one can peruse correlation globes estimated from the NHANES here: http://www.chiragjpgroup.org/exposome_correlation/html/. We claim the exposome globe is a first step toward identifying putative mixtures of exposures associated with disease. These mixtures may be a result of common routes of exposure or behaviors (e.g., foods are mixtures of nutrients or smoking behavior can result in a mixture of hydrocarbons and heavy metals). These systematic correlations may also help identify shared characteristics of exposures; for example, chlorinated persistent pollutants were all densely correlated with one another perhaps due to shared routes of exposure, but also because they happen to be lipophilic and have similar metabolic fates.

Discussion

As described above, EWAS may facilitate many different ways of screening for factors. We describe extensions that might be used off-the-shelf to accommodate longitudinal data and statistical learning methods that consider the entire matrix of dependent variables at once.

Longitudinal Data

As discussed, environmental factors are dynamic. One way to capture the dynamic relationship between environmental factors and a phenotype of interest includes repeatedly measuring individuals over time. An example includes a longitudinal cohort study, in which a cohort is followed for a certain amount of time beginning prior to disease onset, such as childhood or adolescence. This type of study design might lessen the bias of reverse causality, but not completely.

For a binary dependent variable, the Cox proportional hazard model is a common analytic model that can accommodate both time-dependent independent and dependent variables. With this model, we simply substitute line 4 of Algorithm 1 with the Cox model that inputs time-dependent variables. For both continuous and dependent variables, hierarchical modeling techniques such as generalized estimating equations may be utilized. The EWAS as described by Algorithm 1 depends on the computation of individual p-values and effect size for the environmental factor, and statistical tests for these modeling techniques provide this requirement. Calculation of the empirical FDR proceeds also in the same way.

Feature Selection: Shrinkage Methods

The EWAS screening method considers each environmental factor in a separate linear model iteratively (Algorithm 1). This makes feasible the screening and interpretation of many variables and not over-fitting the linear model (i.e., $p \ll n$, where p is the number of predictors, n is the number of individuals). However, this falsely assumes independence between environmental factors. Statistical learning methods, such as "shrinkage" methods, enable one to model the dependent variables simultaneously in the "over-determined" ($p \geq n$) setting.

Two such popular shrinkage methods include the "Lasso" (Tibshirani 1996) and "elastic net" (Zou and Hastie 2005). These methods are extensions of multivariate regression and have some relation to tree "boosting" methods (James et al. 2013) and are applicable over the generalized linear model family, including Cox proportional hazards for longitudinal data. Both the lasso and elastic net are able to fit an overdetermined model by constraining the size of coefficients ("shrinking"). Because these methods consider the entire set of independent variables simultaneously (i.e., multiple regression), Algorithm 1 is supplanted with the shrinkage procedure. Further, k-fold cross-validation is utilized to select features that have the lowest prediction variability on k number of datasets held out of the model building process (James et al. 2013).

Feature selection operates through optimizing prediction accuracy of the dependent variable and not through ordering of test-statistics of individual coefficients used in inference. Thus, we must reconfigure parts of Stage 1 (FDR estimation) and Stage 2 (Validation) to accommodate this. Reconfiguring Stage 1, we use one cohort as the "discovery" cohort, applying the shrinkage method to find factors associated with the phenotype. Within this cohort, k-fold cross-validation is applied in order to optimize prediction accuracy with prediction cohort. Thereafter, the top factors found through this method are "validated" individually in additional validation cohorts using common tools for inference (e.g., GLM). Successful validation requires low nominal p and FDR values for the validation analyses.

Of course, "classical" methods for feature selection exist in the linear regression domain, such as "forward-stepwise" and "backward-stepwise." These methods may be used to select environmental factors, but we opt out of discussion of these methods due to their high variability in subset selection due to the stepwise procedure, ultimately reducing their prediction accuracy (Vittinghoff et al. 2005). The shrinkage methods discussed above avoid this problem.

In this chapter, we have presented a straightforward and generalizable way to associate environmental variables toward identification of multiple factors in disease. Furthermore, we present a way of ranking what variables we may want to pursue for further study through computation of the FDR. Because of its proposed utility, the method has become a center point of discussion and debate. We also discussed that observational data are fraught with biases, such as confounding, and the numerous variables possibly associated with phenotype, challenge discovery, and increase the chances of false positives association, which is probably rampant today in environmental health sciences (Ioannidis 2005; Ioannidis et al. 2011). Methods such as multiplicity control and assessment of the vibration of effects (Patel et al. 2015a) are ways of mitigating such challenges, and we anticipate that with emerging data, larger sample sizes, and advanced analytic methods, we may be able to use this new and exciting data modality to discover multiple factors causally related to disease. The next challenge includes how we—health professionals, caregivers, and individuals alike—may intervene on not one or a few environmental factors, but the entirety of the exposome itself!

References

Bartell SM, Griffith WC, Faustman EM (2004) Temporal error in biomarker-based mean exposure estimates for individuals. J Expo Anal Environ Epidemiol 14:173–179

Benjamini Y, Hochberg Y (1995) Controlling the false discovery rate: a practical and powerful approach to multiple testing. J R Stat Soc Series B Stat Methodology 57(1):289–300

Butte AJ, Kohane IS (2000) Mutual information relevance networks: functional genomic clustering using pairwise entropy measurements. Pac Symp Biocomput 5:418–429

Centers for Disease Control and Prevention (CDC), and National Center for Health Statistics (NCHS) (2013a) National Health and Nutrition Examination Survey Data, 1999–2000. U.S. Department of Health and Human Services, Centers for Disease Control and Prevention, Hyattsville http://www.cdc.gov/nchs/nhanes/nhanes99_00.htm

Centers for Disease Control and Prevention (CDC), and National Center for Health Statistics (NCHS) (2013b) National Health and Nutrition Examination Survey Data, 2001–2002. U.S. Department of Health and Human Services, Centers for Disease Control and Prevention, Hyattsville http://www.cdc.gov/nchs/nhanes/nhanes01-02.htm

Centers for Disease Control and Prevention (CDC), and National Center for Health Statistics (NCHS) (2013c) National Health and Nutrition Examination Survey Data, 2005–2006. U.S. Department of Health and Human Services, Centers for Disease Control and Prevention, Hyattsville http://www.cdc.gov/nchs/nhanes/nhanes2005-2006/nhanes05_06.htm

Centers for Disease Control and Prevention (CDC), and National Center for Health Statistics (NCHS) (2013d) National Health and Nutrition Examination Survey Data, 2003–2004. U.S. Department of Health and Human Services, Centers for Disease Control and Prevention, Hyattsville http://www.cdc.gov/nchs/nhanes/nhanes2003-2004/nhanes03_04.htm

Davey Smith G, Ebrahim S (2005) What can mendelian randomisation tell us about modifiable behavioural and environmental exposures? BMJ 330(7499):1076–1079

Dennis KK, Marder E, Balshaw DM, Cui Y, Lynes MA, Patti GJ, Rappaport SM, Shaughnessy DT, Vrijheid M, Barr DB (2017) Biomonitoring in the era of the exposome. Environ Health Perspect 125(4):502

Efron B (2010) Large-scale inference. Cambridge University Press, Cambridge

Eisen MB, Spellman PT, Brown PO, Botstein D (1998) Cluster analysis and display of genome-wide expression patterns. Proc Natl Acad Sci U S A 95(25):14863–14868

Frayling T, Timpson N, Weedon M, Zeggini E, Freathy R, Lindgren C, Perry J et al (2007) A common variant in the FTO gene is associated with body mass index and predisposes to childhood and adult obesity. Science 316(5826):889–894

Gibson G (2008) The environmental contribution to gene expression profiles. Nat Rev Genet 9 (8):575–581

Goldstein D (2009) Common genetic variation and human traits. N Engl J Med 360(17):1696–1698

Greenland S, Morgenstern H (2011) Confounding in health research. Annu Rev Public Health 22:189–212

Hardy J, Singleton A (2009) Genomewide association studies and human disease. N Engl J Med 360(17):1759–1768

Hindorff LA, Sethupathy P, Junkins HA, Ramos EM, Mehta JP, Collins FS, Manolio TA (2009) Potential etiologic and functional implications of genome-wide association loci for human diseases and traits. Proc Natl Acad Sci U S A 106:9362–9367

Hooper L, Ness AR, Smith GD (2001) Antioxidant strategy for cardiovascular diseases. Lancet 357:1705–1706

International HapMap, Consortium (2005) A haplotype map of the human genome. Nature 437 (7063):1299–1320
Ioannidis JP, Ntzani EE, Trikalinos TA, Contopoulos-Ioannidis DG (2001) Replication validity of genetic association studies. Nat Genet 29(3):306–309
Ioannidis JPA (2005) Why most published research findings are false. PLoS Med 2(8):e124
Ioannidis JPA (2016) Exposure-wide epidemiology: revisiting Bradford hill. Stat Med 35 (11):1749–1762
Ioannidis JPA, En YL, Poulton R, Chia KS (2009) Researching genetic versus nongenetic determinants of disease: a comparison and proposed unification. Sci Transl Med 1(7):7ps8
Ioannidis JPA, Tarone R, McLaughlin JK (2011) The false-positive to false-negative ratio in epidemiologic studies. Epidemiology 22(4):450–456
James G, Witten D, Hastie T, Tibshirani R (2013) An introduction to statistical learning: with applications in R. Springer Texts in Statistics 103. Springer, New York
Krzywinski M, Schein J, Birol I, Connors J, Gascoyne R, Horsman D, Jones SJ, Marra MA (2009) Circos: an information aesthetic for comparative genomics. Genome Res 19(9):1639–1645
Langfelder P, Horvath S (2008) WGCNA: an R package for weighted correlation network analysis. BMC Bioinformatics 9(1):559
Louis B, Germaine M, Sundaram R (2012) Exposome: time for transformative research. Stat Med 31(22):2569–2575
Manly BFJ (2007) Randomization, bootstrap and Monte Carlo methods in biology, 3rd edn. Chapman and Hall/CRC, Boca Raton
Manolio TA, Brooks LD, Collins FS (2008) A HapMap harvest of insights into the genetics of common disease. J Clin Invest 118(5):1590–1605
Manolio TA, Collins FS, Cox NJ, Goldstein DB, Hindorff LA, Hunter DJ, Mccarthy MI et al (2009) Finding the missing heritability of complex diseases. Nature 461(7265):747–753
Manrai AK, Cui Y, Bushel PR, Hall M, Karakitsios S, Mattingly CJ, Ritchie M et al (2017) Informatics and data analytics to support exposome-based discovery for public health. Annu Rev Public Health 38(1):279–294
McGinnis DP, Brownstein JS, Patel CJ (2016) Environment-wide association study of blood pressure in the national health and nutrition examination survey (1999–2012). Sci Rep 6:30373
Miller GW, Jones DP (2014) The nature of nurture: refining the definition of the exposome. Toxicol Sci 137(1):1–2
NCI-NHGRI Working Group on Replication in Association Studies, Chanock SJ, Manolio T, Boehnke M, Boerwinkle E, Hunter DJ, Thomas G et al (2007) Replicating genotype-phenotype associations. Nature 447(7145):655–660
Noble WS (2009) How does multiple testing correction work? Nat Biotechnol 27(12):1135–1137
Omenn GS, Goodman GE, Thornquist MD, Balmes J, Cullen MR, Glass A, Keogh JP et al (1996) Effects of a combination of beta carotene and vitamin a on lung cancer and cardiovascular disease. N Engl J Med 334:1150–1155
Patel CJ, Ioannidis JPA (2014a) Studying the elusive environment in large scale. JAMA 311 (21):2173–2174
Patel CJ, Ioannidis JPA (2014b) Placing epidemiological results in the context of multiplicity and typical correlations of exposures. J Epidemiol Community Health 68(11):1096–1100
Patel CJ, Manrai AK (2015) Development of exposome correlation globes to map out environment-wide associations. Pac Symp Biocomput 20:231–242
Patel CJ, Bhattacharya J, Butte AJ (2010) An environment-wide association study (EWAS) on type 2 diabetes mellitus. PLoS One 5(5):e10746
Patel CJ, Cullen MR, Ioannidis JPA, Butte AJ (2012) Systematic evaluation of environmental factors: persistent pollutants and nutrients correlated with serum lipid levels. Int J Epidemiol 41 (3):828–843
Patel CJ, Rehkopf DH, Leppert JT, Bortz WM, Cullen MR, Chertow GM, Ioannidis JPA (2013) Systematic evaluation of environmental and behavioural factors associated with all-cause

mortality in the United States national health and nutrition examination survey. Int J Epidemiol 42(6):1795–1810
Patel CJ, Yang T, Zhongkai H, Wen Q, Sung J, El-Sayed YY, Cohen H et al (2014) Investigation of maternal environmental exposures in association with self-reported preterm birth. Reprod Toxicol 45:1–7
Patel CJ, Burford B, Ioannidis JPA (2015a) Assessment of vibration of effects due to model specification can demonstrate the instability of observational associations. J Clin Epidemiol 68:1046–1058
Patel CJ, Ioannidis JPA, Cullen MR, Rehkopf DH (2015b) Systematic assessment of the correlations of household income with infectious, biochemical, physiological, and environmental factors in the United States, 1999–2006. Am J Epidemiol 181(3):171–179
Patel CJ, Ji J, Sundquist J, Ioannidis JPA, Sundquist K (2016a) Systematic assessment of pharmaceutical prescriptions in association with cancer risk: a method to conduct a population-wide medication-wide longitudinal study. Sci Rep 6:31308
Patel CJ, Manrai AK, Corona E, Kohane IS (2016b) Systematic correlation of environmental exposure and physiological and self-reported behaviour factors with leukocyte telomere length. Int J Epidemiol 46(1):44–56. https://doi.org/10.1093/ije/dyw043
Patel CJ, Pho N, McDuffie M, Easton-Marks J, Kothari C, Kohane IS, Avillach P (2016c) A database of human exposomes and phenomes from the US national health and nutrition examination survey. Sci Data 3:160096
Pearson TA, Manolio TA (2008) How to interpret a genome-wide association study. JAMA 299 (11):1335–1344
Peto R, Doll R, Buckley JD, Sporn MB (1981) Can dietary beta-carotene materially reduce human cancer rates? Nature 290:201–208
Polderman TJC, Benyamin B, de Leeuw CA, Sullivan PF, van Bochoven A, Visscher PM, Posthuma D (2015) Meta-analysis of the heritability of human traits based on fifty years of twin studies. Nat Genet 47:702–709
Rappaport SM (2012) Discovering environmental causes of disease. J Epidemiol Community Health 66:99–102
Rappaport SM, Barupal DK, Wishart D, Vineis P, Scalbert A (2014) The blood exposome and its role in discovering causes of disease. Environ Health Perspect 122(8):769–774
Schwartz D, Collins F (2007) MEDICINE: environmental biology and human disease. Science 316:695–696
Tibshirani R (1996) Regression shrinkage and selection via the lasso. J R Stat Soc Series B Stat Methodology 58(1):267–288
Tzoulaki I, Patel CJ, Okamura T, Chan Q, Brown IJ, Miura K, Ueshima H et al (2012) A nutrient-wide association study on blood pressure. Circulation 126(21):2456–2464
Visscher PM, Brown MA, McCarthy MI, Yang J (2012) Five years of GWAS discovery. Am J Hum Genet 90(1):7–24
Vittinghoff E, Glidden D, Shiboski S, McCulloch C (2005) Regression methods in biostatistics: linear, logistic, survival, and repeated measures models. Springer, New York
Welter D, MacArthur J, Morales J, Burdett T, Hall P, Junkins H, Klemm A et al (2014) The NHGRI GWAS catalog, a curated resource of SNP-trait associations. Nucleic Acids Res 42:D1001–D1006
Westfall PH, Stanley Young S (1993) Resampling-based multiple testing. Wiley, New York
Wild CP, Scalbert A, Herceg Z (2013) Measuring the exposome: a powerful basis for evaluating environmental exposures and cancer risk. Environ Mol Mutagen 54(7):480–499
Wild CP (2012) The exposome: from concept to utility. Int J Epidemiol 41(1):24–32
Wood AR, Esko T, Yang J, Vedantam S, Pers TH, Gustafsson S, Chu AY et al (2014) Defining the role of common variation in the genomic and biological architecture of adult human height. Nat Genet 46(11):1173–1186
Zou H, Hastie T (2005) Regularization and variable selection via the elastic net. J R Stat Soc Series B Stat Methodology 67:301–320

Part V
Exposome Characterization Around the Globe

Chapter 13
HERCULES: An Academic Center to Support Exposome Research

Megan M. Niedzwiecki and Gary W. Miller

Abstract The HERCULES Exposome Research Center was first funded under the Environmental Health Sciences Core Center P30 Program by the National Institute of Environmental Health Sciences in 2013. The Core Center Program is designed to support infrastructure for research in environmental health. Emory took a unique approach and focused their proposal on the singular topic of the exposome. The idea was to build intellectual and physical infrastructure that would facilitate exposome research. Cores were developed to expand analytical capabilities in targeted and untargeted mass spectrometry and to provide support for data analysis. A major goal of HERCULES was to promote the exposome concept, which has been accomplished through a series of workshops, seminars, and courses. HERCULES has supported the development of new research centers that use the exposome approach and anticipates continued expansion of exposome research at the home institutions.

Keywords Hercules · Exposome research · Data analysis tools · Targeted and untargeted mass spectrometry

The NIEHS Core Centers Program

The National Institute of Environmental Health Sciences (NIEHS), one of 27 institutes and centers of the National Institutes of Health (NIH) in the United States, was established to discover the impacts of the environment on human health. To advance these goals, the NIEHS provides funding for Environmental Health Sciences (EHS) Core Centers, which are centralized facilities to support environmental health research efforts at academic institutions. As of 2017, there are 20 EHS Core Centers across the United States. The NIEHS states that the EHS Core Centers should support innovation and be on the cutting edge of science. Research should cross a

M. M. Niedzwiecki · G. W. Miller (✉)
Department of Environmental Health, Rollins School of Public Health, Emory University, Atlanta, GA, USA
e-mail: gary.miller@emory.edu

S. Dagnino, A. Macherone (eds.), *Unraveling the Exposome*,
https://doi.org/10.1007/978-3-319-89321-1_13

variety of disciplines, and the activities of a Core Center should produce a synergistic effect that result in greater depth, breadth, quality, innovation, and productivity beyond what individual researchers would attain by working independently.

The Exposome in Environmental Health Research

The emphasis on innovative, interdisciplinary research set forth by the NIEHS is exemplified in the concept of the *exposome*. First introduced by Christopher Wild in 2005, the exposome encompasses complex environmental exposures—ranging from a variety of internal and external exposures, including built environment and social factors—throughout the life course (Wild 2005, 2012). This definition was refined in late 2013 by Miller and Jones to include the biological responses to these exposures, as well as explicitly include chemicals generated from endogenous processes (Miller and Jones 2014; Miller 2014). A complement to the genome, the focus of exposome research is to study the impacts of complex environmental exposures on human health. One major goal is to improve methods to detect, quantify, and analyze exposures for a better understanding of the causative impacts of complex exposures on biological systems.

HERCULES: An NIEHS Core Center Focused on the Exposome

The HERCULES Exposome Research Center at Emory University in Atlanta, Georgia—which was originally an acronym for the *Health and Exposome Research Center: Understanding Lifetime Exposures*—was the first NIEHS P30 Core Center with an emphasis on the exposome concept. First funded by NIEHS in 2013 and renewed in 2017, HERCULES seeks to place the environment in its proper place in how we study, prevent, and treat human disease.

The vision of the HERCULES Center is to demonstrably advance the role of environmental health sciences in clinical and public health using the platform of the exposome. Healthcare and biomedical research have become increasingly genome-centric, which is partly due to the impressive achievements and technological advances in the genomics field. The exposome, which embraces a strategy and scale similar to genomic research, is poised to elevate the prominence of the environment in health and disease research. In the initial application for the P30 Core Center, Emory proposed that the grand vision of the exposome would create an intellectual climate that fostered innovation, collaboration, and progress. For this task, the Center has 12 action-oriented goals:

1. *Serve* as an intellectual hub for environmental health research.
2. *Foster* environmental health research.

3. *Support* technical innovation.
4. *Promote* collaboration.
5. *Discover* links between the environment and human health.
6. *Develop* an infrastructure for environmental health data.
7. *Build* partnerships with the community.
8. *Disseminate* science through innovation modalities.
9. *Provide* a vision for environmental health at Emory and beyond.
10. *Contribute* to the development of the exposome.
11. *Facilitate* children's health research using the exposome framework.
12. *Advance* the goals of the NIEHS Strategic Plan.

Cores of the HERCULES Center

To address these 12 goals, HERCULES is organized into five distinct Cores—the Administrative Core, Integrated Health Sciences Facility Core (IHFSC), Environmental Health Data Sciences Core (EHDSC), Community Outreach and Engagement Core (COEC), and Pilot Core—that provide services for the development, application, and promotion of exposome research throughout the institution.

Administrative Core

Headed by Director Gary W. Miller, Ph.D., a toxicologist, and Deputy Director Paige Tolbert, Ph.D., an environmental epidemiologist, the Administrative Core serves critical leadership, organizational structure, and identity roles for HERCULES. The goals of the Core are to (1) provide strategic vision for HERCULES, (2) promote the mission of HERCULES via effective and dynamic leadership, (3) effectively manage center resources and evaluate outcomes, (4) oversee the recruitment, appointment, evaluation, training, and development of Center members, and (5) disseminate knowledge resulting from HERCULES activities to the scientific community and HERCULES partners. The Administrative Core assures that HERCULES remains on a path of excellence by providing inspirational vision, scientific oversight, fiscal management, resource allocation, and critical appraisal of activities.

The Executive Committee, consisting of the Directors of the five Cores and an elected representative from the Stakeholder Advisory Board (SAB; *described in the COEC subsection*), meets regularly to discuss the functioning of the Center, as well as developments at Emory, at NIEHS, or in the broader scientific community that are likely to have an impact on the functioning of the Center. The Executive Committee interfaces with the Internal Advisory Board, External Advisory Board, and SAB, who provide feedback on how the Center can best serve the needs of individuals at Emory and in the larger community.

Integrated Health Sciences Facility Core

The IHFSC facilitates the translation and integration of exposure research across the research spectrum, from basic research through clinical- and population-based studies. The Core emphasizes research on the impact of complex, time-varying exposures on biological systems by providing support in three major areas: targeted exposure analysis, untargeted exposure analysis, and translational/population-based research consultation.

Targeted Exposure Analysis

Headed by Dana Boyd Barr, Ph.D., an internationally recognized expert in analytical chemistry and biomarker development, the IHSFC provides targeted analysis of environmental chemicals in a variety of biological matrices using extensively tested, gold-standard procedures with authentic standards for each chemical of interest. The laboratory contains multiple types of mass spectrometers (e.g., triple quadrupole mass spectrometers, time-of-flight [TOF] mass spectrometers, ion trap mass spectrometers) with different interfaces and ionization techniques (e.g., high-performance liquid chromatography [HPLC], gas chromatographs [GC], electrospray ionization [ESI], chemical ionization [EI], matrix-assisted laser desorption interfaces [MALDI]). The service also provides instruction on procedures for sample collection and storage and hands-on instrumentation training for investigators as needed.

Untargeted Exposure Analysis

The untargeted exposure service, termed “high-resolution metabolomics” (HRM), is headed by director Dean Jones, Ph.D. The service builds upon an environmental metabolomics program that Dr. Jones developed with support from NIEHS and other sources. HRM enables the measurement of thousands of small-molecule metabolites in biologic samples, providing an innovative capability for cell culture, animal models, and human studies. By providing information on metabolites in metabolic pathways, hundreds of known environmental chemicals, and thousands of unidentified chemicals, the platform represents an exposome “reference platform” linking exposures and biological responses (Jones 2016). HRM analyses are conducted on a range of high-sensitivity mass spectrometers, including the Thermo Scientific™ Q-Exactive HF, Orbitrap-LTQ Velos, and Orbitrap-LTQ-FT. A Thermo Scientific™ GC-Orbitrap provides additional capabilities for environmental chemicals that do not ionize well with ESI or atmospheric-pressure chemical ionization.

The service also focuses on the development of novel approaches to process and interpret untargeted metabolomics data, particularly in the context of the exposome.

Recent improvements include new mass spectral data extraction methods for low-abundance chemicals (Yu et al. 2009), optimized combinations of ionization modes and chromatography, and enhanced bioinformatics methods (e.g., data analysis workflows (Uppal et al. 2013), pathway and network analysis software (Li et al. 2013), and probability-based feature identification (Uppal et al. 2017)). Notably, the Core developed a protocol to provide quantification of environmental chemicals in blood and urine using a reference standardization approach (Go et al. 2015). With these capabilities in place, the service provides users exposome-level data that are reproducible, biologically interpretable, and cost effective.

Translational/Population-Based Research Consultation

Headed by epidemiologist and biologist Carmen Marsit, Ph.D., and physician-scientists Thomas Ziegler, M.D., and Miriam Vos, M.D., this service promotes the translation of basic science concepts into clinical/population research designs, specifically for the incorporation of the exposome into clinical medicine and human health research. Services include (1) discussions with investigators of research goals and needs/benefits from clinical translation, (2) organization of informal topic-specific workshops on clinical translation of basic and population research, (3) assistance with clinical research design and access to clinical and population-based research resources, (4) access to clinical populations and samples from well-phenotyped cohorts and healthy control data to compare disease-specific populations, and (5) training and education through organized symposia and workshops.

Environmental Health Data Sciences Core

Exposome-level questions require a "data ecosystem" of interconnected, mutually dependent systems and elements. The EHDSC consolidates the interdisciplinary expertise to facilitate big data analysis in a more comprehensive fashion than what could be accomplished by independent project teams. Under the codirection of Eberhard Voit, Ph.D., a systems biologist at the Georgia Institute of Technology and Lance Waller, Ph.D., and biostatistician at Emory, the goals of the EHDSC are to expand and foster (1) a data ecosystem, (2) an analytic ecosystem, and (3) a collaborative ecosystem for exposome-related research. For these tasks, the EHDSC focuses on the continued development of a data science pipeline for exposome data generation, integration, and analysis.

Data Ecosystem

Exposome research involves complex, interrelated data sets from familiar (e.g., time series of speciated pollutant levels), maturing (e.g., metabolomics), and emerging (e.g., microbiome) data types, as well as data streams from sensors and electronic health records. The diverse and heterogeneous nature of exposome data demands flexible and adaptable support systems for data collection, integration, management, query, linkage, and analysis, while maintaining rigor and reproducibility. Building on the Big Data to Knowledge (BD2K) (Margolis et al. 2014) initiative from the NIH, the EHDSC has developed two specific aspects of the data ecosystem to address these demands—databases/informatics and a hybrid local-cloud computing structure. The EHDSC develops, implements, and monitors plans for long-term data storage, tracking access by investigators. To best address the needs of HERCULES researchers, the EHDSC is developing a system with local computing where feasible and cloud-based solutions for large-scale, complex analyses and back-up storage for large datasets. Ultimately, the EHDSC hopes to lead an effort to create an "Exposome Workbench"—a cloud-based virtual database and toolbox for exposome research—analogous to the Metabolomics Workbench (Sud et al. 2016) hosted by the University of California, San Diego.

Analytic Ecosystem

The EHDSC provides expertise in methodological development and implementation for exposome research, for hypothesis generation via targeted and untargeted analytics (e.g., data summarization, categorization, visualization, and exploration), as well as validation and assessment of predefined hypotheses (e.g., mathematical and statistical modeling, parameter estimation, simulation, and prediction). For these tasks, the EHDSC focuses on analytics/visualization, biostatistics, bioinformatics, computational toxicology, machine learning, and systems biology. The EHDSC has particular strengths in integrative, high-dimensional data analyses to describe complex systems, e.g., multiscale–omics analyses, physiologically based toxicokinetic (PBTK) modeling, and nonparametric modeling of nonlinear compartment systems.

Collaborative Ecosystem

The collaborative ecosystem encompasses data sources, data analysts, and those interested in results, ranging from community stakeholders to lab-based scientists. The EHDSC offers proposal development assistance for HERCULES investigators, career development training and support for junior investigators and those new to exposome research, and communication and interpretation of results for colleagues and the community. Additionally, the EHDSC participates in regular HERCULES

Data Club meetings and sponsors workshops and courses to introduce investigators to new tools, data products, and study designs.

Community Outreach and Engagement Core

Directed by Michelle Kegler, Dr.P.H. and Melanie Pearson, Ph.D., the COEC develops partnerships with community leaders, government agencies, and researchers at other local institutions to better understand the environmental health concerns faced by the larger community. The actions of the COEC are centered on the principles of community-based participatory research (CBPR) (O'Fallon and Dearry 2002) to ensure equitable input from all stakeholders. The goals of the COEC are to (1) maintain and expand multi-directional dialogue with the Atlanta community, (2) implement and evaluate a community grant program to enhance community capacity to address local environmental health concerns and to partner in community-engaged research projects, and (3) guide and support HERCULES scientists in community engagement and outreach.

These goals are largely achieved through an actively engaged Stakeholder Advisory Board (SAB) that not only serves as a platform for multi-directional communication about environmental health issues and research, but also makes decisions about how the COEC should advance its agenda. The SAB, which consists of 30 members, includes concerned citizens; representatives from small and national nonprofit organizations; academic partners; and representatives from city, county, state, and federal government agencies. The SAB conducts regular meetings with HERCULES faculty, attends HERCULES Center retreats, and organizes forums to interact with the local community. The COEC fosters community-based efforts and awards mini-grants to community organizations to enable citizens to develop and lead programs within their own communities. In addition, the COEC promotes community-based participatory research that has led to the funding of projects under the Pilot Core (see below).

Pilot Core

Pilot projects are a primary mechanism by which the exposome concept is propagated throughout Emory University and the greater community. Under the direction of Edward Morgan, Ph.D., the Pilot Core aims to (1) expand and integrate exposome-related research at Emory and Georgia Tech with an emphasis on cutting-edge technologies, (2) stimulate interdisciplinary collaboration by encouraging collaborative exposome-related research projects, (3) promote the development of junior investigators, (4) increase the number of R-level grants from NIEHS by generating new preliminary data and building strong interdisciplinary research teams, and (5) promote translational research in environmental health sciences.

Notably, CBPR is an area of emphasis, which strengthens the bidirectional links of the COEC and the broader community to the HERCULES center.

A total of 19 pilot projects (40,000 USD each) were funded in the first 4-year cycle of HERCULES. In keeping with the goals of the Pilot Core, nine of the pilot award awardees were junior investigators; eight were new to Environmental Health Sciences; awardees planned to implement cutting-edge technologies, with seven of the projects using the metabolomics core, two using the systems biology core, and four using the targeted analyses core; and awardees embraced translational research approaches, with 13 projects involving human clinical or epidemiological study designs. Moving forward, the Pilot Core has identified several key exposome-related areas for growth (epigenetics, systems biology, machine learning, bioinformatics, high-throughput toxicology, and spatial and temporal statistical models), and pilot proposals in these areas will be strongly encouraged.

Promoting the Exposome

Workshops, Seminars, Courses, and Symposiums

The Center has developed, hosted, and participated in a series of programs to teach exposome-related concepts and approaches. For example, in recognition of the microbiome as an emerging area in exposome research, the Center co-hosted the Emory Microbiome Symposium in May 2015 with nearly 300 attendees. HERCULES Core Directors have helped develop a big data symposium series, termed Thinking Big, that addresses issues in the health sciences. Several Center investigators played important roles in the 2-day NIEHS Exposome Workshop in January 2015, which discussed the state of the science in exposome research (Dennis et al. 2016, 2017). The Center also developed an academic course on the exposome entitled "Genome, Exposome, and Health," as well as several exposome-focused lectures within existing courses, which introduced over 1500 undergraduate students at Emory to the exposome paradigm. Center Director Dr. Gary Miller authored the book *The Exposome: A Primer* (Miller 2014) in 2014 (published by Academic Press, Elsevier), which was used in the exposome semester course and was also provided to the COEC SAB members and Center investigators.

Emory Exposome Summer Course

Building on the "Genome, Exposome, and Health" course, the Center launched the first Emory Exposome Summer Course (Niedzwiecki and Miller 2017) from June 12 to 17, 2016. The Course, which drew 150 attendees from around the globe, provided a comprehensive and intensive overview of the emerging science of the exposome. Leading experts from top institutions (Johns Hopkins, Harvard, UCLA,

Berkeley, EPA, Emory, NIEHS, National Center for Advancing Translational Sciences [NCATS]) presented on exposome-related concepts, and interactive laboratory sessions allowed participants to test cloud-based programs to analyze exposome-related datasets. The Center plans to make the Summer Course a recurring event (planned for 2018), and additionally plans to assemble the content from the semester course and the Summer Course to create a modified massive open online course (MOOC) on the exposome for an even broader reach.

Nontraditional Dissemination Initiatives

In addition to traditional scientific dissemination outlets, HERCULES supports several dissemination initiatives aimed at the public. The Human Exposome Project (humanexposomeproject.com) and the Emory HERCULES (emoryhercules.com) websites provide information on research related to the exposome and activities of our center in clear, easily interpretable language. The online promotion of the exposome concept by HERCULES has led thousands of people to learn about the topic: the Center Twitter feeds have received hundreds of thousands of impressions, and the Exposome Wikipedia page, which was created by the Center, continues to receive a high level of traffic. Offline, the Center has developed brochures on the exposome that have been used by community members to educate members of the general public about environmental health, as well as distributed at various academic and industry conferences.

The Future of HERCULES

The successful renewal of HERCULES provides five more years of support for the Center. We anticipate continued expansion of exposome-related research across Emory in the upcoming years. Two notable examples of the influence of HERCULES include the recently awarded Children's Health Exposure Analysis Resource (CHEAR) and the Children's Health, the Environment, the Microbiome, and Metabolomics (C-CHEM^2) centers at Emory, both of which were built upon the foundation established by HERCULES. Several other research grants and programs have taken advantage of HERCULES resources to incorporate environmental factors into human disease research, including partnerships with major centers supported by the National Cancer Institute and NCATS have been developed. At Emory, the presence of an NIH-funded Core to foster exposome-related research has led to a significant expansion of efforts to link environmental exposures to human health.

References

Dennis KK, Auerbach SS, Balshaw DM, Cui Y, Fallin MD, Smith MT, Spira A, Sumner S, Miller GW (2016) The importance of the biological impact of exposure to the concept of the exposome. Environ Health Perspect 124(10):1504–1510. https://doi.org/10.1289/ehp140

Dennis KK, Marder E, Balshaw DM, Cui Y, Lynes MA, Patti GJ, Rappaport SM, Shaughnessy DT, Vrijheid M, Barr DB (2017) Biomonitoring in the era of the exposome. Environ Health Perspect 125(4):502–510. https://doi.org/10.1289/ehp474

Go YM, Walker DI, Liang Y, Uppal K, Soltow QA, Tran V, Strobel F, Quyyumi AA, Ziegler TR, Pennell KD, Miller GW, Jones DP (2015) Reference standardization for mass spectrometry and high-resolution metabolomics applications to exposome research. Toxicol Sci 148(2):531–543. https://doi.org/10.1093/toxsci/kfv198

Jones DP (2016) Sequencing the exposome: a call to action. Toxicol Rep 3:29–45. https://doi.org/10.1016/j.toxrep.2015.11.009

Li S, Park Y, Duraisingham S, Strobel FH, Khan N, Soltow QA, Jones DP, Pulendran B (2013) Predicting network activity from high throughput metabolomics. PLoS Comput Biol 9(7): e1003123. https://doi.org/10.1371/journal.pcbi.1003123

Margolis R, Derr L, Dunn M, Huerta M, Larkin J, Sheehan J, Guyer M, Green ED (2014) The National Institutes of Health's Big Data to Knowledge (BD2K) initiative: capitalizing on biomedical big data. J Am Med Inform Assoc 21(6):957–958. https://doi.org/10.1136/amiajnl-2014-002974

Miller GW (2014) The exposome: a primer. Academic Press, Waltham, MA

Miller GW, Jones DP (2014) The nature of nurture: refining the definition of the exposome. Toxicol Sci 137(1):1–2. https://doi.org/10.1093/toxsci/kft251

Niedzwiecki MM, Miller GW (2017) The exposome paradigm in human health: lessons from the Emory Exposome Summer Course. Environ Health Perspect 125(6):064502. https://doi.org/10.1289/ehp1712

O'Fallon LR, Dearry A (2002) Community-based participatory research as a tool to advance environmental health sciences. Environ Health Perspect 110(Suppl 2):155–159

Sud M, Fahy E, Cotter D, Azam K, Vadivelu I, Burant C, Edison A, Fiehn O, Higashi R, Nair KS, Sumner S, Subramaniam S (2016) Metabolomics Workbench: an international repository for metabolomics data and metadata, metabolite standards, protocols, tutorials and training, and analysis tools. Nucleic Acids Res 44(Database issue):D463–D470. https://doi.org/10.1093/nar/gkv1042

Uppal K, Soltow QA, Strobel FH, Pittard WS, Gernert KM, Yu T, Jones DP (2013) xMSanalyzer: automated pipeline for improved feature detection and downstream analysis of large-scale, non-targeted metabolomics data. BMC Bioinformatics 14(1):15. https://doi.org/10.1186/1471-2105-14-15

Uppal K, Walker DI, Jones DP (2017) xMSannotator: an R package for network-based annotation of high-resolution metabolomics data. Anal Chem 89(2):1063–1067. https://doi.org/10.1021/acs.analchem.6b01214

Wild CP (2005) Complementing the genome with an "exposome": the outstanding challenge of environmental exposure measurement in molecular epidemiology. Cancer Epidemiol Biomark Prev 14(8):1847–1850. https://doi.org/10.1158/1055-9965.epi-05-0456

Wild CP (2012) The exposome: from concept to utility. Int J Epidemiol 41(1):24–32. https://doi.org/10.1093/ije/dyr236

Yu T, Park Y, Johnson JM, Jones DP (2009) apLCMS—adaptive processing of high-resolution LC/MS data. Bioinformatics 25(15):1930–1936. https://doi.org/10.1093/bioinformatics/btp291

Chapter 14
EXPOsOMICs: Meet-in-the-Middle and Network Perturbation

Christiana A. Demetriou, Davide Degli Esposti, Kristi Pullen Fedinick, and Paolo Vineis

Abstract Systems biology has been driven by technology (the development of omics) and by statistical modelling and bioinformatics. We aim to bring biological thinking back. We suggest that three traditions of thought need to be considered: (a) causality in epidemiology, for example the "sufficient-component-cause framework", and causality in other sciences, for example the Salmon and Dowe approach; (b) new acquisitions about disease pathogenesis, for example the "branched evolution model" in cancer, and the role of biomarkers in this process; (c) the burgeoning of omic research, with a large number of "signals" that need to be interpreted. To address the new challenges of epidemiology, the concept of the "exposome" has been proposed. We show examples from recent projects in the field, namely, new omic approaches applied to epidemiological studies; and in particular, the identification of hallmarks of cancer as intermediate steps between exposure to carcinogens and the cancer phenotype, according to the "meet-in-the-middle" concept. We use examples derived from the study of mutational spectra in tumours and benzo(a) pyrene and bisphenol A as model carcinogens. We suggest conceptualising the detection and tracing of signals in terms of information transmission.

Electronic supplementary material The online version of this chapter (https://doi.org/10.1007/978-3-319-89321-1_14) contains supplementary material, which is available to authorized users.

C. A. Demetriou
Neurology Clinic D, The Cyprus Institute of Neurology and Genetics, Nicosia, Cyprus

D. D. Esposti
Ecotoxicology Team, UR Riverly, National Research Institute of Science and Technology for Environment and Agriculture (Irstea), Lyon-Villeurbanne, France

K. P. Fedinick
Natural Resources Defense Council, Washington, DC, USA

P. Vineis (✉)
Imperial College London, London, UK

HuGeF Foundation, Torino, Italy
e-mail: p.vineis@imperial.ac.uk

S. Dagnino, A. Macherone (eds.), *Unraveling the Exposome*,
https://doi.org/10.1007/978-3-319-89321-1_14

Keywords Epidemiology · Adductomics · Evidential pluralism · Information transition

Introduction: The Meet-in-the-Middle Concept as a Tool to Reinforce Causality

As discussed in other chapters of this book, there are two broad interpretations of the exposome and they are complementary. One, called "top-down", is mainly interested in identifying new causes of disease by an *agnostic* approach based on omic technologies, similar to what has been applied in genetics with the GWAS design. This first approach is sometimes called "EWAS", or "exposome-wide association study" (Rappaport 2016), and utilizes tools such as metabolomics or adductomics to generate new hypotheses on disease etiology. The second general approach is called "bottom-up" and starts with a set of exposures or environmental compartments to determine the pathways or networks by which such exposures lead to disease, that is which pathways/networks are perturbed. We have used the latter approach in the EXPOsOMICS investigation (Vineis et al. 2017) as we explain below. Another frequent distinction is between untargeted (agnostic) and targeted investigations, the latter involving specific molecules based on a priori hypotheses.

Whether our investigation starts from disease in search of a cause (top-down) or it starts from the other end (bottom-up), we need in any case to connect an external exposure with a disease outcome through a set of intermediate biomarkers. The simplest approach to do this is the "meet-in-the-middle" (MITM) approach. This consists in measuring intermediate biomarkers (often with an agnostic omic investigation) and relating them retrospectively to measurements of external exposure and prospectively to a health outcome (disease, or aging, or other outcomes) in the context of a longitudinal investigation. If the same set of markers are robustly associated with both ends of the exposure to disease continuum, this is strong validation of a causal hypothesis. As we show below drawing from our experience in EXPOsOMICS, this approach is made possible both by the technological developments in omics and external measurements, and by the existence of long-term longitudinal population cohorts with biological samples stored for many years. This approach has been rarely used before: As we show below with the example of coronary heart disease, previous investigations either analyzed biomarkers in relation to exposure to air pollution, or they analyzed biomarkers of disease risk, but did not look at both associations in the same set of individuals with a longitudinal design.

The EXPOsOMICS Project and Its Goals

EXPOsOMICS is a European Commission-funded bottom-up project on the exposome of air pollution and water contamination. By comprehensively addressing the integration of the external and the internal exposomes at the individual level, EXPOsOMICS:

- Pooled and integrated information from short-term, experimental human studies and long-term epidemiological cohorts—including adults, children, and newborns—to enable focused investigations to refine environmental exposure assessment based on the concept of *life-course epidemiology*.
- Characterised the exposome by (a) measuring the external component of the exposome at different critical life stages by employing novel tools and drawing on experience gained in existing initiatives (personal exposure monitoring, databases coupled with GIS, remote sensing), with a focus on air and water pollution; and (b) measuring biomarkers of the internal exposome (xenobiotics and metabolites), using omic technologies (adductome, metabolome, transcriptome, epigenome and proteome).
- Integrated external and internal exposure measures to comprehensively model and assess exposure to air pollution and water contamination in large population cohorts, through novel statistical modelling.

  At the end of the EXPOsOMICS project, we have measured five types of omics for between 2000 and 3000 subjects, and we have thus created a large database that can have multiple uses:

  - Generation of new hypotheses on disease etiology by agnostic perusal of the database
  - Exploration of pathways and cross-omic networks
  - Follow-up of subjects with omic data (e.g. several 1000 with both methylome and metabolome) for mortality and other outcomes
  - New nested case–control studies within the database cohort to collect further information on exposures
  - Additional omic measurements to allow complementarity with current measurements.

As a final goal, this might become an open resource for other researchers for future exposome work. Further details on the project design have been published in Vineis et al. (2017).

Meet-in-the-Middle: The Example of Air Pollution and Coronary Heart Disease

To exemplify how the EXPOsOMICS data can be used to perform a MITM analysis, we have considered the mediation exerted by protein and methylation markers on the relationship between air pollution and coronary heart disease. Coronary heart diseases (CHDs) are among the major causes of death and disability worldwide. Exposure to ambient air pollution has been linked to a wide range of adverse health effects, including mortality and morbidity due to respiratory diseases and CHD (Wolf et al. 2015). Several epidemiological studies have reported an increased risk of mortality from all-causes, cardiovascular disease and respiratory disease associated with long-term exposure to fine

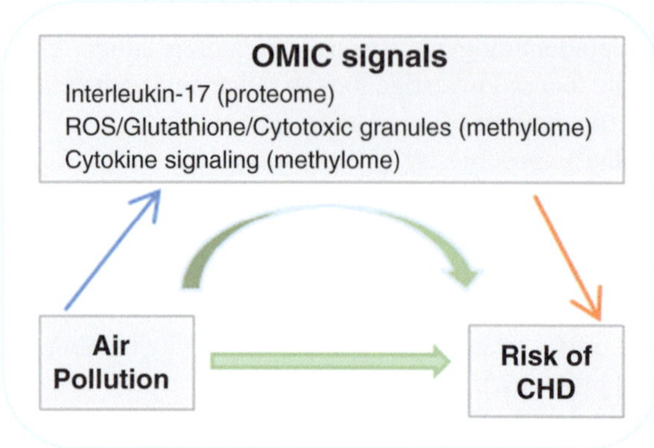

Fig. 14.1 Exemplification of meet-in-the-middle. From Fiorito et al. (2017): Oxidative stress-induced inflammation mediates the effect of air pollution on CHD: a prospective study. One inflammatory protein (Interleukin-17) and two DNA methylation inflammatory pathways ("ROS/Glutathione/Cytotoxic granules" and "Cytokine signaling") were significantly associated with both exposure to air pollution and the risk of CHD, fulfilling the "meet-in-the-middle" hypothesis

and coarse particulate matters ($PM_{2.5}$ and PM_{10}), nitrogen oxide (NO_x), nitrogen dioxide (NO_2), and elemental carbon. Mechanistic explanations have been put forward to explain the effect of air pollution on the cardiovascular system, particularly oxidative stress and inflammation (Uzoigwe et al. 2013; Newby et al. 2015). To assess causality, ideally one needs evidence from a MITM approach, that is one in which biomarkers are associated retrospectively to exposure and prospectively to disease in the same subjects (Vineis et al. 2013). In our EXPOsOMICS nested case–control study (in non-smokers), inflammatory and oxidative pathways in DNA methylation were associated with both air pollution measures and CHD, in particular "Adhesion-Extravasation-Migration", "G-Protein coupled receptor signaling", "MAPK signaling", and "ROS/Glutathione/Cytotoxic granules". Also involvement of proteins indicated inflammation as a link between exposure and disease. Our study was the first to use a "MITM" design to investigate longitudinally the mechanistic chain linking air pollution to CHD risk and suggests the involvement of oxidative stress and inflammatory pathways (Fiorito et al. 2017) (Fig. 14.1). The same approach can now be used with other outcomes in both children and adults.

An Extension of the Meet-in-the-Middle: Hallmarks of Cancer and Key Characteristics of Carcinogens

Network Perturbation: Concept

The idea of MITM can be further expanded to investigate networks and their perturbation. This can be particularly useful to strengthen causal assessments in the context of hazard identification. Carcinogens have been argued to lead to the perturbation of one or more "Adverse Outcome Pathways" (AOPs) which consist of a sequence of events including molecular initiating events, biochemical responses, cellular responses, tissue organ responses, individual responses and lastly population responses (NAS 2017). AOP is a term used in a specific context and we prefer to refer more generically to "pathway perturbation" or even better "network perturbation", since carcinogenesis does not proceed linearly along a chain of events but in multiple directions according to the network metaphor.

One conceptual proposal comes from the integration of two different frameworks, one based on the "branched evolution" theory of carcinogenesis (knowledge on the role of *mutagenic* and *selectogenic* events), and the other is the *sufficient-component-cause framework*, that has been used in epidemiology for decades and can be applied also to pathway/network perturbation. These two frameworks are summarised below.

Causal Frameworks in Carcinogenesis

There have been several important changes in the molecular interpretation of carcinogenesis in recent decades that can be summarised in the discovery of the role of epigenetic changes (defined as functional changes in DNA that do not involve a change in the nucleotide sequence), in the acknowledgement of the importance of selection of mutated or epimutated cells and in the unifying concept of "branched evolution". The latter concept has been summarised as follows: "Recent data increasingly demonstrate that evolution often occurs in a branched manner in several tumour types, leading to intratumour diversity, with subclones differing genetically and functionally. (. . .) the selective advantage of any genotype is dependent on the environment. Environment must be defined broadly to include the cell's repertoire of gene and protein expression, the mutations and epigenetic changes already present in that cell, the local external microenvironment, distantly acting factors (such as hormones), and factors external to the host (such as carcinogens and exogenous therapies). (. . .) For some driver mutations, environmental variation may matter less because such mutations almost invariably confer a selective advantage" (Gerlinger et al. 2014).

How do these concepts connect to external exposures? We know that about 40% of cancers would be preventable today, based on epidemiological evidence. These cancers are explained by external (or endogenous) exposures that are unlikely to act in isolation: for example, even a "necessary" cause of cancer, HPV, is itself not sufficient to cause cervical cancer in an individual. On a population scale, HPV is probably able to explain 100% of cervical cancer cases, but the individual case is not entirely explained by HPV alone. The model of causation that applies to single individuals is called the "sufficient-component-cause framework".

The "sufficient-component-cause framework" considers exposures or hits that together lead to the outcome under consideration. This framework corresponds to what has been called by the philosopher of science John Mackie the INUS model, that is "Insufficient Non-Redundant Components of an Unnecessary Sufficient Causal Complex". The model provides a way to account for how multiple factors, whether environmental exposures or genes, combine to result in disease in an individual or population. The INUS approach to causality underlies the model of component causes put forward by Kenneth Rothman that is widely utilised in epidemiology. In this model, sufficient causal complexes are represented by pies with each slice being a single INUS, that is an insufficient component (Rothman and Greenland 2005). Generally, none of the components is necessary. In fact, for cancer and other non-communicable diseases, there are only a few instances of apparently necessary causes: for example, HPV and cervical cancer and perhaps asbestos and asbestiform fibers and mesothelioma.

Each constellation of component causes in a "pie" represents a sufficient causal complex that, if completed, results in the outcome, and there is no redundant component cause, so that the removal of any one component cause will result in complete outcome/disease prevention. Each component cause is a non-redundant part of that specific causal complex but a particular factor may play a role in multiple causal complexes.

Causal Frameworks and Meet-in-the-Middle Approaches to Carcinogenesis

With respect to the molecular interpretation of carcinogenesis, pathway or network perturbation events can function as different components in a "causal pie", with a minimum number of components required for cancer initiation. Pathway perturbation events can provide a selective advantage either directly (mutagens) or indirectly (selectogens). As a result, the order of pathway perturbation events following exposures can enhance understanding and evaluation of carcinogenicity.

Therefore, in the attempt to integrate mechanistic data in carcinogen evaluations, based on the meet-in-the-middle approach, knowledge of the temporal sequence of events in carcinogenesis would be precious. However, currently a sequence-oriented assessment of molecular evidence or mechanistic data is hindered by the lack of an established or even hypothesised order of events in carcinogenesis. In fact, no attempt has been made to date to put the mechanistic changes of cancer in a temporal sequence.

In the following section, we attempt to gather evidence on the potential temporal sequence of events in carcinogenesis, referring to the causal frameworks described above. We refer to carcinogenesis because the types of mechanistic events involved are better understood than for other non-communicable diseases, through the seminal paper on the Hallmarks of Cancer by Hanahan and Weinberg (2011) (Table 14.1).

Table 14.1 The hallmarks of cancer (from Hanahan and Weinberg 2011)

Genome instability and mutation
Inflammation
Sustaining proliferative signalling
Evading growth suppressors
Resisting cell death
Enabling replicative immortality
Inducing angiogenesis
Activating invasion and metastasis
Reprogramming of energy metabolism
Evading immune destruction

In addition to the temporal order of events in carcinogenesis, a practical application of the meet-in-the-middle approach in hazard assessment also rests on the distinction and interplay between the hallmarks of cancer and the "key characteristics" of carcinogens as described by Smith et al. (2016b). The key characteristics of carcinogens are reported in Table 14.2. Clearly, there is a substantial overlap, but the two concepts have a different meaning and purpose, since the former have been proposed as markers of the cancer phenotype, while the latter are related to the modes of action of carcinogens. We propose here to establish the relationship between the middle-to-outcome nature of the hallmarks with the bottom-to-middle approaches of the key characteristics to be used in carcinogen evaluations (Fig. 14.2), and whether these converge toward a suggested temporal sequence of events. However, for simplicity and to allow an overlap with the hallmarks of cancer, we refer to the latter in the evaluation of the carcinogens we have selected (rather than applying the key characteristics as described in Table 14.2). Therefore, we will exclusively use the framework of hallmarks of cancer throughout this chapter.

The Temporal Order of Genetic Alterations in Carcinogenesis

This review provides evidence of the temporal nature of the hallmarks of cancer, a missing component identified in the Hanahan and Weinberg 2011 paper, in order to provide more of a pathway- and eventually network- based way to understand the relationships between the various mechanisms by which cells can become cancerous.

The review deals with the temporal sequence of hallmarks of cancer first from the point of view of pathological specimens of cancer (essentially branched mutations), and then considering the role played by some relevant carcinogens (bisphenol A and benzo(a)pyrene, chosen on the basis of different modes of action).

Table 14.2 Key characteristics of carcinogens

Is electrophilic or can be metabolically activated
Parent compound or metabolite with an electrophilic structure (e.g. epoxide or quinone), formation of DNA and protein adducts
Is genotoxic
DNA damage (DNA-strand breaks, DNA–protein cross-links or unscheduled DNA synthesis), intercalation, gene mutations, cytogenetic changes (e.g. chromosome aberrations or micronuclei)
Alters DNA repair or causes genomic instability
Alterations of DNA replication or repair (e.g. topoisomerase II, base-excision or double-strand break repair)
Induces epigenetic alterations
DNA methylation, histone modification, microRNA expression
Induces oxidative stress
Oxygen radicals, oxidative stress, oxidative damage to macromolecules (e.g. DNA or lipids)
Induces chronic inflammation
Increased white blood cells, myeloperoxidase activity, altered cytokine, or chemokine production
Is immunosuppressive
Decreased immunosurveillance, immune-system dysfunction
Modulates receptor-mediated effects
Receptor activation or inactivation (e.g. ER, PPAR, AhR) or modulation of endogenous ligands (including hormones)
Causes immortalization
Inhibition of senescence, cell transformation
Alters cell proliferation, cell death or nutrient supply
Increased proliferation, decreased apoptosis, changes in growth factors, energetics and signalling pathways related to cellular replication or cell cycle

Any characteristic could interact with any other (such as oxidative stress, DNA damage and chronic inflammation), and a combination provides stronger evidence of a cancer mechanism than one would alone. Sources: Smith et al. (2016b).

Fig. 14.2 The components of the Meet-in-the-middle approach: Bottom to Middle refers to the action of carcinogenic agents, Middle to Top to the appearance of the cancer phenotype according to hallmarks of cancer

Tumour Branched Evolution

The order in which genetic events become established in tumours can be very insightful, particularly with respect to enlightening the critical events in carcinogenesis. This sequential order of genetic alterations has been shown to vary between and within tumours. The following is an overview of studies which have examined the genomic alterations in and the branched evolution of different tumours and/or different sections within a single tumour.

The premise of these studies is that if a gene is found mutated with a high frequency in multiple tumours from many different patients, or in many samples from the same tumour, it is less likely to be a passenger mutation and more likely to provide the cell with a selective advantage, permitting it to expand and eventually dominate the population. Also, ubiquitous genetic alterations between several samples from the same tumour are more likely to have occurred early in the evolution of the cancer, whereas later alterations, present only in some tumour sections, most likely contributed to tumour branching.

The evidence in the reviewed studies is explored in an attempt to place the hallmarks of cancer in a temporal order of appearance. All the background information is reported in *Appendix*, where the studies are grouped per type of cancer examined.

From the studies we have reviewed, it occurs that even though tumour cells can accumulate many mutations as they evolve, not all of these mutations are driving mutations (i.e. play a causal role). Some mutations are just a by-product of the increased genetic instability which characterises the process of tumourigenesis.

Studies have shown that around 20 driver genes are usually mutated per tumour (Sjöblom et al. 2006), but which specific genes will be mutated in different tumours might vary. In addition, cells from different sections of the same tumour might have a diverse mutational landscape, suggesting that different mutations might converge toward the same phenotypic result.

This is because the intracellular pathways often deregulated in cancer (pathways that drive carcinogenesis) incorporate many genes which operate as an information transfer cascade. Mutations in any one of the genes can have a similar effect, that is stopping the cascade and the transfer of information. This view is supported by the observation that multiple hits in different genes of the same pathway in individual tumours are less frequent than expected (Vogelstein and Kinzler 2004) and in different tumours different members of each pathway may be affected in a mutually exclusive manner (Parsons et al. 2008). Furthermore, different cancer specimens of the same tumour type carry their individual cancer driver gene mutational signature, which seems to be mutually exclusive as the number of common mutant cancer driver genes between tumours is limited (Sjöblom et al. 2006).

Even though the sequential order of driving genetic alterations can vary between and within tumours, and even though in different tumours different members of each pathway may be affected in a mutually exclusive manner, the main cancer pathways affected are almost ubiquitous and follow a generally common sequence. For the

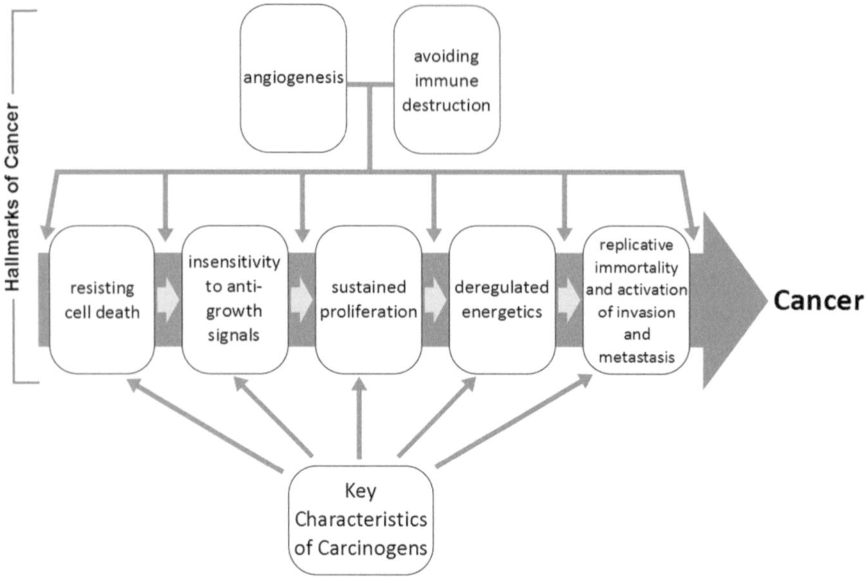

Fig. 14.3 Tentative reconstruction of a sequence of hallmarks based on mutations in biopsies (see text). It should be noted that the first three steps can occur in varied order. Also, we do not consider in the fugure the fact that both mutations and clone selection operate in carcinogenesis

evidence base for this proposed sequence, see *Appendix*. These cancer pathways and their sequential alterations enable the suggestion of a temporal sequence for the appearance of cancer hallmarks: *resisting cell death, insensitivity to anti-growth signals, sustained proliferation, deregulated energetics, replicative immortality and activation of invasion and metastasis*. The first three hallmarks can be regarded as almost simultaneous as their exact order is switched in different investigations. Angiogenesis and avoiding immune destruction are perhaps the only hallmarks with a varying position in the above sequence but they are most often evidenced to follow deregulated energetics and to precede replicative immortality, and invasion and metastasis.

In conclusion, the evidence on tumour biopsies proposes a sequence for 5 of the hallmarks, with the other two potentially residing in different places along the path, as indicated in Fig. 14.3.

Perturbation of Hallmarks of Cancer by Carcinogens: Examples

As a second step, we have summarised the evidence on the hallmarks of cancer activated by specific carcinogens (as roughly equivalent to the carcinogens' key characteristics). The carcinogens have been chosen because of their different levels of evidence in humans—strong evidence for carcinogenicity for benzo(a)pyrene and weak for BPA—and different modes of action. We have followed the same scheme proposed above to organise the evidence on how hallmarks of cancer are affected by carcinogens.

Detailed information about the two chemicals is given in *Appendix*. Here we briefly summarise the main findings.

BaP—A recent review provided in the context of the Halifax project (Engström et al. 2015) described benzo(a)pyrene (B*a*P) as a prototypical multi-disruptor influencing sustained proliferation during carcinogenesis. Interestingly, they found that the mechanisms leading to this phenotype could not be entirely explained by the genotoxicity of its metabolites. In particular, for PAHs mixtures, mechanisms involving the chemical–receptor interaction, notably with estrogen receptor and aryl hydrocarbon receptor, have been described (Baird et al. 2005). Moreover, other than nuclear receptor activation, B*a*P was also shown to induce epigenetic alterations, affecting DNA methylation, histone modification and non-coding RNA expression (Chappell et al. 2016). Finally, similar to the conclusion based on our literature review, B*a*P was found associated with most key characteristics of carcinogens (Engström et al. 2015), but there was no evidence to support a temporal sequence of events. In conclusion, B*a*P may exert its multipotent and multi-tissue carcinogenic effects by affecting all hallmarks of cancer. Even at low concentrations and as part of chemical mixtures, B*a*P and PAHs perturb multiple signalling pathways, including hormonal pathways and those regulating energy metabolism, depending on the status of the target cell.

BPA—Based on our literature review, BPA was found directly or indirectly associated with all hallmarks of cancer (*Table 1 in the Supplementary material* includes all studies investigating direct effects of BPA on at least one of the hallmarks). However, for some hallmarks such as deregulating cellular energetics and replicative immortality, the evidence is not sufficiently direct or consistent.

Similarly, a recent review assessing the carcinogenic potential of low dose exposures to chemical mixtures in the environment has examined BPA and its disruptive action on pathways related to the hallmarks of cancer (Goodson et al. 2015). This review confirmed BPA's effect on pathways involved in tissue invasion and metastasis, resistance to cell death, sustained proliferative signalling, tumour promoting inflammation, immune system evasion, evasion of anti-growth signalling, tumour microenvironment and genome instability.

In conclusion, even though the evidence is not conclusive to support a carcinogenic role, BPA, even at low concentrations, may increase cancer risk by affecting most known hallmarks of cancer/key characteristics. The majority of the evidence supports the following two hallmarks for BPA: Alters cell proliferation, cell death, or nutrient supply and Modulates receptor-mediated effects.

However, existing evidence cannot provide information regarding the temporal order of events in BPA's mode of action and thus cannot corroborate the sequence of hallmarks hypothesised from tumour biopsies. Nevertheless, in the future, establishment of a temporal order of the key characteristics of carcinogens can strengthen evaluations regarding BPA's carcinogenicity as more evidence becomes available. In the same manner, it will facilitate incorporation of mechanistic data in carcinogenicity evaluations of other chemicals/exposures.

Conclusions: Suggestions for Exposome Research

We have proposed the MITM approach as one way to reinforce causality in exposome research. First, we have shown the (so far limited) application of the approach to the EXPOsOMICS data on cardiovascular disease (Fig. 14.1). Then we broadened the perspective by considering a more complex causal interpretation based on network perturbation and the sufficient-component-causal model. We used carcinogenesis (and its branched evolution model) to exemplify how an enlarged meet-in-the-middle approach to causality could work. We referred to the hallmarks of cancer to describe the cancer phenotype, that is we summarised the evidence on a potential temporal sequence in the hallmarks. Then we referred conceptually to the key characteristics of carcinogens to tackle the other end of the continuum, exposure to selected carcinogens. In fact, for simplicity and their substantial equivalence, also in this exercise we used the hallmarks of cancer. The overlap between the two concepts is only partial, and future exercises should probably treat hallmarks and key characteristics separately as belonging to the middle-to-outcome segment and the bottom-to-middle segments of the causal chain, respectively. The present was only a first simplified attempt.

The lesson we can draw is that both the meet-in-the-middle concept and its extension to encompass networks can be extremely powerful tools in exposome research, but more evidence is needed. Our review of hallmarks of cancer and their temporal sequence, based on mutational spectra in different cancer sites, allowed us to propose a hypothetical temporal sequence of the hallmarks: *resisting cell death, insensitivity to anti-growth signals, sustained proliferation, deregulated energetics, replicative immortality and activation of invasion and metastasis.* On the other side, it was not possible to isolate a clear sequence of hallmarks for the two model carcinogens we have chosen. Also, it was not possible to find a clear overlay of hallmarks in the cancer sites we have examined and the hallmarks associated with the two carcinogens. In particular, it seems that both carcinogens have a multiplicity of mechanisms or modes of action, affecting several of the characteristics.

In spite of the limitations, we believe that what we propose may be a promising approach, both for the summary and evaluation of the existing literature on environmental causes of disease, and for novel exposome research.

Appendix: Exposomics: Meet-in-the-Middle and Network Perturbation

Evidence of Temporal Sequence of Hallmarks of Cancer

Colorectal Cancer

Colorectal cancer is the most frequently examined tumour with respect to its genetic alterations and branched evolution.

In 95 colorectal cancer samples, using the Hidden Conjunctive Bayesian Networks (H-CNB) model, researchers identified APC mutations as initiating events (accumulation rate (AR) 0.39/year—consistent for an early driving role), followed by KRAS (AR = 0.12/year), PIK3CA (AR = 0.009/year), and other mutations such as EVC2, FBXW7, EPHA3 and TCF7L2 (Gerstung et al. 2011). In parallel, TP53 (AR = 0.06/year) mutations could be placed either before or after APC and KRAS mutations since their presence was independent of the presence of APC and KRAS mutations (Gerstung et al. 2011).

The model places key signalling pathways in the following order: Small GTPase pathway, Apoptosis/Wnt-Notch signalling, and homophilic cell adhesion. These alterations are often found before alterations in KRAS and TGF-β signalling pathways and before alterations in G1/S phase control. DNA damage control and JNK are altered at later stages, probably through TP53 mutations. Integrin signalling and invasion are not as common, suggesting roles at later stages of colon carcinogenesis.

Translating the pathways into hallmarks, the evidence above places evasion of growth suppressors, cell proliferation and resisting cell death prior to avoiding immune destruction and invasion and metastasis. Angiogenesis occurs simultaneously with initiation of invasion, whereas immortality occurs in parallel with the hallmarks prior to invasion and metastasis.

The early presence of the APC and KRAS mutations, as well as their high mutation accumulation rates (ARs: 0.39 and 0.12 per year Gerstung et al. 2011), is consistent with an early driving role.

These findings are in agreement with the Fearon and Vogelstein model of colorectal carcinogenesis (Fearon and Vogelstein 1990) where APC, a tumour suppressor gene, is firstly inactivated, allowing the normal epithelium to become hyperplastic and eventually form an early adenoma. Activation of KRAS then contributes to sustained proliferation and the formation of an intermediate adenoma, whereas further loss of tumour suppressors, such as Smad4, contributes to the formation of a late adenoma. Loss of p-53 then allows resistance to cell death and immortality. Additional genetic alterations, such as the accumulated loss of suppressor genes on additional chromosomes, correlate with the ability of the carcinomas to metastasise and cause death. However, Fearon and Vogelstein emphasise the importance of the accumulation of these changes, rather than their temporal order which may vary.

Box
Technical Definitions:

- Mutation Accumulation Rate (AR):
  Estimated yearly accumulation rate for a particular mutation. The higher the accumulation rate, the higher the frequency of occurrence of the mutation in different tumour samples
- Driver mutations:
  Mutations involved causally in the neoplastic process, conferring a selective advantage, thus being positively selected for during tumorigenesis.
- Passenger mutations provide no positive or negative selective advantage to the tumor but are retained by chance during repeated rounds of cell division and clonal expansion.
- Passenger Mutation Rates:
  Estimated through the quantification of synonymous (silent) missense mutations, because such mutations are expected to be biologically inert and can therefore exert no positive or negative selective advantage
- Passenger Probability Scores:
  These gene-specific scores are based on a Likelihood Ratio Test (LRT) for the null hypothesis that, for the gene under consideration, the mutation rates are the same as the passenger mutation rates. Small probability scores support rejection of the null hypothesis and acceptance that mutation rates are in fact higher than the passenger mutation rates. Therefore, small passenger probability scores support a "driver" role for mutations in a particular gene.

Rosenberg et al. (2009) also identified mutations in the APC and KRAS genes as an early event in rodent colon cancer models. Even in early premalignant lesions, increased proliferative activity, growth factor signalling and K-RAS mutations were evident.

In addition, Wood et al. (2007) in their analysis of 11 colorectal tumours identified APC, KRAS and TP53 as the most frequent driving mutations, followed by mutations in PIK3CA, FBXW7, CSMD3, TNN, NAV3, SMAD4 and many more. All these genes have passenger probability scores of <0.0001.

Independently, Beerenwinkel et al. (2007) also identify APC, TP53 and KRAS as the most often mutated genes in their analysis of 78 candidate cancer genes in 35 tumour (colon adenoma and carcinoma) samples. Therefore, evasion of growth suppressors, resisting cell death and sustained proliferation are again evidenced as the first cancer hallmarks to appear.

With respect to the temporal order of angiogenesis, studies corroborate its simultaneous occurrence with the initiation of invasion. Evidence from Takahashi et al. (2003) places the angiogenic switch between mucosal and submucosal invasive cancer. Hanahan and Folkman (1996) also place angiogenesis prior to solid tumour formation but following hyperproliferation. In addition, Zhang et al. (2001) evidence angiogenesis in early premalignant stages of tumour development. Furthermore, in benign colorectal adenomas, VEGF protein and RNA levels exceed those of normal colonic mucosa (Ono and Miki 2000; Wong et al. 1999). Lastly, the presence of VEGF A and B at the adenoma stage (Hanrahan et al. 2003) further supports the order of the angiogenic switch sometime after hyperproliferation and prior to invasion.

Lastly, even though APC, TP53 and KRAS mutations were identified at a high frequency in different sections from the primary tumour of a colorectal cancer patient (Kogita et al. 2015), only one section from this tumour contained PIK3CA mutations, at a low frequency. Interestingly, the metastatic tumour from the same patient harboured PIK3CA mutations at a high frequency. This evidence supports the temporal order of replicative immortality just prior to invasion and metastasis.

Pancreatic Cancer

Using the H-CNB to model 90 cases of pancreatic cancer, mutations in KRAS (AR > 100/year, prevalence = 100%) appear to initialise progression followed by TP53 (AR = 0.34/year), CDKN2A (AR = 0.013/year) and MLL3 (AR = 0.00066). SMAD4 is also mutated independently (AR = 0.015/year) (Gerstung et al. 2011).

Extrapolating to pathway level, apoptosis, G1/S transition, Hedgehog and TGF-β signalling pathways were ubiquitous, indicating their alteration at the earliest stages of pancreatic carcinogenesis. Then, alterations in small GTPase-dependent signalling and KRAS signalling arise independently, followed by alterations in DNA damage control, JNK and Wnt/Notch signalling. These alterations also contribute to replicative immortality. Integrin signalling is altered at late stages after homophilic cell adhesion, further contributing to senescence. Lastly, invasion pathways were found unaltered in these samples, suggesting roles in even later stages of carcinogenesis.

This model indicates that in pancreatic cancer, the hallmarks of insensitivity to anti-growth signals, cell proliferation, resisting cell death and angiogenesis are followed by deregulating cellular energetics, which in turn precedes immune system evasion, cell migration, invasion and metastasis. The model does not allow temporal placing of immortality.

Interestingly, quantitative analysis of the timing of the genetic evolution of pancreatic cancer, based on cell proliferation rates of normal pancreatic tissue and pancreatic metastatic tumours, suggested that at least a decade passes between the initial driving mutation and the birth of the parental, non-metastatic founder cell and at least five more years are needed for the acquisition of metastatic ability (Yachida et al. 2010). However, these numbers are mathematical estimates based on a small number of tumour samples; therefore, the evidence for the timing is weak.

Primary Glioblastoma

In 78 primary glioblastoma tumours, TP53 (AR = 0.015/year), PTEN (AR = 0.012/year), EGFR (AR = 0.0026/year), NF1 (AR = 0.0043/year), PI3CA (AR = 0.033/year), IDH1 (AR = 0.0087/year), PIK3R1 (AR = 0.0037) and RB1(AR = 0.011) mutations were identified as the most commonly mutated genes (Gerstung et al. 2011).

The H-CNB model identified TP53 as the first mutation, with NF1, PTEN and EGFR being mutated in parallel. Mutations then proceed in the following order: PIK3CA, PIK3R1 and RB1. However, the low accumulation rates of these individual mutations indicate a low probability for a specific gene alteration in primary glioblastomas.

This mutation order highlights apoptosis and small GTPase pathways as the first and most commonly affected pathways. G1/S phase transition, Wnt/Notch signalling and KRAS signalling were also mutated early. Alterations in these pathways are followed by alterations in DNA damage control, JNK signalling, homophilic cell adhesion and integrin signalling, the latter two occurring independently of the former two pathways. No alterations in the invasion pathway were identified.

Therefore, translating the pathways into hallmarks, in primary glioblastomas, cell proliferation and resistance to cell death are the first hallmarks identified, followed by deregulated cellular energetics, evasion of growth suppressors and angiogenesis. TP53 alterations are also likely to initiate the process of replicative immortality which is completed at later stages with PIK3CA and RB1 mutations. Invasion and metastasis is the last hallmark to occur, after avoiding immune destruction.

Independently, Parsons et al. (2008) investigated the frequency of mutations in 105 samples of glioblastoma tumours. They found CDKN2A, TP53, EGFR, PTEN, NF1, CDK4, RB1, IDH1, PIK3CA and PIK3R1 as the most frequently altered genes, in this order. With the exception of PIK3CA, and PIK3R1, all other gene alterations have a passenger probability of <0.01, indicating their role as driving mutations.

Based on these genes' functions and assuming that the most prevalent/driving alterations are necessary for initiation of carcinogenesis, whereas less frequent alterations have later roles in the carcinogenesis cascade, then in glioblastoma tumours, cancer hallmarks are expected to appear in the following order: evasion of cell death, insensitivity to anti-growth signals, sustained cell proliferation, deregulated energetics (Warburg effect), immortality, and invasion and metastasis. Unfortunately, this list does not allow estimation of the temporal position of angiogenesis and evasion of immune destruction in glioblastoma tumours.

Renal Carcinoma

Gerlinger et al. (2012) studied intratumour branched evolution in 2 renal cell carcinomas by multi-region sequencing and identified that VHL, SETD2, PTEN and KDM5C underwent multiple, distinct and spatially separated inactivating mutations.

In the first patient, VHL was identified as the first driving mutation, followed by SETD2 mutations which break the phylogenetic tree in two branches. In one branch (core biopsy samples), heterogeneity is further propagated by KDM5C and mTOR mutations. In the second branch (metastases samples), SETD2 mutations are also followed by KDM5C mutations.

VHL, the gene identified as firstly mutated, is involved in cell division and formation of new blood vessels, among other functions, identifying cell proliferation and perhaps initiation of angiogenesis as early, widespread events in renal carcinogenesis. The SETD2 gene normally trimethylates histone 3 lysine 36 at sites of active transcription. Its mutation results in silencing of transcription instead of an active chromatin conformation. This leads to altered nucleosome dynamics and DNA replication stress, as well as to failure in loading lens epithelium-derived growth factor and the Rad51 homologous recombination repair factor at DNA breaks (Kanu et al. 2015). Therefore, growth factor evasion follows cellular proliferation. KMD5C, which is mutated downstream, is involved in regulation of transcription through transcriptional repression, which most likely contributes to maintaining genetic instability. Lastly, mTOR is a target for cycle arrest, and immunosuppressive effects, and when mutated it is involved in deregulated metabolism, evasion of anti-growth signals, genetic instability and replicative immortality (Carnero et al. 2015), suggesting that immune system evasion and replicative immortality are later events.

In the second patient, VHL and PBRM1 were mutated in all tumour specimens followed by SETD2 mutations which break the phylogenetic tree in two branches. In one branch, the metastatic branch, SEDT2 mutations are accompanied by P53 mutations. In the other branch, the core tumour branch, PTEN mutations lead to further branching.

PBRM1 is necessary for ligand-dependent transcriptional activation and is involved in chromatin organisation, whereas PTEN is a tumour suppressor enzyme which regulates cell division, apoptosis, cell movement, adhesion and angiogenesis. P53 is one of the most well-established tumour suppressors associated with all cancer hallmarks, namely, increased cancer metabolism, angiogenesis, genetic instability, immune evasion, resistance to cell death, replicative immortality, sustained proliferative signalling, invasion and metastasis (Nahta et al. 2015). This branching further confirms that cell proliferation, new blood vessel formation and transcriptional activation are widespread in tumours and occur early. They are followed by repression of anti-growth signals, through chromatin inactivation (SEDT2 mutations) and anti-growth factor inactivation (PTEN and P53 mutations). Lastly this branching suggests that other hallmarks such as replicative immortality, invasion and metastasis, as mediated by P53, follow at later stages of renal carcinogenesis.

Melanoma

Multi-region sequencing in 41 multiple melanoma biopsies from 8 individual tumours (Harbst et al. 2016) found that the three melanoma driver genes, namely, BRAF, NRAS and NF1, all key components of the MAPK pathway, were ubiquitously mutated in a mutually exclusive pattern. The ubiquitous expression of these genes is consistent with an early role of this pathway (involved in cell proliferation and insensitivity to anti-growth signals) in melanoma formation. On the contrary, mutations in the PI3K pathway were heterogeneous indicating that such mutations occur later in metastatic melanoma evolution. Since the PI3K is important for many cell activities, including cell growth and division (proliferation), movement (migration) of cells and cell survival, its appearance later in carcinogenesis suggests later appearance of the hallmarks of resistance to cell death, and invasion and metastasis. IDH1, which has an essential role in glucose metabolism, is also mutated heterogeneously in some melanoma tumours, suggesting later appearance of the Warburg effect.

Breast Carcinoma

In an analysis of 11 breast cancer tumours, Wood et al. (2007) identified TP53 and PIK3CA as the most frequent, driving mutations. The two genes have passenger probability scores of <0.0001, indicating their important driving roles in carcinogenesis. Assuming that the most prevalent/driving alterations are necessary for initiation of carcinogenesis, the high mutation frequency of these genes highlights the early appearance of the resisting cell death, insensitivity to anti-growth signals and cell proliferation cancer hallmarks.

Combined Evidence on Different Tumor Types

Looking at colon, pancreatic and glioblastoma tumours together, the H-CNB model (Gerstung et al. 2011) identified the following order of pathway alterations as the most likely: Apoptosis, TGF-β signalling, small GTPase-dependent signalling (other than KRAS), Wnt/Notch signalling, control of G1/S phase transition, KRAS signalling, Hedgehog signalling, DNA damage control, JNK, homophilic cell adhesion, integrin signalling and invasion.

Based on the above, the following order of hallmarks emerges: resistance to cell death, insensitivity to anti-growth signals, sustaining proliferative signalling, deregulated cellular energetics, inducing angiogenesis, avoiding immune destruction, enabling replicative immortality, and invasion and metastasis.

Evidence for Angiogenesis in Premalignant Lesions

Raica et al. (2009) reviewed angiogenesis in premalignant lesions and concluded that tumour angiogenesis is not necessarily a characteristic of invasive tumour but may occur prior to malignancy, as defined by invasion and metastasis. In their review, they gather evidence that microvessel density (MVD) was significantly increased in a relatively large spectrum of premalignant squamous cell lesions, such as in the oral mucosa, skin, uterine cervix, vulva and anal canal. Interestingly, for a number of these lesions, MVD was found to correlate with a major pro-angiogenic factor, namely, vascular endothelial growth factor (VEGF).

Premalignant lesions of glandular epithelia, including gastric metaplasia and dysplasia, atypical adenoma of the colon, atypical hyperplasia and carcinoma in situ of the breast also exhibited VEGF overexpression.

Evidence of Temporal Sequence of Key Characteristics for Selected Carcinogens

Benzo[*a*]pyrene and the Hallmarks of Cancer

Benzo[a]pyrene (B*a*P) is a ubiquitous contaminant, belonging to the large group of organic compounds with two or more fused aromatic (benzene) rings, namely, polycyclic aromatic hydrocarbons (PAHs) (IARC 2012; WHO 2010). PAHs and B*a*p formed during incomplete combustion and their major sources include tobacco smoke, residential and commercial heating with wood or coal, motor-vehicle exhaust and industrial emission (IARC 2012). Occupational exposures occur in aluminum production, roofing and paving involving coal-tar pitch, coal liquefaction, coal-tar distillation, wood impregnation, chimney sweeping and power plants (IARC 2010). While no epidemiological studies on B*a*P alone are available, B*a*P produced tumours in multiple organs and tissues in all animal species tested following exposures by many different routes (IARC 2012). Mechanistic evidences from in vitro and in vivo studies, including human exposures, showed that B*a*P is clearly genotoxic following its metabolization into highly reactive species that form DNA adducts leading to sister chromatid exchange, chromosomal aberrations, micronuclei or DNA damage (IARC 2012). Based on these multiple experimental evidences, B*a*P may be considered a well-established model carcinogen. Thus, we decided to review the literature available on B*a*P and its capacity to affect the different hallmarks of cancer.

Hallmark 1. Sustaining Proliferative Signalling

B*a*P is a very well-known chemical able to immortalise human cells, such as breast epithelial cells (Gudjonsson et al. 2004). One of the earliest mechanistic evidence of

B*a*P effects on cell proliferation was its ability to form DNA adducts and recurrent mutations were observed in the genes of the *RAS* superfamily both in humans and in mice (Meng et al. 2010; Hu et al. 2003; Wei et al. 1999). Beyond the mutagenic effects on genes associated to cell proliferation, Kometani et al. (2009) showed that B*a*P was able to promote proliferation of human lung cancer cells after 24 weeks of exposure by activating the EGFR pathway through induction of EGFR ligands, amphiregulin and epiregulin. Moreover, B*a*P may promote cell proliferation through the activation of nuclear receptor signalling pathways, such as aryl hydrocarbon receptor (AhR) and estrogen receptor (ER) in humans, mice and other non-mammalian organisms (Andrysik et al. 2007; Charles et al. 2000; Tian et al. 2013; Wen and Pan 2015). In conclusion, B*a*P is able to induce cell proliferation through both genotoxic and non-genotoxic receptor-induced mechanisms in humans and various experimental models.

Hallmark 2. Evading Growth Suppressors

B*a*P exposure allows cells to evade G1 arrest and induces cell abnormal proliferation (Du et al. 2006). Most studies reported an activation of the ERK pathway in different cell types (Wang et al. 2015a; Hamouchene et al. 2011; Du et al. 2006). Interestingly, a short exposure of 24 h was sufficient to induce a dose-related activation of MAPK in normal human embryo lung diploid fibroblasts (Du et al. 2006), and apparently cellular response to B*a*P exposure was also dependent on the growth kinetics within a target cell population (Hamouchene et al. 2011), suggesting different susceptibility based on cell state and differentiation.

Hallmark 3. Resisting Cell Death

A recent exome-wide mutation profile on immortal human mammary epithelial cells exposed to B*a*P showed that genes involved in various biological processes, including regulation of cell death, harbor mutations predicted to impact protein function (Severson et al. 2014). However, other in vitro reports showed that exposure to B*a*P or its DNA-reactive metabolite anti-benzo[a]pyrene-7,8-diol-9,10-epoxide (BPDE) induces or sensitizes human cells to either receptor-mediated or mitochondrial-mediated apoptosis (Stolpmann et al. 2012; Sang et al. 2012). Similar results were obtained using a reverse-phase protein array comparing mouse primary liver tumours induced by B*a*P to their normal adjacent tissues (Phillips et al. 2015). The results showed both down-regulation (e.g. cleaved caspase 7, caspase 3) and up-regulation (Bax, Bad, Bcl-xL) of some pro-apoptotic proteins in tumour tissues. In conclusion, B*a*P may exert either a pro-apoptotic effect or an anti-apoptotic effect. In vivo data on the temporal sequence of these effects are lacking, although it has been suggested that the carcinogenic effect of B*a*P or PAHs mixtures may be also mediated by chronic inflammation and cell death subsequent to the exposure (Engström et al. 2015).

Hallmark 4. Inducing Angiogenesis

It has been recently shown that 1 month exposure to different B*a*P concentrations increased in a dose-dependent manner the capacity of hepatoma cell lines, BEL-7404, to recruit vascular endothelial cells and promote angiogenesis through increased secretion of vascular endothelial growth factor (VEGF) (Ba et al. 2014). Similarly, low, non-cytotoxic concentration of B*a*P induced hypoxia-inducible factor-1a, responsible for the adaptation to hypoxic conditions and the promotion of angiogenesis (Mavrofrydi and Papazafiri 2012). Interestingly, B*a*P and its metabolites may have a distinct and opposite effect on VEGF expression (Li et al. 2015), suggesting that tissue-specific or genetic inter-individual differences in the CYP450 expression may play a role in determining the overall effect on angiogenesis induction by B*a*P and PAHs in general.

Hallmark 5. Enabling Replicative Immortality

B*a*P has been shown to induce efficiently immortalization of Syrian hamster normal dermal cells through its mutagenic activity causing the direct inactivation of p53 and INK4 alterations (Yasaei et al. 2013; Newbold et al. 1980). Similarly, p53 mutations in Hupki cells (embryonic murine fibroblast with human p53 gene) exposed to B*a*P were correlated with p53 mutations in human lung tumours, supporting the direct role of B*a*P in causing smokers lung tumour p53 mutations (Liu et al. 2005). In conclusion, B*a*P seems to enable replicative immortality mainly through its genotoxic properties.

Hallmark 6. Activating Invasion and Metastasis

There is evidence that B*a*P can promote cell migration, invasion and metastasis (Ochieng et al. 2015). Recently, some studies provided mechanistic clues showing that B*a*P may contribute to lung cancer cell invasion and metastasis by up-regulating pro-inflammatory chemokines (IL8, CCL-2, CCL-3) and one of the master regulators of the epithelial-to-mesenchymal transition, Twist (Zhang et al. 2016a; Wang et al. 2015c). Moreover, B*a*P was able to induce cell migration in triple-negative breast cancer MDA-MB-231 cells through a lipoxygenase- and Src-dependent pathway, notably by increasing the secretion of metalloproteinase MMP-2 and MMP-9 (Castillo-Sanchez et al. 2013). B*a*P-treatment was also able to increase the metastatic potential of hepatocellular carcinoma cell lines in a mouse model, likely through an activation of both angiogenesis and NF-kB pathway (Ba et al. 2014). In conclusion, B*a*P is able to promote cell migration and invasion, contributing to increase the metastatic potential of several epithelial cells in different experimental settings.

Hallmark 7. Deregulating Cellular Energetics

Recent studies showed that B*a*P is able to alter the function of mitochondria, cellular organelles having a key role in cellular energetics as well as in programmed cell death. In particular, B*a*P (range, 0–500 mM) was able to lower mtDNA content in TK6 cells, a human lymphoblastoid cell line (Pieters et al. 2013). Interestingly, in the same study, indoor exposure to PAHs was associated with decreased mtDNA content in the blood (Pieters et al. 2013). Similarly, B*a*P induced mitochondrial damage in vivo in mice cervical tissues, strongly associated with increased oxidative stress (Gao et al. 2010). Interestingly, a short-term exposure (24–48 h) to B*a*P induced an increased expression of several components of the mitochondrial respiratory chain (Salazar et al. 2004), suggesting an adaptative process that could cause a mitochondria-derived increased oxidative stress to the exposed cells and tissues. In conclusion, while a more metabolism-focused research on the effects of B*a*P is needed, some evidence exists that it could directly impact cellular metabolism and energy production, notably through an alteration of the mitochondrial function.

Hallmark 8. Avoiding Immune Destruction

Early studies in the 1980s showed B*a*P has immunotoxic effects in mice, by reducing antibody production and inducing DNA adducts in splenic leukocytes (Dean et al. 1983; Ginsberg et al. 1989). A more recent study showed that B*a*P at a dose as low as 10 mg/kg b.w. (generally considered as non-toxic) was able to induce changes in thymus weight and spleen B-cell populations (De Jong et al. 1999).

In conclusion, while no evidence was available to test the hypothesis that B*a*P may affect cancer cells immune recognition, B*a*P is able to exert immunotoxicity, thus contributing to avoiding immune destruction by cancer cells.

Enabling Characteristics: Genome Instability and Mutation and Tumour-Promoting Inflammation

The acquisition of the hallmarks of cancer is made possible by two enabling characteristics, namely, genome instability (which allows accumulation of random mutations) and the inflammatory state accompanying many premalignant and malignant lesions (Hanahan and Weinberg 2011). As a genotoxic compound, B*a*P and its metabolites are a well-known mutagen for both human and experimental animals. Interestingly, a recent study involving next-generation sequencing of exomes in immortal human mammary epithelial cells showed that DNA repair genes were among the genes harboring B*a*P-induced mutation with a potential functional impact, suggesting a potential impact on genome stability of the mutated cells (Severson et al. 2014). Moreover, B*a*P increased the number of oxidatively induced clustered DNA lesions in normal primary breast-derived cells which were correlated with the number of chromosomal aberrations (Sigounas et al. 2010). These lesions were associated with a decrease of antioxidant defense capacity and an increased ROS and DNA repair gene transcription (Sigounas et al. 2010), suggesting that oxidative stress and DNA damage and repair response are strictly correlated following B*a*P exposure.

Important cross talks between chemicals and immune system in carcinogenesis have been recently reviewed (Kravchenko et al. 2015). Interestingly, interleukin 6 (IL-6) and tumour necrosis factor alpha (TNF-a) produced by macrophages have been shown to be critical to promote malignant transformation of human bronchial epithelial cells in a bionic airway chip culture system and in an animal model (Li et al. 2015). Similarly, it was observed that TNF-a strongly augmented the formation of stable Bap diol epoxide-DNA adducts in alveolar type II epithelial cells (Umannová et al. 2011). Thus, the inflammatory response to PAHs containing BaP and the immunosuppressive ability of BaP toward acquired immunity cells previously described may synergistically cooperate in the carcinogenicity of BaP aPAHs mixtures.

Bisphenol A and the Hallmarks of Cancer

Bisphenol A (BPA) is a chemical compound, produced in large quantities for use primarily in the production of polycarbonate plastics and epoxy resins. Human exposure to BPA is widespread. More specifically, in a study conducted by the Centers for Disease Control and Prevention (CDC) found detectable levels of BPA in 93% of 2517 urine samples from people 6 years and older in the United States ("Bisphenol A (BPA)" 2016). BPA exhibits estrogen mimicking, hormone-like properties that raise health concerns about its suitability in some consumer products and food containers. However, the ability of BPA to act as a carcinogen has been a matter of great debate.

Some epidemiological studies implicate BPA exposure in cancer incidence but the direction of the association is not consistent. In a case–control study in China, urine BPA concentration was positively associated with meningioma diagnosis (Duan et al. 2013). BPA was also shown to be positively associated with established cancer risk factors such as visceral obesity, waist circumference, glucose homeostasis and concentrations of inflammatory markers. Conversely, in a case–control study in Tokyo, serum BPA was found to be lower in premenopausal women with complex endometrial hyperplasia with malignant potential compared to controls and was also significantly lower in women with endometrial cancer than in controls (Hiroi et al. 2004). Also, despite not demonstrating an association between probable occupational BPA exposure and breast cancer, a population case–control study in the USA found BPA exposure to be more common in controls rather than cases (Aschengrau et al. 1998).

In contrast to the epidemiological studies, mechanistic evidence from in vitro and in vivo studies showed that BPA impacts on several of the cancer hallmarks and may thus act as a carcinogen. Here, we review the available experimental evidence of BPA effects on the different hallmarks of cancer.

Hallmark 1. Sustaining Proliferative Signalling

Proliferative effects of BPA are supported by a wealth of in vitro and in vivo studies. Most studies evidence a direct effect of BPA on cell proliferation (Sheng et al. 2013; Sengupta et al. 2013; Park and Choi 2014; Newbold et al. 2007; Nakagawa and Suzuki 2001; Moral et al. 2008; Jung et al. 2011; Han et al. 2001; Ge et al. 2014a; Colerangle and Roy 1997; Betancourt et al. 2010a; Liu et al. 2015; Zhu et al. 2009; Pisapia et al. 2012; Lam et al. 2015; Smith et al. 2016a, b; Zhang et al. 2012; Schafer et al. 1999; Mlynarcikova et al. 2013; Recchia et al. 2004; Ibrahim et al. 2016; Park et al. 2009; Murray et al. 2007; Kang et al. 2012; Ayyanan et al. 2011; Lee et al. 2012a; Wang et al. 2015b, 2014a; Jenkins et al. 2009, 2011). Yet other studies imply proliferative effects by evidencing BPA's ability to mimic estrogen and have estrogenic effects (Katchy et al. 2014; Hall and Korach 2013; Gould et al. 1998; Dong et al. 2011; Chun and Gorski 2000; Lee et al. 2014; Maruyama et al. 1999; Kim et al. 2003; Hwang et al. 2011; Zhang et al. 2012; Terasaka et al. 2004) and by studying the alterations in gene expression conferred by BPA exposures (Hall and Korach 2013; Hess-Wilson et al. 2006). Interestingly, BPA exposure was shown to have proliferative effects in tissues, particularly mammary tissue, of mice/rat offspring even when exposed pre- or perinatally through maternal exposure during pregnancy or lactation (Murray et al. 2007; Newbold et al. 2009; Markey et al. 2001; Wang et al. 2014b; Betancourt et al. 2010a; Mandrup et al. 2016; Ayyanan et al. 2011; Ichihara et al. 2003).

Many of these studies indicate that the proliferative effects of BPA are dose-dependent, with higher doses invoking greater rates of proliferation in the nanomolar range (Moral et al. 2008; Lin et al. 2013; Colerangle and Roy 1997; Wang et al. 2013). However, in the micromolar concentration ranges, BPA effects were shown to be reversed, causing decreased proliferation in Sertoli TM4 cells (Ge et al. 2014b).

The signal transduction pathways activated by BPA through which it exerts its effects have also been extensively investigated. BPA has been shown to bind to ERα receptors (Nakagawa and Suzuki 2001; Miyakoshi et al. 2009) but other studies show BPA effects to be G protein–coupled receptor (GPCR) dependent, with particular involvement of the GPER receptor (Pupo et al. 2012; Song et al. 2015, p. 2; Bouskine et al. 2009). BPA was also shown to confer androgen-independent stimulation of the androgen receptor AR-T877A (Wetherill et al. 2002).

BPA induces specific gene expression changes (Park and Choi 2014) targeting proteins that regulate cell cycle progression such as the p38 MAPK proteins in human mammary cancer and ovarian adenocarcinoma cell lines (Lee et al. 2014; Kang et al. 2013), CD4K and cyclin D1 proteins in human neuroblastoma cells in vitro and neuroblastoma tumours in rats in vivo (Zhu et al. 2009), PCNA, bcl2 and bax proteins in ovarian cancer cell lines (Mlynarcikova et al. 2013; Yu et al. 2004), STAT3 in a human mammary cancer cell line (Zhang et al. 2012) and p16 and cyclin

E in a human mammary epithelial cell line (Qin et al. 2012). More specifically, BPA has been shown to increase expression of cyclin D1, responsible for the G1/S cycle transition, and to decrease expression of p21, a potent CDK inhibitor that arrests cell cycle in the G1 phase in human mammary and ovarian cancer cell lines (Hwang et al. 2013; Kang et al. 2012; Lee et al. 2012a). BPA also activates both cAMP-dependent protein kinase and cGMT-dependent protein kinase pathways, causing phosphorylation of cAMP responsive element-binding protein (CREB) and Rb as shown in human testicular seminoma cells (Bouskine et al. 2009). In addition, the PTEN/AKT/p53 axis was evidenced to be involved in BPA proliferative effects. Specifically, BPA induced up-regulation of oncogenic miR-19a and miR-19b and caused dysregulated expression of miR-19-related downstream proteins, including PTEN, p-AKT, p-MDM2, p53 in human mammary cancer cells (Li et al. 2015). The IGF-1 signalling pathway is also activated by BPA, resulting in increased proliferation (Klotz et al. 2000). Perhaps more importantly, induction of key genes and proteins in the PI3K-mTOR pathway—AKT1, RPS6 and 4EBP1 and a concurrent reduction in the tumour suppressor, that is phosphatase and tensin homolog gene protein—has been shown in benign breast cells following BPA exposure. Lastly, other changes conferred by BPA, such as elevated phospho-AKT, c-Raf, and phospho-ERKs1 and 2 and decreased TGFβ, have been shown to predispose to chemically induced cancer in adult rats, even when exposure was prenatal (Betancourt et al. 2010b).

On the other hand, far fewer studies suggest that BPA exposure is not sufficient to exert proliferative effects, and that co-exposures such as phytoestrogens and IGF 1 (Katchy et al. 2014; Ishido 2004), or co-activators and particular conditions such as oestrogen depletion, or selective mutation of the ER (Hess-Wilson et al. 2006) or carcinogenic exposures such as 3,2′-dimethyl-4-aminobiphenyl (DMAB) (Ichihara et al. 2003) or catechol (Oikawa 2005) are needed.

Still other studies demonstrate no proliferative effects of BPA on membrane estrogen receptor-alpha-enriched GH3/B6/F10 pituitary tumour cells (Kochukov et al. 2009), in rat liver epithelial WB-F344 cells (Dong et al. 2014), in a human breast cancer cell line MCF-7 (Diel et al. 2002), in the human endometrial cell line, ECC-1 (Bergeron et al. 1999) and in adult female prenatally exposed mice (Yoshida et al. 2004).

Lastly, few studies demonstrate decreased proliferation in lung cancer cells (Andreescu et al. 2005) or in murine-derived multipotent neural progenitor cells (NPC) (Kim et al. 2009) as exposure concentrations increased but both studies used fairly high concentrations in the micromolar range.

Overall, the majority of studies support proliferative effects of BPA in nanomolar concentrations, which are also the most relevant with regard to human environmental exposure.

Hallmark 2. Evading Growth Suppressors

As already discussed, BPA induces cell abnormal proliferation by allowing cells to evade G1 arrest. By disrupting cell cycle regulatory proteins, BPA also allows a cell to evade growth suppression. More specifically, BPA down-regulates p16 and p21 (Hwang et al. 2013; Kang et al. 2012; Lee et al. 2012a; Qin et al. 2012), both of which play important roles in regulating/halting transition from G1 to S phase. In addition, by phosphorylating Rb (Bouskine et al. 2009), BPA negates Rb's regulatory role in cell cycle progression and thus its tumour suppressive properties. Furthermore, BPA deregulates PTEN and p53 (Li et al. 2015) and decreases TGFβ (Betancourt et al. 2010b), thus inhibiting them from exerting their tumour suppressive properties and allowing cells to evade their growth regulatory roles.

Lastly, BPA reduces expression of the tumour suppressor, phosphatase and tensin homolog gene protein, in human high-risk donor breast epithelial cells (Goodson et al. 2011).

Hallmark 3. Resisting Cell Death

Apoptotic effects of BPA seem to be contradictory.

BPA at 10^{-7} M in epithelial stromal co-cultures derived from the contralateral tissue of patients with breast cancer induced gene expression patterns that facilitate apoptosis evasion (Dairkee et al. 2008). In addition, in a human MCF-7 breast cancer cell line, BPA reduced cell apoptosis at all concentrations higher than 10^{-8} M, with the lowest rate of apoptosis observed at a dose of 10^{-6} M (Diel et al. 2002). In ovarian cancer cells, BPA co-exposure with leptin was shown to have anti-apoptotic properties by inhibiting caspase 3 expression and activity by modulating STAT3 and ERK1/2 signalling pathways (Ptak et al. 2013). Similarly, in ovarian cancer cells, BPA suppressed expression of pro-apoptotic genes, increased expression of pro-survival genes and decreased caspase 3 activity (Ptak et al. 2011). In vivo, BPA decreased apoptosis in neonatal/prepubertal rats exposed to BPA through lactation and this decrease was associated with changes in PR-A, SRC1–3, erbB3 and Akt activity (Jenkins et al. 2009). Lastly, in cells treated with tamoxifen, pretreatment with BPA evidenced dose-dependent apoptotic evasion (Goodson et al. 2011).

Yet other studies indicate that BPA associated cytotoxicity can decrease cell viability and induce apoptosis in a dose-dependent manner (S. Terasaka et al. 2004; Jenkins et al. 2011). In ER-positive breast cancer cells (MCF-7: WS8) and in insulin-secreting cell lines (INS-1), BPA at higher concentrations ($2\text{x}10^{-8}$ to 10^{-6} M) promotes apoptosis (Sengupta et al. 2013; Lin et al. 2013) through the release of cytochrome C. In Sertoli cells, apoptosis was also induced but by the transmission of apoptotic signals to the mitochondria via the CAM-CAMKII-ERK1/2 pathways (Qian et al. 2014). Expression changes such as up-regulation of PCNA and bcl-2 mRNA expression and down-regulation of the bax mRNA expression, following exposure to BPA (32×10^{-7}M), suggest involvement of this pathway in apoptosis in ovarian cancer cells (Yu et al. 2004). BPA-induced caspase 3 increase was also implied to mediate pro-apoptotic effects in the mammary gland of adult albino rats (Ibrahim et al. 2016). Caspase activation, phosphorylation of ERK and AKT and involvement of both intrinsic and extrinsic apoptotic pathways were also evidenced in leukemia cells (Bontempo et al. 2009).

Hallmark 4. Inducing Angiogenesis

A few studies evidenced BPA to have pro-angiogenic effects by investigating mechanisms involved in this hallmark. BPA increased microvessel density and VEGF expression in neuroblastoma tumours from mice (BPA: 200 mg/kg per day), and VEGF expression was also increased in cultured neuroblastoma cells in vitro (2 μg/mL) (Zhu et al. 2009). In breast cancer cell lines, BPA (1^{-10} μM) was shown to regulate VEGF expression by an ER-dependent mechanism, and suppression of VEGF effects via use of inhibitors suggested involvement of MEK, p38 kinase and PI3K pathways in BPA's pro-angiogenic effects (Buteau-Lozano et al. 2008). Yet, in bovine aortic endothelial cells (BAECs), BPA's (100 nM) pro-angiogenic effects were exerted by up-regulation of XIAP expression via GPER activation and not ERα or ERβ activation (Liu et al. 2015). Lastly, BPA (1, 40 and 100 nM) stimulated VEGF-R2 expression in both cancerous and non-cancerous cell lines and also stimulated VEGF-A only in non-cancerous cells (Ptak and Gregoraszczuk 2015). Given VEGF-A's role in angiogenesis, these studies imply that BPA can have pro-angiogenic results.

Hallmark 5. Enabling Replicative Immortality

BPA was shown to successfully increase the levels of both p16 and cyclin E and to induce cellular senescence in human mammary epithelial cells, as assessed by the number of human heterochromatin protein-1γ positive cells (Qin et al. 2012). Even though BPA was not directly shown to induce senescence in other studies, its exposure effects such as phosphorylation of Rb (Bouskine et al. 2009), deregulation of PTEN and p53 (Li et al. 2015), decrease in TGFβ (Betancourt et al. 2010b) and other tumour suppressors, as well as induction of key genes and proteins in the PI3K-mTOR pathway (Li et al. 2015) can contribute toward replicative immortality.

Hallmark 6. Activating Invasion and Metastasis

Several studies evidence that BPA can promote cell migration, invasion and metastasis (Zhang et al. 2016b; Kim et al. 2015a; Chen et al. 2015; Liu et al. 2015), and epithelial-to-mesenchymal transition (EMT) was shown to play an important role in the acquisition of this hallmark (Chen et al. 2015; Wang et al. 2015b).

Treatment of SW840 colorectal cancer cells promoted metastasis via induction of EMT characterised by acquiring mesenchymal spindle-like morphology and increase in N-cadherin. In parallel, E-cadherin was decreased and Snail transcription factor was increased (Chen et al. 2015). BPA was also shown to stimulate migration through the up-regulation of migration-related factors MMP2, MMP3 and MMP9 in addition to N-cadherin in neuroblastoma and ovarian cancer cells (Ptak et al. 2014; Zhu et al. 2010). These effects were shown to be dose sensitive and dependent on the MAPK and PI3K/Akt signalling pathways (Ptak et al. 2014). Up-regulation of MMPs following BPA expression was also evident in lung cancer cells but in these cells BPA treatment induced the up-regulation through rapid activation of ERK1/2 via GPER/EGFR (Zhang et al. 2014). Similarly, in MDA-MB-231 breast cancer cells, BPA induced GPER-dependent activation of migration, and an invasion process involving AP-1/NFκB-DNA binding activity through an Src- and ERK2-dependent pathway (Castillo Sanchez et al. 2016). The IKK-β/NF-κB signalling pathway was also elsewhere shown to be implicated in BPA-induced cervical cancer cell migration and invasion (Ma et al. 2015). Another mechanism implicated in BPA-induced cell migration was the Erβ-mediated integrin B1/MMP9 pathway (Shi et al. 2016; Kim et al. 2015b).

Despite the implicated mechanisms, all studies above evidenced BPA effects on migration. However, it is important to note that one study evidenced that BPA (0.1 and 1 μM concentration) reduced cell motility of rat liver epithelial WB-F344 cells (Dong et al. 2014).

Hallmark 7. Deregulating Cellular Energetics

One study investigating primarily the effects of BPA exposure on apoptosis of INS-1 cells evidenced that BPA-induced apoptosis was associated with mitochondrial defects, including depletion of ATP, release of cytochrome c, loss of mitochondrial mass and membrane potential, and alterations in expression of genes involved in mitochondrial function and metabolism. These mitochondrial defects may confer dysregulation of cellular energetics. However, the direct effects of BPA on reprogramming cells to favour glycolysis (Warburg effect) have not been investigated.

Hallmark 8. Avoiding Immune Destruction

Even though no studies have evidenced a direct capacity of BPA to affect recognition of cancer cells by immune cells, or immunosuppressive factor secretion from cancer cells, several studies suggest adverse effects on the functioning of the immune system that result from exposure to BPA (immunotoxicity).

Firstly, BPA exposure-related gene expression changes indicated an enrichment of genes involved in immune system response and regulation (Fic et al. 2015). BPA was also shown to affect the regulation of the immune system by reducing NO and TNFα production, through inhibition of NF-κβ mediated by ER (Kim and Jeong 2003). NF-κB is a first responder to harmful cellular stimuli such as reactive oxygen species (ROS), tumour necrosis factor alpha (TNFα) and interleukin 1-beta (IL-1β).

By inhibiting NF-κβ, the organism's ability to regulate immune response is compromised and this might enable immune destruction evasion of cancer cells.

Enabling Characteristics: Genome Instability and Mutation and Tumour-Promoting Inflammation

In an update of their original publication on the hallmarks of cancer (Hanahan and Weinberg 2000), Hanahan and Weinberg recently identified two enabling characteristics which promote malignancy and the establishment of the cancer hallmarks. These characteristics are genomic instability (which facilitates accumulation of mutations) and tumour-promoting inflammation (Hanahan and Weinberg 2011).

Several publications support a genotoxic role of BPA in vivo and in vitro. BPA's cytotoxicity was evidenced in rat INS-1 cells where it increased DNA strand breaks, expression of DNA damage associated genes (p53 and CHK2) and ROS concentration (Xin et al. 2014). Increased ROS production and compromised mitochondrial function were also evidenced following BPA exposure in sperm and brain tissue from mice (Ooe et al. 2005). In addition, at high doses ($\geq$100 μM) BPA induces numerical chromosomal changes almost in the diploid range and associated cellular transformation (Tsutsui et al. 1998). Improper chromosome segregation in response to BPA exposure can also cause micronuclei formation (Kabil et al. 2008) and BPA can also cause induction of DNA fragmentation (H. Terasaka et al. 2005; Ptak et al. 2011). BPA's cytotoxic and genotoxic effects are supported by studies showing BPA's pro-apoptotic effects, especially at larger exposures including p53 activation (H. Terasaka et al. 2005; Sengupta et al. 2013; Lin et al. 2013; Ptak et al. 2011).

Furthermore, BPA-treated cells were evidenced to have increased expressions of genes involved in DNA repair in order to overcome the DNA damage caused by the chemical (Fernandez et al. 2012).

In addition to BPA itself, BPA metabolites were also shown to have genotoxic effects. BPA-Q, when not properly detoxified, can cause depurination of DNA (Qiu et al. 2004) and many other BPA metabolites can cause formation of DNA adducts in vivo (liver and mammary cells of female CD-1 mice, male Sprague-Dawley rats and Swiss ICR (CD-1) mice) (Izzotti et al. 2009, 2010; De Flora et al. 2011).

BPA can contribute to genomic instability also through epigenetic and transcriptional changes. In adult mice exposed in utero, epigenetic effects of BPA included increased histone-3-trimethylation which is usually associated with repressed gene expression (Doherty et al. 2010; Dhimolea et al. 2014; Wong et al. 2015) and in breast cancer cell lines, BPA was shown to alter the miRNA profiles, including expression of miR-21 (Tilghman et al. 2012). Furthermore, BPA exposure increased transcription and chromatin modification factor recruitment in key promoters such as HOXB9 (Deb et al. 2016), HOXC6 (Hussain et al. 2015) and HOTAIR long non-coding RNA (Bhan et al. 2014). Lastly, BPA can induce altered methylation in several regions of the genome, including imprinting regions, in the embryo and placenta after maternal exposure (Hanna et al. 2012; Susiarjo et al. 2013; Jorgensen et al. 2016).

These epigenetic changes can explain the aberrant transcriptomic profiles observed after BPA exposure in breast neoplasms and endometriosis (Roy et al. 2015), human MCF-7 breast cancer cells and T47D carcinoma cells (Tilghman et al. 2012; Vivacqua et al. 2003; Buterin et al. 2006), human ovarian cancer cell lines (Hayes et al. 2016), endometrial cancer cells including estrogen receptor (ER)-proficient Ishikawa plus and ER-deficient Ishikawa minus endometrial cancer cells (Boehme et al. 2009; Gertz et al. 2012), a mouse Sertoli cell line (TTE3 cells) (Tabuchi et al. 2002), rodent prostate cancer (PCa) models (Ho et al. 2015) and female Sprague-Dawley offspring born to mothers exposed to BPA during gestation (Grassi et al. 2016). Genes aberrantly expressed following BPA exposure include but are not limited to CYP19A1, EGFR, ESR2, FOS and IGF1 (Roy et al. 2015), and many growth- and development-related genes such as HOXC1 and C6, Wnt5A, Frizzled, TGFbeta-2 and STAT inhibitor 2 (Singleton et al. 2006).

In turn, transcriptome changes confer aberrant proteomic profiles involving several key cellular processes such as cell proliferation, inhibition of apoptosis, tissue remodeling, inflammation, stress response and glutathione synthesis in vitro (primary organotypic cultures of the mouse mammary gland and non-cancerous human high-risk donor breast epithelial cell (HRBEC) cultures) (Williams et al. 2016; Dairkee et al. 2013) and in vivo (rats exposed prenatally or prepubertally and Swiss ICR (CD-1) mice) (Betancourt et al. 2012, 2014; Izzotti et al. 2010; Lee et al. 2012b; Tang et al. 2012). Proteins whose expression is altered in response to BPA exposure have important roles in protein metabolism, signal transduction, development and cell cycle regulation (Lamartiniere et al. 2011) and can therefore make cells more susceptible to carcinogenesis.

In addition to genetic and epigenetic effects, BPA can also confer instability by antagonistically binding to cellular regulator receptors (Moriyama et al. 2002; Greathouse et al. 2012). Lastly, BPA directly affects DNA repair through impairment of the double-strand break repair machinery (Allard and Colaiácovo 2010) and through binding of key players of DNA repair such as DNA-PKcs (Ito et al. 2008). On the other hand, it might also prevent initiation of repair of oxidised base lesions by the base excision repair pathway (Gassman et al. 2015).

In addition, to genomic instability, BPA also contributes to tumour promoting inflammation, especially in the tumour microenvironment. In adult noble rats, BPA induced marked infiltration of CD4+ and CD8+ T cells into the prostatic intraepithelial neoplasia (PIN) (Lam et al. 2015). This immune cell infiltration is an integral part of the tumour microenvironment and can be used to a tumour's advantage since it may be "utilized" in sustaining angiogenesis, stimulating proliferation, facilitating tissue invasion and supporting metastatic dissemination (Hanahan and Weinberg 2011).

In addition, treatment of THP1 macrophages and primary human macrophages with BPA increased pro-inflammation cytokines tumour necrosis factor-α (TNF-α) and interleukin-6 (IL-6) production but decreased anti-inflammation cytokines interleukin-10 (IL-10) and transforming growth factor-β (TGF-β) production (Liu et al. 2014). BPA alters expression of both anti- and pro-inflammatory modulators in mammary tissues after in-utero exposure of CD1 mice, creating a supporting microenvironment for tumour formation. Some of these changes include decreased expression of members of the chemokine CXC family, interleukin 1 (Il1) gene family (Il1β and Il1rn), interleukin 2 gene family (Il7 receptor) and interferon gene family (interferon regulatory factor 9 (Irf9)), as well as immune response gene 1 (Irg1) (Fischer et al. 2016).

References

Allard P, Colaiácovo MP (2010) Bisphenol A impairs the double-strand break repair machinery in the germline and causes chromosome abnormalities. Proc Natl Acad Sci U S A 107 (47):20405–20410. https://doi.org/10.1073/pnas.1010386107

Andreescu S, Sadik OA, McGee DW (2005) Effect of natural and synthetic estrogens on A549 lung cancer cells: correlation of chemical structures with cytotoxic effects. Chem Res Toxicol 18 (3):466–474. https://doi.org/10.1021/tx0497393

Andrysík Z, Vondrácek J, Machala M, Krcmár P, Svihálková-Sindlerová L, Kranz A, Weiss C, Faust D, Kozubík A, Dietrich C (2007) The aryl hydrocarbon receptor-dependent deregulation of cell cycle control induced by polycyclic aromatic hydrocarbons in rat liver epithelial cells. Mutat Res 615:87–97. https://doi.org/10.1016/j.mrfmmm.2006.10.004

Aschengrau A, Coogan PF, Quinn M, Cashins LJ (1998) Occupational exposure to estrogenic chemicals and the occurrence of breast cancer: an exploratory analysis. Am J Ind Med 34 (1):6–14

Ayyanan A, Laribi O, Schuepbach-Mallepell S, Schrick C, Gutierrez M, Tanos T, Lefebvre G, Rougemont J, Yalcin-Ozuysal Ö, Brisken C (2011) Perinatal exposure to bisphenol A increases adult mammary gland progesterone response and cell number. Mol Endocrinol 25 (11):1915–1923. https://doi.org/10.1210/me.2011-1129

Ba Q, Li J, Huang C, Qiu H, Li J, Chu R, Zhang W, Xie D, Wu Y, Wang H (2014) Effects of benzo [a]pyrene exposure on human hepatocellular carcinoma cell angiogenesis, metastasis, and NF-κB signaling. Environ Health Perspect 123:246–254. https://doi.org/10.1289/ehp.1408524

Baird WM, Hooven LA, Mahadevan B (2005) Carcinogenic polycyclic aromatic hydrocarbon-DNA adducts and mechanism of action. Environ Mol Mutagen 45:106–114. https://doi.org/10.1002/em.20095

Beerenwinkel N, Antal T, Dingli D, Traulsen A, Kinzler KW, Velculescu VE, Vogelstein B, Nowak MA (2007) Genetic progression and the waiting time to cancer. PLoS Comput Biol 3 (11):e225. https://doi.org/10.1371/journal.pcbi.0030225

Bergeron RM, Thompson TB, Leonard LS, Pluta L, Gaido KW (1999) Estrogenicity of bisphenol A in a human endometrial carcinoma cell line. Mol Cell Endocrinol 150(1–2):179–187. https://doi.org/10.1016/S0303-7207(98)00202-0

Betancourt AM, Eltoum IA, Desmond RA, Russo J, Lamartiniere CA (2010a) In utero exposure to bisphenol A shifts the window of susceptibility for mammary carcinogenesis in the rat. Environ Health Perspect 118(11):1614–1619. https://doi.org/10.1289/ehp.1002148

Betancourt AM, Mobley JA, Russo J, Lamartiniere CA (2010b) Proteomic analysis in mammary glands of rat offspring exposed in utero to bisphenol A. J Proteome 73(6):1241–1253. https://doi.org/10.1016/j.jprot.2010.02.020

Betancourt AM, Wang J, Jenkins S, Mobley J, Russo J, Lamartiniere CA (2012) Altered carcinogenesis and proteome in mammary glands of rats after prepubertal exposures to the hormonally active chemicals bisphenol A and genistein. J Nutr 142(7):1382S–1388S. https://doi.org/10.3945/jn.111.152058

Betancourt A, Mobley JA, Wang J, Jenkins S, Chen D, Kojima K, Russo J, Lamartiniere CA (2014) Alterations in the rat serum proteome induced by prepubertal exposure to bisphenol A and genistein. J Proteome Res 13(3):1502–1514. https://doi.org/10.1021/pr401027q

Bhan A, Hussain I, Ansari KI, Bobzean SA, Perrotti LI, Mandal SS (2014) Bisphenol-A and diethylstilbestrol exposure induces the expression of breast cancer associated long noncoding RNA HOTAIR in vitro and in vivo. J Steroid Biochem Mol Biol 141:160–170. https://doi.org/10.1016/j.jsbmb.2014.02.002

"Bisphenol A (BPA)" (2016). https://www.niehs.nih.gov/health/topics/agents/sya-bpa/. Accessed 19 Sept

Boehme K, Simon S, Mueller SO (2009) Gene expression profiling in Ishikawa cells: a fingerprint for estrogen active compounds. Toxicol Appl Pharmacol 236(1):85–96. https://doi.org/10.1016/j.taap.2009.01.006

Bontempo P, Mita L, Doto A, Miceli M, Nebbioso A, Lepore I, Franci GL et al (2009) Molecular analysis of the apoptotic effects of BPA in acute myeloid leukemia cells. J Transl Med 7:48. https://doi.org/10.1186/1479-5876-7-48

Bouskine A, Nebout M, Brücker-Davis F, Benahmed M, Fenichel P (2009) Low doses of bisphenol A promote human seminoma cell proliferation by activating PKA and PKG via a membrane G-protein-coupled estrogen receptor. Environ Health Perspect 117(7):1053–1058. https://doi.org/10.1289/ehp.0800367

Buteau-Lozano H, Velasco G, Cristofari M, Balaguer P, Perrot-Applanat M (2008) Xenoestrogens modulate vascular endothelial growth factor secretion in breast cancer cells through an estrogen receptor-dependent mechanism. J Endocrinol 196(2):399–412. https://doi.org/10.1677/JOE-07-0198

Buterin T, Koch C, Naegeli H (2006) Convergent transcriptional profiles induced by endogenous estrogen and distinct xenoestrogens in breast cancer cells. Carcinogenesis 27(8):1567–1578. https://doi.org/10.1093/carcin/bgi339

Carnero A, Blanco-Aparicio C, Kondoh H, Lleonart ME, Martinez-Leal JF, Mondello C, Scovassi AI et al (2015) Disruptive chemicals, senescence and immortality. Carcinogenesis 36(Suppl 1): S19–S37. https://doi.org/10.1093/carcin/bgv029

Castillo Sanchez R, Gomez R, Perez Salazar E (2016) Bisphenol A induces migration through a GPER-, FAK-, Src-, and ERK2-dependent pathway in MDA-MB-231 breast cancer cells. Chem Res Toxicol 29(3):285–295. https://doi.org/10.1021/acs.chemrestox.5b00457

Castillo-Sanchez R, Villegas-Comonfort S, Galindo-Hernandez O, Gomez R, Salazar EP (2013) Benzo-[a]-pyrene induces FAK activation and cell migration in MDA-MB-231 breast cancer cells. Cell Biol Toxicol 29:303–319. https://doi.org/10.1007/s10565-013-9254-1

Chappell G, Pogribny IP, Guyton KZ, Rusyn I (2016) Epigenetic alterations induced by genotoxic occupational and environmental human chemical carcinogens: a systematic literature review. Mutat Res Rev Mutat Res 768:27–45. https://doi.org/10.1016/j.mrrev.2016.03.004

Charles GD, Bartels MJ, Zacharewski TR, Gollapudi BB, Freshour NL, Carney EW (2000) Activity of benzo[a]pyrene and its hydroxylated metabolites in an estrogen receptor-alpha reporter gene assay. Toxicol Sci 55:320–326

Chen Z-J, Yang X-L, Liu H, Wei W, Zhang K-S, Huang H-B, Giesy JP, Liu H-L, Du J, Wang H-S (2015) Bisphenol A modulates colorectal cancer protein profile and promotes the metastasis via induction of epithelial to mesenchymal transitions. Arch Toxicol 89(8):1371–1381. https://doi.org/10.1007/s00204-014-1301-z

Chun TY, Gorski J (2000) High concentrations of bisphenol A induce cell growth and prolactin secretion in an estrogen-responsive pituitary tumor cell line. Toxicol Appl Pharmacol 162 (3):161–165. https://doi.org/10.1006/taap.1999.8840

Colerangle JB, Roy D (1997) Profound effects of the weak environmental estrogen-like chemical bisphenol A on the growth of the mammary gland of noble rats. J Steroid Biochem Mol Biol 60 (1–2):153–160

Dairkee SH, Seok J, Champion S, Sayeed A, Mindrinos M, Xiao W, Davis RW, Goodson WH (2008) Bisphenol A induces a profile of tumor aggressiveness in high-risk cells from breast cancer patients. Cancer Res 68(7):2076–2080. https://doi.org/10.1158/0008-5472.CAN-07-6526

Dairkee SH, Gloria Luciani-Torres M, Moore DH, Goodson WH (2013) Bisphenol-A-induced inactivation of the p53 axis underlying deregulation of proliferation kinetics, and cell death in non-malignant human breast epithelial cells. Carcinogenesis 34(3):703–712. https://doi.org/10.1093/carcin/bgs379

De Flora S, Micale RT, La Maestra S, Izzotti A, D'Agostini F, Camoirano A, Davoli SA et al (2011) Upregulation of clusterin in prostate and DNA damage in spermatozoa from bisphenol A-treated rats and formation of DNA adducts in cultured human prostatic cells. Toxicol Sci 122(1):45–51. https://doi.org/10.1093/toxsci/kfr096

De Jong WH, Kroese ED, Vos JG, Van Loveren H (1999) Detection of immunotoxicity of benzo[a]pyrene in a subacute toxicity study after oral exposure in rats. Toxicol Sci 50:214–220

Dean JH, Luster MI, Boorman GA, Lauer LD, Leubke RW, Lawson L (1983) Selective immunosuppression resulting from exposure to the carcinogenic congener of benzopyrene in B6C3F1 mice. Clin Exp Immunol 52:199–206

Deb P, Bhan A, Hussain I, Ansari KI, Bobzean SA, Pandita TK, Perrotti LI, Mandal SS (2016) Endocrine disrupting chemical, bisphenol-A, induces breast cancer associated gene HOXB9 expression in vitro and in vivo. Gene 590(2):234–243. https://doi.org/10.1016/j.gene.2016.05.009

Dhimolea E, Wadia PR, Murray TJ, Settles ML, Treitman JD, Sonnenschein C, Shioda T, Soto AM (2014) Prenatal exposure to BPA alters the epigenome of the rat mammary gland and increases the propensity to neoplastic development. PLoS One 9(7):e99800. https://doi.org/10.1371/journal.pone.0099800

Diel P, Olff S, Schmidt S, Michna H (2002) Effects of the environmental estrogens bisphenol A, O, p′-DDT, P-tert-octylphenol and coumestrol on apoptosis induction, cell proliferation and the expression of estrogen sensitive molecular parameters in the human breast cancer cell line MCF-7. J Steroid Biochem Mol Biol 80(1):61–70. https://doi.org/10.1016/S0960-0760(01)00173-X

Doherty LF, Bromer JG, Zhou Y, Aldad TS, Taylor HS (2010) In utero exposure to diethylstilbestrol (DES) or bisphenol-A (BPA) increases EZH2 expression in the mammary gland: an epigenetic mechanism linking endocrine disruptors to breast cancer. Horm Cancer 1(3):146–155. https://doi.org/10.1007/s12672-010-0015-9

Dong S, Terasaka S, Kiyama R (2011) Bisphenol A induces a rapid activation of Erk1/2 through GPR30 in human breast cancer cells. Environ Pollut 159(1):212–218. https://doi.org/10.1016/j.envpol.2010.09.004

Dong Y, Araki M, Hirane M, Tanabe E, Fukushima N, Tsujiuchi T (2014) Effects of bisphenol A and 4-nonylphenol on cellular responses through the different induction of LPA receptors in liver epithelial WB-F344 cells. J Recept Signal Transduct Res 34(3):201–204. https://doi.org/10.3109/10799893.2013.876040

Du HJ, Tang N, Liu BC, You BR, Shen FH, Ye M, Gao A, Huang C (2006) Benzo[a]pyrene-induced cell cycle progression is through ERKs/cyclin D1 pathway and requires the activation of JNKs and p38 mapk in human diploid lung fibroblasts. Mol Cell Biochem 287:79–89. https://doi.org/10.1007/s11010-005-9073-7

Duan B, Xuebin H, Zhao H, Qin J, Luo J (2013) The relationship between urinary bisphenol A levels and meningioma in chinese adults. Int J Clin Oncol 18(3):492–497. https://doi.org/10.1007/s10147-012-0408-6

Engström W, Darbre P, Eriksson S, Gulliver L, Hultman T, Karamouzis MV, Klaunig JE, Mehta R, Moorwood K, Sanderson T, Sone H, Vadgama P, Wagemaker G, Ward A, Singh N, Al-Mulla F, Al-Temaimi R, Amedei A, Colacci AM, Vaccari M, Mondello C, Scovassi AI, Raju J, Hamid RA, Memeo L, Forte S, Roy R, Woodrick J, Salem HK, Ryan EP, Brown DG, Bisson WH (2015) The potential for chemical mixtures from the environment to enable the cancer hallmark of sustained proliferative signalling. Carcinogenesis 36(Suppl 1):S38–S60. https://doi.org/10.1093/carcin/bgv030

Fearon ER, Vogelstein B (1990) A genetic model for colorectal tumorigenesis. Cell 61(5):759–767

Fernandez SV, Huang Y, Snider KE, Zhou Y, Pogash TJ, Russo J (2012) Expression and DNA methylation changes in human breast epithelial cells after bisphenol A exposure. Int J Oncol 41 (1):369–377. https://doi.org/10.3892/ijo.2012.1444

Fic A, Jurković Mlakar S, Juvan P, Mlakar V, Marc J, Sollner Dolenc M, Broberg K, Peterlin Mašič L (2015) Genome-wide gene expression profiling of low-dose, long-term exposure of human osteosarcoma cells to bisphenol A and its analogs bisphenols AF and S. Toxicol In Vitro 29 (5):1060–1069. https://doi.org/10.1016/j.tiv.2015.03.014

Fiorito G, Vlaanderen J, Polidoro S, Gulliver J, Galassi C, Ranzi A, Krogh V, Grioni S, Agnoli C, Sacerdote C, Panico S, Tsai MY, Probst-Hensch N, Hoek G, Herceg Z, Vermeulen R, Ghantous A, Vineis P, Naccarati A, EXPOsOMICS Consortium (2017) Oxidative stress and inflammation mediate the effect of air pollution on cardio- and cerebrovascular disease: a prospective study in nonsmokers. Environ Mol Mutagen 59(3):234–246. https://doi.org/10.1002/em.22153

Fischer C, Mamillapalli R, Goetz LG, Jorgenson E, Ilagan Y, Taylor HS (2016) Bisphenol A (BPA) exposure in utero leads to immunoregulatory cytokine dysregulation in the mouse mammary gland: a potential mechanism programming breast cancer risk. Horm Cancer 7(4):241–251. https://doi.org/10.1007/s12672-016-0254-5

Gao M, Long J, Li Y, Shah W, Fu L, Liu J, Wang Y (2010) Mitochondrial decay is involved in BaP-induced cervical damage. Free Radic Biol Med 49:1735–1745. https://doi.org/10.1016/j.freeradbiomed.2010.09.003

Gassman NR, Coskun E, Stefanick DF, Horton JK, Jaruga P, Dizdaroglu M, Wilson SH (2015) Bisphenol A promotes cell survival following oxidative DNA damage in mouse fibroblasts. PLoS One 10(2):e0118819. https://doi.org/10.1371/journal.pone.0118819

Ge L-C, Chen Z-J, Liu H-Y, Zhang K-S, Liu H, Huang H-B, Zhang G et al (2014a) Involvement of activating ERK1/2 through G protein coupled receptor 30 and estrogen receptor α/β in low doses of bisphenol A promoting growth of sertoli TM4 cells. Toxicol Lett 226(1):81–89. https://doi.org/10.1016/j.toxlet.2014.01.035

Ge L-C, Chen Z-J, Liu H, Zhang K-S, Su Q, Ma X-Y, Huang H-B et al (2014b) Signaling related with biphasic effects of bisphenol A (BPA) on sertoli cell proliferation: a comparative proteomic analysis. Biochim Biophys Acta 1840(9):2663–2673. https://doi.org/10.1016/j.bbagen.2014.05.018

Gerlinger M, McGranahan N, Dewhurst SM, Burrell RA, Tomlinson I, Swanton C (2014) Cancer: evolution within a lifetime. Annu Rev Genet 48:215–236. https://doi.org/10.1146/annurev-genet-120213-092314

Gerlinger M, Rowan AJ, Horswell S, Larkin J, Endesfelder D, Gronroos E, Martinez P et al (2012) Intratumor heterogeneity and branched evolution revealed by multiregion sequencing. N Engl J Med 366(10):883–892. https://doi.org/10.1056/NEJMoa1113205

Gerstung M, Eriksson N, Lin J, Vogelstein B, Beerenwinkel N (2011) The temporal order of genetic and pathway alterations in tumorigenesis. PLoS One 6(11):e27136. https://doi.org/10.1371/journal.pone.0027136

Gertz J, Reddy TE, Varley KE, Garabedian MJ, Myers RM (2012) Genistein and bisphenol A exposure cause estrogen receptor 1 to bind thousands of sites in a cell type-specific manner. Genome Res 22(11):2153–2162. https://doi.org/10.1101/gr.135681.111

Ginsberg GL, Atherholt TB, Butler GH (1989) Benzo[a]pyrene-induced immunotoxicity: comparison to DNA adduct formation in vivo, in cultured splenocytes, and in microsomal systems. J Toxicol Environ Health 28:205–220. https://doi.org/10.1080/15287398909531341

Goodson WH, Lowe L, Carpenter DO, Gilbertson M, Manaf Ali A, Lopez de Cerain Salsamendi A et al (2015) Assessing the carcinogenic potential of low-dose exposures to chemical mixtures in the environment: the challenge ahead. Carcinogenesis 36(Suppl 1):S254–S296

Goodson WH, Luciani MG, Aejaz Sayeed S, Jaffee IM, Moore DH, Dairkee SH (2011) Activation of the mTOR pathway by low levels of xenoestrogens in breast epithelial cells from high-risk women. Carcinogenesis 32(11):1724–1733. https://doi.org/10.1093/carcin/bgr196

Gould JC, Leonard LS, Maness SC, Wagner BL, Conner K, Zacharewski T, Safe S, McDonnell DP, Gaido KW (1998) Bisphenol A interacts with the estrogen receptor alpha in a distinct manner from estradiol. Mol Cell Endocrinol 142(1–2):203–214

Grassi TF, da Silva GN, Bidinotto LT, Rossi BF, Quinalha MM, Kass L, Muñoz-de-Toro M, Barbisan LF (2016) Global gene expression and morphological alterations in the mammary gland after gestational exposure to bisphenol A, genistein and indole-3-carbinol in female Sprague-Dawley offspring. Toxicol Appl Pharmacol 303:101–109. https://doi.org/10.1016/j.taap.2016.05.004

Greathouse KL, Bredfeldt T, Everitt JI, Lin K, Berry T, Kannan K, Mittelstadt ML, Ho S-m, Walker CL (2012) Environmental estrogens differentially engage the histone methyltransferase EZH2 to increase risk of uterine tumorigenesis. Mol Cancer Res 10(4):546–557. https://doi.org/10.1158/1541-7786.MCR-11-0605

Gudjonsson T, Villadsen R, Rønnov-Jessen L, Petersen OW (2004) Immortalization protocols used in cell culture models of human breast morphogenesis. Cell Mol Life Sci 61(19-20):2523–2534. https://doi.org/10.1007/s00018-004-4167-z

Hall JM, Korach KS (2013) Endocrine disrupting chemicals promote the growth of ovarian cancer cells via the ER-CXCL12-CXCR4 signaling axis. Mol Carcinog 52(9):715–725. https://doi.org/10.1002/mc.21913

Hamouchene H, Arlt VM, Giddings I, Phillips DH (2011) Influence of cell cycle on responses of MCF-7 cells to benzo[a]pyrene. BMC Genomics 12:333. https://doi.org/10.1186/1471-2164-12-333

Han D, Tachibana H, Yamada K (2001) Inhibition of environmental estrogen-induced proliferation of human breast carcinoma MCF-7 cells by flavonoids. In Vitro Cell Dev Biol Anim 37(5):275–282

Hanahan D, Folkman J (1996) Patterns and emerging mechanisms of the angiogenic switch during tumorigenesis. Cell 86(3):353–364

Hanahan D, Weinberg RA (2000) The hallmarks of cancer. Cell 100(1):57–70

Hanahan D, Weinberg RA (2011) Hallmarks of cancer: the next generation. Cell 144(5):646–674

Hanna CW, Bloom MS, Robinson WP, Kim D, Parsons PJ, vom Saal FS, Taylor JA, Steuerwald AJ, Fujimoto VY (2012) DNA methylation changes in whole blood is associated with exposure to the environmental contaminants, mercury, lead, cadmium and bisphenol A, in women undergoing ovarian stimulation for IVF. Hum Reprod 27(5):1401–1410. https://doi.org/10.1093/humrep/des038

Hanrahan V, Currie MJ, Gunningham SP, Morrin HR, Scott PAE, Robinson BA, Fox SB (2003) The angiogenic switch for vascular endothelial growth factor (VEGF)-A, VEGF-B, VEGF-C, and VEGF-D in the adenoma-carcinoma sequence during colorectal cancer progression. J Pathol 200(2):183–194. https://doi.org/10.1002/path.1339

Harbst K, Lauss M, Cirenajwis H, Isaksson K, Rosengren F, Torngren T, Kvist A et al (2016) Multi-region whole-exome sequencing uncovers the genetic evolution and mutational heterogeneity of early-stage metastatic melanoma. Cancer Res 76(16):4765–4774. https://doi.org/10.1158/0008-5472.CAN-15-3476

Hayes L, Weening A, Morey LM (2016) Differential effects of estradiol and bisphenol A on SET8 and SIRT1 expression in ovarian cancer cells. Dose Response 14(2):1559325816640682. https://doi.org/10.1177/1559325816640682

Hess-Wilson JK, Boldison J, Weaver KE, Knudsen KE (2006) Xenoestrogen action in breast cancer: impact on ER-dependent transcription and mitogenesis. Breast Cancer Res Treat 96(3):279–292. https://doi.org/10.1007/s10549-005-9082-y

Hiroi H, Tsutsumi O, Takeuchi T, Momoeda M, Ikezuki Y, Okamura A, Yokota H, Taketani Y (2004) Differences in serum bisphenol A concentrations in premenopausal normal women and women with endometrial hyperplasia. Endocr J 51(6):595–600

Ho S-M, Cheong A, Lam H-M, Wen-Yang H, Shi G-B, Zhu X, Chen J et al (2015) Exposure of human prostaspheres to bisphenol A epigenetically regulates SNORD family noncoding RNAs

via histone modification. Endocrinology 156(11):3984–3995. https://doi.org/10.1210/en.2015-1067
Hu W, Feng Z, Tang M-S (2003) Preferential carcinogen-DNA adduct formation at codons 12 and 14 in the human K-ras gene and their possible mechanisms. Biochemistry 42:10012–10023. https://doi.org/10.1021/bi034631s
Hussain I, Bhan A, Ansari KI, Deb P, Bobzean SAM, Perrotti LI, Mandal SS (2015) Bisphenol-A induces expression of HOXC6, an estrogen-regulated homeobox-containing gene associated with breast cancer. Biochim Biophys Acta 1849(6):697–708. https://doi.org/10.1016/j.bbagrm.2015.02.003
Hwang K-A, Park S-H, Yi B-R, Choi K-C (2011) Gene alterations of ovarian cancer cells expressing estrogen receptors by estrogen and bisphenol A using microarray analysis. Lab Anim Res 27(2):99–107. https://doi.org/10.5625/lar.2011.27.2.99
Hwang K-A, Kang N-H, Yi B-R, Lee H-R, Park M-A, Choi K-C (2013) Genistein, a soy phytoestrogen, prevents the growth of BG-1 ovarian cancer cells induced by 17β-estradiol or bisphenol A via the inhibition of cell cycle progression. Int J Oncol 42(2):733–740. https://doi.org/10.3892/ijo.2012.1719
IARC Working Group on the Evaluation of Carcinogenic Risks to Humans (2010) Some non-heterocyclic polycyclic aromatic hydrocarbons and some related exposures. IARC Monogr Eval Carcinog Risks Hum 92:1–853
Ibrahim MAA, Elbakry RH, Bayomy NA (2016) Effect of bisphenol A on morphology, apoptosis and proliferation in the resting mammary gland of the adult albino rat. Int J Exp Pathol 97 (1):27–36. https://doi.org/10.1111/iep.12164
Ichihara T, Yoshino H, Imai N, Tsutsumi T, Kawabe M, Tamano S, Inaguma S, Suzuki S, Shirai T (2003) Lack of carcinogenic risk in the prostate with transplacental and lactational exposure to bisphenol A in rats. J Toxicol Sci 28(3):165–171
International Agency for Research on Cancer, Weltgesundheitsorganisation (Ed) (2012) IARC monographs on the evaluation of carcinogenic risks to humans, volume 100 F, chemical agents and related occupations: this publication represents the views and expert opinions of an IARC Working Group on the Evaluation of Carcinogenic Risks to Humans, which met in Lyon, 20–27 October 2009. IARC, Lyon
Ishido M (2004) Transient inhibition of synergistically insulin-like growth factor-1- and bisphenol A-induced poliferation of estrogen receptor alpha (ERalpha)-positive human breast cancer MCF-7 cells by melatonin. Environ Sci 11(3):163–170
Ito Y, Koessler T, Ibrahim AEK, Rai S, Vowler SL, Abu-Amero S, Silva A-L et al (2008) Somatically acquired hypomethylation of IGF2 in breast and colorectal cancer. Hum Mol Genet 17(17):2633–2643. https://doi.org/10.1093/hmg/ddn163
Izzotti A, Kanitz S, D'Agostini F, Camoirano A, De Flora S (2009) Formation of adducts by bisphenol A, an endocrine disruptor, in DNA in vitro and in liver and mammary tissue of mice. Mutat Res 679(1–2):28–32. https://doi.org/10.1016/j.mrgentox.2009.07.011
Izzotti A, Longobardi M, Cartiglia C, D'Agostini F, Kanitz S, De Flora S (2010) Pharmacological modulation of genome and proteome alterations in mice treated with the endocrine disruptor bisphenol A. Curr Cancer Drug Targets 10(2):147–154
Jenkins S, Raghuraman N, Eltoum I, Carpenter M, Russo J, Lamartiniere CA (2009) Oral exposure to bisphenol A increases dimethylbenzanthracene-induced mammary cancer in rats. Environ Health Perspect 117(6):910–915. https://doi.org/10.1289/ehp.11751
Jenkins S, Wang J, Eltoum I, Desmond R, Lamartiniere CA (2011) Chronic oral exposure to bisphenol A results in a nonmonotonic dose response in mammary carcinogenesis and metastasis in MMTV-erbB2 mice. Environ Health Perspect 119(11):1604–1609. https://doi.org/10.1289/ehp.1103850
Jorgensen EM, Alderman MH, Taylor HS (2016) Preferential epigenetic programming of estrogen response after in utero xenoestrogen (bisphenol-A) exposure. FASEB J 30(9):3194–3201. https://doi.org/10.1096/fj.201500089R

Jung J-W, Park S-B, Lee S-J, Seo M-S, Trosko JE, Kang K-S (2011) Metformin represses self-renewal of the human breast carcinoma stem cells via inhibition of estrogen receptor-mediated OCT4 expression. PLoS One 6(11):e28068. https://doi.org/10.1371/journal.pone.0028068

Kabil A, Silva E, Kortenkamp A (2008) Estrogens and genomic instability in human breast cancer cells--involvement of Src/Raf/Erk signaling in micronucleus formation by estrogenic chemicals. Carcinogenesis 29(10):1862–1868. https://doi.org/10.1093/carcin/bgn138

Kang NH, Hwang KA, Kim TH, Hyun SH, Jeung EB, Choi KC (2012) Induced growth of BG-1 ovarian cancer cells by 17β-estradiol or various endocrine disrupting chemicals was reversed by resveratrol via downregulation of cell cycle progression. Mol Med Rep 6(1):151–156. https://doi.org/10.3892/mmr.2012.887

Kang NH, Hwang KA, Lee HR, Choi DW, Choi KC (2013) Resveratrol regulates the cell viability promoted by 17β-estradiol or bisphenol A via down-regulation of the cross-talk between estrogen receptor α and insulin growth factor-1 receptor in BG-1 ovarian cancer cells. Food Chem Toxicol 59:373–379. https://doi.org/10.1016/j.fct.2013.06.029

Kanu N, Grönroos E, Martinez P, Burrell RA, Yi Goh X, Bartkova J, Maya-Mendoza A et al (2015) SETD2 loss-of-function promotes renal cancer branched evolution through replication stress and impaired DNA repair. Oncogene 34(46):5699–5708. https://doi.org/10.1038/onc.2015.24

Katchy A, Pinto C, Jonsson P, Nguyen-Vu T, Pandelova M, Riu A, Schramm K-W et al (2014) Coexposure to phytoestrogens and bisphenol A mimics estrogenic effects in an additive manner. Toxicol Sci 138(1):21–35. https://doi.org/10.1093/toxsci/kft271

Kim JY, Jeong HG (2003) Down-regulation of inducible nitric oxide synthase and tumor necrosis factor-alpha expression by bisphenol A via nuclear factor-kappaB inactivation in macrophages. Cancer Lett 196(1):69–76

Kim K, Son TG, Park HR, Kim SJ, Kim HS, Kim HS, Kim TS, Jung KK, Han SY, Lee J (2009) Potencies of bisphenol A on the neuronal differentiation and hippocampal neurogenesis. J Toxic Environ Health A 72(21–22):1343–1351. https://doi.org/10.1080/15287390903212501

Kim KB, Seo KW, Kim YJ, Park M, Park CW, Kim PY, Kim JI, Lee SH (2003) Estrogenic effects of phenolic compounds on glucose-6-phosphate dehydrogenase in MCF-7 cells and uterine glutathione peroxidase in rats. Chemosphere 50(9):1167–1173

Kim YS, Choi KC, Hwang KA (2015a) Genistein suppressed epithelial-mesenchymal transition and migration efficacies of BG-1 ovarian cancer cells activated by estrogenic chemicals via estrogen receptor pathway and downregulation of TGF-β signaling pathway. Phytomedicine 22(11):993–999. https://doi.org/10.1016/j.phymed.2015.08.003

Kim Y-S, Hwang K-A, Hyun S-H, Nam K-H, Lee C-K, Choi K-C (2015b) Bisphenol A and nonylphenol have the potential to stimulate the migration of ovarian cancer cells by inducing epithelial–mesenchymal transition via an estrogen receptor dependent pathway. Chem Res Toxicol 28(4):662–671. https://doi.org/10.1021/tx500443p

Klotz DM, Hewitt SC, Korach KS, Diaugustine RP (2000) Activation of a uterine insulin-like growth factor I signaling pathway by clinical and environmental estrogens: requirement of estrogen receptor-alpha. Endocrinology 141(9):3430–3439. https://doi.org/10.1210/endo.141.9.7649

Kochukov MY, Jeng Y-J, Watson CS (2009) Alkylphenol Xenoestrogens with varying carbon chain lengths differentially and potently activate signaling and functional responses in GH3/B6/F10 somatomammotropes. Environ Health Perspect 117(5):723–730. https://doi.org/10.1289/ehp.0800182

Kogita A, Yoshioka Y, Sakai K, Togashi Y, Sogabe S, Nakai T, Okuno K, Nishio K (2015) Inter- and intra-tumor profiling of multi-regional colon cancer and metastasis. Biochem Biophys Res Commun 458(1):52–56. https://doi.org/10.1016/j.bbrc.2015.01.064

Kometani T, Yoshino I, Miura N, Okazaki H, Ohba T, Takenaka T, Shoji F, Yano T, Maehara Y (2009) Benzo[a]pyrene promotes proliferation of human lung cancer cells by accelerating the epidermal growth factor receptor signaling pathway. Cancer Lett 278:27–33. https://doi.org/10.1016/j.canlet.2008.12.017

Kravchenko J, Corsini E, Williams MA, Decker W, Manjili MH, Otsuki T, Singh N, Al-Mulla F, Al-Temaimi R, Amedei A, Colacci AM, Vaccari M, Mondello C, Scovassi AI, Raju J, Hamid RA, Memeo L, Forte S, Roy R, Woodrick J, Salem HK, Ryan EP, Brown DG, Bisson WH, Lowe L, Lyerly HK (2015) Chemical compounds from anthropogenic environment and immune evasion mechanisms: potential interactions. Carcinogenesis 36(Suppl 1):S111–S127. https://doi.org/10.1093/carcin/bgv033

Lam H-M, Ho S-M, Chen J, Medvedovic M, Tam NNC (2015) Bisphenol A disrupts HNF4-α-regulated gene networks linking to prostate preneoplasia and immune disruption in noble rats. Endocrinology 157(1):207–219. https://doi.org/10.1210/en.2015-1363

Lamartiniere CA, Jenkins S, Betancourt AM, Wang J, Russo J (2011) Exposure to the endocrine disruptor bisphenol A alters susceptibility for mammary cancer. Horm Mol Biol Clin Invest 5 (2):45–52. https://doi.org/10.1515/HMBCI.2010.075

Lee HR, Hwang KA, Park MA, Yi BR, Jeung EB, Choi KC (2012a) Treatment with bisphenol A and methoxychlor results in the growth of human breast cancer cells and alteration of the expression of cell cycle-related genes, cyclin D1 and p21, via an estrogen receptor-dependent signaling pathway. Int J Mol Med 29(5):883–890. https://doi.org/10.3892/ijmm.2012.903

Lee H-S, Pyo M-Y, Yang M (2012b) Set, a putative oncogene, as a biomarker for prenatal exposure to bisphenol A. Asian Pac J Cancer Prev 13(6):2711–2715

Lee H-S, Park E-J, Oh J-H, Moon G, Hwang M-S, Kim S-Y, Shin M-K et al (2014) Bisphenol A exerts estrogenic effects by modulating CDK1/2 and p38 MAP kinase activity. Biosci Biotechnol Biochem 78(8):1371–1375. https://doi.org/10.1080/09168451.2014.921557

Li E, Xu Z, Zhao H, Sun Z, Wang L, Guo Z, Zhao Y, Gao Z, Wang Q (2015) Macrophages promote benzopyrene-induced tumor transformation of human bronchial epithelial cells by activation of NF-KB and STAT3 signaling in a bionic airway chip culture and in animal models. Oncotarget 6(11):8900–8913

Lin Y, Sun X, Qiu L, Wei J, Huang Q, Fang C, Ye T, Kang M, Shen H, Dong S (2013) Exposure to bisphenol A induces dysfunction of insulin secretion and apoptosis through the damage of mitochondria in rat insulinoma (INS-1) cells. Cell Death Dis 4:e460. https://doi.org/10.1038/cddis.2012.206

Liu J, Jin X, Zhao N, Ye X, Ying C (2015) Bisphenol A promotes X-linked inhibitor of apoptosis protein-dependent angiogenesis via G protein-coupled estrogen receptor pathway. J Appl Toxicol 35(11):1309–1317. https://doi.org/10.1002/jat.3112

Liu Y, Mei C, Liu H, Wang H, Zeng G, Lin J, Xu M (2014) Modulation of cytokine expression in human macrophages by endocrine-disrupting chemical bisphenol-A. Biochem Biophys Res Commun 451(4):592–598. https://doi.org/10.1016/j.bbrc.2014.08.031

Liu Z, Muehlbauer K-R, Schmeiser HH, Hergenhahn M, Belharazem D, Hollstein MC (2005) p53 mutations in benzo(a)pyrene-exposed human p53 knock-in murine fibroblasts correlate with p53 mutations in human lung tumors. Cancer Res 65:2583–2587. https://doi.org/10.1158/0008-5472.CAN-04-3675

Ma X-F, Zhang J, Shuai H-L, Guan B-Z, Luo X, Yan R-L (2015) IKKβ/NF-κB mediated the low doses of bisphenol A induced migration of cervical cancer cells. Arch Biochem Biophys 573:52–58. https://doi.org/10.1016/j.abb.2015.03.010

Mandrup K, Boberg J, Isling LK, Christiansen S, Hass U (2016) Low-dose effects of bisphenol A on mammary gland development in rats. Andrology 4(4):673–683. https://doi.org/10.1111/andr.12193

Markey CM, Luque EH, Munoz De Toro M, Sonnenschein C, Soto AM (2001) In utero exposure to bisphenol A alters the development and tissue organization of the mouse mammary gland. Biol Reprod 65(4):1215–1223

Maruyama S, Fujimoto N, Yin H, Ito A (1999) Growth stimulation of a rat pituitary cell line MtT/E-2 by environmental estrogens in vitro and in vivo. Endocr J 46(4):513–520

Mavrofrydi O, Papazafiri P (2012) Hypoxia-inducible factor-lα increase is an early and sensitive marker of lung cells responding to benzo[a]pyrene. J Environ Pathol Toxicol Oncol 31:335–347. https://doi.org/10.1615/JEnvironPatholToxicolOncol.v31.i4.40

Meng F, Knapp GW, Green T, Ross JA, Parsons BL (2010) K-Ras mutant fraction in A/J mouse lung increases as a function of benzo[a]pyrene dose. Environ Mol Mutagen 51:146–155. https://doi.org/10.1002/em.20513

Miyakoshi T, Miyajima K, Takekoshi S, Osamura RY (2009) The influence of endocrine disrupting chemicals on the proliferation of ERalpha knockdown-human breast cancer cell line MCF-7; new attempts by RNAi technology. Acta Histochem Cytochem 42(2):23–28. https://doi.org/10.1267/ahc.08036

Mlynarcikova A, Macho L, Fickova M (2013) Bisphenol A alone or in combination with estradiol modulates cell cycle- and apoptosis-related proteins and genes in MCF7 cells. Endocr Regul 47 (4):189–199

Moral R, Wang R, Russo IH, Lamartiniere CA, Pereira J, Russo J (2008) Effect of prenatal exposure to the endocrine disruptor bisphenol A on mammary gland morphology and gene expression signature. J Endocrinol 196(1):101–112. https://doi.org/10.1677/JOE-07-0056

Moriyama K, Tagami T, Akamizu T, Usui T, Saijo M, Kanamoto N, Hataya Y, Shimatsu A, Kuzuya H, Nakao K (2002) Thyroid hormone action is disrupted by bisphenol A as an antagonist. J Clin Endocrinol Metab 87(11):5185–5190. https://doi.org/10.1210/jc.2002-020209

Murray TJ, Maffini MV, Ucci AA, Sonnenschein C, Soto AM (2007) Induction of mammary gland ductal hyperplasias and carcinoma in situ following fetal bisphenol A exposure. Reprod Toxicol 23(3):383–390. https://doi.org/10.1016/j.reprotox.2006.10.002

Nahta R, Al-Mulla F, Al-Temaimi R, Amedei A, Andrade-Vieira R, Bay SN, Brown DG et al (2015) Mechanisms of environmental chemicals that enable the cancer hallmark of evasion of growth suppression. Carcinogenesis 36(Suppl 1):S2–S18. https://doi.org/10.1093/carcin/bgv028

Nakagawa Y, Suzuki T (2001) Metabolism of bisphenol A in isolated rat hepatocytes and oestrogenic activity of a hydroxylated metabolite in MCF-7 human breast cancer cells. Xenobiotica 31(3):113–123. https://doi.org/10.1080/00498250110040501

National Academy of Sciences (US) (2017) Using 21st century science to improve risk-related evaluations. National Academies Press, Washington, DC

Newbold RF, Warren W, Medcalf AS, Amos J (1980) Mutagenicity of carcinogenic methylating agents is associated with a specific DNA modification. Nature 283:596–599

Newbold RR, Jefferson WN, Padilla-Banks E (2007) Long-term adverse effects of neonatal exposure to bisphenol A on the murine female reproductive tract. Reprod Toxicol 24 (2):253–258. https://doi.org/10.1016/j.reprotox.2007.07.006

Newbold RR, Jefferson WN, Padilla-Banks E (2009) Prenatal exposure to bisphenol A at environmentally relevant doses adversely affects the murine female reproductive tract later in life. Environ Health Perspect 117(6):879–885. https://doi.org/10.1289/ehp.0800045

Newby DE, Mannucci PM, Tell GS, Baccarelli AA, Brook RD, Donaldson K et al (2015) Expert position paper on air pollution and cardiovascular disease. Eur Heart J 36(2):83–93b

Ochieng J, Nangami GN, Ogunkua O, Miousse IR, Koturbash I, Odero-Marah V, McCawley LJ, Nangia-Makker P, Ahmed N, Luqmani Y, Chen Z, Papagerakis S, Wolf GT, Dong C, Zhou BP, Brown DG, Colacci AM, Hamid RA, Mondello C, Raju J, Ryan EP, Woodrick J, Scovassi AI, Singh N, Vaccari M, Roy R, Forte S, Memeo L, Salem HK, Amedei A, Al-Temaimi R, Al-Mulla F, Bisson WH, Eltom SE (2015) The impact of low-dose carcinogens and environmental disruptors on tissue invasion and metastasis. Carcinogenesis 36(Suppl 1):S128–S159. https://doi.org/10.1093/carcin/bgv034

Oikawa S (2005) Sequence-specific DNA damage by reactive oxygen species: implications for carcinogenesis and aging. Environ Health Prev Med 10(2):65–71. https://doi.org/10.1007/BF02897995

Ono T, Miki C (2000) Factors influencing tissue concentration of vascular endothelial growth factor in colorectal carcinoma. Am J Gastroenterol 95(4):1062–1067. https://doi.org/10.1111/j.1572-0241.2000.01909.x

Ooe H, Taira T, Iguchi-Ariga SM, Ariga H (2005) Induction of reactive oxygen species by bisphenol A and abrogation of bisphenol A-induced cell injury by DJ-1. Toxicol Sci 88 (1):114–126. https://doi.org/10.1093/toxsci/kfi278
Park M-A, Choi K-C (2014) Effects of 4-nonylphenol and bisphenol A on stimulation of cell growth via disruption of the transforming growth factor-β signaling pathway in ovarian cancer models. Chem Res Toxicol 27(1):119–128. https://doi.org/10.1021/tx400365z
Park S-H, Kim K-Y, An B-S, Choi J-H, Jeung E-B, Leung PCK, Choi K-C (2009) Cell growth of ovarian cancer cells is stimulated by xenoestrogens through an estrogen-dependent pathway, but their stimulation of cell growth appears not to be involved in the activation of the mitogen-activated protein kinases ERK-1 and p38. J Reprod Dev 55(1):23–29
Parsons DW, Jones S, Zhang X, Lin JC-H, Leary RJ, Angenendt P et al (2008) An integrated genomic analysis of human glioblastoma multiforme. Science 321(5897):1807
Phillips TD, Richardson M, Cheng Y-SL, He L, McDonald TJ, Cizmas LH, Safe SH, Donnelly KC, Wang F, Moorthy B, Zhou G-D (2015) Mechanistic relationships between hepatic genotoxicity and carcinogenicity in male B6C3F1 mice treated with polycyclic aromatic hydrocarbon mixtures. Arch Toxicol 89:967–977. https://doi.org/10.1007/s00204-014-1285-8
Pieters N, Koppen G, Smeets K, Napierska D, Plusquin M, De Prins S, Van De Weghe H, Nelen V, Cox B, Cuypers A, Hoet P, Schoeters G, Nawrot TS (2013) Decreased mitochondrial DNA content in association with exposure to polycyclic aromatic hydrocarbons in house dust during wintertime: from a population enquiry to cell culture. PLoS One 8:e63208. https://doi.org/10.1371/journal.pone.0063208
Pisapia L, Del Pozzo G, Barba P, Caputo L, Mita L, Viggiano E, Russo GL et al (2012) Effects of some endocrine disruptors on cell cycle progression and murine dendritic cell differentiation. Gen Comp Endocrinol 178(1):54–63. https://doi.org/10.1016/j.ygcen.2012.04.005
Ptak A, Gregoraszczuk EL (2015) Effects of bisphenol A and 17β-Estradiol on vascular endothelial growth factor a and its receptor expression in the non-cancer and cancer ovarian cell lines. Cell Biol Toxicol 31(3):187–197. https://doi.org/10.1007/s10565-015-9303-z
Ptak A, Wróbel A, Gregoraszczuk EL (2011) Effect of bisphenol-A on the expression of selected genes involved in cell cycle and apoptosis in the OVCAR-3 cell line. Toxicol Lett 202 (1):30–35. https://doi.org/10.1016/j.toxlet.2011.01.015
Ptak A, Rak-Mardyła A, Gregoraszczuk EL (2013) Cooperation of bisphenol A and leptin in inhibition of caspase-3 expression and activity in OVCAR-3 ovarian cancer cells. Toxicol In Vitro 27(6):1937–1943. https://doi.org/10.1016/j.tiv.2013.06.017
Ptak A, Hoffmann M, Gruca I, Barć J (2014) Bisphenol A induce ovarian cancer cell migration via the MAPK and PI3K/Akt signalling pathways. Toxicol Lett 229(2):357–365. https://doi.org/10.1016/j.toxlet.2014.07.001
Pupo M, Pisano A, Lappano R, Santolla MF, De Francesco EM, Abonante S, Rosano C, Maggiolini M (2012) Bisphenol A induces gene expression changes and proliferative effects through GPER in breast cancer cells and cancer-associated fibroblasts. Environ Health Perspect 120 (8):1177–1182. https://doi.org/10.1289/ehp.1104526
Qian W, Zhu J, Mao C, Liu J, Wang Y, Wang Q, Liu Y, Gao R, Xiao H, Wang J (2014) Involvement of CaM-CaMKII-ERK in bisphenol A-induced sertoli cell apoptosis. Toxicology 324:27–34. https://doi.org/10.1016/j.tox.2014.06.001
Qin X-Y, Fukuda T, Yang L, Zaha H, Akanuma H, Zeng Q, Yoshinaga J, Sone H (2012) Effects of bisphenol A exposure on the proliferation and senescence of normal human mammary epithelial cells. Cancer Biol Ther 13(5):296–306. https://doi.org/10.4161/cbt.18942
Qiu S-X, Yang RZ, Gross ML (2004) Synthesis and liquid chromatography/tandem mass spectrometric characterization of the adducts of bisphenol A O-quinone with glutathione and nucleotide monophosphates. Chem Res Toxicol 17(8):1038–1046. https://doi.org/10.1021/tx049953r
Raica M, Cimpean AM, Ribatti D (2009) Angiogenesis in pre-malignant conditions. Eur J Cancer 45(11):1924–1934. https://doi.org/10.1016/j.ejca.2009.04.007
Rappaport SM (2016) Genetic factors are not the major causes of chronic diseases. PLoS One 11(4): e0154387. https://doi.org/10.1371/journal.pone.0154387

Recchia AG, Vivacqua A, Gabriele S, Carpino A, Fasanella G, Rago V, Bonofiglio D, Maggiolini M (2004) Xenoestrogens and the induction of proliferative effects in breast cancer cells via direct activation of oestrogen receptor α. Food Addit Contam 21(2):134–144. https://doi.org/10.1080/02652030310001641177

Rosenberg DW, Giardina C, Tanaka T (2009) Mouse models for the study of colon carcinogenesis. Carcinogenesis 30(2):183–196. https://doi.org/10.1093/carcin/bgn267

Rothman KJ, Greenland S (2005) Causation and causal inference in epidemiology. Am J Public Health 95(Suppl 1):S144–S150

Roy D, Morgan M, Yoo C, Deoraj A, Roy S, Yadav VK, Garoub M, Assaggaf H, Doke M (2015) Integrated bioinformatics, environmental epidemiologic and genomic approaches to identify environmental and molecular links between endometriosis and breast cancer. Int J Mol Sci 16 (10):25285–25322. https://doi.org/10.3390/ijms161025285

Salazar I, Pavani M, Aranda W, Maya JD, Morello A, Ferreira J (2004) Alterations of rat liver mitochondrial oxidative phosphorylation and calcium uptake by benzo[a]pyrene. Toxicol Appl Pharmacol 198:1–10. https://doi.org/10.1016/j.taap.2004.02.013

Sang H, Zhang L, Li J (2012) Anti-benzopyrene-7,8-diol-9,10-epoxide induces apoptosis via mitochondrial pathway in human bronchiolar epithelium cells independent of the mitochondria permeability transition pore. Food Chem Toxicol 50:2417–2423. https://doi.org/10.1016/j.fct.2012.04.041

Schafer TE, Lapp CA, Hanes CM, Lewis JB, Wataha JC, Schuster GS (1999) Estrogenicity of bisphenol A and bisphenol A dimethacrylate in vitro. J Biomed Mater Res 45(3):192–197

Sengupta S, Obiorah I, Maximov PY, Curpan R, Jordan VC (2013) Molecular mechanism of action of bisphenol and bisphenol A mediated by oestrogen receptor alpha in growth and apoptosis of breast cancer cells. Br J Pharmacol 169(1):167–178. https://doi.org/10.1111/bph.12122

Severson PL, Vrba L, Stampfer MR, Futscher BW (2014) Exome-wide mutation profile in benzo[a] pyrene-derived post-stasis and immortal human mammary epithelial cells. Mutat Res Genet Toxicol Environ Mutagen 775–776:48–54. https://doi.org/10.1016/j.mrgentox.2014.10.011

Sheng Z-G, Huang W, Liu Y-X, Zhu B-Z (2013) Bisphenol A at a low concentration boosts mouse spermatogonial cell proliferation by inducing the G protein-coupled receptor 30 expression. Toxicol Appl Pharmacol 267(1):88–94. https://doi.org/10.1016/j.taap.2012.12.014

Shi T, Zhao C, Li Z, Zhang Q, Jin X (2016) Bisphenol a exposure promotes the migration of NCM460 cells via estrogen receptor-mediated integrin β1/MMP-9 pathway. Environ Toxicol 31 (7):799–807. https://doi.org/10.1002/tox.22090

Sigounas G, Hairr JW, Cooke CD, Owen JR, Asch AS, Weidner DA, Wiley JE (2010) Role of benzo[alpha]pyrene in generation of clustered DNA damage in human breast tissue. Free Radic Biol Med 49:77–87. https://doi.org/10.1016/j.freeradbiomed.2010.03.018

Singleton DW, Feng Y, Yang J, Puga A, Lee AV, Khan SA (2006) Gene expression profiling reveals novel regulation by bisphenol-A in estrogen receptor-alpha-positive human cells. Environ Res 100(1):86–92. https://doi.org/10.1016/j.envres.2005.05.004

Sjöblom T, Jones S, Wood LD, Parsons DW, Lin J, Barber TD et al (2006) The consensus coding sequences of human breast and colorectal cancers. Science 314(5797):268–274

Smith LC, Ralston-Hooper KJ, Lee Ferguson P, Sabo-Attwood T (2016a) The G protein-coupled estrogen receptor agonist G-1 inhibits nuclear estrogen receptor activity and stimulates novel phosphoproteomic signatures. Toxicol Sci 151(2):434–446. https://doi.org/10.1093/toxsci/kfw057

Smith MT, Guyton KZ, Gibbons CF, Fritz JM, Portier CJ, Rusyn I, DeMarini DM, Caldwell JC, Kavlock RJ, Lambert PF, Hecht SS, Bucher JR, Stewart BW, Baan RA, Cogliano VJ, Straif K (2016b) Key characteristics of carcinogens as a basis for organizing data on mechanisms of carcinogenesis. Environ Health Perspect 124(6):713–721

Song H, Zhang T, Yang P, Li M, Yang Y, Wang Y, Du J, Pan K, Zhang K (2015) Low doses of bisphenol A stimulate the proliferation of breast cancer cells via ERK1/2/ERRγ signals. Toxicol In Vitro 30(1 Pt B):521–528. https://doi.org/10.1016/j.tiv.2015.09.009

Stolpmann K, Brinkmann J, Salzmann S, Genkinger D, Fritsche E, Hutzler C, Wajant H, Luch A, Henkler F (2012) Activation of the aryl hydrocarbon receptor sensitises human keratinocytes for CD95L- and TRAIL-induced apoptosis. Cell Death Dis 3:e388. https://doi.org/10.1038/cddis.2012.127

Susiarjo M, Sasson I, Mesaros C, Bartolomei MS (2013) Bisphenol a exposure disrupts genomic imprinting in the mouse. PLoS Genet 9(4):e1003401. https://doi.org/10.1371/journal.pgen.1003401

Tabuchi Y, Zhao Q-L, Kondo T (2002) DNA microarray analysis of differentially expressed genes responsive to bisphenol A, an alkylphenol derivative, in an in vitro mouse sertoli cell model. Jpn J Pharmacol 89(4):413–416

Takahashi Y, Ellis LM, Mai M (2003) The angiogenic switch of human colon cancer occurs simultaneous to initiation of invasion. Oncol Rep 10(1):9–13

Tang W-y, Morey LM, Cheung YY, Birch L, Prins GS, Ho S-m (2012) Neonatal exposure to estradiol/bisphenol A alters promoter methylation and expression of Nsbp1 and Hpcal1 genes and transcriptional programs of Dnmt3a/B and Mbd2/4 in the rat prostate gland throughout life. Endocrinology 153(1):42–55. https://doi.org/10.1210/en.2011-1308

Terasaka H, Kadoma Y, Sakagami H, Fujisawa S (2005) Cytotoxicity and apoptosis-inducing activity of bisphenol A and hydroquinone in HL-60 cells. Anticancer Res 25(3B):2241–2247

Terasaka S, Aita Y, Inoue A, Hayashi S, Nishigaki M, Aoyagi K, Sasaki H et al (2004) Using a customized DNA microarray for expression profiling of the estrogen-responsive genes to evaluate estrogen activity among natural estrogens and industrial chemicals. Environ Health Perspect 112(7):773–781

Tian S, Pan L, Sun X (2013) An investigation of endocrine disrupting effects and toxic mechanisms modulated by benzo[a]pyrene in female scallop Chlamys farreri. Aquat Toxicol 144–145:162–171. https://doi.org/10.1016/j.aquatox.2013.09.031

Tilghman SL, Bratton MR, Chris Segar H, Martin EC, Rhodes LV, Li M, McLachlan JA, Wiese TE, Nephew KP, Burow ME (2012) Endocrine disruptor regulation of microRNA expression in breast carcinoma cells. PLoS One 7(3):e32754. https://doi.org/10.1371/journal.pone.0032754

Tsutsui T, Tamura Y, Yagi E, Hasegawa K, Takahashi M, Maizumi N, Yamaguchi F, Barrett JC (1998) Bisphenol-A induces cellular transformation, aneuploidy and DNA adduct formation in cultured Syrian hamster embryo cells. Int J Cancer 75(2):290–294

Umannová L, Machala M, Topinka J, Schmuczerová J, Krčmář P, Neca J, Šujanová K, Kozubík A, Vondracek J (2011) Benzo[a]pyrene and tumor necrosis factor-α coordinately increase genotoxic damage and production of proinflammatory mediators in aleolar epithelial type II cells. Toxicol Lett 206:121–129. https://doi.org/10.1016/j.toxlet.2011.06.029

Uzoigwe JC, Prum T, Bresnahan E, Garelnabi M (2013) The emerging role of outdoor and indoor air pollution in cardiovascular disease. N Am J Med Sci 5(8):445–453

Vineis P, van Veldhoven K, Chadeau-Hyam M, Athersuch TJ (2013) Advancing the application of omics-based biomarkers in environmental epidemiology. Environ Mol Mutagen 54(7):461–467

Vineis P, Chadeau-Hyam M, Gmuender H, Gulliver J, Herceg Z, Kleinjans J, Kogevinas M, Kyrtopoulos S, Nieuwenhuijsen M, Phillips DH, Probst-Hensch N, Scalbert A, Vermeulen R, Wild CP, EXPOsOMICS Consortium (2017) The exposome in practice: design of the EXPOsOMICS project. Int J Hyg Environ Health 220(2 Pt A):142–151

Vivacqua A, Recchia AG, Fasanella G, Gabriele S, Carpino A, Rago V, Di Gioia ML et al (2003) The food contaminants bisphenol A and 4-nonylphenol act as agonists for estrogen receptor alpha in MCF7 breast cancer cells. Endocrine 22(3):275–284. https://doi.org/10.1385/ENDO:22:3:275

Vogelstein B, Kinzler KW (2004) Cancer genes and the pathways they control. Nat Med 10 (8):789–799

Wang B-Y, Wu S-Y, Tang S-C, Lai C-H, Ou C-C, Wu M-F, Hsiao Y-M, Ko J-L (2015a) Benzo[a] pyrene-induced cell cycle progression occurs via ERK-induced Chk1 pathway activation in human lung cancer cells. Mutat Res 773:1–8. https://doi.org/10.1016/j.mrfmmm.2015.01.009

Wang K-H, Kao A-P, Chang C-C, Lin T-C, Kuo T-C (2015b) Bisphenol A-induced epithelial to mesenchymal transition is mediated by cyclooxygenase-2 up-regulation in human endometrial carcinoma cells. Reprod Toxicol 58:229–233. https://doi.org/10.1016/j.reprotox.2015.10.011

Wang Y, Zhai W, Wang H, Xia X, Zhang C (2015c) Benzo(a)pyrene promotes A549 cell migration and invasion through up-regulating Twist. Arch Toxicol 89:451–458. https://doi.org/10.1007/s00204-014-1269-8

Wang D, Gao H, Bandyopadhyay A, Wu A, Yeh IT, Chen Y, Zou Y et al (2014a) Pubertal bisphenol A exposure alters murine mammary stem cell function leading to early neoplasia in regenerated glands. Cancer Prev Res (Phila) 7(4):445–455. https://doi.org/10.1158/1940-6207.CAPR-13-0260

Wang J, Jenkins S, Lamartiniere CA (2014b) Cell proliferation and apoptosis in rat mammary glands following combinational exposure to bisphenol A and genistein. BMC Cancer 14:379. https://doi.org/10.1186/1471-2407-14-379

Wang K-H, Kao A-P, Chang C-C, Lin T-C, Kuo T-C (2013) Bisphenol A at environmentally relevant doses induces cyclooxygenase-2 expression and promotes invasion of human mesenchymal stem cells derived from uterine myoma tissue. Taiwan J Obstet Gynecol 52(2):246–252. https://doi.org/10.1016/j.tjog.2013.04.016

Wei SJ, Chang RL, Merkler KA, Gwynne M, Cui XX, Murthy B, Huang MT, Xie JG, Lu YP, Lou YR, Jerina DM, Conney AH (1999) Dose-dependent mutation profile in the c-Ha-ras proto-oncogene of skin tumors in mice initiated with benzo[a]pyrene. Carcinogenesis 20:1689–1696

Wen J, Pan L (2015) Short-term exposure to benzo[a]pyrene disrupts reproductive endocrine status in the swimming crab Portunus trituberculatus. Comp Biochem Physiol C Toxicol Pharmacol 174–175:13–20. https://doi.org/10.1016/j.cbpc.2015.06.001

Wetherill YB, Petre CE, Monk KR, Puga A, Knudsen KE (2002) The xenoestrogen bisphenol A induces inappropriate androgen receptor activation and mitogenesis in prostatic adenocarcinoma cells. Mol Cancer Ther 1(7):515–524

Williams KE, Lemieux GA, Hassis ME, Olshen AB, Fisher SJ, Werb Z (2016) Quantitative proteomic analyses of mammary organoids reveals distinct signatures after exposure to environmental chemicals. Proc Natl Acad Sci U S A 113(10):E1343–E1351. https://doi.org/10.1073/pnas.1600645113

Wolf K, Stafoggia M, Cesaroni G, Andersen ZJ, Beelen R, Galassi C et al (2015) Long-term exposure to particulate matter constituents and the incidence of coronary events in 11 european cohorts. Epidemiology 26(4):565–574

Wong MP, Cheung N, Yuen ST, Leung SY, Chung LP (1999) Vascular endothelial growth factor is up-regulated in the early pre-malignant stage of colorectal tumour progression. Int J Cancer 81(6):845–850

Wong RLY, Wang Q, Treviño LS, Bosland MC, Chen J, Medvedovic M, Prins GS, Kannan K, Ho S-M, Walker CL (2015) Identification of secretaglobin Scgb2a1 as a target for developmental reprogramming by BPA in the rat prostate. Epigenetics 10(2):127–134. https://doi.org/10.1080/15592294.2015.1009768

Wood LD, Parsons DW, Jones S, Lin J, Sjöblom T, Leary RJ, Shen D et al (2007) The genomic landscapes of human breast and colorectal cancers. Science 318(5853):1108–1113. https://doi.org/10.1126/science.1145720

World Health Organization (Ed) (2010) Who guidelines for indoor air quality: selected pollutants. WHO, Copenhagen

Xin F, Jiang L, Liu X, Geng C, Wang W, Zhong L, Yang G, Chen M (2014) Bisphenol A induces oxidative stress-associated DNA damage in INS-1 cells. Mutat Res Genet Toxicol Environ Mutagen 769:29–33. https://doi.org/10.1016/j.mrgentox.2014.04.019

Yachida S, Jones S, Bozic I, Antal T, Leary R, Baojin F, Kamiyama M et al (2010) Distant metastasis occurs late during the genetic evolution of pancreatic cancer. Nature 467(7319):1114–1117. https://doi.org/10.1038/nature09515

Yasaei H, Gilham E, Pickles JC, Roberts TP, O'Donovan M, Newbold RF (2013) Carcinogen-specific mutational and epigenetic alterations in INK4A, INK4B and p53 tumour-suppressor

genes drive induced senescence bypass in normal diploid mammalian cells. Oncogene 32:171–179. https://doi.org/10.1038/onc.2012.45
Yoshida M, Shimomoto T, Katashima S, Watanabe G, Taya K, Maekawa A (2004) Maternal exposure to low doses of bisphenol A has no effects on development of female reproductive tract and uterine carcinogenesis in donryu rats. J Reprod Dev 50(3):349–360
Yu Z, Zhang L, Desheng W (2004) Effects of three environmental estrogens on expression of proliferation and apoptosis-associated genes in PEO4 cells. Wei Sheng Yan Jiu 33(4):404–406
Zhang J, Chang L, Jin H, Xia Y, Wang L, He W, Li W, Chen H (2016a) Benzopyrene promotes lung cancer A549 cell migration and invasion through up-regulating cytokine IL8 and chemokines CCL2 and CCL3 expression. Exp Biol Med (Maywood) 241:1516–1523. https://doi.org/10.1177/1535370216644530
Zhang X-L, Liu N, Weng S-F, Wang H-S (2016b) Bisphenol A increases the migration and invasion of triple-negative breast cancer cells via oestrogen-related receptor gamma. Basic Clin Pharmacol Toxicol 119(4):389–395. https://doi.org/10.1111/bcpt.12591
Zhang K-S, Chen H-Q, Chen Y-S, Qiu K-F, Zheng X-B, Li G-C, Yang H-D, Wen C-J (2014) Bisphenol A stimulates human lung cancer cell migration via upregulation of matrix metalloproteinases by GPER/EGFR/ERK1/2 signal pathway. Biomed Pharmacother 68 (8):1037–1043. https://doi.org/10.1016/j.biopha.2014.09.003
Zhang W, Fang Y, Shi X, Zhang M, Wang X, Tan Y (2012) Effect of bisphenol A on the EGFR-STAT3 pathway in MCF-7 breast cancer cells. Mol Med Rep 5(1):41–47. https://doi.org/10.3892/mmr.2011.583
Zhang X, Gaspard JP, Chung DC (2001) Regulation of vascular endothelial growth factor by the Wnt and K-Ras pathways in colonic neoplasia. Cancer Res 61(16):6050–6054
Zhu H, Xiao X, Zheng J, Zheng S, Dong K, Yong Y (2009) Growth-promoting effect of bisphenol A on neuroblastoma in vitro and in vivo. J Pediatr Surg 44(4):672–680. https://doi.org/10.1016/j.jpedsurg.2008.10.067
Zhu H, Zheng J, Xiao X, Zheng S, Dong K, Liu J, Wang Y (2010) Environmental endocrine disruptors promote invasion and metastasis of SK-N-SH human neuroblastoma cells. Oncol Rep 23(1):129–139

Chapter 15
Building an Early Life Exposome by Integrating Multiple Birth Cohorts: HELIX

Martine Vrijheid and Lea Maitre

Abstract The exposome has conceptually been described to comprise three overlapping domains: (1) a general external environment including factors such as the urban environment, climate factors, social capital, stress; (2) a specific external environment including specific contaminants, diet, physical activity, tobacco, and (3) an internal environment including internal biological factors such as metabolism, gut microflora, inflammation, and oxidative stress. Here, we aim to illustrate how these three domains and their interrelations may be studied in an epidemiological study design, using the HELIX (Human Early Life Exposome) project as an example. HELIX takes pregnancy and childhood periods ("early life") as a starting point. In six existing birth cohort studies in Europe, HELIX estimated prenatal and postnatal exposures. Exposure models for the outdoor exposome (air pollutants, noise, meteorological factors, and natural and built environment characteristics) were developed for a total of 30,000 mother–child pairs. Exposure biomarkers (for persistent organic pollutants, metals, phthalate metabolites, phenolic compounds and organophosphate pesticides) and omics markers (metabolites, proteins, mRNA, miRNA, DNA methylation) were measured in a subset of 1200 children. Nested repeat-sampling panel studies ($N = 150$) collected data on variability in personal exposure to air pollution and built environment measures, in biomarkers for non-persistent chemicals (phthalates and phenolic compounds) and in all omics techniques. Outcome examinations were carried out using common protocols in the six cohorts. We will discuss some first results of the HELIX project, including a description of the correlation structure of multiple exposure data.

Keywords Early-life exposome · Birth cohorts · Pre-natal and post-natal exposures

M. Vrijheid (✉) · L. Maitre
ISGlobal, Institute for Global Health, Barcelona, Spain
e-mail: martine.vrijheid@isglobal.org; lea.maitre@isglobal.org

S. Dagnino, A. Macherone (eds.), *Unraveling the Exposome*,
https://doi.org/10.1007/978-3-319-89321-1_15

The Early Life Exposome

An important feature of the exposome is its longitudinal and dynamic nature, in contrast to the static and fixed genome. Over the course of their lifetime an individual is exposed to levels of different environmental factors that vary on an hourly to yearly basis, with certain exposures being greatest during certain periods of life such as occupational exposures during the working portion of a lifetime. Since continuous exposome monitoring would be impossible, it is envisaged that the longitudinal or "life course exposome" would be built up from cross-sectional snapshots at key life periods such as in utero, early childhood, adolescence, adulthood, and old age (Wild 2012). The choice of time point at which the exposome will be surveyed is crucial since the health effects of exposures vary over time and sensitivity to the exposome will vary according to age and developmental stage.

The developing fetus, infant, and child may be especially vulnerable to effects of environmental exposures since these are periods of rapidly growing and developing organs, immature metabolism, and the received dose relative to its bodyweight may be greater than later in life. In utero and early life exposure to environmental stressors during critical windows can disrupt developmental processes and permanently alter body structure, metabolism, and physiology, leading to chronic pathologies in later life (Barouki et al. 2012; Heindel et al. 2015). There is now moderate to good evidence for the effects of prenatal exposure to environmental contaminants, including air pollutants, polychlorinated biphenyls (PCBs), lead, mercury and organophosphate pesticides on fetal growth, neurological development, and the respiratory and immune systems (Vrijheid et al. 2016). Evidence is also growing for effects on childhood growth, obesity, and metabolic signaling (Vrijheid et al. 2016). Under The Developmental Origin of Health and Disease hypothesis such effects may be expected to have a lifetime impact (Godfrey et al. 2010; Van den Bergh 2011). At the same time it is clear that, up to now, the environment and child health field has almost uniquely focused on single exposure-health effect relationships; there is no global view of how various types of exposures coexist and jointly impact on health. Therefore the "early life exposome" is key, both as a starting point to develop a lifetime exposome and due to the heightened impact of the exposome at this time.

Here, we aim to illustrate how the exposome concept may be implemented in an epidemiological study design, using the Human Early Life Exposome (HELIX) project as an example (full study design described in Vrijheid et al. 2014). The project takes pregnancy and childhood periods (early life) as the starting point for putting the exposome into practice. The HELIX project measures a wide range of chemical and physical environmental hazards from food, consumer products, water, air, noise, climate, and the built environment among mother–child pairs, during pregnancy and childhood, and links these to the internal exposome using a range of omics platforms. The project then associates the external and internal parts of the exposome with childhood health and development outcomes, namely pre- and postnatal growth, obesity, neurodevelopment, and asthma, lung function, and allergies, each thought to have an important environmental component and each related to later adult disease.

Integrating Multiple Birth Cohorts to Measure the Exposome

Six existing longitudinal population-based birth cohort studies in Europe form the basis of the HELIX project (Vrijheid et al. 2014): the Born in Bradford (BiB) study in Bradford, UK, the Étude des Déterminants pré et postnatals du développement et de la santé de l'Enfant (EDEN) study in France, the INfancia y Medio Ambiente (INMA) cohort in Spain, the Kaunus cohort (KANC) in Lithuania, the Norwegian Mother and Child Cohort Study (MoBa), and the RHEA Mother Child Cohort study in Crete, Greece. These cohorts were selected because (1) they each have a large amount of existing longitudinal data from early pregnancy through childhood; (2) they were able to implement new follow-up examinations of the children at similar ages (6–11 years), old enough for accurate measurement of the phenotypes of interest for HELIX; and (3) they were able to integrate new questionnaires, biosampling, and clinical examinations using a common protocol. The selection of cohorts followed a strategy to obtain data in different regions of Europe.

In general, exposure estimates can be obtained in cohort studies for very large numbers of subjects by exposure models, for example geospatial models, and questionnaires, whereas exposure and omics biomarkers can, for cost reasons, only be obtained in smaller numbers of subjects. Exposure misclassification remains a key concern of exposure assessment, and care must be taken that the exposome analyses do not multiply these problems. HELIX addresses this issue with repeat biological sampling to fully capture exposure to short-lived nonpersistent pollutants (a class within which new varieties are continuously produced by the chemical industries), personal exposure sampling to validate models used for larger population, and the integration of detailed information on how mothers move through their external environment. This type of intensive data collection is feasible in an even smaller number of subjects. Therefore, to construct the exposome, HELIX used a multilevel study design, drawing on nested study populations from the cohorts for different levels of data collection (Fig. 15.1).

1. The outdoor exposome in a large study population.

At the first level, a total of around 30,000 mother–child pairs across the 6 HELIX cohorts were available for study with already collected data on certain exposures (e.g., tobacco smoke), key covariates (e.g., diet, health behaviors, markers of socioeconomic status), and health outcomes (e.g., et pourqu: neurodevelopment, growth, respiratory health—Table 15.1). These existing datasets were harmonized across the cohorts. Based on residential address history and geostatistical models, HELIX has estimated exposures to a range of outdoor exposures (the "outdoor exposome"), including ambient air pollution, noise pollution, UV radiation, temperature, green and blue spaces, and the built environment for all 30,000 mother–child

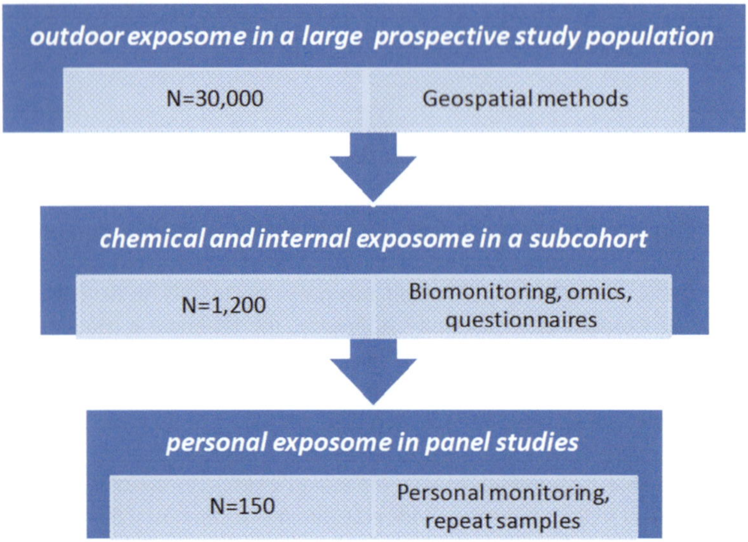

Fig. 15.1 Nested study design of the HELIX study

pairs *during pregnancy and childhood* (Table 15.2). Data from existing regulatory monitors and remote sensing data were used to inform ambient spatial exposure models. Exposure estimates were obtained for pregnancy trimesters, the entire pregnancy period, the first year of life, and during childhood for different time periods, including 1 day, 1 week, 1 month, and 1 year before outcome assessment. The harmonized data on outcomes, exposures, and covariates are held centrally in the HELIX data warehouse located at ISGlobal.

2. The chemical and internal exposome in a subcohort.

At the second level, HELIX selected a subcohort of around 1200 mother–child pairs from these entire cohorts, to be fully characterized for a broad suite of environmental chemical exposures during pregnancy and childhood (Table 15.2) and "omics" biomarkers during childhood (Table 15.3), and to be clinically examined (Table 15.1). The nested design means that for the subcohort subjects, the outdoor exposome has been characterized following the methods outlined in the preceding section. During the subcohort follow-up examination, new biological samples were collected suitable for all planned biomarker and omics analyses. Two urine samples (one before bedtime and one first morning void) were collected to better capture short-lived biomarker metabolites and provide more stable metabolome coverage than would be achieved with one spot urine sample. Stored maternal samples, collected during the pregnancy, were used to measure the pregnancy biomarkers.

In the subcohort, biomarkers of exposure were measured in appropriate biological samples collected from the children at age 6–11 years and previously collected from mothers during pregnancy. Exposure biomarkers include: Persistent Organic

Table 15.1 Child health outcomes measured and harmonized across the HELIX cohorts—Birth, 0–5 years, 6–11 years

Health/development outcomes	Methods	BiB	EDEN	INMA	KANC	MoBa	RHEA
Birth							
Birth weight	Measurements	√	√	√	√	√	√
Gestational duration	Records	√	√	√	√	√	√
1–5 years							
Repeated weight and height, BMI	Measurements and records	√	√	√	√	√	√
Blood pressure (4–5 years)	Measurements	√	√	√			√
Neurodevelopment—cognition	Psychology tests and questionnaires		√	√		√	√
Neurodevelopment—motor skills, language	Psychology tests and questionnaires	√	√	√		√	√
Neurodevelopment—behavior	Questionnaires	√	√	√	√	√	√
Asthma, wheeze, respiratory infection, allergy	Questionnaires	√	√	√	√	√	√
Lung function (4–5 years)	Spirometry	√	√	√			√
6–11 years: common HELIX protocol							
Weight, height, BMI	Measurements	√	√	√	√	√	√
Waist circumference	Measurements	√	√	√	√	√	√
Skinfolds	Measurements	√	√	√	√	√	√
Fat mass	Bioimpedance	√	√	√	√	√	√
Blood pressure	Measurements	√	√	√	√	√	√
Asthma, wheeze, respiratory infection, allergy	Questionnaire	√	√	√	√	√	√
Lung function	Spirometry	√	√	√	√	√	√
Neurodevelopment—cognition, language	Computer tests	√	√	√	√	√	√
Neurodevelopment—behavior	Questionnaires	√	√	√	√	√	√

Pollutants (POPs) and perfluoroalkyl substances (PFAS) in blood samples, metals in whole blood, and nonpersistent chemicals (phthalates, phenols, organophosphate pesticides) and cotinine in urine samples.

During the follow-up examination, trained nurses carried out health examinations of the children to measure weight, height, waist circumference, skin folds, bioimpedance, blood pressure, and lung function. Interviews with the mothers collected information on indoor exposures, environmental tobacco smoke, physical activity, time activity, diet, social factors, stress, sleep, exposure-related habits, and

Table 15.2 Environmental exposures assessed in HELIX during pregnancy and childhood

Exposure group	Typical sources of exposure	Exposure assessment methods
Ambient air pollutants: NO_2, $PM_{2.5}$, PM_{10}, $PM_{2.5}$ absorbance	Traffic emissions and industrial sources	Land use regression spatial models combined with routine monitoring and OMI satellite data for temporal variability
Noise	Traffic	Existing municipal noise maps to obtain spatial estimates. Address-based modeling of noise home and school
UV	Sun	Remote sensing (satellite) UV radiation maps
Temperature, humidity, pressure	Climate	Remote sensing (satellite) temperature (from thermal infrared band), humidity and pressure maps, and data from local meteorological stations
Built environment/green spaces	Built environment	Normalized Difference Vegetation Index (NDVI) from satellite. Building density, walkability score, accessibility, bike lanes, etc., derived from GIS
Persistent organic pollutants (POPs): PCBs, DDE, HCB, and PBDEs	Mainly diet, breastfeeding. PBDEs used as flame-retardants. Production eliminated or restricted	Serum analysis: Solid phase extraction and determination on GC-MS/MS
Perfluoroalkyl substances (PFASs): PFOS, PFOA, PFNA, PFUnDA, PFHxS	Many consumer products including surfactants; surface protection for textiles, carpets, and upholstery; nonstick cookware, food packaging	Plasma analysis: On line column switching LC-MS/MS
Metals: Mercury, cadmium, lead, arsenic, others	Natural and many industrial sources	Whole blood analysis: ICP-MS
Phthalates: 10 phthalate metabolites originating from 6 parent phthalates	Plasticizers in consumer products, including building materials, clothing, food containers, adhesives, and personal care products	Urine analysis in pool of last night and first morning void: On line column switching LC-MS/MS
Phenols: BPA, triclosan, 4 parabens (methyl-, ethyl-, propyl-, butyl paraben), benzophenone 3	BPA: found in coatings of food cans and plastic bottles. Other phenols: many household and personal products	Urine analysis in pool of last night and first morning void: On line column switching LC-MS/MS
Organophosphate (OP) pesticides: 6 nonspecific DAPs, metabolites of OP pesticides	Many agricultural and household uses. Exposure mainly through diet and from hand-to-mouth contact with surfaces treated with OPs insecticides	Urine analysis in pool of last night and first morning void: Solid phase extraction on 96-well plates and LC-QTOF

(continued)

Table 15.2 (continued)

Exposure group	Typical sources of exposure	Exposure assessment methods
Environmental tobacco smoke and cotinine	Tobacco smoke	Questionnaire and cotinine data during and after pregnancy and at age 6–11 years. urine analysis in pool of last night and first morning void
Indoor air pollutants: NO_2, $PM_{2.5}$, BTEX	Indoor cooking, heating, cleaning, smoking. Outdoor air pollutants	Prediction modeling using questionnaire data on indoor sources including cooking, heating, cleaning, and ventilation and passive BTEX and NO_2 sampling and active $PM_{2.5}$ sampling in the home in panel studies
Water disinfection by-products	Drinking water chlorination	Estimates from water companies combined with questionnaires on water consumption, showering, bathing, and swimming

UV ultraviolet, *PCBs* polychlorinated biphenyls, *DDE* dichlorodiphenyl-dichloroethylene, *HCB* hexachlorobenzene, *PBDE* polybrominated diphenyl ethers, *PFOS* perfluorooctane sulfonate, *PFOA* perfluoro-octanoate, *PFNA* perfluoro-nonanoic acid, *PFHxS* perfluorohexane sulfonate, *PFUnDA* perfluoroundecanoate, *BPA* bisphenol A, *DAPs* Dialkylphosphates, *NO_2* nitrogen dioxide, *PM* particulate matter (including $PM_{2.5}$ and PM_{10})

Table 15.3 Omics measured in HELIX at 6–11 years

	Omics platform used
Metabolomics	Untargeted 1H NMR analysis in urine Targeted Biocrates AbsoluteIDQ p180 Kit analysis in serum
Proteomics	Targeted Luminex Technologies of up to 50 proteins including cytokines, adipokines, and apolipoproteins in plasma
Transcriptomics	GeneChip® Human Transcriptome Array 2.0 (Affymetrix) analysis of coding mRNAs and long noncoding transcripts in child whole blood samples SurePrint Human miRNA Microarray rel 21 from Agilent of whole blood miRNA
DNA methylation	Infinium Human Methylation 450 BeadChip for genome-wide methylation analysis of DNA extracted from child whole blood samples

asthma and allergy. Neurodevelopmental outcomes were assessed through a battery of internationally standardized, nonlinguistic, and culturally blind computer tests. Parents completed the Conner's and Child Behavior Checklist (CBCL) questionnaire to assess child behavioral problems.

3. The personal exposome in panel studies.

At the third level, intensive panel studies collected data on short-term temporal variability in exposure biomarkers and omics biomarkers, individual behaviors (physical activity, mobility, time activity), and personal and indoor exposures (Fig. 15.1). The children panel study included children from the HELIX subcohort ($n = 150$). A pregnant women panel study includes 150 pregnant women from outside the cohorts; mothers from the cohorts could not be used for this purpose because their pregnancies were several years ago. Subjects in the children and pregnant women panel study were followed for 1 week in two seasons. From these subjects daily urine samples were collected and at the end of each monitoring week, blood samples were collected following the same procedures as for the subcohort. Blood samples were used to measure repeat omics signals (see below). The daily urine samples were used to measure repeat biomarkers of the nonpersistent chemicals (phthalates, phenols, organophosphate pesticides, cotinine); these data are used to characterize inter- and intra-individual variability in the urine biomarkers, and where possible, correct for the uncertainties in the larger cohort.

Smartphones were worn by the participants in the panel studies to provide information on how children and pregnant women move through their external environment, and on their physical activity levels. The specially developed ExpoApp integrates data from the in-built accelerometer and GPS for this purpose. The panel study subjects also wore electronic wristband UV dosimeters, active $PM_{2.5}$ Cyclone samplers, and MicroAthelometers for continuous black carbon monitoring. Personal exposure estimates from these sensors are used to characterize uncertainties in the spatial exposure models. Indoor air pollution was measured using passive samplers for nitrogen dioxide (NO_2) and BTEX (benzene, toluene, ethylene, and xylene), and active $PM_{2.5}$ samplers, installed in the home; these data are used to develop indoor air pollution models in the larger subcohort.

Linking the Exposome to Child Health

Having collected multilevel data on numerous environmental exposures, it is apparent that different statistical approaches will be needed for the different stages of exposome data analyses, which include describing exposure mixtures and exposome patterns (as most normally encountered in reality), identifying the most important exposures with respect to adverse health effects (crucial for regulatory policy), and identifying synergistic or interactive effects, which mainly due to issues of power has been relatively unexplored in the epidemiological literature. To determine how the external exposome impacts on child health and development, HELIX has developed a multistep statistical analysis approach that is based on several tools and methods (reviewed by Billionnet et al. 2012; Chadeau-Hyam et al. 2013; Sun et al. 2013), including an agnostic exposure-wide association study (ExWAS) approach, reduction dimension and variable selection techniques, and clustering techniques. Simulation studies as part of the project have evaluated the performance of these models with respect to false discovery proportion (FDP) and sensitivity, and indicate the statistical methods are limited in their ability to efficiently differentiate true predictors from correlated covariates in a realistic exposome context, emphasizing that correlations between exposures are the main challenge for exposome research (Agier et al. 2016). It is therefore important that exposome studies interpret exposure effects in light of their correlations to other exposures (Robinson et al. 2015) and select their statistical methods carefully (Agier et al. 2016; Slama and Vrijheid 2015).

Integrating the Internal Exposome

The integration of "omics" into the exposome in HELIX allows us to measure the molecular "fingerprint" or signature of environmental exposures in early life. Each child is made up of a unique molecular profile at the methylome, transcriptome, proteome, or metabolome levels, as the result of the interaction between its genome and early life events partially captured in the external exposome. These molecular signatures are a fine indication of the personal internal exposure to environmental factors. These omics signatures can also reveal early markers of later effects. Environmental exposures may be associated with oxidative stress or inflammation during childhood, without any clinical manifestations until adulthood as chronic diseases. Children may also display differences in susceptibility to their environment and in the era of personalized medicine, a personal exposome assessment should consider molecular susceptibility. For example, the toxicity of arsenic, an ubiquitous metal whose exposure occurs mainly through consumption of fish and crustaceans, will depend heavily on the capacity of the liver and potentially gut microbiota to methylate arsenic species (Hsueh et al. 2016; Claus et al. 2016).

The HELIX project has measured metabolite profiles in blood serum and urine, proteins in plasma, coding and small noncoding RNAs (including miRNAs), and DNA methylation in whole blood. These measurements were performed on the new samples collected from the subcohort of 1200 children at age 6–11 years and in the two repeat panel study periods. The repeat panel study measurements are used to assess the likely sources of variability within and between individuals. The use of new samples for all omics measurements ensures comparability between techniques and cohorts. The use of a similar time point for all omics techniques also allows integration of the different techniques during data analysis. The omics platforms employed in HELIX are detailed in Table 15.3.

The omics data analyses in HELIX aim to evaluate the molecular signatures associated with the external exposome detailed in previous sections. Analyses will focus both on screening all exposures and on the more in-depth analysis of a priori defined exposures for which long-term pre- and postnatal exposure estimates are available and for which there is some first literature (tobacco smoke, exposure to POPs, and air pollution). In the screening analysis of all exposures we will first assess the association between a single exposure and a single omic marker. In this ExWAS analysis, we will select a list of significant hits for each omics based on a p-value threshold accounting for multiple testing. This threshold will be based on the exact number of tests that will be estimated by using exposome correlation matrix.

Secondly, we will define the variability explained by the exposome within omics data and potential common exposure clusters defined by their omics signatures using a dimension reduction approach: the Two-way Orthogonal PLS (O2PLS) model.

In order to interpret the large amounts of information generated by individual omics techniques, HELIX will systematically identify common biological pathways across omics data and link it to health based on public libraries. Each omics signature associated with an exposure can be linked to gene-related pathways (Reactome DB, GO/KEGG), to a biological function, to a disease (DisGeNET), or to a specific anatomical compartment (Bgee DB). It may be difficult to link omics together due to differences in their library identifiers, in particular proteins and metabolites. However, recent solutions (such as BridgeDB) allow the mapping of all the omics used in HELIX. Coupling real human data and in silico analysis allows the identification of the missing link for connecting the responses at different levels of biological organization and thus, linking environmental exposure to disease. This work will eventually allow a better classification of exposures by their biological pathways and potentially prioritize the regulation of the most toxic ones. Similar to developments in the fields of toxicology and pharmacology, the identification of perturbed pathways of well-characterized exposures may allow predicting the public health burden of more recent, less characterized exposures.

More in-depth analysis of a priori defined exposures will include hypothesis driven analyses of the omics data. The impact of different time windows, short- and long-term effects, or pre- and postnatal exposures will also be assessed. Finally, meaningful "hits" will be linked to health outcomes, similar to the "meet-in-the middle" approach to biomarker discovery (Chadeau-Hyam et al. 2011).

Conclusions

The exposome concept is a broad concept that has generated much discussion of how it should be interpreted. It is part of a wider trend in related fields such as exposure assessment, toxicology, systems biology, and public health policy to take a more holistic approach that is more reflective of the exposure mixtures we encounter day to day. Although implementing exposome research remains challenging, characterization of the early life stages of the exposome is a logical and feasible area to focus efforts due to its lifetime importance and defined exposure window.

Separately from HELIX, several other ongoing studies are also characterizing parts of the external exposome and internal exposome at various biological levels (Robinson and Vrijheid 2015). To move towards a unified exposome approach, efforts are needed to piece these parts together across locations and ages, ultimately resulting in a more complete life-course exposome picture. One current focus is on linking internal exposome profiles to single or a few external exposures, and it may turn out that some of the internal exposome responses emerging are more generalizable to whole classes of external exposures, which may eventually minimize the need for comprehensive external exposure assessment. Further efforts in the investigation of omics and molecular pathways using large-scale agnostic molecular technologies will be needed to identify biological pertubations and early signs of damage from environmental agents that can be used to improve causal inference and predict later chronic disease risk. Alternatively, parts of the internal exposome could be considered as important exposures in their own right rather than external exposome mediators, and "exposome-wide" health analyses could analyze these domains side by side, with emphasis on the assessment of multiple and unknown exposures and their interactions through an agnostic approach. This would permit the systematic identification of the most important environmental causes of disease and quantification of the scale of the associated disease burden. A final promise of the exposome is that the application of personal external and internal exposome assessment tools, in combination with susceptibility analysis of genetic and epigenetic predisposition, will allow for a better characterization of personal environmental health risks and targeted prevention; early life may prove particularly efficient for this given that effective prevention and intervention at this time can give life-long returns on investment.

In conclusion, the exposome provides an important new framework to integrate and improve knowledge on the environmental component of disease etiology, with the early life period crucial to full understanding of the human exposome. A more holistic approach, incorporating information on how the various components of the early life exposome coexist and interact, will improve our knowledge of the causes of the complex and multifactorial developmental disorders of childhood, some of which are becoming increasingly common, and ultimately lead to better prevention strategies.

References

Agier L, Portengen L, Chadeau-Hyam M, Basagana X, Giorgis-Allemand L, Siroux V, Robinson O, Vlaanderen J, González JR, Nieuwenhuijsen MJ, Vineis P, Vrijheid M, Slama R, Vermeulen R (2016) A systematic comparison of linear regression-based statistical methods to assess exposome-health associations. Environ Health Perspect 124(12):1848–1856

Barouki R, Gluckman PD, Grandjean P, Hanson M, Heindel JJ (2012) Developmental origins of non-communicable disease: implications for research and public health. Environ Health 11:42

Billionnet C, Sherrill D, Annesi-Maesano I (2012) Estimating the health effects of exposure to multi-pollutant mixture. Ann Epidemiol 22:126–141

Chadeau-Hyam M, Athersuch TJ, Keun HC, De Iorio M, Ebbels TM, Jenab M, Sacerdote C, Bruce SJ, Holmes E, Vineis P (2011) Meeting-in-the-middle using metabolic profiling - a strategy for the identification of intermediate biomarkers in cohort studies. Biomarkers 16(1):83–88

Chadeau-Hyam M, Campanella G, Jombart T, Bottolo L, Portengen L, Vineis P, Liquet B, Vermeulen RC (2013) Deciphering the complex: methodological overview of statistical models to derive OMICS-based biomarkers. Environ Mol Mutagen 54(7):542–557

Claus Henn B, Ettinger AS, Hopkins MR, Jim R, Amarasiriwardena C, Christiani DC, Coull BA, Bellinger DC, Wright RO (2016) Prenatal arsenic exposure and birth outcomes among a population residing near a mining-related superfund site. Environ Health Perspect 124(8):1308–1315

Godfrey KM, Gluckman PD, Hanson MA (2010) Developmental origins of metabolic disease: life course and intergenerational perspectives. Trends Endocrinol Metab 21:199–205
Heindel JJ, Balbus J, Birnbaum L, Brune-Drisse MN, Grandjean P, Gray K, Landrigan PJ, Sly PD, Suk W, Cory Slechta D, Thompson C, Hanson M (2015) Developmental origins of health and disease: integrating environmental influences. Endocrinology 156(10):3416–3421
Hsueh YM, Chen WJ, Lee CY, Chien SN, Shiue HS, Huang SR, Lin MI, Mu SC, Hsieh RL (2016) Association of arsenic methylation capacity with developmental delays and health status in children: a prospective case-control trial. Sci Rep 6:37287
Robinson O, Vrijheid M (2015) The pregnancy exposome. Curr Environ Health Rep 2(2):204–213
Robinson O, Basagaña X, Agier L, de Castro M, Hernandez-Ferrer C, Gonzalez JR, Grimalt JO, Nieuwenhuijsen M, Sunyer J, Slama R, Vrijheid M (2015) The pregnancy exposome: multiple environmental exposures in the INMA-Sabadell Birth Cohort. Environ Sci Technol 49 (17):10632–10641
Slama R, Vrijheid M (2015) Some challenges of studies aiming to relate the exposome to human health. Occup Environ Med 72(6):383–384
Sun Z, Tao Y, Li S, Ferguson KK, Meeker JD, Park SK et al (2013) Statistical strategies for constructing health risk models with multiple pollutants and their interactions: possible choices and comparisons. Environ Health 12:85. https://doi.org/10.1186/1476-069X-12-85
Van den Bergh BR (2011) Developmental programming of early brain and behaviour development and mental health: a conceptual framework. Dev Med Child Neurol 53(Suppl 4):19–23
Vrijheid M, Slama R, Robinson O, Chatzi L, Coen M, van den Hazel P et al (2014) The human early-life exposome (HELIX): project rationale and design. Environ Health Perspect 122:535–544
Vrijheid M, Casas M, Gascon M, Valvi D, Nieuwenhuijsen M (2016) Environmental pollutants and child health-a review of recent concerns. Int J Hyg Environ Health 219(4–5):331–342
Wild CP (2012) The exposome: from concept to utility. Int J Epidemiol 41:24–32

Chapter 16
The HEALS Project

D. A. Sarigiannis

Abstract Exposome appears as a very promising tool for better understanding the complexity of interactions between genome and environment, especially when investigating large population studies. The HEALS project aims at identifying the complex links among genes, environment, and human disease on allergies and asthma, neurodevelopmental/neurodegenerative and metabolic disorders based on individual exposome characterization and how should this be implemented in large cohorts. HEALS relies on the re-analysis of existing cohort studies and the deployment of a Pilot European Exposure and Health Examination Survey. Although the analysis will start from the collection of biomonitoring data, a wide array of omics technologies (completed by confirmatory in vitro testing) will be employed. Lifetime exposure assessment will involve novel technologies such as sensors and agent- based modelling. Mapping the different omics responses onto regulatory networks and disease pathways will allow understanding the intermediate stages from exposure to disease at individual as well as population level. HEALS is expected to provide additional insights into the way to synthesize different data and methodological tools for assessing the internal and external exposome overall aiming to a better understanding of both the potential mechanisms and the origin of disease. This includes (1) how different environmental factors contribute cumulatively to disease and (2) the common nodes of exposure and molecular events resulting in phenomenally different health outcomes. HEALS is a comprehensive methodological advance aiming to provide the way of linking interdisciplinary research towards the understanding of genome and lifetime environmental interaction at individual and population level.

Keywords Life-time exposure assessment · Agent-based modelling

D. A. Sarigiannis (✉)
HERACLES Research Center on the Exposome and Health, Center for Interdisciplinary Research and Innovation, Thessaloniki, Greece

Environmental Engineering Laboratory, School of Chemical Engineering, Aristotle University of Thessaloniki, Thessaloniki, Greece

University School for Advanced Study (IUSS), Pavia, Italy
e-mail: denis@eng.auth.gr

S. Dagnino, A. Macherone (eds.), *Unraveling the Exposome*,
https://doi.org/10.1007/978-3-319-89321-1_16

Introduction and Overview of Coordinated Exposome Efforts

The exposome concept came as a complement to the human genome; although the decoding of the human genome (Schmutz et al. 2004) increased our understanding of the underlying causes of disease, genomic information explains only a part of the population burden of disease. Thus, it is evident that environmental factors are equally or eventually more important and what is critical is the interaction of environmental factors with the biological systems. Towards a better understanding of the causal links among genome, environment, and disease, unraveling the exposome implies that both environmental exposures and genetic variation are reliably measured simultaneously.

Unraveling the exposome is daunting, particularly in the light of the enormous amount of information that needs to be integrated. Because of dedicated actions and projects following the European Commission's (EC) Environment & Health Action Plan 2004–2010, various harmonization efforts have occurred. Projects such as COPHES (harmonization of HBM), EHES (harmonization of Health Surveys), EU-menu (harmonization of data collection on food consumption) or CHICOS (harmonization of child cohort studies) or U-BIOPRED (unbiased biomarkers in prediction of respiratory disease outcomes) all aim at providing common ground for the often-disparate information which was scattered across Europe. In addition, European twin registries have collected biological material and longitudinal phenotypic and exposure data on tens of thousands of twins providing a valuable resource for investigating the development of complex phenotypes and their underlying biology. The HEALS (Health and Environment-wide Associations based on Large Population Surveys) project is a logical progression from many of the achievements of the Environment and Health Action Plan (EHAP) 2004–2010. Making optimal use of the availability of harmonized data across Europe, HEALS introduces significant advances to environmental and health data fusion, including assimilation of data from satellite remote sensing for direct measurement of environmental exposure to airborne pollutants such as particulate matter (PM) and for providing accurate spatially resolved estimates of population exposed to environmental pollutants. Two other projects comprise the EU exposome initiative, namely Exposomics (led by Imperial College, in London, UK) and HELIX (led by CREAL in Barcelona, Spain). Exposomics focuses on the development of a systematic way to measure the influence of environmental exposures on health. Towards this aim, Exposomics deals with the development of a personal exposure monitoring (PEM) system (including sensors, smartphones, geo-referencing, satellites) for collecting data on the individual external exposome as well as on analyzing biological samples (internal markers of external exposures) using multiple "omic" technologies. Relationships between external exposures (as measured by PEM) and global profiles of molecular features (as measured by omics) in the same individuals constitute the overall methodology (Vineis et al. 2013), opening the way to "exposome-wide association studies" (EWAS). The ultimate goal is to use the new tools in risk assessment and in the estimation of the burden of environmental disease with a special focus on the molecular epidemiology of cancer (Chadeau-Hyam et al. 2013; Vineis and Wild 2014). The Human Early-Life Exposome (HELIX) project (Vrijheid et al. 2014) is a collaborative research project that aims to implement novel exposure assessment and biomarker methods to characterize early-life exposure to multiple environmental factors and associate these with omics biomarkers and child health outcomes, characterizing thus the early-life exposome. HELIX uses a multilevel study design, drawing on nested study populations on four different levels of data collection, including data from existing, as well as from an originally designed sub-cohort for HELIX that includes 1200 mother–child pairs (Vrijheid et al. 2014). Similarly to Exposomics, a wide array of external and internal exposome assessment strategies will be applied.

A fourth project, CROME (Cross-Mediterranean Environment and Health Network) is also part of the EU-funded exposome projects albeit with a regional focus on the Mediterranean basin. CROME focuses on integrating human biomonitoring measurements into the exposome construction process as a means to shed more light into the links between environmental and dietary exposures to toxic metals and persistent organic chemicals and adverse health outcomes with emphasis on cancer (Sarigiannis et al. 2015) and neurotoxicity.

In the USA, two are the main exposome-related initiatives funded by NIEHS, namely the HERCULES center at Emory University in Georgia and the Exposome Research Center at the University of California at Berkeley. HERCULES takes a multi-faceted approach to the exposome, attempting to estimate the allostatic load from environmental exposures over an individual's life course. The UC Berkeley Center instead focuses on refining untargeted metabolomics as the main technological means towards unraveling the individual exposome and linking it to human disease adopting a top-down approach (Rappaport and Smith 2010).

HEALS (Health and Environment-wide Associations based on Large Population Surveys) brings together a comprehensive array of novel technologies, data analysis, and modelling tools that support the efficient design and execution of large-scale exposome studies. The HEALS approach brings together and organizes environmental, socio-economic, exposure, biomarker, and health effect data; in addition, it includes all the procedures and computational sequences necessary for applying advanced bioinformatics coupling advanced data mining, biological and exposure modelling so as to ensure that environmental exposure-health associations are studied comprehensively. The overall approach will be verified in a series of population studies across Europe, tackling various levels of environmental exposure, age windows and gender differentiation of exposure, and socio-economic and genetic variability. The main objective of HEALS is the refinement of an integrated methodology and the application of the corresponding analytical and computational tools for performing environment-wide association studies in support of EU-wide environment and health assessments. The HEALS approach will be refined based on pre-existing population data and then it will be applied in a pilot environment and health examination survey covering 18 EU Member States. The lessons learned will be translated into scientific advice towards the development of protocols and guidelines for the setting up of a European environment and health examination survey.

Methodological Tools Relevant to HEALS

Overall Methodological Concept and Innovation Elements

The overall methodological concept of HEALS and the different arrays involved is graphically illustrated in Fig. 16.1. This includes a wide array of state-of-the-art technologies across all major disciplines of the environmental exposure, biochemistry, molecular biology, toxicology, bioinformatics, and epidemiology arena. The exposome is a methodological tool to functionally integrate the different type of information and to canalize this through a specific workflow, aiming at providing additional insights on the environmental causes of diseases. In simple word, environmental exposure put the pressure to the living organism. Genome defines how the organism will respond under the continuously changing environmental pressures. Biomarkers and especially -omics biomarkers describe a snapshot of the result of this dynamic interaction of the living organism against the environmental pressures at different levels of biological organization. Should the environmental pressure defeat the homeostatic mechanisms, disease phenotypes will appear as well. Thus, every piece of information involved in this process is equally important; methodological oversimplification or skip of data in any of the intermediate steps will downgrade the aims from causal link to a loose association. Under this prism, the different technological arrays used in HEALS and their functional integration are presented more in detail. HEALS brings together a comprehensive array of novel technologies, data

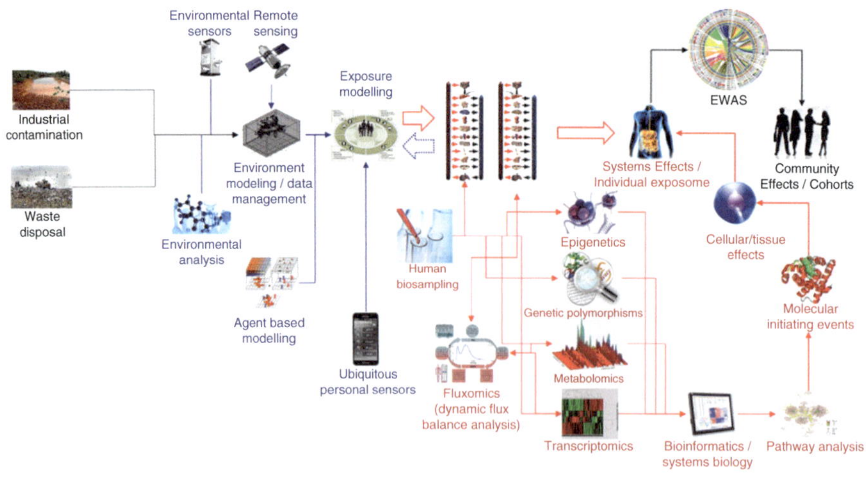

Fig. 16.1 Assessing health impacts of hazardous waste using the exposome paradigm

analysis, and modelling tools that support the efficient design and execution of large-scale exposome studies. The HEALS approach brings together and organizes environmental, socio-economic, exposure, biomarker and health effect data; in addition, it includes all the procedures and computational sequences necessary for applying advanced bioinformatics coupling advanced data mining, biological and exposure modelling to ensure that environmental exposure-health associations are studied comprehensively. The overall approach will be verified in a series of population studies across Europe, tackling various levels of environmental exposure, age windows and gender differentiation of exposure, and socio-economic and genetic variability. The main objective of HEALS is the refinement of an integrated methodology and the application of the corresponding analytical and computational tools for performing environment-wide association studies in support of EU-wide environment and health assessments. The HEALS approach will be refined based on pre-existing population data and then it will be applied in a pilot environment and health examination survey covering 18 EU Member States. The lessons learned will be translated into scientific advice towards the development of protocols and guidelines for the setting up of a European environment and health examination survey.

Exposome studies require novel tools to address the complexity of emerging environmental health issues. Critical for success will be the ability to bring together existing geospatial, environmental, health and socio-economic data, and to collect new high-resolution data using innovative environmental micro-sensors, remote sensing or other community and omics/systems biology based approaches to describe the exposome for, e.g., endocrine disruption-related syndromes and sex-related changes (menopause), neurodegenerative or respiratory diseases. Focus will be on susceptibility windows during growth (including pregnancy) and development, and on the unequal distribution of the burden of epigenetically active food and environment-related disease to vulnerable populations such as the young, elderly, socio-economic disadvantaged, gender and ethnic minorities. Mapping the entire lifecycle of an individual may not be necessary if critical lifetime events where an individual's geospatial lifeline crosses a noteworthy environmental event (Sabel et al. 2009) are recognized and understood. Hence, the most relevant exposure episodes in an individual's life could be reconstructed and linked to socio-economic conditions at critical life stages such as prenatal exposure, puberty, or the reproductively active period. Whereas exposure during all life stages may entail adverse effects, children, pregnant women, and the elderly are particularly susceptible; thus, these population subgroups will be the focus of this project. Modelling the mobility patterns of the population at risk at the individual level is challenging. There are considerable conceptual and computational difficulties involved in intersecting data on the distributions of pollutants, and/or the patterns of movements of recipient individuals or groups, reflecting the limitations of available data on environmental conditions and human distributions. With the advent of geographic information systems (GIS), GPS to track individuals, and personal environmental monitoring, undertaking such analyses throughout an individual's lifetime is now possible.

For the first time, HEALS tries to reverse the paradigm of "nature versus nurture" and adopt one defined by complex and dynamic interactions between DNA sequence, epigenetic DNA modifications, gene expression and environmental factors that all combine to influence disease phenotypes. HEALS will start from analysis of data collected in ongoing epidemiological EU studies involving mother/infant pairs, children, or adults including the elderly to evidence relevant environmental exposure/health outcome associations. These associations will aid in designing pilot surveys using an integrated approach, where the selection of biomarkers of exposure, effects and individual susceptibility result in integrated risk assessment. In the context of this new paradigm, a relevant contribution for a better understanding of the diseases comes also from twin studies. Reviewed evidence makes it clear that substantial phenotypic differences in monozygotic (MZ) twins, that are genetically identical, arise from both non-shared environment and epigenetic DNA modification patterns responsible for incomplete penetrance of genes. The following key issues will be addressed to characterize the individual exposome:

- *Human biomonitoring (HBM)* and *biobanking* are a central theme. Sources and levels of exposure change over time and it is envisioned that individual exposomes can be (re)constructed by analyzing toxicants in (preferentially non-invasively collected) human matrices obtained during critical life stages.
- *Understanding of the interaction between HBM and exposure modelling (EM) or estimation* is another key factor for elucidating the exposome. HBM only has a limited ability to identify specific exposure sources, and may not be applicable to all relevant environmental stressors (e.g., PMx or noise). EM is therefore an essential part of any strategy aimed at unraveling the exposome. Integrating HBM data, EM, and environmental monitoring data will lead to a more conclusive picture of the exposome and related health outcomes and will be of use in future large-scale population studies.
- *Lifestyle/behavior patterns* (such as time-activity-location, food consumption, use of consumer products, etc.) are needed to understand individual and population-based geospatial lifelines.
  - Spatial information and initiatives to harmonize their collection (INSPIRE, GMES) can transform the way scientists and policy makers think about exposure to environmental stressors.
  - At the same time, behavioral information functions as the most accessible and direct way for policy makers and risk assessors to understand and manage an individual's exposure patterns.
  - These exposure factors can be used to derive aggregate and cumulative exposure models, leading to probabilistic exposure assessments. In this context, data from the European Social Survey (ESS) a biennial multi-country survey started in 2002 covering over 30 nations (http://ess.nsd.uib.no/) will provide contextual variables on change and continuity in a wide range of social variables including well-being, health and security, human values, demographics, and socio-economics.
- *Innovations in sensor technology* create possibilities to collect environmental data at unprecedented depth and breadth. Finding the right balance between limited

amounts of high quality data from standardized environmental monitoring campaigns (e.g., EIONET, AIRBase) and large amounts of moderate quality data by sensor networks will transform the way we understand and interact with our environment.

- Due to the substantial technical and ethical hurdles involved in collecting real individual space-time movement data for whole populations, we propose simulating movement and interaction behavior using *agent-based models* (ABM) informed by sensor technologies. ABM aim to simulate and organize social behaviors to understand the dynamics of real-world systems. A style of computational modelling, ABM, focuses on simulating individuals (agents) and their interactions with other agents and their environments, as part of a larger complex system. The use of ABM will enable us to better understand the behavior of individuals and populations in social and evolutionary settings, and to "fill-in" the gaps in the exposome currently not available from real-world monitoring and sensor data.
- Current toxicological state-of-the-art couples estimations of *biologically effective dose (BED)* with early biological events to derive dose-effect models, which can be used in combination with the probabilistic exposure estimates (Georgopoulos et al. 2008; Sarigiannis et al. 2016) to derive biomarkers of exposure and/or effect. Combined use of epidemiological, clinical, and genetic/epigenetic analysis data will shed light on the effect of risk modifying factors such as lifestyle choices and DNA polymorphisms and methylation (Sarigiannis et al. 2009). Distinguishing exposures of limited duration or periods of specific relevance (e.g., in utero) from chronic exposures may render new biomarkers more specific. *Metabolomics and adductomics* are key to this analytical and data interpretation process. They will be functionally integrated with *transcriptomics* and *proteomics* to provide the mechanistic underpinning for establishing causality in the association between health status and exposure to environmental stressors.

Detailed Methodological Elements

Assessment of External Exposure

A major objective of HEALS is to integrate the existing dataset and to fill data gaps to unravel the external exposome of individuals and population subgroups to multiple stressors via different pathways. Towards this aim, an extensive data collection/data mining scheme is employed. Data are mined from different information sources including past and ongoing research and survey projects both at country and European level and EU-wide monitoring systems specifically in the areas of interest to the population studies foreseen in the project. They have been brought together into a comprehensive database of environmental information in Europe, the European Data Management System (EDMS) of HEALS. EDMS is designed in such a way to be fully compatible with the IPCheM information system on chemical contamination in Europe that is being developed by the European Commission's Joint Research Centre.

Environmental data, combined with exposure models, provide the basis for the development of a methodological and computational framework for estimating the external exposures of selected population groups to multiple stressors via different exposure routes (i.e., inhalation, ingestion, and dermal). Depending on data availability, geospatial analysis and multimedia modelling are used to estimate concentrations of the analyzed toxic substances in "microenvironments" and food. Methods for data integration as well as for handling missing data are being exploited by using data fusion techniques aiming at maximizing the information available through "intelligent" merging of the disparate environmental data available and the agent-based and other exposure modelling techniques such as personal sensor technologies. Probabilistic exposure modelling methodologies based on Markov-Chain Monte Carlo techniques are used to estimate external exposure for selected population groups across time and space and to integrate uncertainty associated with the measurement and modelling of the various agent-specific (individual) exposure doses. Thus, based on "individual" exposure estimates, estimations for wider and coherent population subgroups can be derived.

To unravel the individual external exposome, HEALS uses several types of personal sensors (Loh et al. 2017) which can be grouped according to the type of data they provide: passive pollution measuring sensors to measure the pollution levels encountered in different locations where users spend their days, tracking location and physical activity sensors to gather information about the spatial patterns of user location and physical activity. Direct reading monitors help to identify whether peak exposures are more important than average exposure values, identify specific exposure pathways that dominate in critical time windows over an individual's lifetime, and finally build individual exposure profiles.

This highly novel approach gives access to an unprecedented amount of "individualized exposure data," which could greatly improve our understanding of exposure and health associations but which are worthless without interpretation (e.g., human behavior recognition). This requires statistical advances, sophisticated data mining techniques, computing power as well as a careful sharing of data sources while also maintaining privacy protections for personal data. Big data is difficult to be used with classical relational databases, desktop statistics, and traditional visualization packages. What is common for big data treatment is that it is not just about storing huge amounts of data; it is the ability to mine and integrate data, extracting new knowledge from it. Appling this innovative framework to construct the individual exposome in the pilot EU-wide Exposure and Health survey (EXHES) as well as in the existing cohorts, HEALS will bring advances in this area to overcome the current limiting factors related to the analysis and the interpretation of the enormous wealth of data generated necessary to move the current approach from a population to a personalized level.

According to the HEALS paradigm, using data fusion techniques, traditional health and exposure data derived from fixed monitoring networks are supplemented by a range of emerging novel techniques and technologies such as agent-based models (ABM), mobile phone apps, environmental sensor-webs, micro-sensors, and satellite remote sensing. HEALS uses the ICAROS and SMAQ horde of Earth observation data processing tools to derive high resolution estimates of particulate matter loading in the areas covered by the project case studies. The satellite-derived signal is processed through an indirect method that is based on the estimation of aerosol optical depth by comparing the Mie scattering coefficient of atmospheric aerosol between a polluted and a "clean" image of the study domain taken as reference. The total column signal derived by this method is corrected based on the horizontal profile of the atmospheric mixing height to estimate the portion of atmospheric turbidity (expressed in terms of aerosol optical depth) that is due to the particulate matter that is contained in the lowermost part of the troposphere that is close to the Earth's surface. This correction is a key feature that distinguishes the approaches used in HEALS from other techniques that have been proposed as valid alternatives. Comparison of the particulate matter loading over a spatial grid at 30×30 m^2 against ground air quality monitors showed that the residual error is on the order of 4–5%, a result that enhances significantly the availability of air pollution data at the individual residence or workplace level at a minimal cost.

In addition, HEALS will considerably improve exposure modelling and phenotype identification using deterministic and probabilistic approaches, and applying new epidemiological and statistical methods to relate modelled exposure to health outcomes. ABMs are informed by data relating to an individual's behavior within his/her environment (such as movement data within specific microenvironments) and between individuals exploring interactions around health related behaviors and risk determinants such as low SES. Using these parameters and the evolution of agents, simulations produce detailed information relating to the emulated systems, data that can be used to fill in the gaps that exist in traditional datasets.

Assessment of the Internal Exposome

Human Biomonitoring

Human Biomonitoring (HBM) data allow for a new approach to exposure assessment even when the quantity and quality of external exposures are unknown or ambiguous. Biomonitoring data can be used to compare exposures of the general population with special subpopulations and with toxicological animal data. They can also be used in risk assessment and risk management. For risk assessment, biomarker measurements are used to estimate dose, which can then be compared with toxicological parameters normally obtained from animal studies. A key task in human biomonitoring is their interpretation so as to put them in perspective exposure data with presumed toxic doses. Typically, large-scale human biomonitoring surveys gather biomarker measurements from single time points. These data are then used to make inferences about longer periods of toxicant intake, if biomarker values are representative of steady-state conditions. However, this may not be a justified assumption, and may require additional investigation. By repeated sampling of individuals with particularly high and particularly low biomarker values (e.g., $>P_{95}$ and $<P_5$), more insight in the toxicant exposure can be obtained. The time-lag between sampling depends on the chemical half-life and the matrix used for biomarker determination.

Quantitative interpretation of biomonitoring data can best be accomplished by linking PBBK modelling with exposure pathway modelling in a probabilistic framework. This is carried out by combining PBBK models and HBM data to make inferences about environmental exposure scenarios for biomarker data collected in population-based studies (Mosquin et al. 2009). PBBK models when used in combination with Markov chain Monte Carlo (MCMC) techniques may allow disentangling the contribution of different environmental compartments through exposure reconstruction within a probabilistic framework. Ultimately, as biomarker data also reflect individual accumulation, distribution, metabolism, and excretion (ADME) characteristics of chemicals, HBM data offer an excellent opportunity for the validation of the PBBK models.

Multi-Omics, In Vitro Confirmatory Analysis and Exposure Biology Workflow

The HEALS approach to the internal exposome relies on evaluating the maximum available information from multiple omics data. Towards this aim, a detailed exposure biology workflow has been developed and it is being applied in HEALS, entailing a stepwise procedure moving from untargeted to more targeted analysis. This includes the main omics technologies including in particular:

- Transcriptomics: it allows us to identify how gene expression changes under different exposure conditions. The analysis of gene expression in its entirety allows detection of broad coordinated trends which cannot be discerned by more targeted assays. The aim is to identify the different transcription "fingerprints" related to the different combination of exposure and to identify the different toxic mechanisms, associated with transcriptional changes in gene expression that are causally linked to or downstream of the toxicity. Transcriptomics analysis is carried out using the Microarray Scanner Agilent SureScan on SurePrint G3 Human Geb Exp v3 Array Kit 8x60K microarrays. The results of microarray analysis are validated using quantitative real-time polymerase chain reaction (RT-PCR). Significance of differentially expressed transcripts scored by a change in abundance greater than or equal to a threshold of $\pm$ twofold relative to abundance in controls are considered as significant. Significant effects ($p < 0.05$) are determined by one-way analyses of variance (ANOVA). Additional info related to top interacting genes are obtained from the Comparative Toxicogenomics Database (North Carolina State University, USA).
- Epigenomics: Recent work has shown that there is a strong interplay between miRNAs and epigenetic regulators. Their expression and mutual interactions are often highly coordinated, suggesting that epigenetic regulation can be fully

understood by only considering epigenetic regulators and miRNAs as partners of a combined regulatory network. Epigenomics analysis is carried out using the Microarray Scanner Agilent SureScan on SurePrint G3 unrestricted miRNA 8x60K microarrays. Total RNA samples will be spiked using the Agilent MicroRNA Spike-In Kit to assess the labeling and hybridization efficiencies. The labeled miRNA will be hybridized to human miRNA. Fluorescent signal intensities will then be detected on an Agilent Microarray Scanner and extracted for further data interpretation.

- Proteomics analysis is done in a sub-set of human blood plasma samples ($n = 300$) and in extracts from cell cultures. Protein changes can be closer to the relevant functional impact of a studied stimulus. Although mRNA and protein expression are tightly linked through translation, their correlation is limited, and mRNA transcript levels only explain about 50% of the variation of protein levels. This is because of the additional levels of protein regulation including their rate of translation and degradation. Moreover, regulation of protein activity does not stop at its expression level; rather it is often further controlled through posttranslational modification such as phosphorylation. MS-based protein patterns will be acquired in accurate, reproducible, and high sensitive manner by using an LC-MS/MS Thermo Orbitrap Velos. Our approach is based on the application of untargeted label-free proteomics, which allows the identification of many proteins enabling detection of small changes in protein patterns. Absolute intensity-based quantification (Schwanhausser et al. 2011) is be performed to estimate the amount of protein in each sample and statistical analysis will reveal significant changes ($p < 0.05$) in protein abundance.
- Metabolomics: Metabolomics represents an effective approach not only for determining the presence of multiple AOPs that may be occurring in parallel with an already identified AOP, but also for providing physiological support and expanding knowledge on these AOPs relevant to complex exposures. Metabolomics has the potential to inform and support the definition of AOPs based on physiology-related data that incorporate key events. This includes the identification of previously undefined key events by measuring changes in endogenous metabolites involved in a wide variety of biochemical pathways, and associate these changes with exposure to specific combinations of exogenous compounds. Starting from untargeted metabolomics, potential pathways of toxicity will be hypothesized. Further targeted metabolomics analysis will be used for pathway analysis. Targeted metabolomics will also provide quantitative detail on the dynamic metabolic fluxes and thus support mechanistically the dose-time-response modelling, relevant to the toxic responses. Protocols for plasma and urine samples, as well as supernatant extracts of the mini-brains have been already developed and applied, using state-of-the-art equipment that includes the GC/MS-MS/QToF Agilent 7200A, the LC/MS-MS/QToF Agilent 1290–6540 UHD and the LC/MS-MS Thermo Orbitrap.

The reasoning behind the joint analysis of multi-omics is that not all perturbations at the level of gene expression should be considered because they would result in adaptive, not toxic responses. Thus, gene expression-based molecular response pathways should be coupled to the prevalent pathways identified from bioinformatic analysis of metabolite profiles. Metabolomics data provide a closer link to potential phenotypic/clinical observations. Coupled transcriptomics and metabolomics data analysis will allow us to identify among the perturbed pathways the more likely ones to be associated with adverse outcomes. This approach touches directly the main weakness of transcriptomics, namely the failure to distinguish between adaptive and toxic responses. The further combined analysis that includes also epigenetics and proteomics will provide an even more comprehensive picture of the actual toxic responses.

Internal Dose Modelling

The use of internal dose modelling aims at integrating exposure data and modelling output with HBM data. Its goals are to (1) provide the time history of the internal exposure profile, focusing on susceptible developmental stages; (2) assimilate the biomonitoring data related to the cohorts to estimate the individual exposome in quantitative terms; and (3) derive reliable Biologically Effective Dose (BED) values for the compounds of interest so that they can be associated to observed health outcomes. The key component of the above is the development of a lifetime (including gestation and breastfeeding) generic PBBK model (Sarigiannis et al. 2016) incorporating mixtures interaction (Sarigiannis and Gotti 2008) and a framework for biomonitoring data assimilation (Georgopoulos et al. 2008). Aiming to expand the applicability of the generic PBBK model to cover as much as possible the chemical space, parameterization of the model for known and new chemicals with limited information is done through the development of QSAR models (Sarigiannis et al. 2017; Papadaki et al. 2017). The generic PBBK model will also be used to reconstruct exposure from HBM data (Andra et al. 2015). PBBK modelling will also be used to estimate the internal doses of xenobiotics that exceed levels associated with biological pathway alterations (Judson et al. 2011) and, eventually, health risk. The latter can involve the use of specific omics results (metabolomics analysis) and associations of BED to early biological responses. In addition, BED would be used to quantify the effect of compound-induced extracellular perturbations on metabolic states, to directly couple the PBBK model with metabolic regulatory networks. Direct coupling defines a feedback loop that connects clearance and metabolite production rates to metabolism regulation (Eissing et al. 2011) via dynamic flux balance analysis (FBA) (Krauss et al. 2012), thus linking to the regulatory networks identified in the bioinformatics analysis carried out in HEALS.

Bioinformatics

Another key aspect of HEALS is the carving of innovative bioinformatics strategies for biomarker prediction. The bioinformatics tools currently available for biomarker detection and analysis range from statistical approaches to data mining. The latter can be grouped as descriptive or predictive. The first describes the dataset in a concise and brief manner and presents general properties of the data. The second constructs one or a set of models, performs inference on the available dataset, and tries to predict the outcome on new datasets. Data mining techniques have been carried out on different types of data, including clinical (Exarchos et al. 2006), biological (Exarchos et al. 2009), and environmental (Manrai et al. 2017), exhibiting constantly excellent results.

In HEALS, bioinformatics further advances the concept of exposome, by integrating the omics responses of different omics levels, through mapping of regulatory networks and disease pathways. Functional integration of different omics results using bioinformatics tools aiming at development of adverse outcome pathways (AOPs) (Gutsell and Russell 2013) for the endpoints addressed in HEALS. Different omics data derived both from existing on-line databases and from in vitro laboratory experimentation are mapped onto regulatory pathways. They are analyzed using network visualization environments such as Agilent GeneSpring, Thompson-Reuters MetaCoreTM and Reactome/Functional Interaction network plug-in for Cytoscape, so as to create systems toxicology hypotheses from human data, related to the endpoints addressed in HEALS, as well as to identify common nodes across several pathways of exposure to different compounds of relevance. This is especially important under the prism of cumulative exposure; metabolomic data, in combination with the transcriptome data, show that the toxicologic mechanism for the combined toxicity of several compounds exerting interactions beyond additivity occur at a network level (Roede et al. 2014).

Cohort Studies

Use of Big Data and Re-Analysis of Previous Studies

The HEALS approach and tools are put to test through their application in many population studies (incl. Twins' studies) across different exposure settings tackling key health endpoints of the SCALE initiative and the Parma Declaration for both children and the elderly. The overall population size involved in these studies is up to ca. 50,000 individuals covering different age, gender, and socio-economic status groups. The cohorts involved are dispersed across Europe to provide sufficient geographic coverage to facilitate drawing conclusions at the EU-wide scale. These cohorts have been carefully selected to comprise twins. The classical twin methods combined with novel technologies represent a powerful approach towards identifying and understanding the mechanisms and pathways that underlie complex traits. In this way, epigenetic modification of both exposure to and effect of co-exposure to chemical, physical, and biological environmental stressors throughout a person's lifetime will be easier to consider when deducing environmental exposure and health associations. HEALS makes use of environmental and health data including bio-banks from pre-existing twin registries from Denmark (Danish Twin Registry:

14,000+ twins born 1931–1969), Finland (Finnish Twin Cohort Study: 12,966 MZ and DZ twin pairs (25,932 individuals) with both members currently alive), Italy (Italian Twin Register: 25000 twins), Netherlands (Netherlands Twin Register: 87,000 twins), Norway (Norwegian Twin Registry: 31,440 twins, MoBa Study: 1900 twins), Sweden (Swedish Twin Registry: 20,000 MZ pairs, 25,000 same-sex DZ pairs, and 30,000 opposite-sex DZ pairs, respectively), and the UK (EpiTwins: 5000 twins followed-up 20 years).

EXHES Study

Based on the results of these population studies, a pilot exposure and health survey (EXHES) has been organized in 10 EU countries to test the applicability of the HEALS approach for EU-wide large population surveys. EXHES will combine a longitudinal and a nested case-control phase to allow for better definition of environmental exposures and better characterization of disease and risk phenotypes over the limited duration of the project, whilst setting the foundation for post-project follow-up. The technological and computational integration proposed in HEALS will be tested through EXHES about both technical feasibility and cost-effectiveness. The lessons drawn from the EXHES pilot survey will provide the basis for drafting scientific advice, protocols and, eventually, guidance for the setting up of a European Health and Exposure Survey, paving the way to EU-wide assessments.

The objectives of the HEALS pilot European Exposure and Health Examination Survey (EXHES) are:

1. To collect *new harmonized and standardized exposure and health data in European twins and matched singletons*, describe for the first time the normal and pathological development of twins and matched singletons in different European centers, and to make comparisons within and between countries;
2. To obtain *baseline measures for assessment of future trends* in environmental exposures and major chronic diseases and to provide a framework for further etiological research into lifestyle, environmental, epigenetic, genetic and medical care factors affecting health;
3. To determine *the contribution of the exposome to later health and development compared to other factors* operating during childhood. To confirm that the plasticity of the fetal and early life period may lead to programming of body functions by environmental conditions as early as in-utero including low level exposures to environmental toxics;
4. To *provide insight into the potential mechanisms at the origin of diseases*.

The EXHES survey focuses on children health and development: growth and development of the adipose tissue, neurodevelopment, as well as immune, respiratory, and metabolic functions that constitute intermediate phenotypes/endotypes. However, it is also expected that the most frequent conditions such as respiratory diseases, asthma, allergies, overweight (disease phenotypes) will be observed in the first years of life. Risk and protective markers and factors considered include mother-and-child diet, pre- and post-natal environmental exposure, social factors, and -omics markers.

Overall Methodology for Assessing Environmental Burden of Disease: Lifetime Risks for Individuals

HEALS introduces a novel approach towards defining causal associations between health status and environmental stressors through the integrated use of advanced statistical tools for environment-wide association studies (EWAS). Environmental factors that are correlated are not considered confounders; rather they are covariates, which are in "linkage disequilibrium" with each other. EWAS findings could then be used to identify further factors that may be in "disequilibrium," for further detailed measurement and causal identification. Internal doses will be coupled to health impacts on the local population through advanced statistical methods to derive the dose–response functions which account for differences in exposure patterns, susceptibility differences, and inter-individual variation in health response. The approach starts from the biomarker values measured in different biological matrices (urine and peripheral blood) to estimate through the application of the lifetime generic PBBK model the biological effective dose in the target tissue, which is consistent with the biomarker level measured. To estimate the health impact HEALS uses a statistical approach based on survey-weighted logistic multivariate regression adjusted for different covariates (age, sex, socio-economic status, etc.) linking internal doses with health effects or intermediate biological events that can be associated to health perturbations through pathway analysis considering the interdependence of the covariates (using as metric an analogy of the "linkage disequilibrium" metric used in genome-wide association studies). The general formulation of the approach is based on the mathematical linkage of health end points (expressed in terms of odds ratio, p) with different covariates (age sex, SES, lifestyle choices such as smoking, etc.) and the internal dose in the target tissue (X_{Factor}) such as the expression below:

$$\text{logit}\left({}^{p}/_{1-p}\right)=\alpha+\beta_0\cdot\text{cov}_1+\beta_1.\text{cov}_1+\beta_2\cdot X_{\text{factor}}\ldots+\beta_n.\text{cov}_n$$

where cov represents the different covariates used in the model and α and β are the regression coefficients which consider the interdependence between covariates. Considering that adjustment-variable selection in epidemiology can be broadly grouped into background knowledge-based and statistics-based approaches, the use of *Directed Acyclic Graphs* (DAGs) has come to be a core tool in the background-knowledge approach. DAGs present assumed relationships between variables graphically and, based on these assumptions, identify variables to adjust for confounding and other biases. The enhancement in DAGs construction in epidemiology that includes arrow-on-arrow representations for effect modification proposed by Weinberg (2007) boosted their applicability, including the identification of the direction of unmeasured confounding bias (VanderWeele et al. 2008), adjustment for socio-economic status in occupational cancer studies (Fleischer and Diez Roux 2008), identifying variables that need to be adjusted for high resolution spatial epidemiology, as well as metabolic syndrome confounders (Shahar 2010). *Bayesian inference* based modelling has recently been used for the minimization of covariates and determinants in epidemiology, considering that such models offer a more scientifically defensible framework for epidemiologic analysis than the fixed-effects models now prevalent in epidemiology (Greenland 2000). Their applicability domain includes Bayesian regression models for analyzing longitudinal binary process data with emphasis on dealing with missing data (Su and Hogan 2008), Bayesian adjustment for covariate measurement errors (Hossain and Gustafson 2009), and more specifically for health association studies as Bayesian sensitivity analysis for ill-measured and unobserved confounders (Gustafson et al. 2010). In HEALS, the tools mentioned above are combined to provide an integrative framework for identifying the proper covariates that affect the health endpoints of relevance under the light of the health surveys, as well as to constrain the magnitude and direction of bias parameters. In practice, HEALS ushers in "enviromics," the study of a wide array of environmental factors in relation to health and biology.

Conclusions

HEALS introduces a new exposome based paradigm for interdisciplinary scientific work in environment and health. This denotes an approach that builds on the exploration of the interconnections between the co-existence of multiple stressors and the different scales of biological organization that together produce the final adverse health effect. This marks a clear departure from the conventional paradigm, which seeks to shed light on the identification of singular cause–effect relationships between stressors and health outcomes. It entails creating a new way of combining health-relevant information coming from different disciplines, including (but not limited to) environmental science, epidemiology, toxicology, physiology, molecular biology, biochemistry, mathematics and computer science. In HEALS approach, all factors affecting internal and external exposome are treated as covariates, rather than just as co-founders. The functional integration of these different information classes into a unique framework will result in understanding the complex interaction between genome and exposure to environmental factors. This is being verified through the re-evaluation of existing cohorts (including re-analysis of samples) and the deployment of a new cohort, which in turn acts as pilot for the design of future relevant studies.

References

Andra SS, Charisiadis P, Karakitsios S, Sarigiannis DA, Makris KC (2015) Passive exposures of children to volatile trihalomethanes during domestic cleaning activities of their parents. Environ Res 136(0):187–195. https://doi.org/10.1016/j.envres.2014.10.018

Chadeau-Hyam M, Campanella G, Jombart T, Bottolo L, Portengen L, Vineis P, Liquet B, Vermeulen RCH (2013) Deciphering the complex: methodological overview of statistical models to derive OMICS-based biomarkers. Environ Mol Mutagen 54(7):542–557. https://doi.org/10.1002/em.21797

Eissing T, Kuepfer L, Becker C, Block M, Coboeken K, Gaub T, Goerlitz L, Jaeger J, Loosen R, Ludewig B, Meyer M, Niederalt C, Sevestre M, Siegmund HU, Solodenko J, Thelen K, Telle U, Weiss W, Wendl T, Willmann S, Lippert J (2011) A computational systems biology software platform for multiscale modeling and simulation: integrating whole-body physiology, disease biology, and molecular reaction networks. Front Physiol 2:4

Exarchos TP, Papaloukas C, Fotiadis DI, Michalis LK (2006) An association rule mining-based methodology for automated detection of ischemic ECG beats. IEEE Trans Biomed Eng 53 (8):1531–1540

Exarchos TP, Tsipouras MG, Papaloukas C, Fotiadis DI (2009) An optimized sequential pattern matching methodology for sequence classification. Knowl Inf Syst 19(2):249–264

Fleischer NL, Diez Roux AV (2008) Using directed acyclic graphs to guide analyses of neighbourhood health effects: an introduction. J Epidemiol Community Health 62(9):842–846

Georgopoulos PG, Sasso AF, Isukapalli SS, Lioy PJ, Vallero DA, Okino M, Reiter L (2008) Reconstructing population exposures to environmental chemicals from biomarkers: challenges and opportunities. J Expo Sci Environ Epidemiol 19(2):149–171

Greenland S (2000) When should epidemiologic regressions use random coefficients? Biometrics 56(3):915–921

Gustafson P, McCandless LC, Levy AR, Richardson S (2010) Simplified Bayesian sensitivity analysis for mismeasured and unobserved confounders. Biometrics 66(4):1129–1137

Gutsell S, Russell P (2013) The role of chemistry in developing understanding of adverse outcome pathways and their application in risk assessment. Toxicol Res 2(5):299–307

Hossain S, Gustafson P (2009) Bayesian adjustment for covariate measurement errors: a flexible parametric approach. Stat Med 28(11):1580–1600

Judson RS, Kavlock RJ, Setzer RW, Cohen Hubal EA, Martin MT, Knudsen TB, Houck KA, Thomas RS, Wetmore BA, Dix DJ (2011) Estimating toxicity-related biological pathway altering doses for high-throughput chemical risk assessment. Chem Res Toxicol 24(4):451–462

Krauss M, Schaller S, Borchers S, Findeisen R, Lippert J, Kuepfer L (2012) Integrating cellular metabolism into a multiscale whole-body model. PLoS Comput Biol 8(10):e1002750

Loh M, Sarigiannis D, Gotti A, Karakitsios S, Pronk A, Kuijpers E, Annesi-Maesano I, Baiz N, Madureira J, Oliveira Fernandes E, Jerrett M, Cherrie J (2017) How sensors might help define the external exposome. Int J Environ Res Public Health 14(4):434

Manrai AK, Cui Y, Bushel PR, Hall M, Karakitsios S, Mattingly CJ, Ritchie M, Schmitt C, Sarigiannis DA, Thomas DC, Wishart D, Balshaw DM, Patel CJ (2017) Informatics and data analytics to support exposome-based discovery for public health. Annu Rev Public Health 38:279–294. https://doi.org/10.1146/annurev-publhealth-082516-012737

Mosquin PL, Licata AC, Liu B, Sumner SCJ, Okino MS (2009) Reconstructing exposures from small samples using physiologically based pharmacokinetic models and multiple biomarkers. J Expo Sci Environ Epidemiol 19(3):284–297

Papadaki K, Sarigiannis DA, Karakitsios SP (2017) Modeling of adipose/blood partition coefficient for environmental chemicals. Food Chem Toxicol 110c:274–285

Rappaport SM, Smith MT (2010) Environment and disease risks. Science 330(6003):460–461

Roede JR, Uppal K, Park Y, Tran V, Jones DP (2014) Transcriptome–metabolome wide association study (TMWAS) of maneb and paraquat neurotoxicity reveals network level interactions in

toxicologic mechanism. Toxicol Rep 1(0):435–444. https://doi.org/10.1016/j.toxrep.2014.07.006
Sabel CE, Boyle P, Raab G, Löytönen M, Maasilta P (2009) Modelling individual space-time exposure opportunities: a novel approach to unravelling the genetic or environment disease causation debate. Spat Spatiotemporal Epidemiol 1(1):85–94
Sarigiannis D, Marafante E, Gotti A, Reale GC (2009) Reflections on new directions for risk assessment of environmental chemical mixtures. Int J Risk Assess Manag 13(3–4):216–241
Sarigiannis D, Karakitsios S, Handakas E, Simou K, Solomou E, Gotti A (2016) Integrated exposure and risk characterization of bisphenol-a in Europe. Food Chem Toxicol 98:134–147. https://doi.org/10.1016/j.fct.2016.10.017
Sarigiannis D, Papadaki K, Kontoroupis P, Karakitsios SP (2017) Development of QSARs for parameterizing physiology based ToxicoKinetic models. Food Chem Toxicol 106 (Pt A):114–124. https://doi.org/10.1016/j.fct.2017.05.029
Sarigiannis DA, Gotti A (2008) Biology-based dose-response models for health risk assessment of chemical mixtures. Fresenius Environ Bull 17(9 B):1439–1451
Sarigiannis DA, Karakitsios SP, Zikopoulos D, Nikolaki S, Kermenidou M (2015) Lung cancer risk from PAHs emitted from biomass combustion. Environ Res 137:147–156. https://doi.org/10.1016/j.envres.2014.12.009
Schmutz J, Wheeler J, Grimwood J, Dickson M, Yang J, Caoile C, Bajorek E, Black S, Chan YM, Denys M, Escobar J, Flowers D, Fotopulos D, Garcia C, Gomez M, Gonzales E, Haydu L, Lopez F, Ramirez L, Retterer J, Rodriguez A, Rogers S, Salazar A, Tsai M, Myers RM (2004) Quality assessment of the human genome sequence. Nature 429(6990):365–368
Schwanhausser B, Busse D, Li N, Dittmar G, Schuchhardt J, Wolf J, Chen W, Selbach M (2011) Global quantification of mammalian gene expression control. Nature 473(7347):337–342. https://doi.org/10.1038/nature10098
Shahar E (2010) Metabolic syndrome? A critical look from the viewpoints of causal diagrams and statistics. J Cardiovasc Med 11(10):772–779
Su L, Hogan JW (2008) Bayesian semiparametric regression for longitudinal binary processes with missing data. Stat Med 27(17):3247–3268
VanderWeele TJ, Hernán MA, Robins JM (2008) Causal directed acyclic graphs and the direction of unmeasured confounding bias. Epidemiology 19(5):720–728
Vineis P, Wild CP (2014) Global cancer patterns: causes and prevention. Lancet 383 (9916):549–557. https://doi.org/10.1016/s0140-6736(13)62224-2
Vineis P, van Veldhoven K, Chadeau-Hyam M, Athersuch TJ (2013) Advancing the application of omics-based biomarkers in environmental epidemiology. Environ Mol Mutagen 54(7):461–467. https://doi.org/10.1002/em.21764
Vrijheid M, Slama R, Robinson O, Chatzi L, Coen M, van den Hazel P, Thomsen C, Wright J, Athersuch TJ, Avellana N, Basagaña X, Brochot C, Bucchini L, Bustamante M, Carracedo A, Casas M, Estivill X, Fairley L, van Gent D, Gonzalez JR, Granum B, Gražuleviciene˙ R, Gutzkow KB, Julvez J, Keun HC, Kogevinas M, McEachan RRC, Meltzer HM, Sabidó E, Schwarze PE, Siroux V, Sunyer J, Want EJ, Zeman F, Nieuwenhuijsen MJ (2014) The human early-life exposome (HELIX): project rationale and design. Environ Health Perspect 122 (6):535–544. https://doi.org/10.1289/ehp.1307204
Weinberg CR (2007) Can DAGs clarify effect modification? Epidemiology 18(5):569–572

Part VI
Conclusion

Chapter 17
Unravelling the Exposome: Conclusions and Thoughts for the Future

Sonia Dagnino

Abstract Since it was first defined by Christopher Wild, to today, the concept of the exposome has evolved into a valuable tool to evaluate human exposure and health. In the chapters of this book the definition of the exposome paradigm and concept is discussed. The most recent techniques based on OMICs, targeted and untargeted analysis for its characterization are described. The challenges arising from the amount of data that needs to be processed to understand the complex mechanism behind exposure and health outcome, as well as the latest statistical and data analysis methods have been discussed. Finally, multiple projects across the globe have or are currently tackling the application of the exposome concept to real studies, and these studies have been presented in this book. In this section, we will provide a chapter summary, and describe how the exposome paradigm has advanced since its definition. We will also explore question such as: what have we learned so far? Does the exposome paradigm provide valuable information? And what needs to be improved? Finally, we will discuss the possibility of future applications of the exposome in disease causality and personalized medicine.

Keywords The exposome paradigm · The future of exposomics

The Evolution of the Exposome Through Time

Since the completion of the Human Genome Project, the study of the genome was thought to be the key to understanding the causes of chronic diseases and mortality. Thus, the extensive studies carried on GWAS (genome-wide association study) have shown that this is not the case. It has now been widely accepted that only a small proportion of non-communicable diseases can be explained by the study of the Genome (Lichtenstein et al. 2000; Saracci and Vineis 2007; Rappaport 2016), and

S. Dagnino (✉)
Department of Epidemiology and Biostatistics, MRC-PHE Centre for Environment and Health, School of Public Health, Imperial College, London, UK
e-mail: sonia.dagnino@imperial.ac.uk

S. Dagnino, A. Macherone (eds.), *Unraveling the Exposome*,
https://doi.org/10.1007/978-3-319-89321-1_17

that the study of the exposome is a key element in the understanding of disease etiology. Since the first appearance of the terminology, the definition of the exposome has evolved. Wild (2005) described the exposome as the ensemble of factors that "encompasses life-course environmental exposures (including lifestyle factors), from the prenatal period onwards" (Wild 2005). For the first time, the environmental exposures were considered as a whole that needs to be assessed simultaneously in order to understand its influence on the body. This novel idea was further expanded upon in 2010–2011 by Martin Smith and Stephen Rappaport (Rappaport 2011; Rappaport and Smith 2010). Therein, they highlighted that the current state of epidemiological knowledge pertaining to the impact of environmental exposures on human health was extremely poor. They further posited that epidemiologist should shift their views from narrow hypothesis considering one or two environmental risk factors to a more comprehensive view considering total exposures throughout individuals' lifetime (Rappaport 2011).

In 2012, Wild stratified the exposome into three domains (Wild 2012):

1. The general external exposome: including the living environment, climate, social class, education.
2. A specific external exposome: including diet, smoking habits, physical activity, occupation, etc.
3. And finally, an internal exposome: which includes internal and biological factors, defined by the metabolome, microbiome, inflammation, oxidative stress, and ageing.

This new description of the exposome stressed the need to carry transdisciplinary research, going from "molecular mechanisms, biotechnology, bioinformatics, biostatistics, epidemiology, social sciences and clinical research" (Wild 2012) in order to tackle the exposome in its whole.

More recently, the definition was expanded by Miller and Jones to include behavior. This definition goes beyond lifestyle as it also includes interactions with our surroundings (relationships, stress, physical emotions, etc.) and considers endogenous processes that potentially modify exposures (Miller and Jones 2014). The revised and comprehensive complexity of the exposome is shown in Fig. 17.1.

Although it is still in its infancy, the concept has been growing and several attempts have been made to define it and characterize it. The number of citations of the "exposome" in PubMed has risen from 17 in 2011 to 76 in 2017 (Fig. 17.2), indicating a growing interest in the subject.

The reason for its slow development relies on the difficulty of the measurements. Measuring the totality of all exposures at a given time point already presents itself with huge challenges. Additionally, characterization of the exposome requires longitudinal measurements—all exposures through lifetime—creating a dense, 3-dimensional array of data and meta-data. This rich dataset must be analyzed with sophisticated bioinformatics tools to identify associations between exposures and disease.

The first attempt to define measuring the exposome was put forward by Rappaport (2011), who proposed two complementary methods to tackle the exposome. The first method was called a bottom-up approach. This consists of measuring all external exposures by analyzing environmental samples (air, water,

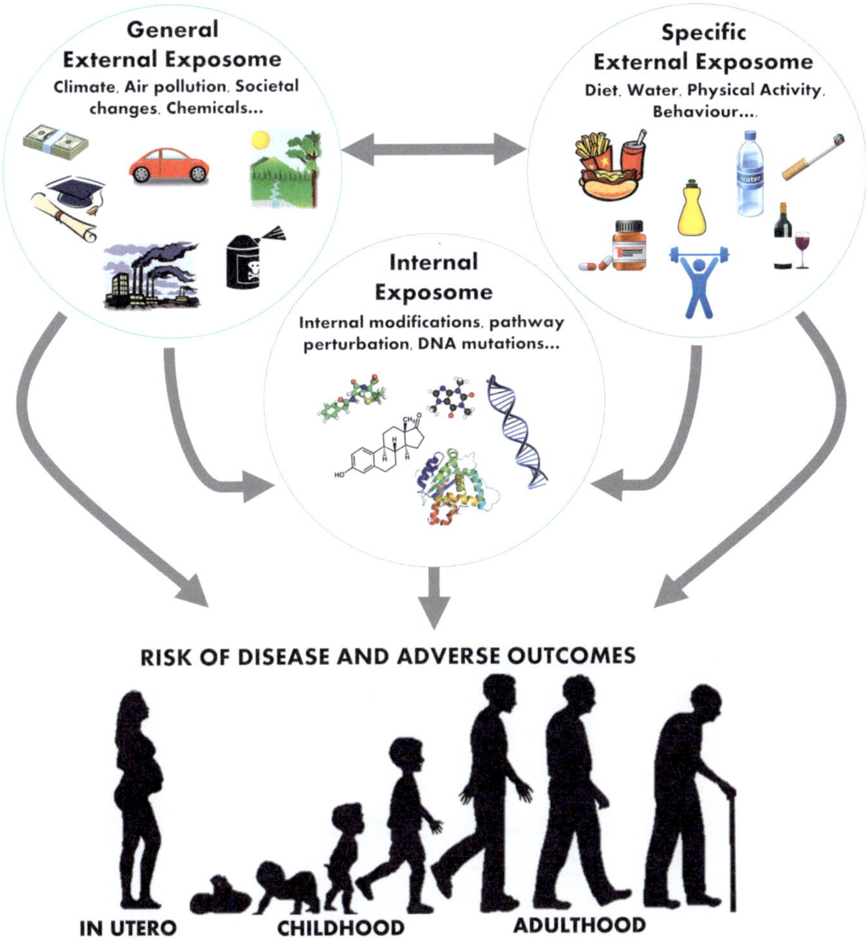

Fig. 17.1 The three layers of factors that can influence the exposome from conception onwards through our life course

food, etc.) and linking them to health status. The second approach is top-down. This approach consists of measuring exposures and internal processes by sampling individuals' blood, urine or other body fluids and linking them to health outcomes. Both methods have limitations, the bottom-up approach does not consider internal processes, such as metabolism, oxidative-stress, and inflammation but it does provide a link to the exposure source. In the top-down approach, the analyses consider all exposures as well as metabolites, by-products, and internal effects, but it does not identify the source of the exposure. In 2010, the national academy of science organized a symposium "The Exposome: A Powerful Approach for Evaluating Environmental Exposures and Their Influences on Human Disease," the discussion

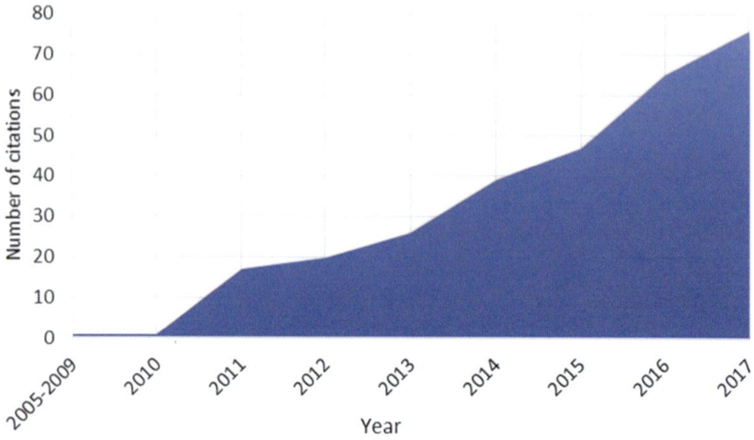

Fig. 17.2 Number of citations for "exposome" in Pubmed (data extracted on 8-11-2017)

concluded that the top-down approach would make more sense, as it allows to "focus on specific toxicants that have a known effect on human health" (National Academy of Science 2010). The discussion also suggested the use of OMICs technologies, which had already proven efficient for biomarker discovery, and should be applied to the study of the exposome. These thoughts lead to the birth of several large international projects for the study of the exposome.

The Study of the Exposome: International Projects

Over the past decade, multiple exposome-related projects and centers have been created. The main projects and centers are summarized in Table 17.1. Most of these projects have been described in detail in Part V of this book.

In Europe, the REA (Research Executive Agency) has devoted a significant attention to the early understanding of the exposome. Financed by the Seventh Framework Program for Research and Technological Development (FP-7), three major projects have been created: EXPOsOMICs, HELIX, and HEALS. EXPOsOMICs has been the first large project created for the understanding of the exposome (Vineis et al. 2017). The project has focused on a top-down approach, focusing on air pollution and water contamination. Although this project has now officially ended, a great amount of data has been produced and analysis is still ongoing. Outputs of EXPOsOMICs include the production of a European database combining OMICs and environmental exposure measurements on over 3000 individuals. The outcome of EXPOsOMICs will further guide future projects and research in the exposome.

The HELIX project (Vrijheid et al. 2014) is devoted to understanding how the exposome influences health at an early stage. The in utero and early-life exposure are

Table 17.1 Project and research centers devoted to the study and understanding of the EXPOSOME across the world

Project	Objectives of the project	Funder	Start date	End date
EXPOsOMICs	Develop new approaches to assess environmental exposures and to link them to biochemical and molecular changes in the body, with focus on air pollution and water contaminants	EU FP-7	2012	2017
HELIX	Combining all environmental hazards that mothers and children are exposed to and link it to health, growth, and development of children	EU FP-7	2013	2018
HEALS	Functional integration of omics derived data and biochemical biomonitoring to create the internal exposome at the individual level	EU FP-7	2013	2018
HERCULES	Provide key infrastructure and expertise to develop and refine new tools and technologies for the study of the exposome	NIEHS	2013	2022
CHEAR	Advance understanding about how the environment impacts children's health and development	NIEHS	2015	
JECS	Identify harmful factors in the environment affecting children's growth and health, and to investigate the relationship between such factors and children's health condition	Japan	2011	2032
I_3 care	International Exposome Center	Netherlands-Hong-Kong-Canada		

critical as they can influence health outcomes further in life. Exposure science had, until now, focused primarily on evaluating single exposure-health effect relationships. HELIX provides the framework to assess the effect of multiple exposures in these critical stages. The HELIX project is still ongoing but will end in 2018.

The HEALs project has a focus on creating new analytical and computational methods to integrate exposome-wide association studies (EWAS) (HEALS 2017). The aim of HEALS is to introduce new methodologies for data analysis and modeling tools to tackle large-scale exposome studies. Similarly to HELIX, HEALS will end in 2018.

In the USA, significant funding has been provided by the National Institute of Environmental Health Sciences (NIEHS) to create the HERCULES exposome Research Center (HERCULES 2017). The aim and project of the center have been described in detail in Chap. 13 of this book. In brief, HERCULES provides key infrastructure and funding for exposome-related projects. The creation of the center has already led to the birth of the Children's Health Exposure Analysis Resource (CHEAR) project (NIEHS 2017), which, similarly to HELIX, focusses on the exposome of critical early-stages of life.

In Japan, a substantial effort has been put in the creation of the Japan Environment and Children's study (Kawamoto et al. 2014), involving 100,000 parent–child pairs followed from birth to 13 years of age. The study will periodically measure exposure to chemicals by direct analysis and questionnaires as well as lifestyle and health outcomes.

Other efforts across the globe include the creation of the International Exposome Center, the I_3 Care Centre, which is a global collaboration between the Chinese University of Hong Kong, the University of Utrecht, and the University of Toronto. The Center aims at "expanding the exposome concept through the development of new investigative tools to discover and quantify modifiable risk factors" (I_3 Care Center 2017) and provides a framework for the development of new international projects on the understanding of the exposome.

Although these projects offered great advances in the description and understanding of the exposome, we are still far from understanding the full mechanisms and implications of exposome science. Further projects and investments will be needed to pursue advances in this domain.

Are Omics Tools Suitable for the Evaluation of the Internal Exposome?

Characterization and measurement of the exposome is multi-disciplinary and relies on multiple analytical tools and technologies. The implementation of omics (big-data) technologies has yielded significant advances for the study of the internal exposome. As discussed in this book, omics provide useful tools to assess effects of exposure at certain time-points. Metabolomics, Epigenomics, and Transcriptomics have been extensively described in Part III of this book. To these, we must add other very important omics, such as Genomics, Proteomics, Adductomics (Rappaport et al. 2012), and the Microbiome—which should also be considered when assessing the exposome. These technologies allow the acquisition of many measurements, and they are often based on untargeted approaches, allowing an agnostic analysis of markers. When associated to exposures and outcomes, omics data can help identify biomarkers of exposure.

Overall, omics technologies have several advantages that make them very suitable to assess the exposome. Omics measurements can be applied on freshly obtained human samples, but also on samples from existing biorepositories within large cohort studies. Furthermore, as mentioned by Vineis et al. (2013), it is possible to "investigate platform-specific markers playing a role in the biological pathway linking exposure to disease risk", each marker can give valuable information on mechanism and pathways involved in disease etiology and exposure. Finally, measurements on different omics platforms can be applied on the same samples, allowing the potential of cross-omics studies. If a signal belonging to a same

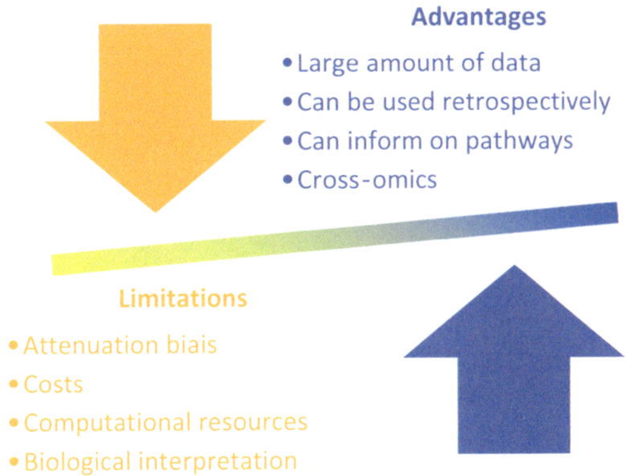

Fig. 17.3 Advantages and Limitations of using OMICs technologies for the exposome

biologically relevant pathway is found in multiple omics, the causal link from exposure to health outcome is reinforced (Fig. 17.3).

Although very useful for the characterization of the exposome, omics measurements also present several limitations. The first is that omics markers can vary through time. Therefore, when they are measured on only one sample, bias can be introduced. Perrier et al. (2016) reported that using multiple samples pooled together from one specimen can reduce the attenuation bias without increasing analytical costs. Undeniably, cost is another limitation for omics measurements, although over time these are being reduced, it has been estimated that the cost of a full suite of omics measurement can go from €50,000 to €100,000 for a study with 500 subjects (Siroux et al. 2016). The development of new cost-effective technologies in the future will be necessary to perform larger studies.

Other limitations include the amount of data produced by omics measurements. The management of these large data sets is a great challenge. Firstly, it requires a significant computational investment both for data processing and data storage. Secondly, as illustrated in Part IV of this book, the amount of data produced requires new statistical methodologies to interpret the significant signals. Finally, another important limitation of omics technologies relies on biological interpretation. Thousands of measured signals cannot always be interpreted and annotated due to technical limitations, making it sometimes impossible to form conclusions based on the findings. This limitation can be reduced by the creation of databases of signals and the sharing of data among researchers which can contribute to the identification of unknowns.

Apart from these limitations, omics are still considered the best option for the characterization of the internal exposome. Moreover, the continuous development of new technologies increases the capabilities of evaluating internal signals.

What Advances Have Been Made for the External Exposome?

As reviewed in Chap. 2 and Part III of this book, the assessment of the external exposome can rely on many different tools and technologies. Environmental exposure and health data can now be obtained by using models based on geographical information systems (GIS). Data obtained by geo-localization technologies can be used to build models for exposure assessment. These models can be applied to large populations, and further be refined with direct measurements obtained by personal sensing on smaller sub-sets of individuals. A great example of this technique has been applied as part of EXPOsOMICs. GIS based models have been successfully applied to air pollution estimates in multiple cities in Europe (Gulliver et al. 2017) and models have been refined with data collected by personal monitoring devices. Other methods such as remote sensing, smartphone based sensors and apps, questionnaires, photos as well as personal dosimeters and sensors are currently being used in multiple projects, as reviewed in Chap. 10 of this book. Additionally, new easy-to-use and low costs sensors are being developed, such as silicone wristbands. These have been applied as passive sampling devices for the evaluation of exposure of 30 individuals where 49 compounds were assessed with a simple and easy-to-use method (O'Connell et al. 2014). Smartphone and sensor-based technologies based on wireless devices present a great advantage, in fact, the information stored by these devices can be made available through internet portals through collaborations with app developers. The collected data can provide information on many subjects without the need of setting up a specific study. Additionally, continuously collected data can significantly improve the estimation of long-term and chronic exposures by reducing the gaps in data collection. These technologies are well suited to evaluate exposures related to location such as air pollution, noise, green-spaces etc., or what is more widely defined as the "urban exposome."

One of the greatest limitations of these technologies, similarly to omics, is the amount of data produced. Great computing power is required to store, manage and to perform statistical analysis on these large datasets. Other issues include privacy and ethical issues, particularly related to data collected through apps and stored on the internet. Researchers must ensure that consent of participants is obtained and that proper and secure data management and storage insuring privacy is maintained.

The analysis of the external exposome also relies on direct measurements. As explained in Part II of this book, high throughput technologies based on mass-spectrometry allow the analysis of thousands of compounds in food, air, water, dust, etc. to evaluate exposure to pollutants. As mentioned in Chap. 8, limitations of these technologies rely today on the fact that they are often semi-quantitative and can only be applied to a small number of samples. Further development in technologies for analysis and sample preparation are required to apply these technologies to a large number of samples with greater precision.

Unravelling the Exposome: Where Do We Go from Here?

Over the past decade, research studies and government funded projects have offered guidance for the future needs and goals of the exposome. These precursor projects provide great advances in the field and will serve as examples and guides for future research. In this section we will highlight a few points that need to be considered in future exposome research.

Collaborations and Data Sharing

In 2012, Wild offered suggestions to transfer the exposome from concept to utility (Wild 2012). He pointed out that one of the main difficulties for the exposome would be the attempt from single groups of researchers to tackle the concept in its whole, he recommended that greater success would be achieved through collaboration, focusing on "defined goals and shared expertise" and sharing the information in the public domain. This approach has been proven successful by multiple collaborative projects such as EXPOsOMICs and HELIX (Vineis et al. 2017; Vrijheid et al. 2014).

The concept of the exposome is broad, and therefore requires the collaboration across multiple disciplines: epidemiology, biology, statistics, computer science, exposure science, etc. In 2014, the National Academy of Sciences, Engineering and Medicine held a workshop to establish which were the obstacles of data sharing in environmental research (Principles and Obstacles for Sharing Data from Environmental Health Research: Workshop Summary 2016). As part of this discussion, it was noted that the quality of the data is key to the sharing. Quality controls and common scientific language must be applied to make sharing and collaborations successful. Epidemiologist and statistician need to collaborate with experts in mass spectrometry, GIS and more, to ensure that proper study designs are applied, and that data is produced with proper methodologies and analyzed correctly. Similar conclusions were discussed in 2015 at the NIEHS Exposome workshop, during which several recommendations were made for future research in exposome. One of the recommendations was based on the need to develop databases and coordinating centers for "collaborative data storage, access and analysis." Recommendations also included the establishment of language and methodological standards for sample collection, and use of technologies (Dennis et al. 2016). These principles have already been implemented by a recent NIEHS initiative as part of the Children's Health Exposure Analysis Resource (CHEAR) (NIEHS 2017). The CHEAR project will serve as a structure to facilitate the evaluation of early life-exposome. Methods and data collection will be harmonized among platforms to ensure cohesion among selected cohorts.

Development of New Tools and Methods to Measure Biological Responses

Another difficulty to tackle in future exposome research is how to consider all aspects of a life course. Up to now, most projects have focused on snap-shots—measuring the exposome at specific time-points throughout the life-course and associating these with disease. Complete exposome studies need to account for longitudinal, multiple exposures and responses. In this context, the elucidation of important pathways is not simple and will require the development of new methodologies and models. In their review, Stingone et al. (2017) suggested a two-step approach to facilitate the interpretation of the exposome's complexity. The first step is the identification of critical windows of exposure, which will be studied in detail. The second step relies on the creation of solid and repeatable study designs which can facilitate the comparisons across different exposome measurements over time. Stingone et al. recommend the study based on merging different cohorts and databases that have focused on short windows. For this purpose, new methods that integrate individual studies at different specific life-stages, the relationship between multiple exposures, different omics measurements, and health outcomes must be developed. This approach is currently being tested in the EXPOsOMICs project. As illustrated in Chaps. 11 and 12 of this book, new approaches to address the high-dimensional data produced in exposome studies that can link data to pathway analysis and network construction can contribute to converge data from different platforms and guide biological interpretation.

Other Applications, the Exposome, and Precision Medicine

In recent decades, epidemiology has shifted its attention to chronic diseases in developed countries. This shift, as well as the failure of genomics to explain the causative factors of many chronic diseases, has led to a reconsideration of the health impact of environmental exposures. The study of the exposome provides growing evidence that exposures to toxicants (air, water, soil, food, household products, etc.) contributes to many chronic diseases. In fact, many diseases have been linked to environmental exposure such as diabetes (Chevalier and Fenichel 2015), Alzheimer's disease (Genuis and Kelln 2015), reduced fertility (Buck Louis et al. 2013), cardiovascular diseases (Xu et al. 2009), cancer (Cao et al. 2011; Kim et al. 2013; Teitelbaum et al. 2015), and more (Bijlsma and Cohen 2016).

It was estimated that over 4.9 million deaths could be attributed to exposure to environmental chemicals in 2011 (WHO 2015). This figure stresses the importance of considering the effect of exposures on health outcomes. However, little has been done to bring knowledge of the possibilities of including environmental health in prevention and diagnosis to clinicians. Advances in exposome technologies, such as the development of OMICs (including genomics), and personal sensors can provide tools for clinicians to evaluate the role of environmental exposures in patient's health. Classical views of disease onset and progression need to be replaced and considered holistically as caused by a complex interplay of genetic and environmental factors (GxE as referred in Chap. 1). In current medicine, diagnosis is based on clinical practice guidelines which consider the patients not as an individual but as a group (Ziegelstein 2017). Precision medicine is defined by considering unique biological characteristics of each individual and tailoring the diagnosis and therapeutic specifically to the individual's phenotype. In this context, exposome tools could be used to integrate and interpret data about disease causation and prevention. Genomics, metabolomics, proteomics, epigenomics, and other omics can deliver information that can lead to more precise diagnosis (Collins and Varmus 2015). Epigenetic, for instance, can provide evidence of inherited effects and alterations due to environmental exposures which can be implicated in disease causation. The inclusion of regular comprehensive monitoring of omic levels as well as the use of sensor-based technologies could contribute, in the future, to assess differential risk in patients. These recent developments could soon offer a bridge between environmental science and medical research and indicate a new pathway for personalized medicine engaging environmental scientists, epidemiologists, clinicians, and individuals that would help the understanding of the links between the environment and human health.

Acknowledgement This work was supported by Horizon 2020 Marie Skłodowska-Curie fellowship EXACT Identifying biomarkers of EXposure leading to Lung Cancer with AdduCTomics. MSCA Project # 708392.

References

Bijlsma N, Cohen MM (2016) Environmental chemical assessment in clinical practice: unveiling the elephant in the room. Int J Environ Res Public Health 13(2):181. https://doi.org/10.3390/ijerph13020181

Buck Louis GM, Sundaram R, Schisterman EF, Sweeney AM, Lynch CD, Gore-Langton RE, Maisog J, Kim S, Chen Z, Barr DB (2013) Persistent environmental pollutants and couple fecundity: the LIFE study. Environ Health Perspect 121(2):231–236. https://doi.org/10.1289/ehp.1205301

Cao J, Yang C, Li J, Chen R, Chen B, Gu D, Kan H (2011) Association between long-term exposure to outdoor air pollution and mortality in China: a cohort study. J Hazard Mater 186 (2–3):1594–1600. https://doi.org/10.1016/j.jhazmat.2010.12.036

Chevalier N, Fenichel P (2015) Endocrine disruptors: new players in the pathophysiology of type 2 diabetes? Diabetes Metab 41(2):107–115. https://doi.org/10.1016/j.diabet.2014.09.005

Collins FS, Varmus H (2015) A new initiative on precision medicine. N Engl J Med 372 (9):793–795. https://doi.org/10.1056/NEJMp1500523

Dennis KK, Auerbach SS, Balshaw DM, Cui Y, Fallin MD, Smith MT, Spira A, Sumner S, Miller GW (2016) The importance of the biological impact of exposure to the concept of the exposome. Environ Health Perspect 124(10):1504–1510. https://doi.org/10.1289/ehp140

Genuis SJ, Kelln KL (2015) Toxicant exposure and bioaccumulation: a common and potentially reversible cause of cognitive dysfunction and dementia. Behav Neurol 2015:620143. https://doi.org/10.1155/2015/620143

Gulliver J, Morley D, Dunster C, McCrea A, van Nunen E, Tsai MY, Probst-Hensch N, Eeftens M, Imboden M, Ducret-Stich R, Naccarati A, Galassi C, Ranzi A, Nieuwenhuijsen M, Curto A, Donaire-Gonzalez D, Cirach M, Vermeulen R, Vineis P, Hoek G, Kelly FJ (2017) Land use regression models for the oxidative potential of fine particles (PM2.5) in five European areas. Environ Res 160:247–255. https://doi.org/10.1016/j.envres.2017.10.002

HEALS (2017) Health and environment-wide associations based on large population suverys. http://www.heals-eu.eu/. Accessed 10 Nov 2017

HERCULES (2017) HERCULES exposome research center. https://emoryhercules.com/. Accessed 10 Nov 2017

I_3 Care Center (2017) International exposome center. http://exposome.iras.uu.nl/about/. Accessed 08 Nov 2017

Kawamoto T, Nitta H, Murata K, Toda E, Tsukamoto N, Hasegawa M, Yamagata Z, Kayama F, Kishi R, Ohya Y, Saito H, Sago H, Okuyama M, Ogata T, Yokoya S, Koresawa Y, Shibata Y, Nakayama S, Michikawa T, Takeuchi A, Satoh H (2014) Rationale and study design of the Japan environment and children's study (JECS). BMC Public Health 14:25. https://doi.org/10.1186/1471-2458-14-25

Kim KH, Jahan SA, Kabir E, Brown RJ (2013) A review of airborne polycyclic aromatic hydrocarbons (PAHs) and their human health effects. Environ Int 60:71–80. https://doi.org/10.1016/j.envint.2013.07.019

Lichtenstein P, Holm NV, Verkasalo PK, Iliadou A, Kaprio J, Koskenvuo M, Pukkala E, Skytthe A, Hemminki K (2000) Environmental and heritable factors in the causation of cancer—analyses of cohorts of twins from Sweden, Denmark, and Finland. N Engl J Med 343(2):78–85. https://doi.org/10.1056/nejm200007133430201

Miller GW, Jones DP (2014) The nature of nurture: refining the definition of the exposome. Toxicol Sci 137(1):1–2. https://doi.org/10.1093/toxsci/kft251

National Academy of Science (2010) The exposome: a powerful approach for evaluating environmental exposures and their influences on human disease. http://nas-sites.org/emergingscience/files/2011/05/03-exposome-newsletter-508.pdf. Accessed 11 May 2018

NIEHS (2017) Children's health exposure analysis resource (CHEAR). https://www.niehs.nih.gov/research/supported/exposure/chear/index.cfm. Accessed 08 Nov 2017

O'Connell SG, Kincl LD, Anderson KA (2014) Silicone wristbands as personal passive samplers. Environ Sci Technol 48(6):3327–3335. https://doi.org/10.1021/es405022f

Perrier F, Giorgis-Allemand L, Slama R, Philippat C (2016) Within-subject pooling of biological samples to reduce exposure misclassification in biomarker-based studies. Epidimiology 27 (3):378–388. https://doi.org/10.1097/ede.0000000000000460

Principles and Obstacles for Sharing Data from Environmental Health Research: Workshop Summary (2016) 2016 by the National Academy of Sciences, Washington DC. https://doi.org/10.17226/21703

Rappaport SM (2011) Implications of the exposome for exposure science. J Expo Sci Environ Epidemiol 21(1):5–9. https://doi.org/10.1038/jes.2010.50

Rappaport SM (2016) Genetic factors are not the major causes of chronic diseases. PLoS One 11(4): e0154387. https://doi.org/10.1371/journal.pone.0154387

Rappaport SM, Smith MT (2010) Epidemiology. Environment and disease risks. Science (New York, NY) 330(6003):460–461. https://doi.org/10.1126/science.1192603

Rappaport SM, Li H, Grigoryan H, Funk WE, Williams ER (2012) Adductomics: characterizing exposures to reactive electrophiles. Toxicol Lett 213(1):83–90. https://doi.org/10.1016/j.toxlet.2011.04.002

Saracci R, Vineis P (2007) Disease proportions attributable to environment. Environ Health 6:38. https://doi.org/10.1186/1476-069x-6-38

Siroux V, Agier L, Slama R (2016) The exposome concept: a challenge and a potential driver for environmental health research. Eur Respir Rev 25(140):124–129. https://doi.org/10.1183/16000617.0034-2016

Stingone JA, Buck Louis GM, Nakayama SF, Vermeulen RC, Kwok RK, Cui Y, Balshaw DM, Teitelbaum SL (2017) Toward greater implementation of the exposome research paradigm within environmental epidemiology. Annu Rev Public Health 38:315–327. https://doi.org/10.1146/annurev-publhealth-082516-012750

Teitelbaum SL, Belpoggi F, Reinlib L (2015) Advancing research on endocrine disrupting chemicals in breast cancer: expert panel recommendations. Reprod Toxicol 54:141. https://doi.org/10.1016/j.reprotox.2014.12.015

Vineis P, van Veldhoven K, Chadeau-Hyam M, Athersuch TJ (2013) Advancing the application of omics-based biomarkers in environmental epidemiology. Environ Mol Mutagen 54(7):461–467. https://doi.org/10.1002/em.21764

Vineis P, Chadeau-Hyam M, Gmuender H, Gulliver J, Herceg Z, Kleinjans J, Kogevinas M, Kyrtopoulos S, Nieuwenhuijsen M, Phillips DH, Probst-Hensch N, Scalbert A, Vermeulen R, Wild CP (2017) The exposome in practice: design of the EXPOsOMICS project. Int J Hyg Environ Health 220(2 Pt A):142–151. https://doi.org/10.1016/j.ijheh.2016.08.001

Vrijheid M, Slama R, Robinson O, Chatzi L, Coen M, van den Hazel P, Thomsen C, Wright J, Athersuch TJ, Avellana N, Basagana X, Brochot C, Bucchini L, Bustamante M, Carracedo A, Casas M, Estivill X, Fairley L, van Gent D, Gonzalez JR, Granum B, Grazuleviciene R, Gutzkow KB, Julvez J, Keun HC, Kogevinas M, McEachan RR, Meltzer HM, Sabido E, Schwarze PE, Siroux V, Sunyer J, Want EJ, Zeman F, Nieuwenhuijsen MJ (2014) The human early-life exposome (HELIX): project rationale and design. Environ Health Perspect 122(6):535–544. https://doi.org/10.1289/ehp.1307204

WHO (2015) The 10 leading causes of death in the world, 2000 and 2012. http://www.who.int/mediacentre/factsheets/fs310/en/index2.html. Accessed 18 Jan 2015

Wild CP (2005) Complementing the genome with an "exposome": the outstanding challenge of environmental exposure measurement in molecular epidemiology. Cancer Epidemiol Biomark Prev 14(8):1847–1850. https://doi.org/10.1158/1055-9965.epi-05-0456

Wild CP (2012) The exposome: from concept to utility. Int J Epidemiol 41(1):24–32. https://doi.org/10.1093/ije/dyr236

Xu X, Freeman NC, Dailey AB, Ilacqua VA, Kearney GD, Talbott EO (2009) Association between exposure to alkylbenzenes and cardiovascular disease among National Health and Nutrition Examination Survey (NHANES) participants. Int J Occup Environ Health 15(4):385–391. https://doi.org/10.1179/oeh.2009.15.4.385

Ziegelstein RC (2017) Personomics: the missing link in the evolution from precision medicine to personalized medicine. J Pers Med 7(4). https://doi.org/10.3390/jpm7040011

Index

S. Dagnino, A. Macherone (eds.), *Unraveling the Exposome*,
https://doi.org/10.1007/978-3-319-89321-1

MIX
Papier aus verantwortungsvollen Quellen
Paper from responsible sources
FSC® C105338

If you have any concerns about our products,
you can contact us on
ProductSafety@springernature.com

In case Publisher is established outside the EU,
the EU authorized representative is:
Springer Nature Customer Service Center GmbH
Europaplatz 3, 69115 Heidelberg, Germany

Printed by Libri Plureos GmbH
in Hamburg, Germany